T0183625

Lecture Notes in Computer Science 11233

Commenced Publication in 1973
Founding and Former Series Editors:
Gerhard Goos, Juris Hartmanis, and Jan van Leeuwen

More information about this series at http://www.springer.com/series/7409

Hakim Hacid · Wojciech Cellary
Hua Wang · Hye-Young Paik
Rui Zhou (Eds.)

Web Information Systems Engineering – WISE 2018

19th International Conference
Dubai, United Arab Emirates, November 12–15, 2018
Proceedings, Part I

 Springer

Editors
Hakim Hacid (iD)
Zayed University
Dubai, United Arab Emirates

Wojciech Cellary (iD)
Poznan University of Economics
Poznan, Poland

Hua Wang (iD)
University of Victoria
Footscray, VIC, Australia

Hye-Young Paik (iD)
University of New South Wales
Sydney, NSW, Australia

Rui Zhou (iD)
Swinburne University of Technology
Hawthorn, VIC, Australia

ISSN 0302-9743   ISSN 1611-3349 (electronic)
Lecture Notes in Computer Science
ISBN 978-3-030-02921-0   ISBN 978-3-030-02922-7 (eBook)
https://doi.org/10.1007/978-3-030-02922-7

Library of Congress Control Number: 2018958517

LNCS Sublibrary: SL3 – Information Systems and Applications, incl. Internet/Web, and HCI

This Springer imprint is published by the registered company Springer Nature Switzerland AG
The registered company address is: Gewerbestrasse 11, 6330 Cham, Switzerland

Preface

Welcome to the proceedings of the 19th International Conference on Web Information Systems Engineering (WISE 2018), held in Dubai, UAE, during November 12–15, 2018. The series of WISE conferences aims to provide an international forum for researchers, professionals, and industrial practitioners to share their knowledge in the rapidly growing area of Web technologies, methodologies, and applications. The first WISE event took place in Hong Kong, SAR China (2000). Then the trip continued to Kyoto, Japan (2001); Singapore (2002); Rome, Italy (2003); Brisbane, Australia (2004); New York, USA (2005); Wuhan, China (2006); Nancy, France (2007); Auckland, New Zealand (2008); Poznan, Poland (2009); Hong Kong, SAR China (2010); Sydney, Australia (2011); Paphos, Cyprus (2012); Nanjing, China (2013); Thessaloniki, Greece (2014); Miami, USA (2015); Shanghai, China (2016); Puschino, Russia (2017); and this year, WISE 2018 was held in Dubai, UAE, supported by Zayed University.

A total of 209 research papers were submitted to the conference for consideration, and each paper was reviewed by at least three reviewers. Finally, 48 submissions were selected as regular papers (with an acceptance rate of 23% approximately), plus 21 as short papers. The research papers cover the areas of blockchain, security and privacy, social networks, microblog data analysis, graph data, information extraction, text mining, recommender systems, medical data analysis, Web services, cloud computing, data stream, distributed computing, data mining techniques, entity linkage and semantics, Web applications, and data mining applications.

In addition to regular and short papers, the WISE 2018 program also featured four workshops: (1) the 5th WISE Workshop on data quality and trust in big data (QUAT 2018); (2) International Workshop on Edge-Based Computing for Next-Generation Wireless Networks; (3) the Third International Workshop on Information Security and Privacy for Mobile Cloud Computing, Web, and Internet of Things (ISCW 2018); (4) The 1st International Workshop on Cloud Computing Economic Impacts. This year's tutorial program included: (1) Text Mining for Social Media; (2) Towards Privacy-Preserving Identity and Access Management Systems for Web Developers; and (3) From Data Lakes to Knowledge Lakes: The Age of Big Data Analytics.

We also wish to take this opportunity to thank the general co-chairs, Prof. Zakaria Maamar and Prof. Marek Rusinkiewicz; the program co-chairs, Prof. Hakim Hacid, Prof. Wojciech Cellary, and Prof. Hua Wang; the workshop co-chairs, Prof. Michael Sheng and Prof. Tetsuya Yoshida; the tutorial and panel chair, Dr. Hye-Young Helen Paik; the sponsor chair, Dr. Fatma Taher; the finance chair, Prof. Hakim Hacid; the local arrangements co-chairs, Dr. Andrew Leonce, Prof. Huwida Saeed, and Prof. Emad Bataineh; the publication chair, Dr. Rui Zhou; the publicity co-chairs, Dr. Dickson Chiu, Dr. Reda Bouadjenek, and Dr. Vanilson Burégio; the website co-chairs, Mr. Emir Ugljanin and Mr. Emerson Bautista; and the WISE Steering Committee representative, Prof. Yanchun Zhang.

We would like to sincerely thank our keynote speakers:

- Professor Athman Bouguettaya, Professor and Head of School of Information Technologies, University of Sydney, Sydney, Australia
- Professor A. Min Tjoa, Institute of Information Systems Engineering TU Wien, Vienna University of Technology, Vienna, Austria
- Professor Mike P. Papazoglou, Executive Director of the European Research Institute in Services Science (ERISS), University of Tilburg, Tilburg, The Netherlands

In addition, special thanks are due to the members of the international Program Committee and the external reviewers for a rigorous and robust reviewing process. We are also grateful to Zayed University, UAE, Springer Nature Switzerland AG, IBM, and the International WISE Society for supporting this conference. The WISE Organizing Committee is also grateful to the workshop organizers for their great efforts to help promote Web information system research to broader domains.

The local lead organizer, Prof. Hakim Hacid, would also like to thank the following colleagues for their support and dedication at different levels in the preparation of WISE 2018 (alphabetical order): Ahmed Alblooshi, Ahmad Al Rjoub, Alia Sulaiman, Amina El Gharroubi, Ayesha Alsuwaidi, Fatima AlMutawa, Hind AlDosari, Jarita Sebastian, Jimson Lee, Osama Nasr, Rosania Braganza, Sudeep Kumar, Zenelabdeen Alsadig, Gemma Ornedo.

We expect that the ideas that have emerged in WISE 2018 will result in the development of further innovations for the benefit of scientific, industrial, and social communities.

November 2018   Hakim Hacid
                                                                     Wojciech Cellary
                                                                            Hua Wang
                                                                     Hye-Young Paik
                                                                             Rui Zhou

Organization

General Co-chairs

Zakaria Maamar   Zayed University, UAE
Marek Rusinkiewicz   Florida Gulf Coast University, USA

Program Co-chairs

Hakim Hacid   Zayed University, UAE
Wojciech Cellary   Poznań University of Economics and Business, Poland
Hua Wang   Victoria University, Australia

Workshop Co-chairs

Michael Sheng   Macquarie University, Australia
Tetsuya Yoshida   Nara Women's University, Japan

Tutorial and Panel Chair

Hye-Young Helen Paik   University of New South Wales, Australia

Sponsor Chair

Fatma Taher   Zayed University, UAE

Finance Chair

Hakim Hacid   Zayed University, UAE

Local Arrangements Co-chairs

Andrew Leonce   Zayed University, UAE
Huwida Saeed   Zayed University, UAE
Emad Bataineh   Zayed University, UAE

Publication Chair

Rui Zhou   Swinburne University of Technology, Australia

Publicity Co-chairs

Dickson Chiu   The University of Hong Kong, SAR China
Reda Bouadjenek   University of Toronto, Canada
Vanilson Burégio   Federal Rural University of Pernambuco (UFRPE), Recife,
                         Brazil

Website Co-chairs

Emir Ugljanin   State University of Novi Pazar, Serbia
Emerson Bautista   Zayed University, UAE

WISE Steering Committee Representative

Yanchun Zhang   Victoria University, Australia and Fudan University, China

Program Committee

Karl Aberer   EPFL, Switzerland
Marco Aiello   University of Stuttgart, Germany
Mohammed Eunus Ali   Bangladesh University of Engineering and Technology
                         (BUET), Bangladesh
Toshiyuki Amagasa   University of Tsukuba, Japan
Boualem Benatallah   University of New South Wales, Australia
Djamal Benslimane   Lyon 1 University, France
Mohamed Reda   The University of Melbourne, Australia
  Bouadjenek
Athman Bouguettaya   The University of Sydney, Australia
Vanilson Burégio   UFRPE, Brazil
Yi Cai   South China University of Technology, China
Bin Cao   Zhejiang University of Technology, China
Xin Cao   University of New South Wales, Australia
Jinli Cao   Latrobe University, Australia
Barbara Catania   University of Genoa, Italy
Richard Chbeir   University of Pau, France
Cindy Chen   University of Massachusetts Lowell, USA
Lisi Chen   Hong Kong Baptist University, SAR China
Lu Chen   Zhejiang University, China
Jacek Chmielewski   Poznań University of Economics and Business, Poland
Ting Deng   Beihang University, China
Hai Dong   RMIT University, Australia
Schahram Dustar   Vienna University of Technology, Austria
Nora Faci   Lyon 1 University, France
Yunjun Gao   Zhejiang University, China
Dimitrios   Swinburne University of Technology, Australia
  Georgakopoulos

Thanaa Ghanem	Metropolitan State University, USA
Azadeh Ghari Neiat	The University of Sydney, Australia
Claude Godart	Université de Lorraine, France
Daniela Grigori	Université Paris Dauphine, France
Viswanath Gunturi	IIT Ropar, India
Armin Haller	Australian National University, Australia
Tanzima Hashem	Bangladesh University of Engineering and Technology, Bangladesh
Md Rafiul Hassan	King Fahd University of Petroleum and Minerals, Saudi Arabia
Yuh-Jong Hu	National Chengchi University, Taiwan
Hao Huang	Wuhan University, China
Adam Jatowt	Kyoto University, Japan
Dawei Jiang	Zhejiang University, China
Lili Jiang	Umeå University, Sweden
Wei Jiang	Missouri University of Science and Technology, USA
Peiquan Jin	University of Science and Technology of China, China
Eleana Kafeza	Athens University of Economics and Business, Greece
Georgios Kambourakis	University of the Aegean, Greece
Hui Li	Xidian University, China
Jiuyong Li	University of South Australia, Australia
Xiang Lian	Kent State University, USA
Dan Lin	Missouri University of Science and Technology, USA
Sebastian Link	The University of Auckland, New Zealand
Qing Liu	Zhejiang University, China
Cheng Long	Queen's University Belfast, UK
Wei Lu	Renmin University of China, China
Hui Ma	Victoria University of Wellington, Australia
Murali Mani	University of Michigan, USA
Jinghan Meng	University of South Florida, USA
Xiaoye Miao	Zhejiang University, China
Paolo Missier	Newcastle University, UK
Sajib Mistry	The University of Sydney, Australia
Natwar Modani	Adobe Research, India
Mikolaj Morzy	Poznań University of Technology, Poland
Wilfred Ng	Hong Kong University of Science and Technology, SAR China
Kjetil Nørvåg	Norwegian University of Science and Technology, Norway
Mitsunori Ogihara	University of Miami, USA
Min Peng	Wuhan University, China
Francesco Piccialli	University of Naples Federico II, Italy
Olivier Pivert	ENSSAT, France
Tieyun Qian	Wuhan University, China
Lie Qu	The University of Sydney, Australia
Qiang Qu	Shenzhen Institutes of Advanced Technology, China

Additional Reviewers

Abdallah Lakhdari
Ali Hamdi Fergani Ali
Anastasia Douma
Andrei Kelarev
Anila Butt
Balaji Vasan Srinivasan
Bing Huang
Brian Setz
Bulat Nasrulin
Ch. Md. Rakin Haider
Chen Zhan
Demetris Paschalides
Di Yao
Dimitrios Papamartzivanos
Dinesh Pandey
Elio Mansour
Fan Liu
Filippos Giannakas
Gang Chen
Gang Ren
Hao Wu
Hongxu Chen
Ildar Nurgaliev
Jeff Ansah
Jeffery Ansah
Joe Tekli
Karam Bou Chaaya
Le Sun
Liandeng Su

Lili Sun
Luis Sanchez Giraldo
Marios Anagnostopoulos
Masoud Salehpour
Md Saddam Hossain Mukta
Md Saiful Islam
Md Zahidul Islam
Mohammed Bahutair
Panayiotis Smeros
Qinyong Wang
Saisai Ma
Sha Lu
Shahriar Badsha
Shi Zhi
Shiv Kumar Saini
Siuly Siuly
Sunav Choudhary
Sven Hartmann
Tam Nguyen
Thang Duong
Tugrulcan Elmas
Weiqing Wang
Weiyi Huang
Xinghao Li
Xu Yang
Xuechao Yang
Yingnan Shi
Zacharias Georgiou

Contents – Part I

Blockchain

Inter-organizational Business Processes Managed by Blockchain 3
   Hiroaki Nakamura, Kohtaroh Miyamoto, and Michiharu Kudo

Decentralized Voting: A Self-tallying Voting System Using a Smart
Contract on the Ethereum Blockchain 18
   Xuechao Yang, Xun Yi, Surya Nepal, and Fengling Han

Enabling Blockchain for Efficient Spatio-Temporal Query Processing 36
   Ildar Nurgaliev, Muhammad Muzammal, and Qiang Qu

A Robust Spatio-Temporal Verification Protocol for Blockchain 52
   Bulat Nasrulin, Muhammad Muzammal, and Qiang Qu

Towards an End-to-End IoT Data Privacy-Preserving Framework
Using Blockchain Technology 68
   Faiza Loukil, Chirine Ghedira-Guegan, Khouloud Boukadi,
   and Aïcha Nabila Benharkat

Security

i2kit: A Deployment Tool with the Simplicity of Containers
and the Security of Virtual Machines 81
   Pablo Chico de Guzmán, Felipe Gorostiaga, and César Sánchez

Gradient Correlation: Are Ensemble Classifiers More Robust Against
Evasion Attacks in Practical Settings? 96
   Fuyong Zhang, Yi Wang, and Hua Wang

An Improved Lightweight RFID Authentication Protocol
for Internet of Things 111
   Xu Yang, Xun Yi, Yali Zeng, Ibrahim Khalil, Xinyi Huang,
   and Surya Nepal

Dynamic Transitions of States for Context-Sensitive Access
Control Decision 127
   A. S. M. Kayes, Wenny Rahayu, Tharam Dillon, Syed Mahbub,
   Eric Pardede, and Elizabeth Chang

Social Network and Security

CoRank: A Coupled Dual Networks Approach to Trust Evaluation
on Twitter   145
   Peiyao Li, Weiliang Zhao, and Jian Yang

Social Context-Aware Trust Prediction: Methods for Identifying
Fake News   161
   *Seyed Mohssen Ghafari, Shahpar Yakhchi, Amin Beheshti,
and Mehmet Orgun*

Privacy Preserving Social Network Against Dopv Attacks   178
   Yumeng Fu, Wei Wang, Hao Fu, Wu Yang, and Dan Yin

A Hybrid Approach for Detecting Spammers in Online Social Networks   189
   Bandar Alghamdi, Yue Xu, and Jason Watson

DUAL: A Deep Unified Attention Model with Latent Relation
Representations for Fake News Detection.   199
   *Manqing Dong, Lina Yao, Xianzhi Wang, Boualem Benatallah,
Quan Z. Sheng, and Hao Huang*

Social Network

Extracting Representative User Subset of Social Networks Towards User
Characteristics and Topological Features   213
   Yiming Zhou, Yuehui Han, An Liu, Zhixu Li, Hongzhi Yin, and Lei Zhao

Group Identity Matching Across Heterogeneous Social Networks   230
   Hongchao Qin, Ye Yuan, Feida Zhu, and Guoren Wang

NANE: Attributed Network Embedding with Local
and Global Information   247
   Jingjie Mo, Neng Gao, Yujing Zhou, Yang Pei, and Jiong Wang

Topical Authority-Sensitive Influence Maximization   262
   Xiaoqing Xiong, Ruixuan Li, Yuhua Li, Xiwu Gu, and Tianan Liang

Microblog Data Analysis

SensorTree: Bursty Propagation Trees as Sensors
for Protest Event Detection.   281
   Jeffery Ansah, Wei Kang, Lin Liu, Jixue Liu, and Jiuyong Li

Claim Retrieval in Twitter   297
   *Wenjia Ma, Wenhan Chao, Zhunchen Luo,
and Xin Jiang*

PUB: Product Recommendation with Users' Buying Intents
on Microblogs 308
  Xiaoxuan Ren, Tianshu Lyu, and Yan Zhang

Learning Concept Hierarchy from Short Texts Using Context Coherence 319
  Abdulqader Almars, Xue Li, Ibrahim A. Ibrahim, and Xin Zhao

Graph Data

Eliminating Temporal Conflicts in Uncertain Temporal
Knowledge Graphs 333
  Lingjiao Lu, Junhua Fang, Pengpeng Zhao, Jiajie Xu, Hongzhi Yin,
  and Lei Zhao

Renovating Watts and Strogatz Random Graph Generation
by a Sequential Approach 348
  Sadegh Nobari, Qiang Qu, Muhammad Muzammal, and Qingshan Jiang

Which Type of Classifier to Use for Networked Data, Connectivity
Based or Feature Based? 364
  Zan Zhang, Jiuyong Li, Hao Wang, Lin Liu, and Jixue Liu

Diversified and Verbalized Result Summarization for Semantic
Association Search 381
  Yu Gu, Yue Liang, Gong Cheng, Daxin Liu, Ruidi Wei, and Yuzhong Qu

Information Extraction

Main Content Extraction from Heterogeneous Webpages 393
  Julian Alarte, David Insa, Josep Silva, and Salvador Tamarit

Bootstrapped Multi-level Distant Supervision for Relation Extraction 408
  Ying He, Zhixu Li, Guanfeng Liu, Fangfei Cao, Zhigang Chen,
  Ke Wang, and Jie Ma

On the Discovery of Continuous Truth: A Semi-supervised Approach
with Partial Ground Truths 424
  Yi Yang, Quan Bai, and Qing Liu

Web Page Template and Data Separation for Better Maintainability 439
  Chenxu Zhao, Rui Zhang, and Jianzhong Qi

Text Mining

Combining Contextual Information by Self-attention Mechanism
in Convolutional Neural Networks for Text Classification 453
  Xin Wu, Yi Cai, Qing Li, Jingyun Xu, and Ho-fung Leung

Cpriori: An Index-Based Framework to Extract the Generalized
Center Strings.   468
    Shuhan Zhang, Shengluan Hou, and Chaoqun Fei

Topic-Net Conversation Model   483
    *Min Peng, Dian Chen, Qianqian Xie, Yanchun Zhang, Hua Wang,
    Gang Hu, Wang Gao, and Yihan Zhang*

A Hybrid Model Reuse Training Approach for Multilingual OCR   497
    *Zhongwei Xie, Lin Li, Xian Zhong, Luo Zhong, Qing Xie,
    and Jianwen Xiang*

Author Index   513

Contents – Part II

Recommender Systems

SARFM: A Sentiment-Aware Review Feature Mapping Approach
for Cross-Domain Recommendation 3
    Yang Xu, Zhaohui Peng, Yupeng Hu, and Xiaoguang Hong

Integrating Collaborative Filtering and Association Rule Mining
for Market Basket Recommendation 19
    Feiran Wang, Yiping Wen, Jinjun Chen, and Buqing Cao

Unified User and Item Representation Learning for Joint Recommendation
in Social Network 35
    Jiali Yang, Zhixu Li, Hongzhi Yin, Pengpeng Zhao, An Liu,
    Zhigang Chen, and Lei Zhao

Geographical Proximity Boosted Recommendation Algorithms
for Real Estate 51
    Yonghong Yu, Can Wang, Li Zhang, Rong Gao, and Hua Wang

Cross-domain Recommendation with Consistent Knowledge
Transfer by Subspace Alignment............................. 67
    Qian Zhang, Jie Lu, Dianshuang Wu, and Guangquan Zhang

Medical Data Analysis

D-ECG: A Dynamic Framework for Cardiac Arrhythmia Detection from
IoT-Based ECGs.. 85
    Jinyuan He, Jia Rong, Le Sun, Hua Wang, Yanchun Zhang,
    and Jiangang Ma

Jointly Predicting Affective and Mental Health Scores Using Deep Neural
Networks of Visual Cues on the Web 100
    Hung Nguyen, Van Nguyen, Thin Nguyen, Mark E. Larsen,
    Bridianne O'Dea, Duc Thanh Nguyen, Trung Le, Dinh Phung,
    Svetha Venkatesh, and Helen Christensen

Preserving Data Privacy and Security in Australian My Health
Record System: A Quality Health Care Implication................. 111
    Pasupathy Vimalachandran, Yanchun Zhang, Jinli Cao, Lili Sun,
    and Jianming Yong

A Framework for Processing Cumulative Frequency Queries over Medical
Data Streams   121
    Ahmed Al-Shammari, Rui Zhou, Chengfei Liu, Mehdi Naseriparsa,
    and Bao Quoc Vo

Web Services and Cloud Computing

Knowledge-Driven Automated Web Service Composition—An
EDA-Based Approach   135
    Chen Wang, Hui Ma, Aaron Chen, and Sven Hartmann

A CP-Net Based Qualitative Composition Approach for an IaaS Provider . . .   151
    Sheik Mohammad Mostakim Fattah, Athman Bouguettaya,
    and Sajib Mistry

LIFE-MP: Online Virtual Machine Consolidation with Multiple Resource
Usages in Cloud Environments   167
    Deafallah Alsadie, Zahir Tari, Eidah J. Alzahrani,
    and Ahmed Alshammari

Stance and Credibility Based Trust in Social-Sensor Cloud Services   178
    Tooba Aamir, Hai Dong, and Athman Bouguettaya

Data Stream and Distributed Computing

StrDip: A Fast Data Stream Clustering Algorithm Using the Dip Test
of Unimodality   193
    Yonghong Luo, Ying Zhang, Xiaoke Ding, Xiangrui Cai, Chunyao Song,
    and Xiaojie Yuan

Classification and Annotation of Open Internet of Things Datastreams   209
    Federico Montori, Kewen Liao, Prem Prakash Jayaraman,
    Luciano Bononi, Timos Sellis, and Dimitrios Georgakopoulos

Efficient Auto-Increment Keys Generation for Distributed Log-Structured
Storage Systems   225
    Jianwei Huang, Jinwei Guo, Zhao Zhang, Weining Qian,
    and Aoying Zhou

A Parallel Joinless Algorithm for Co-location Pattern Mining Based on
Group-Dependent Shard   240
    Peizhong Yang, Lizhen Wang, Xiaoxuan Wang, and Yuan Fang

Data Mining Techniques

Improving Maximum Classifier Discrepancy by Considering Joint
Distribution for Domain Adaptation.................................... 253
  *Zehang Lin, Zhenguo Yang, Runwei Situ, Feitao Huang, Jianming Lv,
  Qing Li, and Wenyin Liu*

Density Biased Sampling with Locality Sensitive Hashing
for Outlier Detection 269
  *Xuyun Zhang, Mahsa Salehi, Christopher Leckie, Yun Luo, Qiang He,
  Rui Zhou, and Rao Kotagiri*

A Novel Technique of Using Coupled Matrix and Greedy Coordinate
Descent for Multi-view Data Representation........................... 285
  Khanh Luong, Thirunavukarasu Balasubramaniam, and Richi Nayak

Data-Augmented Regression with Generative Convolutional Network 301
  *Xiaodong Ning, Lina Yao, Xianzhi Wang, Boualem Benatallah,
  Shuai Zhang, and Xiang Zhang*

Towards Automatic Complex Feature Engineering 312
  Jianyu Zhang, Françoise Fogelman-Soulié, and Christine Largeron

Entity Linkage and Semantics

Entity Linking Facing Incomplete Knowledge Base.................... 325
  Shaohua Zhang, Jiong Lou, Xiaojie Zhou, and Weijia Jia

User Identity Linkage with Accumulated Information from Neighbouring
Anchor Links 335
  Xiang Li, Yijun Su, Wei Tang, Neng Gao, and Ji Xiang

Mining High-Quality Fine-Grained Type Information from Chinese
Online Encyclopedias 345
  Maoxiang Hao, Zhixu Li, Yan Zhao, and Kai Zheng

Semantics-Enabled Personalised Urban Data Exploration 361
  *Devis Bianchini, Valeria De Antonellis, Massimiliano Garda,
  and Michele Melchiori*

Web Applications

Recommendation for MOOC with Learner Neighbors and Learning Series... 379
  Yanxia Pang, Chang Liao, Wenan Tan, Yueping Wu, and Chunyi Zhou

In-depth Exploration of Engagement Patterns in MOOCs............... 395
  Lei Shi and Alexandra I. Cristea

Topic Evolution Models for Long-Running MOOCs   410
 Arti Ramesh and Lise Getoor

Neuroscientific User Models: The Source of Uncertain User Feedback
and Potentials for Improving Web Personalisation   422
 Kevin Jasberg and Sergej Sizov

Modeling New and Old Editors' Behaviors in Different Languages
of Wikipedia   438
 Anita Chandra and Abyayananda Maiti

Data Mining Applications

A Novel Incremental Dictionary Learning Method for Low Bit Rate
Speech Streaming   457
 Luyao Teng, Yingxiang Huo, Huan Song, Shaohua Teng, Hua Wang,
 and Yanchun Zhang

Identifying Price Index Classes for Electricity Consumers via Dynamic
Gradient Boosting   472
 Vanh Khuyen Nguyen, Wei Emma Zhang, and Quan Z. Sheng

Big Data Exploration for Smart Manufacturing Applications   487
 Ada Bagozi, Devis Bianchini, Valeria De Antonellis,
 and Alessandro Marini

Dark Web Markets: Turning the Lights on AlphaBay   502
 Andres Baravalle and Sin Wee Lee

Author Index   515

Blockchain

Inter-organizational Business Processes Managed by Blockchain

Hiroaki Nakamura[✉], Kohtaroh Miyamoto, and Michiharu Kudo

IBM Research - Tokyo, Tokyo, Japan
{hnakamur,kmiya,kudo}@jp.ibm.com

Abstract. Blockchain technology is highly expected to be a solution to the consistency and trust problems in managing business processes that span across organizational boundaries. However, to execute collaborative business processes, we need a mechanism for enabling entire workflows as a whole, where participants' private processes must agree on the shared inter-organizational processes realized by Blockchain. To address this, we introduce a set of techniques that take business process models as input and transforms them into statecharts for Blockchain and process participants. We also optimize the size of the statechart in order to reduce the number of communications between Blockchain and participants. The statecharts are then used as a basis for generating software artifacts: smart contracts running on Blockchain and Web applications for process participants. Through the evaluation of our solution, we confirmed that our algorithms produce software artifacts that collaboratively work together. By applying the statechart reduction algorithms, we could reduce the number of sending and receiving events by 74% and 65% in two case studies.

Keywords: Process management · Blockchain · Model transformation

1 Introduction

Maintaining consistency and mutual trust in inter-organizational business processes has been a long-lasting challenge for IT industries. In a centralized business process within an organization, information on business activities can be shared and validated, and participants of the process trust each other. In an inter-organizational process, however, when process control is handed over to participants outside of an organization, we cannot validate data accuracy, enforce obligations, or check if conditions are met. As a result, moving control between fragmented processes of different organizations tends to result in inconsistent and untrusted process management.

Blockchain[1] is highly expected to be a solution to the consistency and trust issues in managing inter-organizational business processes [10]. Transactions in

---

[1] In this paper, Blockchain refers to blockchain-related technologies in general.

© Springer Nature Switzerland AG 2018
H. Hacid et al. (Eds.): WISE 2018, LNCS 11233, pp. 3–17, 2018.
https://doi.org/10.1007/978-3-030-02922-7_1

business process networks that span organizations are shared and validated by participants, which is a mechanism that does not require the participants to trust each other. Transactions between participants are also processed by computer programs, called "smart contracts" or "chaincode", that encode business logic agreed on by all process participants. Each participant runs a private process of its own, while the entire process shared by the participants is managed by Blockchain as shown in Fig. 1.

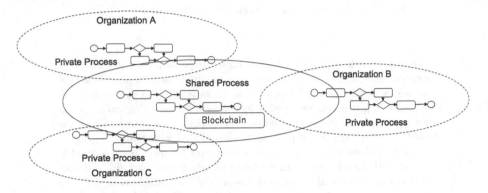

Fig. 1. Inter-organizational collaboration with Blockchain

Although Blockchain technology has the potential to help us shift the level of integration and automation, we still need to address the following challenges in managing inter-organizational business processes using Blockchain.

1. We need to be able to define entire workflows that control processes across organizational boundaries.
2. The private processes of the participants must fulfill the responsibilities imposed by the shared process.
3. Since Blockchain-based solutions still face technological challenges such as throughput, latency, and size and bandwidth limitations [15], the mechanism built on top of Blockchain must be highly optimized.

To resolve the challenges mentioned above, we propose a method for (1) transforming a single business process model into multiple statecharts, (2) optimizing the generated statecharts, and (3) generating smart contracts and participants applications from the statecharts. Since our algorithms allow us to represent an entire workflow that will be automatically transformed into smart contract and Web applications, they are always consistent with each other.

This paper is organized as follows. In Sect. 2, we discuss the background and related works of our approach. Our algorithms for transforming process models into statecharts and for optimizing the generated statecharts are described in Sect. 3. Section 4 presents our method for generating software artifacts, which are smart contract and Web applications. In Sect. 5, we evaluate our approach using two case studies, and Sect. 6 concludes our research with a discussion on future directions.

2   Background and Related Work

Techniques and tools for transforming models, in particular those expressed in
UML and its profiles, have been studied extensively [11], where transformations
from one type of behavior model into a different type of behavior model are
related to our study. In [16], the algebraic properties of the algorithms for transforming UML sequence diagrams into statecharts were shown. Though their
input model type (sequence diagram) is different from ours (process model),
we adopted their framework of representing output statecharts, which focused
on receiving and sending events associated with state transitions. In [16], only
structured models are the target of the transformation, but we need to be able
to handle arbitrarily connected process models. To transform such unstructured
models as input, we used a graph traversing technique described in [13].

In [14], the authors proposed a method for translating process models
described in BPMN into smart contracts in Solidity. The generated smart contracts are either for monitoring the process execution status across all involved
participants or for executing collaborative business processes as well as data
transformation and calculations. Its implementation, called "Caterpillar", was
also presented in [8]. Instead of directly mapping process models with smart
contracts, the use of Petri nets as an intermediary representation allows us to
optimize the generated smart contract [6]. Declarative process notations can also
be used for representing process models that have a corresponding implementation as a smart contract [9].

Our work is different from existing studies in that we use a simple process
diagram in a single pool with swimlanes as input and a statechart as a target
of model transformation. In the case that we want to dig into the details on
how messages are exchanged between participants and how the processes are
supported by software process engines, collaboration diagrams or choreography
diagrams are suitable [3]. However, when we design entire workflows as a whole
without going into detail, the use of a process diagram in a single pool with
swimlanes is a better approach. Statecharts have the following advantages over
other behavior representations.

- We can well represent Blockchain as well as multiple process participants as
  statecharts communicating with each other by receiving and sending events.
- Statecharts have been standardized [12], so developer support, including tool-
  ing and education, is already available.
- Formal aspects of statecharts allow us to define composition and optimization
  algorithms precisely.
- Statecharts are also close to implementation, so the final software artifacts
  can be easily built on the basis of statecharts.

3   Generating Statecharts from Process Models

As shown in Fig. 2, our method first takes a process model as input and
transforms it into statecharts for Blockchain and organizations' applications.

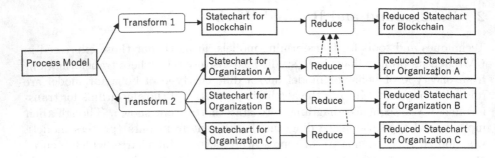

Fig. 2. Transformation and reduction process

Then the organizations' statecharts are reduced into optimized ones, and using the result of reduction computation, the statechart for Blockchain is also reduced.

3.1 Process Models

To define process models, we use the following small set of modeling elements, taken mainly from BPMN [4]. *Start Event* acts as a process trigger. *End Event* represents the result of a process. *Activity* describes work that must be done. *Sequence Flow* shows in which order the activities are performed. *Decision* creates alternative flows, where only one of the paths can be taken. *Merge* combines two alternative sequence flow paths into one. *Swimlane* categorizes activities according to participant roles, where we use only one *Pool* with multiple *Lanes*.

Figure 3 shows an example insurance payment process with two decisions and one merge, where *client*, *insurer*, and *surveyor* are the participants of the process.

After the process is triggered, the *client* prepares a claim and submits it to the *insurer*, who then requests the *surveyor* to survey the claim. The *surveyor*

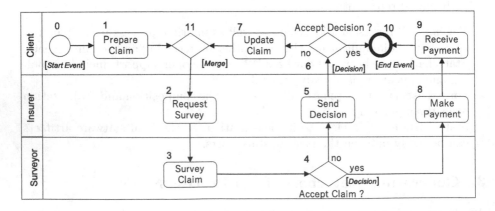

Fig. 3. An example process model of insurance payment

decides whether the claim is accepted or not. If the *surveyor* does not accept it, the *insurer* sends the decision (the claim is not accepted). After receiving the decision, the *client* also decides whether to accept the decision. In the case that the *client* does not accept the decision, she updates the claim and submit it to the *insurer* again; otherwise, the process terminates. If the *surveyor* accepts the claim, the *insurer* makes payment, which will be received by the *client*.

For defining a transformation that takes process models as input, we formalize a process model as a 5-tuple $\langle O, o_0, F, P, R \rangle$ where O is a set of flow objects, including Start and End Events, Activities, Decisions, and Merges, $o_0 \in O$ is the start event, $F \subseteq O \times O \times T$ is the flow relation, P is a set of participants, and $R : O \to P$ is a mappings associating a flow object to a participant. A flow relation (o, o', t) represents an object flow from a object o to another object o' whose flow type is denoted by t. When a flow type can be uniquely derived from a source object and target object, we do not describe it explicitly.

3.2   Statecharts for Representing Shared Processes

We then represent processes shared by multiple organizations by means of a set of statecharts. A Blockchain-based infrastructure can be thought of as a state transition system, where a submitted transaction is recorded as a change in state [5]. For an individual participant, a statechart can also be used for behavior description which provides a basis of interface specification as well as implementation of functions. In addition, interaction between process participants and Blockchain can be modeled as a set of statecharts that send and receive events for their communication. Thus, using statecharts for representing the behavior and interaction of Blockchain and process participants is believed to be a promising approach.

When statecharts are used for representing shared processes, the mapping between the two different types of models must fulfill two requirements: all communications between participants must be intermediated by Blockchain to ensure consistency and trust, and all participants share the same agreed-upon process.

Figure 4 shows a fragment process model with two participants A and B, and corresponding statecharts to represent the behavior of *Activity*. We use three statecharts, one for Blockchain and the others for participants A and B. In this paper, we denote a transition with an arrow from one state s to another state s' with the label $e/a_1, ..., a_n$, where e is an event to receive and $a_1, ..., a_n$ are events to send when the transition occurs. We also describe $/a_1, ..., a_n$ and e in the case that the transition has no receiving event or sending events, respectively.

All statecharts start with the first states s_1, s_1^A, and s_1^B. When an object flow goes into *Activity*, Blockchain sends events b^A and b^B and transitions to state s_2. When b^A and b^B are received, participants A and B transition to states f_2^A and f_2^B, respectively. After *Activity* is finished, participant A transitions to state f_3^A and sends event f^A, which will be received by Blockchain. Blockchain then transitions to state s^3 and sends event f^B, which will be received by participant B. Thus, all communications are intermediated by Blockchain and all participants share the same process.

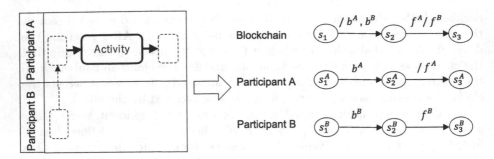

Fig. 4. Process model and corresponding statecharts

We also formalize a statechart as a 5-tuple $\langle S, s_0, F, E, T \rangle$ where S is a set of states, $s_0 \in S$ is the initial state, $F \subseteq S$ is a set of final states, E is a set of events, and $T \subseteq S \times E \times 2^E \times S$ is the transition relation. A transition relation $(s, e, \{a_1, ..., a_n\}, s')$ represents a state transition from one state s to another state s' which occurs when the event e is received and sends a set of events $\{a_1, ..., a_n\}$ as a result of the action associated with the transition.

3.3   Generating Statecharts

We here define algorithms for transforming process models into statecharts as shown in Figs. 5 and 6. The procedure *Transform1* generates a statechart for Blockchain by using sub-procedures *TransformNext1* and *RemoveNoEventTransition*, while *Transform2* generates a statechart for a participant by using sub-procedures *TransformNext2* and *RemoveNoEventTransition*.

TransformNext1 and *TransformNext2* traverse the input process model and create statecharts according to the types of objects in the process model. Activity objects and Merge objects are translated into two state transitions, and Decision objects are translated into three state transitions. Attached to transitions are sending and receiving events, which constitute the communication mechanism described in Sect. 3.2. The algorithms may produce statecharts with transitions that have no receiving or sending events, so we apply *RemoveNoEventTransition* to the statecharts to eliminate such transitions at the end of *Transform1* and *Transform2*.

In these algorithms, we use function $Event : L \times P \times O \rightarrow E$, which takes a label, participant, and object as input and associates with them an event that is unique to the combination of these parameters. For brevity, we denote $Event(l, p, o)$ as l_o^p. This function is shared by *Transform1* and *Transform2*, so they generate statecharts that can communicate with each other by exchanging the events of the same names. We also use an associative memory $Map : O \rightarrow S$ which holds the correspondence between an object in a process model and a state in a statechart. When the algorithms visit an object that has already been visited before, they reuse the state created before and maintained by *Map*.

When the input process model is the one in Fig. 3, the algorithms produce the statecharts shown in Fig. 7.

/* Blockhain */
Procedure $Transform1(M)$
  Input: Process Model $M = \langle O, o_0, F, P, R \rangle$
  Output: Statechart $\langle S, s_0, Fin, E, T \rangle$
  Create the initial state s_0
  $S \leftarrow \{s_0\}; T \leftarrow \emptyset; Fin \leftarrow \emptyset$
  if there exists $(o_0, o) \in F$ **then**
   $TransformNext1(s_0, o)$
  $T \leftarrow RemoveNoEventTransitions(T)$
  return T

Sub-Procedure $TransformNext1(s, o)$
  Input: last state s; current object o
  case $type(o)$ **of**
   End:
    $Fin \leftarrow Fin \cup \{s\}$
   Activity:
    Create a new state $s_1; S \leftarrow S \cup \{s_1\}$
    $A_1 \leftarrow \{b_o^p | p \in P\}$
    $T \leftarrow T \cup \{(s, \emptyset, A_1, s_1)\}$
    Create a new state $s_2; S \leftarrow S \cup \{s_2\}$
    $A_2 \leftarrow \{f_o^p | p \in P \wedge p \neq R(o)\}$
    $T \leftarrow T \cup \{(s_1, f_o^{R(o)}, A_2, s_2)\}$
    if there exists $(o, o') \in F$ **then**
     $TransformNext1(s_2, o')$
   Decision:
    Create a new state $s_1; S \leftarrow S \cup \{s_1\}$
    $A_1 \leftarrow \{b_o^p | p \in P\}$
    $T \leftarrow T \cup \{(s, \emptyset, A_1, s_1)\}$
    Create a new state $s_y; S \leftarrow S \cup \{s_y\}$
    $A_y \leftarrow \{y_o^p | p \in P \wedge p \neq R(o)\}$
    $T \leftarrow T \cup \{(s_1, y_o^{R(o)}, A_y, s_2)\}$
    Create a new state $s_n; S \leftarrow S \cup \{s_n\}$
    $A_n \leftarrow \{n_o^p | p \in P \wedge p \neq R(o)\}$
    $T \leftarrow T \cup \{(s_1, n_o^{R(o)}, A_n, s_2)\}$
    if there exists $(o, o_y, Y) \in F$ **then**
     $TransformNext1(s_y, o_y)$
    if there exists $(o, o_n, N) \in F$ **then**
     $TransformNext1(s_n, o_n)$
   Merge:
    if there exists $s_1 = Map(o)$ **then**
     $T \leftarrow T \cup \{(s, \emptyset, \emptyset, s_1)\}$
    else
     Create a new state $s_1; S \leftarrow S \cup \{s_1\}$
     $T \leftarrow T \cup \{(s, \emptyset, \emptyset, s_1)\}$
     Memorize $Map(o) \triangleq s_1$
     Create a new state $s_2; S \leftarrow S \cup \{s_2\}$
     $T \leftarrow T \cup \{(s_1, \emptyset, \emptyset, s_2)\}$
     if there exists $(o, o') \in F$ **then**
      $TransformNext1(s_2, o')$

/* Participant */
Procedure $Transform2(M, p)$
  Input: Process $M = \langle O, o_0, F, P, R \rangle$
      Participant p
  Output: Statechart $\langle S, s_0, Fin, E, T \rangle$
  Create the initial state s_0
  $S \leftarrow \{s_0\}; T \leftarrow \emptyset; Fin \leftarrow \emptyset$
  if there exists $(o_0, o) \in F$ **then**
   $TransformNext2(s_0, o)$
  $T \leftarrow RemoveNoEventTransitions(T)$
  return T

Sub-Procedure $TransformNext2(s, o)$
  Input: last state s; current object o
  case $type(o)$ **of**
   End:
    $Fin \leftarrow Fin \cup \{s\}$
   Activity:
    Create a new state $s_1; S \leftarrow S \cup \{s_1\}$
    $T \leftarrow T \cup \{(s, b_o^p, \emptyset, s_1)\}$
    Create a new state $s_2; S \leftarrow S \cup \{s_2\}$
    if $p = R(o)$ **then**
     $T \leftarrow T \cup \{(s_1, \emptyset, \{f_o^p\}, s_2)\}$
    else
     $T \leftarrow T \cup \{(s_1, f_o^p, \emptyset, s_2)\}$
    if there exists $(o, o') \in F$ **then**
     $TransformNext2(s_2, o')$
   Decision:
    Create a new state $s_1; S \leftarrow S \cup \{s_1\}$
    $T \leftarrow T \cup \{(s, f_o^p, \emptyset, s_1)\}$
    Create a new state $s_y; S \leftarrow S \cup \{s_y\}$
    Create a new state $s_n; S \leftarrow S \cup \{s_n\}$
    if $p = R(o)$ **then**
     $T \leftarrow T \cup \{(s_1, \emptyset, \{y_o^p\}, s_y)\}$
     $T \leftarrow T \cup \{(s_1, \emptyset, \{n_o^p\}, s_n)\}$
    else
     $T \leftarrow T \cup \{(s_1, y_o^p, \emptyset, s_y)\}$
     $T \leftarrow T \cup \{(s_1, n_o^p, \emptyset, s_n)\}$
    if there exists $(o, o_y, Y) \in F$ **then**
     $TransformNext2(s_y, o_y)$
    if there exists $(o, o_n, N) \in F$ **then**
     $TransformNext2(s_n, o_n)$
   Merge:
    if there exists $s_1 = Map(o)$ **then**
     $T \leftarrow T \cup \{(s, \emptyset, \emptyset, s_1)\}$
    else
     Create a new state $s_1; S \leftarrow S \cup \{s_1\}$
     $T \leftarrow T \cup \{(s, \emptyset, \emptyset, s_1)\}$
     Memorize $Map(o) \triangleq s_1$
     Create a new state $s_2; S \leftarrow S \cup \{s_2\}$
     $T \leftarrow T \cup \{(s_1, \emptyset, \emptyset, s_2)\}$
     if there exists $(o, o') \in F$ **then**
      $TransformNext2(s_2, o')$

Fig. 5. Algorithms for transforming process models into statecharts

Procedure *RemoveNoEventTransitions(T_i)*
  Input: Set of transitions T_i
  Output: Set of transitions T_o
  $T_o \leftarrow T_i$
  repeat until there exists $(s_1, \emptyset, \emptyset, s_2) \in T_o$
    $T_o \leftarrow T_o - \{(s_1, \emptyset, \emptyset, s_2)\}$
    for each $(s_2, E, A, s) \in T_o$
      $T_o \leftarrow T_o - \{(s_2, E, A, s)\}$
      $T_o \leftarrow T_o \cup \{(s_1, E, A, s)\}$
    for each $(s', E, A, s_2) \in T_o$
      $T_o \leftarrow T_o - \{(s', E, A, s_2)\}$
      $T_o \leftarrow T_o \cup \{(s', E, A, s_1)\}$
  return T_o

Fig. 6. Algorithm for removing transitions that do not receive or send events

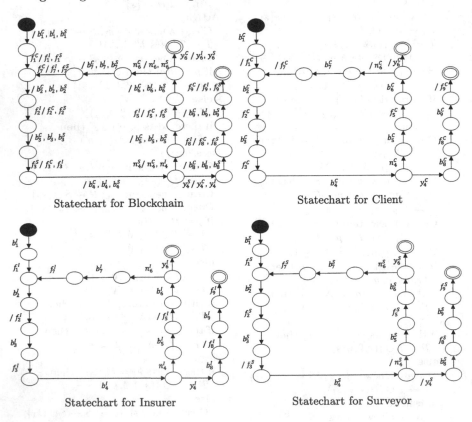

Fig. 7. Generated statecharts

3.4   Reducing Statecharts

The statecharts generated from process models are not optimal in that events that trigger an activity in a process model are sent to all participants even if the participants are not involved in the activity. To reduce the communication overhead in Blockchain-based solutions, we need to eliminate as many events as possible.

We focus on two consecutive transitions that receive events and send no events. In such a case, the path of the state transitions is the same even if we eliminate the receiving event from the first transition. Because no events are sent from the two state transitions, eliminating the receiving event does not change the behavior observed from outside of the statechart.

This principle can be generalized to cases where one state has multiple incoming transitions and/or multiple outgoing transitions. When all outgoing transitions receive events but do not send any events, we can safely eliminate the receiving event of an incoming transition that does not send any events. In Fig. 8(a), s is a state whose outgoing transitions do not sent any events, so we can eliminate e_1 from the transition from s_1 to s (b). Then we remove no-event transition and add new transitions to pass through the removed transition (c). We repeat the elimination steps for all incoming transitions of the target state (d). In the case that the state becomes inaccessible, we remove the state and its outgoing transitions (e).

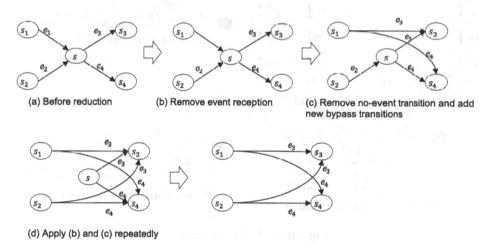

(a) Before reduction   (b) Remove event reception   (c) Remove no-event transition and add
                                                               new bypass transitions

(d) Apply (b) and (c) repeatedly

Fig. 8. Statechart reduction steps

The algorithm in Fig. 9 implements the statechart reduction steps described above. After removing unnecessary receiving events, we can remove the corresponding sending events in other statecharts. The resulting reduced statecharts are shown in Fig. 10.

Procedure *ReduceTransitions*(T_i)
  Input: Set of transitions T_i
  Output: Set of transitions T_o
  $S \leftarrow \{s | (s_1, E_1, A_1, s) \in T_i \land (s, E_2, A_2, s_2) \in T_i\}$
  $T_o \leftarrow T_i$
  for each s in S
    if $A_2 = \emptyset$ for all $(s, E_2, A_2, s_2) \in T_o$ **then**
      for each $(s_1, E_1, \emptyset, s)$ in T_o
        $T_o \leftarrow T_o - \{(s_1, E_1, \emptyset, s)\}$
        for each $(s, E_2, \emptyset, s_2)$ in T_o
          $T_o \leftarrow T_o \cup \{(s_1, E_2, \emptyset, s_2)\}$
      if there exists no $(s_1, E_1, A_1, s) \in T_o$ **then**
        for each $(s, E_2, \emptyset, s_2)$ in T_o
          $T_o \leftarrow T_o - \{(s, E_2, \emptyset, s_2)\}$
  return T_o

Fig. 9. Algorithm for reducing statecharts

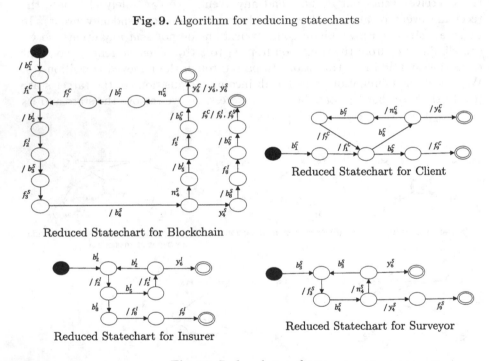

Reduced Statechart for Blockchain

Reduced Statechart for Client

Reduced Statechart for Insurer

Reduced Statechart for Surveyor

Fig. 10. Reduced statecharts

4 Generating Software Artifacts from Statecharts

4.1 Overall Architecture

After obtaining reduced statecharts for Blockchain and process participants, we use these statecharts for generating software artifacts, which are smart contracts and Web applications. Figure 11 shows an overview of the processes for

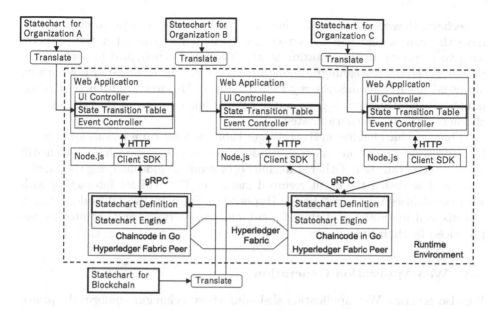

Fig. 11. Overall architecture of the generated artifacts and runtime environment

generating smart contract and Web applications as well as the resulting software artifacts.

We employ Hyperledger Fabric [1], a Blockchain framework project by The Linux Foundation, as an architecture for hosting smart contracts called "chaincode". In our implementation, chaincode in Go language consists of two components: a common statechart engine and a statechart definition. To provide an API for interacting with Hyperledger Fabric, we use its client SDK, which is designed to be used in the Node.js JavaScript runtime. We also created Web applications that consume state transition tables generated from statecharts and interact with Hyperledger Fabric by exchanging events.

4.2   Chaincode Generation

We first output a statechart definition in SCXML [12], an XML-based standardized notation for statechart, and then translate the SCXML representation into JSON data that will be interpreted by our statechart engine running on Hyperledger Fabric. We used SCXML as an intermediate language because supporting tools such as graphical viewers and execution monitors are already available, and it covers complex constructs, such as sub-states and parallel states, found in the Harel statechart [7], which will facilitate the development of our future extensions.

When generating a statechart in SCXML, we have to give specific state names and event names. As for state names, we cannot obtain any clues from the input process models, and so we assign uniquely generated names to states in a statechart. Random names are allowable because state names are closed inside a statechart. However, events are shared by multiple statecharts, so events must have the names that can be agreed on by all participants. For the purpose, we give a specific implementation to the function $Event(l, p, o)$ in Sect. 3.3, so that it generates a meaningful name for an event. One example of a generated name is "Insurer_RequestSurvey_Finish", where "Insurer" is from p (Participant), "RequestSurvey" is from o (Object), "Finish" is from l (Label), and the character strings are concatenated.

The statechart engine on Hyperledger Fabric accepts transactions for requesting statechart operations, such as sending an event to a statechart, which will cause a state transition. Other statechart operations include querying the current state and accessing the event received last time. The API for interacting with such capabilities is provided by the Hyperledger Fabric SDK and Node.js, which installs and instantiates chaincode, submits transactions, and monitor events produced by chaincode.

4.3   Web Application Generation

We also generate Web application skeletons whose behaviors conform the protocols defined by statecharts, which are given in the form of state transition tables. Table 1 is an example that shows the state transition table translated from the reduced statechart for Client in Fig. 10. As we discussed in Sect. 4.2, state names are only unique identifiers while event names must carry their meanings.

Our Web application skeletons are built on the basis of Angular [2], a frontend web application framework in JavaScript. The application has two custom controllers. *UI Controller* first creates clickable buttons for all sending events, enables/disables the buttons according to the current state, and induces a transition when an enabled button is clicked by a Web application user. *Event Controller* sends and receives events through the Node.js server and communicate with the statechart running as a chaincode on Blockchain.

Suppose our statechart for Client is now in state s2. Then an example sequence of execution steps is as follows.

1. Button Controller disables all buttons because the statechart in state s2 has no possible sending events.
2. When Event Controller receives Client_AcceptDecision_Begin event, the statechart transitions to s3.
3. Button Controller enables the buttons for Client_AcceptDecision_No and Client_AcceptDecision_Yes because they are possible sending events when the state is s3.
4. When an application user clicks the button for Client_AcceptDecision_No, the statechart transitions to s4 and Event Controller sends event Client_AcceptDecision_No to Blockchain.
5. Button Controller disables all buttons because the statechart in state s4 has no possible sending events.

Table 1. State transition table for Client

Current state	Receiving event	Sending events	Next state
s0	Client_PrepareClaim_Begin		s1
s1		Client_PrepareClaim_Finish	s2
s2	Client_AcceptDecision_Begin		s3
s2	Client_ReceivePayment_Begin		s6
s3		Client_AcceptDecision_No	s4
s3		Client_AcceptDecision_Yes	s7
s4	Client_UpdateClaim_Begin		s5
s5		Client_UpdateClaim_Finish	s2
s6		Client_ReceivePayment_Finish	s8

5   Evaluation

We evaluated the effectiveness of our approach using two insurance process models: *Marine Insurance* covers the loss of or damage to ships, cargo, and terminals caused by the sea. It involves Importer, Exporter, and Insurer as participants, and the process model contains 22 objects and 25 flows. *Re-Insurance* is "insurance of insurance" which is purchased by an insurance company and allows an insurance company to remain solvent after major claim events. It involves Reinsurer, Cedant (insurance company), and Surveyor as participants, and the process model contains 18 objects and 18 flows.

Tables 2 and 3 summarize the result of our statechart generation and statechart reduction algorithms applied to the Marine Insurance process and Re-Insurance process, respectively. They show the number of states, transitions, receiving events, and sending events before and after statechart reduction.

The statechart generation algorithms produced a set of statecharts that work together to correctly execute process models, but the resulting statecharts for Blockchain required a relatively large number of receiving and sending events (105 events for 22 objects in Marine Insurance, and 93 events for 18 objects in Re-Insurance). By applying the statechart reduction algorithms, we could reduce the number of events by 74% and 65%, which helped us to increase

Table 2. Experimental result of Marine Insurance process

Statechart	Before reduction				After reduction			
	States	Transitions	Receiving events	Sending events	States	Transitions	Receiving events	Sending events
Blockchain	31	35	20	85	31	35	20	17
Importer	31	35	25	10	19	20	10	10
Exporter	31	35	32	3	7	8	5	3
Insurer	31	35	28	7	13	15	8	7

Table 3. Experimental result of Re-Insurance process

Statechart	Before reduction				After reduction			
	States	Transitions	Receiving events	Sending events	States	Transitions	Receiving events	Sending events
Blockchain	31	31	16	77	31	31	16	17
Cedant	31	31	27	4	9	9	5	4
Reinsurer	31	31	20	11	21	21	10	11
Surveyor	31	31	30	1	4	3	2	1

the performance of the entire system. Also the number of state transitions of participants' statecharts was reduced to 68% on average, which makes it easier to build and maintain applications that comply with the process models.

6   Conclusions and Future Work

In this paper, we described a method for resolving the challenges posed by inter-organizational processes with Blockchain. By our method, we can define entire workflows that control processes across organizational boundaries, where the private processes fulfill the responsibilities imposed by the shared process. The resulting software artifacts built on top of Blockchain are highly optimized. We used Hyperledger Fabric as a Blockchain foundation, but our techniques described in this paper can also be applied to other configurations with smart contract capability, such as Ethereum with Solidity smart contract language [5].

Our next goal is to enhance our method to cover more complex process models and wider variety of process execution styles.

Process Fork and Join. In modeling processes, we excluded process forks and joins, which are still important process constructs [17]. Because statecharts support parallel states, mechanisms for representing forks and joins are already available. However, how to incorporate into our framework the mapping of parallel process constructs with statechart constructs is a remaining problem.

Private Processes inside Smart Contract. In this paper, we assumed that participants' private processes are executed outside of Blockchain. Realizing system-driven private processes as a smart contract and combining them with its shared process is another direction of our study.

Acknowledgments. We thank Takaaki Tateishi for providing us with the SCXML translator and statechart engine for Hyperledger Fabric, and Sachiko Yoshihama and Koichi Kamijo for their helpful discussion.

References

1. Androulaki, E., et al.: Hyperledger fabric: a distributed operating system for permissioned blockchains. arXiv preprint arXiv:1801.10228 (2018)
2. Angular (2016). https://angular.io/
3. BPMN 2.0 by Example (2010). https://www.omg.org/cgi-bin/doc?dtc/10-06-02.pdf
4. Business Process Model and Notation Specification Version 2.0.2 (2014). https://www.omg.org/spec/BPMN/2.0.2/PDF/
5. Buterin, V.: A next-generation smart contract and decentralized application platform. White paper (2014)
6. García-Bañuelos, L., Ponomarev, A., Dumas, M., Weber, I.: Optimized execution of business processes on blockchain. In: Carmona, J., Engels, G., Kumar, A. (eds.) BPM 2017. LNCS, vol. 10445, pp. 130–146. Springer, Cham (2017). https://doi.org/10.1007/978-3-319-65000-5_8
7. Harel, D.: Statecharts: a visual formalism for complex systems. Sci. Comput. Program. 8(3), 231–274 (1987)
8. López-Pintado, O., et al.: Caterpillar: a blockchain-based business process management system. In: Proceedings of the BPM Demo Track and BPM Dissertation Award co-located with 15th International Conference on Business Process Modeling (BPM 2017), Barcelona, Spain (2017)
9. Madsen, M.F., et al.: Collaboration among adversaries: distributed workflow execution on a blockchain. In: 2018 Symposium on Foundations and Applications of Blockchain (2018)
10. Mendling, J., et al.: Blockchains for business process management-challenges and opportunities. ACM Trans. Manag. Inf. Syst. (TMIS) 9(1), 4 (2018)
11. Sendall, S., Kozaczynski, W.: Model transformation: the heart and soul of model-driven software development. IEEE Softw. 20(5), 42–45 (2003)
12. State Chart XML (SCXML): State Machine Notation for Control Abstraction (2015). https://www.w3.org/TR/scxml/
13. Tabuchi, N., Sato, N., Nakamura, H.: Model-driven performance analysis of UML design models based on stochastic process algebra. In: Hartman, A., Kreische, D. (eds.) ECMDA-FA 2005. LNCS, vol. 3748, pp. 41–58. Springer, Heidelberg (2005). https://doi.org/10.1007/11581741_5
14. Weber, I., Xu, X., Riveret, R., Governatori, G., Ponomarev, A., Mendling, J.: Untrusted business process monitoring and execution using blockchain. In: La Rosa, M., Loos, P., Pastor, O. (eds.) BPM 2016. LNCS, vol. 9850, pp. 329–347. Springer, Cham (2016). https://doi.org/10.1007/978-3-319-45348-4_19
15. Yli-Huumo, J., et al.: Where is current research on blockchain technology?—a systematic review. PloS One 11(10), e0163477 (2016)
16. Ziadi, T., Helouet, L., Jezequel, J.-M.: Revisiting statechart synthesis with an algebraic approach. In: Proceedings of the 26th International Conference on Software Engineering. IEEE Computer Society (2004)
17. Zur Muehlen, M., Recker, J.: How much language is enough? Theoretical and practical use of the business process modeling notation. In: Bubenko, J., Krogstie, J., Pastor, O., Pernici, B., Rolland, C., Sølvberg, A. (eds.) Seminal Contributions to Information Systems Engineering. Springer, Heidelberg (2013). https://doi.org/10.1007/978-3-642-36926-1_35

Decentralized Voting: A Self-tallying Voting System Using a Smart Contract on the Ethereum Blockchain

Xuechao Yang[1(✉)], Xun Yi[1], Surya Nepal[2], and Fengling Han[1]

[1] School of Science, RMIT University, Melbourne, VIC 3000, Australia
xuechao.yang@rmit.edu.au
[2] CSIRO Data61, Sydney, NSW 2122, Australia

Abstract. Electronic online voting has been piloted in various countries in the recent past. These experiments show that further research is required, to improve the security guarantees of such systems, in terms of vote confidentiality and integrity and validity verification. In this paper we argue that blockchain technology, combined with modern cryptography can provide the transparency, integrity and confidentiality required from reliable online voting. Furthermore, we present a decentralized online voting system implemented as a smart contract on the Ethereum blockchain. The system has no hardwired restrictions on possible vote assignments to candidates, protects voter confidentiality by using a homomorphic encryption system and stores proofs for each element of a vote. To the best of our knowledge, our proposed system is the first decentralized ranked choice online voting system in existence. The underlying Ethereum platform enforces the correct execution of the voting protocol. We also present a security and performance analysis, showing the feasibility of our proposed protocol for real-world voting applications at large scale.

Keywords: Decentralized voting · Ethereum blockchain
Smart contract · Self-tallying

1 Introduction

A blockchain is a public, append-only, immutable ledger maintained by a decentralised peer-to-peer network. Whilst first designed for digital currencies without trusted third parties, blockchain technology has now moved into many fields beyond finance.

In this paper, we focus on blockchain-based online voting. There are a number of existing proposals for such a system, using the blockchain as a public bulletin board to store the voting data, such as FollowMyVote [7] and TIVI [21]. These proposals achieve voter privacy by involving trusted authorities that obfuscate the relation between real-world identities and keys [7], or by shuffling encrypted votes before decrypting [21].

© Springer Nature Switzerland AG 2018
H. Hacid et al. (Eds.): WISE 2018, LNCS 11233, pp. 18–35, 2018.
https://doi.org/10.1007/978-3-030-02922-7_2

We propose a self-tallying online voting system using a smart contract deployed on Ethereum. The system reduces the responsibilities of election authorities to a minimum and allows candidate ranking, instead of just voting for one candidate [14]. The system's voting mechanism is inspired by score voting [23], which enables voters to assign points to different candidates directly without any restrictions apart from the total number of available points specified (Fig. 1).

In most online voting systems, a tallying authority tallies the votes and decrypts the result [1, 22]. Self-tallying was introduced by Kiayisa and Yung [10] and developed by Groth [9], which converts tallying into an open procedure, that allows any voter or a third-party observer to perform the tally computation once all votes are cast. Unfortunately, self-tallying protocols have a fairness drawback, as the last voter can compute the tallying before anyone else. McCorry et al. [14] proposed a self-tallying protocol that avoids this adaptive issue. However, the system requires voters to vote in two rounds. Our system does not have any adaptive issue with only one round of voting.

Ballot	
Alice	2
Bob	2
David	2

(a)

Ballot	
Alice	0
Bob	0
David	6

(b)

Ballot	
Alice	1
Bob	3
David	2

(c)

Fig. 1. In this example six points can be distributed amongst candidates.

Online privacy preservation is one of the most pressing concerns in Internet technology [11–13, 16–20]. Therefore, maintaining the privacy and security of voters is the priority for our online voting system. Our proposed decentralized voting system uses the exponential ElGamal encryption system [3] and an open vote network protocol [14]. The additive homomorphism property of the cryptographic system makes it possible to tally encrypted votes directly without decrypting them. Our proposed system also incorporates cryptographic proofs to ensure the integrity of the voting process and to verify the validity of each vote before it is saved to the blockchain. To the best of our knowledge our voting system is the first decentralized ranked choice online voting system in existence, which meets the following security requirements [22, 23]:

Eligibility of Voters: Only authorized voters can submit their cast votes.

Multiple-Voting Detection: Multiple voting by any one voter is detected and identified.

Privacy of Voters: All submitted votes must be stored securely and secretly and should not reveal voting preferences of the voters.

Integrity of Ballot: No one can modify or duplicate any submitted votes.

Correctness of Tallied Result: Only verified votes are counted to calculate the result.

End-to-End Voter Verifiable: Every voter is able to verify whether his/her vote is posted and counted correctly, and also able to verify the eligibility of other submissions.

The rest of this paper is organized as follow: Sect. 2 demonstrates a simple voting contract that is deployed on Ethereum. Section 3 describes the cryptographic models used in our online voting system. Section 4 presents our proposed online voting system. A security and performance analysis can be found in Sects. 5 and 6, respectively. Finally, Sect. 7 concludes the paper.

2   Preliminaries on Smart Contract Voting

Blockchain-Based Smart Contracts: Blockchain technology was first introduced by the Bitcoin digital currency in 2008 [2]. Bitcoin proposed a solution to securely maintain a decentralized ledger in the presence of a Byzantine failure model [4], in which nodes may act maliciously.

Blockchain ledgers were originally designed to record monetary transactions, but the concept has been widened to provide support for general purpose computing. Ethereum [5] provides a Turing complete platform for decentralized smart contracts. Smart Contracts were first described in 1996 by Nick Szabo [6] and are autonomously executing contracts written in computer code.

The Blockchain provides the following properties which make smart contracts possible:

Transparency. Blockchain transactions are public and can be verified by anyone.

Immutability. The transaction history of a blockchain cannot be altered. As such, the Blockchain can be seen as an append-only database.

Trustless. Participants in blockchain transactions do not have to rely on a trusted third party for their interactions. Trust is provided by the underlying consensus protocol.

Decentralized Voting: Ethereum provides a natural platform for our distributed voting system, in that it provides a decentralized "public bulletin board" to support coordination amongst voters. The execution of the election procedure is enforced by the same consensus mechanisms that secures the Blockchain. The smart contract code is stored on the Blockchain and executed by all peers to reach consensus on its output.

We present a simple voting contract written in Ethereum's Solidity language. The implementation was deployed on Ethereum's Kovan test network and the contract's interface is as follows:

VotingContract(candidateList, voterList, definedPoint): this is the constructor function of the contract. In order for the election administrator to deploy a new contract, there are three parameters that have to be provided: (1) a list of candidates; (2) a list of eligible voters; and (3) total available points. Once the contract is deployed, it is immutable.

submitVote(vote, voterSign): an eligible voter is able to cast and submit a vote via this function. This function calls a contract internal **verifyPendingVote**(vote) function, which verifies the eligibility of the vote. The function returns true (success) or false.

verifyAddedVote(voterID) constant returns (bool): Each voter is able to verify the eligibility of any other voter's submission before self-tallying.

tallyVotes(candidateName) constant returns (uint8): voters can tally any candidate's final received points independently by using this self-tallying function.

The above voting contract submits and stores data in plaintext format. In order to protect the privacy of voters, an encryption system has to be used.

3   Preliminaries on Cryptography

In this section, we introduce the underlying cryptographic building blocks for our proposed online voting system. We combine two cryptographic systems to ensure both voter privacy and verifiability of the result. The two systems involved are the ElGamal Cryptosystem [3] and the distributed encryption protocol described in [8].

ElGamal Cryptosystem: We assume that the cyclic group (G, q, g) is defined. A user has a public key y and private key x. The ElGamal cryptosystem consists of the following algorithms:

Encryption. To encrypt a plaintext message $m \in G$: Randomly choose an integer r from $\mathbb{Z}_q^*$; Computes $c_1 = g^r$; Computes $c_2 = g^m \cdot y^r$. And the encrypted message can be presented as $E(m) = (c_1, c_2)$.

Decryption. A user computes and broadcasts a partially decrypted value, and the final plaintext is revealed. For the ciphertext (c_1, c_2), decryption proceeds as follows: The user with secret key x computes c_1^x and broadcasts to others; Everyone is able to compute $\dfrac{c_2}{c_1^x} = g^m$. Finally, m can be revealed by computing a discrete logarithm.

Homomorphism. ElGamal encryption has an inherited homomorphic property [23], which allows multiplication and exponentiation to be performed on a set of ciphertexts without decrypting them, such as $E(m_1) \times E(m_2) = (g^{r_1}, g^{m_1} \cdot y^{r_1}) \times (g^{r_2}, g^{m_2} \cdot y^{r_2}) = (g^{r_1+r_2}, g^{m_1+m_2} \cdot y^{r_1+r_2}) = E(m_1 + m_2)$.

Distributed Encryption: works as follows: Let G denote a finite cyclic group of prime order q in which the decision Diffie-Hellman problem is intractable. Let g be a generator in G. There are n users, all of whom agree on (G, g). We assume there are n different users $u_1, u_2, \cdots, u_n$. Each user u_i chooses a secret value $x_i \in_R \mathbb{Z}_q$, and computes a public value g^{x_i}, where $1 \leq i \leq n$. Each u_i computes a y_i as below:

$$y_i = \frac{\prod_{j=1}^{i-1} g^{x_j}}{\prod_{j=i+1}^{n} g^{x_j}} \tag{1}$$

which is publicly computable since the computation uses all public values g^{x_i}.

We assume the message for each u_i is m_i, and the encrypted message $E(m_i, y_i, x_i)$ is $(y_i)^{x_i} \cdot g^{m_i}$, where $E(m_i, y_i, x_i)$ denotes the message m_i encrypted using y_i and x_i.

This cryptosystem also provides for **homomorphism addition**. Therefore, anyone can compute the sum of all ciphertexts simple by multiplying all encrypted messages:

$$\prod_{i=1}^{n} E(m_i, y_i, x_i) = \prod_{i=1}^{n}(y_i)^{x_i} g^{m_i} = \prod_{i=1}^{n}(y_i)^{x_i} \times \prod_{i=1}^{n} g^{m_i} = \prod_{i=1}^{n} g^{m_i} = g^{\sum_{i=1}^{n} m_i}$$

where $\prod_{i=1}^{n}(y_i)^{x_i} = 1$, according to [8].

Zero knowledge Proof: Our zero knowledge proof protocol is based on [15]. Given a cyclic group $G = < g > = < h >$ and public knowledge $A = g^x$ and $B = h^x$, the prover wants to convince verifier(s) A and B have the same exponentiation, but the verifier(s) cannot learn the value of x.

    Prover: choose $t \in \mathbb{Z}_q$, computes $T_1 = g^t$, $T_2 = h^t$, $c = Hash(T_1||T_2)$, $s = x \cdot c + t$ sends T_1, T_2, s to **Verifier**

    Verifier: computes $c = Hash(T_1||T_2)$, verifies if $g^s = A^c \cdot T_1$ and if $h^s = B^c \cdot T_2$

If both verifications are passed, the verifier believes the prover knows x, but cannot determine the value of x.

4   Our Proposed Voting System

In this section we present our proposed decentralized, self-tallying, ranked choice, smart contract-based voting system. The basic idea is as follows: The election administrator deploys a voting contract by confirming public parameters (such as the public key of the election). Each voter can then submit a vote via the voting contract, with each vote constituting a transaction of the blockchain system. In case of the vote not being verified as valid by the checks performed in the smart contract, the transaction reverts. After being mined by the blockchain's consensus algorithm, the vote is considered final. Figure 2 presents the stages of our proposed election.

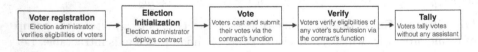

Fig. 2. The five steps of our proposed decentralized online voting system.

The involved participants in our proposed system are:

Election Administrator: An election administrator is required to set the election's parameters, begin the registration stage and add voters to the list of eligible voters. The administrator should also generate a key pair (public key and private key) of the election and contribute the public key to the blockchain. Furthermore, the administrator is responsible for voter registration, generating a candidate list, setting rules of the election, and deploying the voting contract, which cannot be changed once the election started.

Candidates: A list of candidates is generated by the election administrator. Each candidate is a contestant in the election and will receive points from different voters.

Voters: Each voter has a private key and public key. The public key is added to the blockchain after the eligibility of the voter is verified by the election administrator. The voter can submit his/her cast vote via the function provided by the smart contract.

Blockchain Database: A distributed and append-only database. All submissions will be added to the latest block of the chain once they are verified.

Table 1 provides the notations used to explain our protocols.

Table 1. Notations that used in the rest of the paper

n_c:	number of Candidates
n_v:	number of Voters
v_i:	i-th voter; $i \in [1, n_v]$
c_j:	j-th candidate; $j \in [1, n_c]$
$x_{v_i}^{c_j}$:	secret key of i-th voter that is used to vote for c_j candidate; $i \in [1, n_v], j \in [1, n_c]$
pk:	public key of election administrator
sk:	secret key of election administrator
P:	the pre-defined point, where the sum of all assigned points must equal to P
$p_{v_i}^{c_j}$	a point that is assigned by v_i to c_j, $i \in [1, n_v]$, $j \in [1, n_c]$
p^{c_j}	total received point of c_j, $j \in [1, n_c]$
ZPK{...}:	proof of zero knowledge

4.1   Initialization and Voter Registration

Before an election can start the cyclic group (G, p, g) is defined. The election administrator generates an ElGamal key pair (public key pk and secret key sk), and pk is added to the blockchain database, which can be accessed by all voters.

The only rule for defining the election parameters is that the sum of all assigned points must be a fixed number (which we treat it as P), the election administrator defines a list of the candidates and the value of P before the election starts.

In order to register, each voter must select n_c secret keys $x_{v_i}^{c_j} \in Z_p$ and compute the n_c corresponding public keys $g^{x_{v_i}^{c_j}} \pmod{p}$. The voter must register his real-world identification and his/her n_c pubilc keys to the election administrator. Once the eligibility is verified, the voter will be added to the list of eligible voters, and all his/her $g^{x_{v_i}^{c_j}}$ will be added to the blockchain. Once all eligible voters are registered for the election (or the deadline of registration has passed), the election administrator deploys the voting contract.

4.2   Voting Process

The proposed voting system allows voters to assign different scores to different candidates according to their personal preferences. There are three phases in the voting stage: pre-computing, vote casting and proof generation.

We do not remove the connection between the identity of voters and their votes, meaning everyone can see that a voter submitted his/her vote. However, the content of votes is encrypted, meaning no-one is able to reveal the content of any individual vote.

Pre-computing: We assume there are n_v registered voters, and all $g^{x_{v_i}^{c_j}}$ are viewable in the blockchain database. Thus, the pre-computing values $y_{v_i}^{c_j}$ of voters can be computed by using all other $g^{x_{v_i}^{c_j}}$ via Eq. 1. At the end, each v_i has n_c pre-computed values as $y_{v_i}^{c_1}, y_{v_i}^{c_2}, \cdots, y_{v_i}^{c_{n_c}}$, and each value can only be used to vote for the particular c_j.

Vote Casting: Each v_i is able to assign any integer point (from 0 to P) to different candidates, but the sum of all assigned points must equal to P (see Fig. 1), which is the rule each voter has to follow. Because each vote consists of multiple assigned points (according to the number of candidates), those points are treated as private and confidential to the voters. Thus, those scores must be encrypted before submission. In our case, we use $p_{v_i}^{c_j}$ to denote a score that is assigned by voter v_i to candidate c_j, which will be encrypted twice: ElGamal encryption and distributed encryption.

ElGamal Encryption. Each assigned point is encrypted using the public key pk of the election administrator. For example

$$E(p_{v_i}^{c_j}, pk) = g^r, \ g^{(p_{v_i}^{c_j})} \cdot pk^r$$

meaning the score $p_{c_j}^{v_i}$ is encrypted using pk according to the ElGamal encryption.

Distributed Encryption: Once the point is encrypted by ElGamal encryption, the first part (g^r) of the encrypted value will be "encrypted" again by using the private voting key (x_{v_i}) of the voter v_i, such as

$$g^r \rightarrow (y_{v_i}^{c_j})^{x_{v_i}^{c_j}} \cdot g^r$$

where $y_{v_i}^{c_j}$ is computed during the pre-computing phase and publicly accessible.

To summarise, we developed the encryption algorithm based on both the ElGamal encryption and group-based encryption, meaning each assigned point will be encrypted as per Eq. 2

$$E(p_{v_i}^{c_j}, pk, y_{v_i}^{c_j}, x_{v_i}^{c_j}) = (y_{v_i}^{c_j})^{x_{v_i}^{c_j}} g^r, \; g^{(p_{v_i}^{c_j})} \cdot pk^r \qquad (2)$$

where $p_{v_i}^{c_j}$ is encrypted by using pk (public key of the election), $y_{v_i}^{c_j}$ (pre-computed value that is used by v_i to vote c_j) and $x_{v_i}^{c_j}$ (the particular private key of v_i to vote for c_j). Thus, a cast Vote_{v_i} (with n_c candidates) can be presented as:

$$\mathrm{Vote}_{v_i} = \begin{bmatrix} E(p_{v_i}^{c_1}, pk, y_{v_i}^{c_1}, x_{v_i}^{c_1}) \\ \vdots \\ E(p_{v_i}^{c_{n_c}}, pk, y_{v_i}^{c_{n_c}}, x_{v_i}^{c_{n_c}}) \end{bmatrix}$$

Proof Generation: In order to allow anyone to verify the eligibility of each vote without decrypting the cipher text and revealing the content, each voter is required to generate several proofs for his/her vote before submission (ZKP denotes zero knowledge proof):

- ZKP($x_{v_i}^{c_j}$): to prove each encrypted point for the candidate c_j is computed correctly using the voter's private key $x_{v_i}^{c_j}$.
- ZKP(P): to prove that the sum of all encrypted points is equal to P.

The voter v_i has to generate ZKP($x_{v_i}^{c_j}$) for each encrypted point $E(p_{v_i}^{c_j}, pk, y_{v_i}^{c_j}, x_{v_i}^{c_j})$, and ZKP($P$) for the Vote_{v_i}. The summarised processing procedure of the voting stage is shown in Algorithm 1.

Remark 1. The computation details about how to generate the ZKP($x_{v_i}^{c_j}$) and ZKP(P) can be found in **Appendix**.

4.3   Vote Verification Stage

In order to prevent multiple counting of any individual vote into the final result, vote verification is required as follows:

Verify Each Encrypted Point: In order to prevent having any error during tallying all submissions, each encrypted point $E(p_{v_i}^{c_j}, pk, y_{v_i}^{c_j}, x_{v_i}^{c_j})$ has to be confirmed as to have been computed with the correct parameters. The verification can be done by using the corresponding proofs ZKP($x_{v_i}^{c_j}$) that are generated during vote casting.

Verify Sum of All Encrypted Points: According to the rules of the election, each voter cannot assign more than the pre-defined total available point P in his/her cast vote. Using homomorphic addition, anyone is able to compute the sum (encrypted) of all encrypted points and verify the value by using the corresponding proof ZKP(P) that are generated by the voter.

The processing procedure of the verification is shown as Algorithm 2 (The purpose of function **verifyAddedVote** is similar, but the input parameters differ).

Algorithm 1. function **submitVote**

Input : pre-defined point: P, public key pk,
all secret keys of v_i: $x_{v_i}^{c_1}, \cdots, x_{v_i}^{c_{n_c}}$
voting public keys of all voters $g^{x_{v_1}^{c_1}}, \cdots, g^{x_{v_{n_v}}^{n_c}}$

Output: Vote_{v_i}

1 computes $y_{v_i}^{c_j}$, $j \in [1, n_c]$.   ▷ refer to Eq. 1

2 set $\text{Vote}_{v_i} = []$

3 **for** $j \leftarrow 1$ **to** n_c **do**

4 $\quad$ $E(p_{v_i}^{c_j}, pk, y_{v_i}^{c_j}, x_{v_i}^{c_j}) = ((y_{v_i}^{c_j})^{x_{v_i}^{c_j}} \cdot g^r, g^{p_{v_i}^{c_j}} \cdot pk^r)$ ▷ refer to Eq. 2

5 $\quad$ $\text{ZKP}(x_{v_i}^{c_j})$: $\{K_1, K_2, Z_1, Z_2\}$ ▷ **Remark 1**

6 $\quad$ $\text{Vote}_{v_i} = \text{Vote}_{v_i} \cup [E(p_{v_i}^{c_j}, pk, y_{v_i}^{c_j}, x_{v_i}^{c_j}), \text{ZKP}(x_{v_i}^{c_j})]$

7 **end**

8 $\text{ZKP}(P)$: $\{T_1, T_2, s, z\}$ ▷ **Remark 1**

9 $\text{Vote}_{v_i} = \text{Vote}_{v_i} \cup [\text{ZKP}(P)]$

10 $\text{Signature}_{v_i} = \text{Sign}(\text{Vote}_{v_i})$

11 **if** **verifyPendingVote**(Vote_{v_i}) == False **then**

12 $\quad$ **return** False

13 **end**

14 **return** $\text{Vote}_{v_i} = \begin{bmatrix} E(p_{v_i}^{c_1}, pk, y_{v_i}^{c_1}, x_{v_i}^{c_1}), \text{ZKP}(x_{v_i}^{c_1}) \\ \vdots \\ E(p_{v_i}^{c_{n_c}}, pk, y_{v_i}^{c_{n_c}}, x_{v_i}^{c_{n_c}}), \text{ZKP}(x_{v_i}^{c_{n_c}}) \\ \text{ZKP}(P) \end{bmatrix}, \text{Signature}_{v_i}$

Remark 2. The computation details about how to verify the $\text{ZKP}(x_{v_i}^{c_j})$ and $\text{ZKP}(P)$ can be found in **Appendix**.

4.4 Votes Tally Stage

Once all voters have submitted their Vote_{v_i} and the deadline of submission has passed, the election administrator must do the following: (1) compute the tallying result (via homomorphic addition), (2) compute their partially decrypted value and proof; (3) send partially decrypted values (including proofs) to the blockchain.

Each point is encrypted using our developed encryption algorithm (Eq. 2), in which the cipher texts can be computed by homomorphic addition. In this case, we can simply multiply all Vote_{v_i} in the blockchain database as shown below, where we assume there are n_v voters and n_c candidates, and all Vote_{v_i} have been verified as valid.

$$\prod_{i=1}^{n_v} \text{Vote}_{v_i} = \begin{bmatrix} \prod_{i=1}^{n_v} E(p_{v_i}^{c_1} \cdots) \\ \vdots \\ \prod_{i=1}^{n_v} E(p_{v_i}^{c_{n_c}} \cdots) \end{bmatrix} = \begin{bmatrix} \prod_{i=1}^{n_v} (y_{v_i}^{c_1})^{x_{v_i}^{c_1}} g^{r_1}, \ g^{\sum_{i=1}^{n_v} p_{v_i}^{c_1}} pk^{r_1} \\ \vdots \\ \prod_{i=1}^{n_v} (y_{v_i}^{c_{n_c}})^{x_{v_i}^{c_{n_c}}} g^{r_{n_c}}, \ g^{\sum_{i=1}^{n_v} p_{v_i}^{c_{n_c}}} pk^{r_{n_c}} \end{bmatrix}$$

$$(3)$$

Algorithm 2. function **verifyPendingVote**

Input : Vote_{v_i}, g, all $g^{x_{v_i}^{c_j}}$, all $y_{v_i}^{c_j}$
Output: Valid or Invalid

1 **for** $j \leftarrow 1$ **to** n_c **do**
2 $\quad$ sum $* = E(p_{v_i}^{c_j}, \cdots)$
3 $\quad$ //verify $E(p_{v_i}^{c_j}, \cdots)$ using corresonding ZKP$(x_{v_i}^{c_j})$ $\hfill \triangleright$ **Remark 2**
4 $\quad$ $E(p_{v_i}^{c_j}, \cdots) = (c_1, c_2)$ $\hfill \triangleright$ refer to Algorithm 1
5 $\quad$ ZKP$(x_{v_i}^{c_j}) = \{K_1, K_2, Z_1, Z_2\}$ $\hfill \triangleright$ refer to Algorithm 1
6 $\quad$ compute $c = Hash(K_1 \| K_2)$
7 $\quad$ **if** $(y_{v_i}^{c_j})^{Z_1} g^{Z_2} \neq K_1 \times (c_1)^c$ OR $g^{Z_1} \neq K_2 \times (g^{x_{v_i}^{c_j}})^c$ **then**
8 $\quad\quad$ | **return** False
9 $\quad$ **end**
10 **end**
11 //verify sum using corresonding ZKP(P) $\hfill \triangleright$ **Remark 2**
12 assume sum $= (c_1, c_2)$
13 ZKP$(P) = \{T_1, T_2, s, z\}$ $\hfill \triangleright$ refer to Algorithm 1
14 compute $c = Hash(T_1 \| T_2)$
15 **if** $(y_{v_i}^{c_j})^s \neq (\frac{c_1}{z})^c \cdot T_1$ OR $pk^s \neq (\frac{c_2}{g^P})^c \cdot T_2$ **then**
16 $\quad$ | **return** False
17 **end**
18 **return** True

Due to $\prod_{i=1}^{n_v} (y_{v_i}^{c_j})^{x_{v_i}^{c_j}} = 1$ (refer to Sect. 3), $\prod_{i=1}^{n_v} \text{Vote}_{v_i}$ can be treated as n_c ciphertexts by ElGamal encryption, such as $E(\sum_{i=1}^{n_v} p_{v_i}^{c_j}), j \in [1, n_c]$.

The election administrator then has to compute partially decrypted values, such as $(g^{r_1})^{sk}, \cdots, (g^{r_{n_c}})^{sk}$. He/she must also generate the corresponding proof for each partially decrypted value to prove that each value is computed correctly using the secret key sk. Finally, the election administrator broadcasts the partially decrypted values (including the corresponding proofs ZKP(sk)) to the blockchain. The winner of the election can be computed by any voter with Algorithm 3.

Remark 3. The verification of each partial decrypted value can be treated as verifying if $(g^{r_j})^{sk}$ has the same exponentiation as pk, where $pk = g^{sk}$. The procedure is same as the example in Sect. 3.

Because the tallying algorithm is a function of the voting contract, the tallying result can be computed by any voter individually, without any key or decryption function.

5 Security Analysis

This section is devoted to a theoretical security analysis of our system. Noted that none of the previous related papers provided a formal security model, including only a description and an informal security discussion of their systems. Our

---

Algorithm 3. function tallyVotes

Input   : all valid votes $\text{Vote}_{v_1}, \cdots, \text{Vote}_{v_{n_v}}$ in blockchain

all paritial decryption values $(g^{r_1})^{sk}, \cdots, (g^{r_{n_c}})^{sk}$ by election administrator

Output: $p^{c_1}, \cdots, p^{c_{n_c}}$

1 compute $\prod_{i=1}^{n_v} \text{Vote}_{v_i}$ ▷ refer to Eq. 3

2 //verify each partial decryption value using corresponding ZKP(sk)

3 **for** $j \leftarrow 1$ **to** n_c **do**

4 | verify $(g^{r_j})^{sk}$ using corresponding ZKP(sk)   ▷ **Remark 3**

5 **end**

6 //reveal result for all candidate using partial decryption values

7 **for** $j \leftarrow 1$ **to** n_c **do**

8 | $p_{c_j} = \dfrac{\prod_{i=1}^{n_v} g^{p_{v_i}^{c_1}} pk^{r_j}}{(g^{r_j})^{sk}} = \dfrac{g^{\sum_{i=1}^{n_v} p_{v_i}^{c_j}} (g^{sk})^{r_j}}{(g^{r_j})^{sk}} = g^{p_{v_1}^{c_j} + \cdots + p_{v_{n_v}}^{v_{c_j}}}$ ▷ refer to Sect. 3

9 **end**

10 **return** $p_{c_1}, p_{c_2}, \cdots, p_{c_{n_c}}$

---

analysis makes the following assumptions: (1) the election administrator and voters are always identifiable, as all blockchain transactions are signed with sender's private key. (2) Voters will never disclose their private voting keys $x_{v_i}^{c_j}$; (3) the blockchain database is secure and insert-only; (4) Our system relies on several cryptographic protocols, which are presented in Sect. 3 and have reliable published proofs of their security.

Theorem 1. *If the digital signature algorithm (such as DSA) is non-falsifiable, no one is able to submit a ballot by impersonating another voter.*

Proof. In order to prevent adversaries from casting ballots by impersonating authenticated voters, we require each voter to submit with his/her digital signature algorithm. In our proposed system, only eligible voters are added to the voters list by the election administrator once their identities have been verified. The signing_verify key of each verified voter is stored on the blockchain, and the voter is responsible for keeping their signing key secret. Once the election starts, each authorized voter signs their votes by using his/her signing key and submits the vote along with their signature. The smart contract is able to verify if each submission by verifying the digital signature using the corresponding signing_verify key.

Theorem 2. *Only one submission from each voter is accepted as valid.*

Proof. In our proposed system, only the content of a cast vote is encrypted, the identification of the voter (and the digital signature) is in plaintext and can be viewed by everyone. Thus, multiple-voting detection is achieved by our system, as it can always detect whether a voter has previously submitted a vote. Furthermore, depending on the requirements of the particular scenario, our system can accept one submission of each voter or accept multiple submissions for each voter and use the last vote as valid.

Theorem 3. *If the underlying cryptographic systems are semantically secure, then the votes' contents will never be revealed to anyone (including the election administrator).*

Proof. Every vote is encrypted twice before submission. We use the ElGamal cryptosystem and a distributed encryption algorithm, which inherits the homomorphic property from the standard ElGamal system. Both algorithms are semantically secure.

All the submitted votes remain in encrypted form as cipher texts all the time. The homomorphic property makes it possible to add all encrypted votes without decrypting them. Furthermore, there is no relationship between the cipher texts and the corresponding plaintexts since the cryptosystem employed is probabilistic. It applies random numbers, so that the cipher text can take on different values even when the encryption is computed from the same input. Finally, due to each value being encrypted by both the public key of the election administrator and the secret voting key of the voter, the decryption must be done via collaboration of the election administrator and the voter. This means that, if the voter kept his/her secret voting keys as secret all the time (that is also one of our assumptions), even the election administrator cannot reveal anything.

Theorem 4. *Integrity of all cast votes are secured after submission.*

Proof. Firstly, we require voters to sign their cast votes by using their signing keys (refer to Algorithm 1), and we assume voters do not share their signing keys, to ensure that nobody can modify the content of a submission and fake the voter's signature. Secondly, all cast votes will be verified being before being added to the blockchain. Third, all verified votes will be added to the blockchain, being logged in an immutable ledger. Thus, the integrity of all submitted votes is treated as secure.

Theorem 5. *Invalid votes can be detected by any individual voter.*

Proof. Each cast vote is added to the blockchain database with corresponding proofs, generated by using Zero Knowledge Proof. The verification algorithm is public to all voters, which means the voters are able to verify any vote without any assistant.

Theorem 6. *The self-tallying algorithm is proposed public accessible and anyone can use it to tally votes without assistant.*

Proof. Once all votes are verified and added to the blockchain, we require the election administrator to compute the partially decrypted value in order to allow voters to compute the tallied result by themselves. In the meantime, the election administrator must generate corresponding proofs to convince all voters that all partially decrypted values are computed correctly.

Theorem 7. *Voters are able to verify everything of the election.*

Proof. In our system, all content (encrypted votes, proofs and signature) for each submission is broadcasted and added to the blockchain database, where they can be accessed by anyone. We assume that the blockchain database is secure, and it is "append-only". The voters can do the following without any assistant: (1) Voters can verify the blockchain transactions themselves. (2) Voters can verify the integrity of each submission by using the corresponding signing-verify keys from all voters. (3) Voters can verify each partially decrypted value (computed by election administrator) is computed correctly. (4) Voters can self-tally all votes and compute the final result of the election.

6   Performance Analysis

This section discusses the performance of our proposed voting system. The analysis is based on the computation time of each processing step, separated into 3 phases, vote casting performance, votes verification performance and votes tallying performance. In our proposed protocol, each vote is encrypted twice using different keys (common key of election administrator and secret key of the voter, refer to Algorithm 1). All tests were performed using a 512-bit key (p is 512-bit), which provides a higher security level than one-time encryption using a 1024-bit key.

We tested our proposed protocol using a high performance implementation of libgmp via the gmpy2 python module (https://gmpy2.readthedocs.io/en/latest/), on a laptop with the following specifications: 2.8 GHz quad-core Intel Core i7 with 6 MB shared L3 cache and with 16 GB of 1600 MHz DDR3L on-board memory.

We use t to denote the computation time of one exponentiation, where $t = 0.09$ ms. ElGamal encryption requires two exponentiations, and ElGamal decryption requires one exponentiation, where the division can be avoided by using an alternative method (https://wikipedia.org/wiki/ElGamal_encryption). Thus, we use t_E and t_D to denote the computation time of encryption and decryption, respectively, where $t_E = 2t$ and $t_D = t$, approximately. Precomputed values of distributed encryption (refer to Eq. 1) require one exponentiation (the inverse power computation), and encryption also has cost of one exponentiation.

6.1   Vote Casting Performance

The performance can be analysed for the following aspects:

Total Computation Time: According to the Algorithm 1, we use T_{voter} to denote the total time spent before submission (including the proof generation time), where

$$T_{voter} = (t_E \times n_c + t * n_c) + (3 * n_c * t) + (3 * t) = (6n_c + 3)t$$

In this experiment, we tested $T_{v}oter$ in five rounds, varying the number of candidates ($n_c = 3, 5, 10, 15, 20$). The result is shown in (a) Fig. 3. From the results in (a) Fig. 3, we can see the time cost for casting a vote is less than 12 ms even if there are 20 candidates to be ranked.

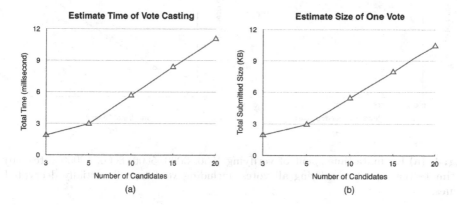

Fig. 3. Performance of voter side when the number of candidates is 3, 5, 10, 15, 20: (a) Time spent encrypting a cast vote, including generation time of all proofs (b) The size of a submission, includes all encrypted values and all proofs.

Total Submission Size: We assume the size of digital signature is 1024-bit (refer to Algorithm 1), and we use S_{vote} to denote the total submission size (bits) for a voter,

$$S_{vote} = (1024 \times n_c) + (2048 \times n_c) + (2048 + 512 + 2 * n_c * 512) + (1024) = 4096 * n_c + 3584$$

The test result is shown in (b) Fig. 3 based on different numbers of candidates ($n_c = 3, 5, 10, 15, 20$). From the result of (b) Fig. 3, we found the submission size of one vote is less than 11 KB even for a 20-candidate ballot.

6.2 Votes Verification Performance

We have also evaluated the performance of the verification time for submissions a member of the public or an independent observer might with to verify. Due to the verification of each voter's identification being equivalent to verifying the digital signature of each submission, this is not computationally expensive. Thus, we concentrated on the performance of Algorithm 2. We use T_{verify} to denote the total time spent verifying votes and n to denote the total number of votes being verified, which can be presented as follows:

$$T_{verify} = \big((5t \times n_c) + (n_c + 1 + 2 * n_c)t + (6t)\big) \times n = (8 * n_c + 7)t$$

We tested T_{verify} in five rounds, varying the numbers of votes verified ($n = 1000, 3000, 5000, 8000, 10000$). In this experiment, we assume the number of candidates is 10 ($n_c = 10$), and the result is shown in Fig. 4(a). From the results in Fig. 4, we found the time spent verifying 10,000 ballots costs less than 1.5 min.

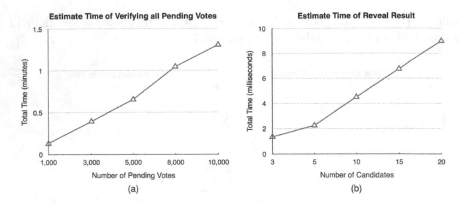

Fig. 4. (a) Estimate time spent of verifying 1000, 3000, 5000, 8000, 10000 votes. (b) Estimate time spent of tallying all votes, including verifying all partially decrypted values.

6.3   Votes Tallying Performance

Our proposed system allows voters to self-tally all submitted votes by using all partially decrypted values from the election administrators. However, before tallying starts, each partially decrypted value must be verified using the corresponding proofs (refer to Sect. 4.4). T_{reveal} is used to denote the total time spent verifying all partially decrypted values, tally all votes, reveal the result (refer to Algorithm 3), which is presented as:

$$T_{reveal} = (4t * n_c) + (n_c * t) = 5 * n_c * t$$

Again, we tested T_{tally} in five rounds varying the number of candidates ($n_c = 3, 5, 10, 15, 20$). The result of this experiment is shown in Fig. 4(b). We found the time spent tallying all votes (including verifying all partial decryption proofs) costs less than 10 ms using the same test machine.

7   Conclusion and Future Work

We have proposed a secure decentralized online voting system using cryptography and a smart contract, which allows the voters to cast their ballots by assigning arbitrary numbers of points to different candidates. This means that the voters can assign equal points to different candidates, or they can assign different points to different candidates. Each cast vote in our system is encrypted before submission and remains encrypted at all times. The additive homomorphic property of the ElGamal cryptosystem enables effective processing of the cipher texts during these procedures. Furthermore, the eligibility of voters and their submissions can be verified by anyone without revealing the contents of the ballots. The security and performance analysis confirm the feasibility of our proposed cryptographic voting contract.

There is a limitation in that our system may suffer abortive issue [10, 14]. We have to assume that all registered voters submit their valid votes since otherwise, any voter can abort the tally without submitting his/her vote. In future work, we will address this issue and consider further generalizations. Furthermore, we will migrate the whole system to an Ethereum network and perform trials of the protocol at a larger scale.

Appendix

In order to protect the privacy of the voters, each assigned point is encrypted (refer to Eq. 2) before submission. However, voters have to generate corresponding proofs to prove their votes are cast correctly by observing the following: 1. each encrypted score is computed correctly using the voter's private key $x_{v_i}^{c_j}$; 2. sum of all encrypted points are equaled to pre-defined total available point P.

Each Encrypted Value is Computed Correctly: We present the proofs generation and verification for an encrypted score in Vote_{v_1}, where we assume v_1 assigned 5 points to c_1, and the encrypted value should be $E(5, pk, y_{v_1}^{c_1}, x_{v_1}) = (c_1, c_2)$, where $c_1 = (y_{v_1}^{c_1})^{x_{v_1}^{c_1}} \cdot g^r$, $c_2 = g^5 \cdot pk^r$:

Prover: generates a random number $k_1, k_2 \in \mathbb{Z}_q$, computes $K_1 = (y_{v_1}^{c_1})^{k_1} \cdot g^{k_2}$, computes $K_2 = g^{k_1}$, $c = Hash(K_1 \| K_2)$, $Z_1 = x_{v_1}^{c_1} c + k_1$, $Z_2 = rc + k_2$. And $ZKP(x_{v_i}^{c_1}) = \{K_1, K_2, Z_1, Z_2\}$

Verifier: compute $c = Hash(K_1 \| K_2)$, verify if $(y_{v_1}^{c_1})^{Z_1} g^{Z_2} = K_1 \times (c_1)^c$, verify if $g^{Z_1} = K_2 \times (g^{x_{v_i}^{c_1}})^c$, where $y_{v_1}^{c_1}$ and $g^{x_{v_i}^{c_1}}$ are public values, and the verifier(s) will never know the value is encrypted from 5.

Sum of All Encrypted Values is Equivalent to Encrypted from P: We present the proofs generation and verification for the sum of all encrypted points in Vote_{v_1}, where we assume $P = 10$, and there are 3 candidates c_1, c_2 and c_3. Voter v_1 cast a vote as:

$$E(5, pk, y_{v_1}^{c_1}, x_{v_1}^{c_1}) = (y_{v_1}^{c_1})^{x_{v_1}^{c_1}} \cdot g^{r_1}, g^5 \cdot pk^{r_1}$$
$$E(2, pk, y_{v_1}^{c_2}, x_{v_1}^{c_2}) = (y_{v_1}^{c_2})^{x_{v_1}^{c_2}} \cdot g^{r_2}, g^2 \cdot pk^{r_2}$$
$$E(3, pk, y_{v_1}^{c_3}, x_{v_1}^{c_3}) = (y_{v_1}^{c_3})^{x_{v_1}^{c_3}} \cdot g^{r_3}, g^3 \cdot pk^{r_3}$$

Prover: multiply them as

$$E(5, pk, y_{v_1}^{c_1}, x_{v_1}^{c_1}) \times E(2, pk, y_{v_1}^{c_2}, x_{v_1}^{c_2}) \times E(3, pk, y_{v_1}^{c_3}, x_{v_1}^{c_3})$$
$$= (y_{v_1}^{c_1})^{x_{v_1}^{c_1}} \cdot (y_{v_1}^{c_2})^{x_{v_1}^{c_2}} \cdot (y_{v_1}^{c_3})^{x_{v_1}^{c_3}} \cdot g^{r_1+r_2+r_3}, g^{5+2+3} \cdot pk^{r_1+r_2+r_3}$$
$$= (y_{v_1}^{c_1})^{x_{v_1}^{c_1}} \cdot (y_{v_1}^{c_2})^{x_{v_1}^{c_2}} \cdot (y_{v_1}^{c_3})^{x_{v_1}^{c_3}} \cdot g^{r_4}, g^{10} \cdot pk^{r_4}$$

We use c_1 and c_2 to denote $(y_{v_1}^{c_1})^{x_{v_1}^{c_1}} \cdot (y_{v_1}^{c_2})^{x_{v_1}^{c_2}} \cdot (y_{v_1}^{c_3})^{x_{v_1}^{c_3}} \cdot g^{r_4}$ and $g^{10} \cdot pk^{r_4}$, respectively.

And then, compute $z = \prod_{j=1}^{n_c}(y_{v_1}^{c_j})^{x_{v_1}^{c_j}}$. In this case, $z = (y_{v_1}^{c_1})^{x_{v_1}^{c_1}} \cdot (y_{v_1}^{c_2})^{x_{v_1}^{c_2}} \cdot (y_{v_1}^{c_3})^{x_{v_1}^{c_3}}$. Selects random numbers $k_1, k_2, k_3 \in \mathbb{Z}_q$, computes $K_1 = (y_{v_1}^{c_1})^{k_1} \cdot (y_{v_1}^{c_2})^{k_2} \cdot (y_{v_1}^{c_3})^{k_3}$, $K_2 = g^{k_1}$, $K_3 = g^{k_2}$, $K_4 = g^{k_3}$, $c = Hash(K_1\|K_2\|K_3\|K_4)$, $Z_1 = x_{v_1}^{c_1}c + k_1$, $Z_2 = x_{v_1}^{c_2}c + k_2$, $Z_3 = x_{v_1}^{c_3}c + k_3$. And ZKP$(z) = \{K_1, K_2, K_3, K_4, Z_1, Z_2, Z_3\}$. Prove $\frac{c_1}{z}(= g^{r_4})$ and $\frac{c_2}{g^{10}}(= pk^{r_4})$ has the same exponentiation (refer to Sect. 3), select $t \in Z$, compute $T_1 = (g)^t$, compute $T_2 = pk^t$, compute $c = Hash(T_1\|T_2)$, compute $s = r_4 \cdot c + t$ (r_4 in this case). ZKP$(P) = \{T_1, T_2, s, z, \text{ZKP}(z)\}$.

Verifier: firstly verify z using ZKP(z), compute $c = Hash(K_1\|K_2\|K_3)$, verify if $(y_{v_1}^{c_1})^{Z_1}(y_{v_1}^{c_2})^{Z_2}(y_{v_1}^{c_3})^{Z_3} = K_1 \times (z)^c$, verify if $(y_{v_1}^{c_1})^{Z_1} = K_1 \times (g^{x_{v_1}^{c_1}})^c$, verify if $(y_{v_1}^{c_2})^{Z_2} = K_2 \times (g^{x_{v_1}^{c_3}})^c$, verify if $(y_{v_1}^{c_3})^{Z_3} = K_3 \times (g^{x_{v_1}^{c_3}})^c$.
Secondly verify P using T_1, T_2, s and z, multiply $E(5, \cdots), E(2, \cdots)$ and $E(3, \cdots)$ as (c_1, c_2), compute $c = Hash(T_1\|T_2)$, verify if $g^s = (\frac{c_1}{z})^c \cdot T_1$, verify if $pk^s = (\frac{c_2}{g^{10}})^c \cdot T_2$. Same as to verify $\frac{c_1}{z}$ and $\frac{c_2}{g^{10}}$ has the same exponentiation r_4.

References

1. Adida, B.: Helios: web-based open-audit voting. In: USENIX Security Symposium, vol. 17, pp. 335–348 (2008)
2. Nakamoto, S.: Bitcoin: a peer-to-peer electronic cash system (2008)
3. ElGamal, T.: A public key cryptosystem and a signature scheme based on discrete logarithms. In: Advances in Cryptology, pp. 10–18 (1984)
4. Lamport, L., Shostak, R., Pease, M.: The Byzantine generals. ACM Trans. Program. Lang. Syst. **4**(3), 382–401 (1982)
5. Buterin, V.: A next-generation smart contract and decentralized application platform (2015). https://github.com/ethereum/wiki/wiki/White-Paper
6. Szabo, N.: Smart contracts: building blocks for digital markets. EXTROPY J. Transhumanist Thought **16** (1996)
7. Followmyvote.com. Introducing a secure and transparent online voting solution for the modern age: Follow My Vote (2016). https://followmyvote.com/
8. Hao, F., Ryan, P.Y., Zieliński, P.: Anonymous voting by two-round public discussion. IET Inf. Secur. **4**(2), 62–67 (2010)
9. Groth, J.: Efficient maximal privacy in boardroom voting and anonymous broadcast. In: International Conference on Financial Cryptography, pp. 90–104 (2004)
10. Kiayias, A., Yung, M.: Self-tallying elections and perfect ballot secrecy. In: International Workshop on Public Key Cryptography, pp. 141–158 (2002)
11. Li, M., Sun, X., Wang, H., Zhang, Y., Zhang, J.: Privacy-aware access control with trust management in web service. World Wide Web **14**(4), 407–430 (2011)
12. Kabir, M.E., Wang, H.: Conditional purpose based access control model for privacy protection. In: Proceedings of the Twentieth Australasian Conference on Australasian Database, vol. 92, pp. 135–142 (2009)
13. Kabir, M.E., Wang, H., Bertino, E.: A role-involved purpose-based access control model. Inf. Syst. Front. **14**(3), 809–822 (2012)
14. McCorry, P., Shahandashti, S.F., Hao, F.: A smart contract for boardroom voting with maximum voter privacy. In: International Conference on Financial Cryptography and Data Security, pp. 357–375 (2017)

15. Schnorr, C.P.: Efficient signature generation by smart cards. J. Cryptol. **4**(3), 161–174 (1991)
16. Sun, X., Li, M., Wang, H.: A family of enhanced (L, α)-diversity models for privacy preserving data publishing. Future Gener. Comput. Syst. **27**(3), 348–356 (2011)
17. Sun, X., Wang, H.: Satisfying privacy requirements before data anonymization. Comput. J. **55**(4), 422–437 (2012)
18. Sun, X., Wang, H., Li, J., Truta, T.M.: Enhanced p-sensitive k-anonymity models for privacy preserving data publishing. Trans. Data Priv. **1**(2), 53–66 (2008)
19. Wang, H., Cao, J., Zhang, Y.: A flexible payment scheme and its role-based access control. IEEE Trans. Knowl. Data Eng. **17**(3), 425–436 (2005)
20. Wang, H., Zhang, Y., Cao, J.: Effective collaboration with information sharing in virtual universities. IEEE Trans. Knowl. Data Eng. **21**(6), 840–853 (2009)
21. Business Wire. Now you can vote online with a selfie. Business Wire (2016). http://www.businesswire.com/news/home/20161017005354/en/Vote-Online-Selfie
22. Yang, X., et al.: A verifiable ranked choice internet voting system. In: Bouguettaya, A., et al. (eds.) WISE 2017. LNCS, vol. 10570, pp. 490–501. Springer, Cham (2017). https://doi.org/10.1007/978-3-319-68786-5_39
23. Yang, X., Yi, X., Nepal, S., Kelarev, A., Han, F.: A secure verifiable ranked choice online voting system based on homomorphic encryption. IEEE Access (2018)

Enabling Blockchain for Efficient Spatio-Temporal Query Processing

Ildar Nurgaliev[1,2], Muhammad Muzammal[2], and Qiang Qu[2(✉)]

[1] Shenzhen College of Advanced Technology,
University of Chinese Academy of Sciences, Shenzhen, China
[2] Shenzhen Institutes of Advanced Technology,
Chinese Academy of Sciences, Shenzhen, China
{ildar,muzammal,qiang}@siat.ac.cn

Abstract. Recent interest in blockchain technology has spurred on a host of new applications in a variety of domains including spatio-temporal data management. The reliability and immutability of blockchain in addition to the decentralized trustless data processing offers promising solutions for modern enterprise systems. However, current blockchain proposals do not support spatio-temporal data processing. Further, a block-based sequential access data structure in the blockchain restricts efficient query processing. Therefore, a blockchain system is desirable that not only supports spatio-temporal data management but also provides efficient query processing. In this work, we propose efficient query processing for spatio-temporal blockchain data. We consider a spatio-temporal blockchain that records both time and location attributes for the transactions. The data storage and integrity is maintained by the introduction of a cryptographically signed tree data structure, a variant of Merkle KD-tree, which also supports fast spatial queries. For the temporal attribute, we consider Bitcoin like near uniform block generation and process temporal queries by a block-DAG data structure without the introduction of temporal indexes. For current position verification, we use Merkle-Patricia-Trie. We also propose a random graph model to generate a block-DAG topology for an abstract peer-to-peer network. A comprehensive evaluation demonstrates the applicability and the effectiveness of the proposed approach.

Keywords: Blockchain · Spatio-temporal queries · Block-DAG

1 Introduction

Blockchain as a paradigm-shift technology emerged originally for the financial transactions [1] and has recently found ways into a variety of application domains. The main idea of a blockchain is a trustless decentralized environment where the reliability and the security of the data is a design requirement. The blockchain technology has demonstrated its applicability for business solutions in

© Springer Nature Switzerland AG 2018
H. Hacid et al. (Eds.): WISE 2018, LNCS 11233, pp. 36–51, 2018.
https://doi.org/10.1007/978-3-030-02922-7_3

sectors including finance, healthcare and others. Consider, for instance, a supply-chain scenario where an item is tracked over the course of transportation. The tracking mechanism requires that not only the spatio-temporal information is continuously updated to the database but the mobility queries about the object are also supported. Typical queries include, for example, 'list all objects at location l at time t', or 'list all objects that moved in a radius r for location l at time interval $[t_1, t_2]$'. The support for such queries is required, for example, for logistic decisions or product monitoring. However, a blockchain implementation of such a business scenario is challenging. For example, spatio-temporal data grows at a much higher rate as compared to the transaction data currently supported by financial blockchain. Further, the consensus protocols for spatio-temporal data require proof-of-location processing. Also, the sequential access mechanism in blockchain does not support efficient query processing.

The value proposition of blockchain over traditional databases is the data integrity through cryptographically signed history [1–3]. For example, financial institutions require an 'append-only' signed data structure that is auditable and traceable. Large enterprise service providers such as Google require spatio-temporal analytics on user data for providing continuous services in a given time and space. Therefore, a spatio-temporal blockchain system design should have two considerations, (i) the scale at which such a system shall be used, and (ii) the kind of query support that will be required by the blockchain system. Note that it is not straightforward to directly adapt database concepts to a blockchain system. A spatio-temporal blockchain system design should consider secure data storage and efficient query processing simultaneously. This work provides a conceptual block design for efficient queries in the block directed acyclic graph (block-DAG). Block-DAG is the blockchain alternative that makes possible to achieve high throughput by way of fast block creation.

In this work, we consider spatio-temporal data processing in a blockchain and propose queries on blockchain without expensive local indexing. The idea is to integrate Merkle-tree with the spatial-index such that spatio-temporal queries are supported without the need of any additional indexes. We assume a spatio-temporal blockchain with an abstract consensus such that the credibility of the data is maintained by the consensus algorithm. We assume that a typical transaction has the following attributes: timestamp, longitude, latitude, and hashed account identifier. Thus, we enable blockchain for efficient query processing without the requirement of additional local indexes.

Our Contribution. We summarize our contribution as follows:

1. We propose the existing block-DAG with pair-wise block order for spatio-temporal data storage and discuss the advantages over a sequential blockchain access. We also propose block header organization such that it supports efficient spatio-temporal queries.
2. For the most recent position verification through mobile clients, we use Merkle-Patricia-trie on peer side and a particular block header in a client side.

3. We propose a random graph model to generate a block-DAG topology for an abstract p2p network.
4. Finally, we demonstrate the efficiency of spatio-temporal queries using 'Pokemon GO' dataset.

The rest of this paper is organized as follows: Sect. 2 is about related concepts and Sect. 3 is an overview of the proposed approach. The details about block construction and authenticated spatial-indexes are in Sect. 4. Section 5 is about spatio-temporal query processing. The evaluation is presented in section 6 and Sect. 7 concludes this work.

2   Preliminaries

The problem of indexing multi-dimensional data that includes time-stamp and spatial location are well studied in the database community [4] but has not been considered for blockchain systems. Recent interest in blockchain technology made most of the companies willing to migrate their database infrastructure to blockchain solutions, one such idea is an open-source system developed by integrating the blockchain with the database where they replicate the SQL statements by blockchain [5]. In this section, we focus on some fundamental concepts that are related to current work.

Spatial Indexes. R-tree was originally proposed by Guttman [6] for spatial indexes and since then many variants including RT-tree [7] and 3D R-tree [8] have been proposed. For efficient range query processing, kd-tree [9] has been proposed that additionally supports efficient k-NN queries for a given point. Nonetheless, kd-tree performs reasonably under the assumption of an initial bulk load and no further updates. This property makes it excellent for cases in which the data is static. The kd-tree exhibits many favorable properties and has been proven to be efficient in practice for low-dimensional data [10].

Another important aspect of data processing is that of data verification. Many authenticated data structures have been proposed for indexing spatial and spatio-temporal data to support verifiable queries such that range queries, k-NN queries, reverse k-NN, skyline queries and others [11]. Authenticated version of spatial index proposals include Merkle kd-tree (Mkd-tree) [12], Merkle R-tree (MR-tree) [13], k-NN-based spatial queries [14] and others.

Blockchain Cryptography and Index basics. A collision-resistant hash function $\mathcal{H}$ maps a string s of any size to a bit vector of a fixed-length, s.t. $\mathcal{H}(s)$ is fast to compute while it is computationally infeasible to find a collision for $s_1 \neq s_2$ s.t. $\mathcal{H}(s_1) = \mathcal{H}(s_2)$ [15]. The Merkle Hash Tree (MHT) has proven to be a base for arbitrary authenticated Directed Acyclic Graph (DAG) structures [16]. The MHT proposes hierarchically organized hashes to verify the integrity and validity of blocks providing a tiny number of hashes or Verification Objects (VO) [17]. These properties facilitate the applicability of MHT in blockchain systems such that MHT is customized for particular usecases [18]. Ethereum [2], for example, uses Merkle Patricia trie to maintain the integrity of the global key-value states where the key is a 32-byte account identifier and the value is the account state.

Blockchain and block-DAG. A blockchain is a datastore where every known participant maintains a copy of the shared data. Bitcoin [1] and Ethereum [2] are the most well-known examples where all the transactions are publicly accessible in an anonymous way. The transaction accuracies are driven by the fact that consensus creates truth. However, throughput in blockchain systems is a bottleneck as the transaction confirmation times are not comparable with, for instance, VISA network. The blockchain performance is strongly tied to a consensus protocol underneath it and hard-coded limitations on computations per block. The authors of [19] propose the block-DAG concept. Its structure or topology is formed from multiple references from every block to its fore-runner blocks with possible conflicting transactions. As a consequence, this leads to changed transaction acceptance rules where the graph topology is directly used as 'votes' that helps to reveal the robust subset of block-DAG without conflicting transactions. Each block references all blocks to which its miner was aware at the time of its creation. Thus, the blocks that take a long time to propagate are prone to be rejected by the system.

Table 1. A list of notation and their meaning.

Symbol	Meaning
B	Block; B_{header}: block header; B_{size}: block size
D	Dataset of points $D = \{x \mid x \in \mathcal{R}^d\}$ where d is space dimensionality
$G(V, E)$	Topology of a block-DAG; V: set of blocks; E: set of references
$\mathcal{D}$	Network delay
$\mathcal{H}$	Cryptographic Hash function; $\mathcal{H}_R$: Merkle hash root function
$\mathcal{T}$	Transaction; $\mathcal{T}_r$: transaction rate per second; $\mathcal{T}_n$: total transactions
k	Number of answer points in k-NN query
q	Hyper-rectangle space range $q = \{x, y\}$ or point $q = \{x\}$ where $x, y \in D$
α	Standard deviation s.t. $\mathcal{N}(\mathcal{T}_r, \alpha^2)$, where $\mathcal{N}$ denotes the normal distribution
β	Time range s.t. $(\beta.start\_time, \beta.end\_time)$
σ	Deterministic search procedure
σ_β	Time range search on $G(V, E)$ by β
σ_q	Range search
$\sigma_{k,q}$	k-NN search
(ϕ, λ)	(Latitude, Longitude)

3   Problem Formulation

In this section, we present related concepts, definitions, and system assumptions. We consider block-DAG paradigm as an alternative to the blockchain and propose a block header that allows efficient spatio-temporal data querying. Table 1 gives a list of useful notation.

Transaction: (formally $\mathcal{T}$) is a tuple, (*longitude, latitude, timestamp* (t), *hashed account identifier* (uid)), that is included in a block. Thus, a transaction is a single cryptographically-signed instruction. We assume that the system records spatio-temporal information only.

Block: (formally $B = (B_{header}, B_{body})$) includes a block header B_{header} and a corresponding list of transactions B_{body} ordered w.r.t. final consequence of leafs in Merkle kd-tree that is organized by a bulk loading algorithm applied to list of transactions of B_{body}.

Block header: (formally B_{header}) contains pieces of information related to transactions and a set of hashes of other block headers. The block header forces to maintain spatial index per block. For the spatial indexes per block, the following properties should hold: (a) header must be authenticated to guarantee that every other ledger holder has similar index structure, (b) supports spatial queries, e.g. range query, nearest neighbors query, etc. and (c) fast access and verification from the last position of a particular item that is associated with its hashed account identifier. To avoid additional indexing of the whole block-DAG, we consider Merkle kd-tree for its bi-functionality, i.e. block integrity verification similar to Bitcoin and fast spatial queries on a block. For simplified current position verification, we consider Merkle-Patricia-trie. The block header contains the following information:

> **blockID**: a unique identifier, $\mathcal{H}(B_{header})$
> **orphanHashes**: a list of hashes of referenced block headers
> **locationRoot**: a 256-bit hash of the root node of the Merkle Patricia trie structure populated with all the account identifiers and associated with the most recent location, time and number of records made per account
> **merkleSpaceRoot**: a 256-bit hash of the root node of the Merkle kd-tree structure populated with each transaction of the block
> **startTime**: a scalar value equal to $\forall \mathcal{T} \in B_{body} : min(\mathcal{T}.t)$
> **endTime**: a scalar value equal to $\forall \mathcal{T} \in B_{body} : max(\mathcal{T}.t)$
> **timestamp**: is the time of block creation
> **nonce**: a 64-bit hash which proves that the sufficient amount of computation has been carried while block creation

Definitions. We denote block-DAG as G. As mentioned already, G enforces a 'causal relation' among blocks which states that if block B_i includes the hash of block B_j, then B_i must have been created after B_j. Although, block-DAG supports fast block creation and short transaction confirmation time, a highly conflicting environment reduces the speed of transaction to be securely confirmed in a local image of G for a particular ledger holder. The SPECTRE protocol introduces $GetRobustAccepted(G)$ that is a function of G and returns a subset of transactions that are securely accepted.

We now present some definitions, properties, and assumptions.

Definition 1 (input(T)). *A subset of confirmed transactions in G are the input of T and belong to the owner of T.*

Definition 2 (conflict). T_1 *and* T_2 *are conflicting iff* $T_1.uid = T_2.uid$ *and* $T_1.t = T_2.t$.

Property 1 (Adjusted Consistency). T_1 is accepted iff $\forall T_i \in input(T_1) : T_i \in GetRobustAccepted(G)$, all $conflicts$ are rejected and the time-stamp of transaction T_1 is not more than δ-away from current time-stamp. δ is system defined.

The Adjusted Consistency property ensures the system includes only fresh T with timestamp satisfied by δ from the real system clock.

Property 2 (Weak Liveness). If transaction T_i is published in G and no conflicting T_j is published, then it is readily included in $GetRobustAccepted(G)$.

Property 3 (Pairwise ordering). When block B is published in G, the system guarantees that G contains blocks published before or exactly after B.

Property 4 (Result Completeness). A response set for a user query must not have any missing results.

Property 5 (Block Soundness). No modification takes place in the B_{body}, neither by adding non-existing transactions nor by updating existing ones.

A miner is expected to maintain the SPECTRE voting protocol to keep the real order between each pair of blocks on their local image of G. The protocol holds these properties and states for fast block creation (at least five blocks per second). The aforementioned properties and a somewhat fixed block creation rate allow temporal queries over block-DAG topology without additional temporal indexes. The system also considers the case of a *Light Node* such as a mobile client which does not store the entire block-DAG [1]. The main idea of the mobile client is to have authenticated tracking for a list of assets.

Remark. The spatio-temporal block-DAG does not require a linear order among all the transactions because the tracking actions of a particular account have no relation to other account actions as that is not same to UTXO from Bitcoin. The linear ordering precedes the great limitation of throughput in a blockchain that comes from low block creation rate that retains the system within reliable parameters. The emerging spatio-temporal data grows fast since a particular user or supplier tracking sends his spatial location on a regular basis. The volume of spatio-temporal data is increasing exponentially that makes it hard to stay on blockchain. The block-DAG is the alternative that makes possible to handle high throughput by fast block creation.

4   Enabling Block-DAG for Spatio-Temporal Queries

A transaction typically includes geodesic coordinates, s.t. *latitude* ϕ and *longitude* λ, to represent a location. Latitude is the angular distance between $-90°$ and $90°$ that represents a location southern or northern from the earth's equator. Longitude angular distance ranges between $-180°$ and $180°$ that represents a place east or west of the imaginary line that is running through *Greenwich*.

4.1   Geo-Spatial Representation and Spatial Index

The k-NN and range queries need a Cartesian coordinate system that is a system of points represented by orthogonal axis. To do so, the 'map projection' is used to convert the geodesic system to two-dimensional coordinates on the map. Nonetheless, it does not matter what coordinate system or what geographic standard is used, as it is not possible to make a projection from a sphere into a rectangle, i.e. a two-axis system, and save all data, preserving angles and distances, at the same time. Haversine is one of the methods to get approximate distance on a sphere (Eq. 1), to use it on Earth with an approximate radius $r_e = 6371km$.

$$dist_{harv} = 2\ r_e\ arcsin\big(min\big[1, \sqrt{sin^2(\frac{\delta\phi}{2}) + cos\phi_i cos\phi_j sin^2(\frac{\delta\lambda}{2})}\big]\big) \quad (1)$$

The k-NN search is a problem where a magnitude-comparable distance can be substituted for the relative distance since the relative ordering of the distances is more important than the actual distances [20]. The tunnel-through distance for points i and j is computed using Eq. 3. For each index, we pre-compute Cartesian coordinates using latitude and longitude and Eq. 2 resulting in 3D Merkle kd-tree.

$$\left.\begin{array}{l} x_i = r_e cos(\lambda_i)cos(\phi_i) \\ y_i = r_e sin(\lambda_i)cos(\phi_i) \\ z_i = r_e sin(\phi_i) \end{array}\right\} \quad (2)$$

$$dist = \sqrt{(x_i - x_j)^2 - (y_i - y_j)^2 - (z_i - z_j)^2)} \quad (3)$$

4.2   Spatial Index per Block

The Merkle Hash Tree, MHT, is generally used as a base for arbitrary authenticated directed acyclic graph structures. The scheme to do verification proposed on MHT is simple. Recompute the hash value incrementally by recreating Merkle tree root for a particular block as per Eq. 4.

$$\mathcal{H}_R(v_i) = \begin{cases} \mathcal{H}(byte(v_i)), & v_i \text{ is a leaf node} \\ \mathcal{H}(v_i, \mathcal{H}_R(v_{i,1}), ..., \mathcal{H}_R(v_{i,n})), & \text{where } \mathcal{H}_R(v_{i,j}) \text{ is the hash value of} \\ & \text{successors of } v_i \text{ in consequent order} \end{cases}$$
$$(4)$$

As mentioned already, the Merkle tree is designed for spatial queries but we want a more general authenticated data structure that is able to process spatial queries. The SPECTRE protocol states that the block creation time could take within one second and the size of that block is expected to be within 40–70 transactions on average. Therefore, having a three-dimensional space, i.e. a sphere in a Cartesian space, and a few points to hold in the structure, we consider a kd-tree. A kd-tree has the advantage that it fits in the main memory and avoids a complicated structure similar to 3D R-tree with additional minimum bounding box coordinates. We use the *Mkd-tree* that is a kd-tree authenticated by Merkle scheme. The kd-tree has a reasonable performance for processing k-NN and range queries. It happens because, at every tree level, the kd-tree has the Euclidean distance in one dimension, that gives a lower bound for pruning while query processing.

The bulk loading of Mkd-tree from a prepared list of transactions T_{list} for a block formation is presented in Algorithm 1, where t is a timestamp of $T \in T_{list}$.

Algorithm 1. BULKLOAD sub-routine for Mkd-tree loading based on T_{list}

Require: $T_{list}, \mu, \mathcal{M} \leftarrow \emptyset$ ▷ μ: recursion depth; $\mathcal{M}$: empty kd-tree
Ensure: $T_{list} \leftarrow \{T.(X, Y, Z) \leftarrow T.(\phi, \lambda) \forall T \in T_{list}\}$;   ▷ T_{list} to Cartesian space
1:   $\sigma \leftarrow \mu \bmod 3$ ▷ Choose splitting point
2:   $T_{list} \leftarrow sorted(T_{list}, by\ T.(X, Y, Z)[\sigma]\ and\ T.t)$ ▷ Sort rest of transactions
3:   $T_{left}, m, T_{right} \leftarrow$ MEDIANSPLIT(T_{list}) ▷ m: root of splitting
4:   $\mathcal{M} \leftarrow m$;
5:   BULKLOAD$(T_{left}, \mu + 1, \mathcal{M})$ ▷ Recursion build left subtree
6:   BULKLOAD$(T_{right}, \mu + 1, \mathcal{M})$ ▷ Recursion build right subtree
7:   **return** $\mathcal{M}$

Lightweight Client. We also consider the possible case of a Light Node, a mobile device, which does not store the entire block-DAG. For the whole block-DAG, The client only needs to query all nodes headers from the peer-to-peer network and select locally the valid part by topological voting procedure over its network that is handled by SPECTRE protocol. The main idea of the mobile client is just to have an authenticated tracking for a list of assets knowing their account identifiers. Nonetheless, the system is able to handle some simple queries and provide verification objects with the results of spatio-temporal queries to a lightweight client. The typical proof of transaction inclusion is handled by Mkd-tree.

4.3   Simplified Last Position Verification

Account. The *account* is a personal data that is associated with a hashed public key of a user's initially generated *pair-key*. The account includes the last record of space-time location and the number of stored transactions per account. The state of all accounts is the state of the whole block-DAG network. In the system, accounts are essential for tracking entities associated with accounts. The presented block-DAG platform is a public transaction-based state machine that takes its start from the genesis empty block (*genesis state*) and incrementally updates the location position of associated anonymous accounts up to its last location (final state).

Account Management. The main challenge of account management in a blockchain is frequent updates of a value per account, therefore, an authenticated data-structure that holds all the account information actually differs from managing transaction history in binary trees. We utilize Merkle Patricia-trie, MPT, that reflects associations between each account identifier and their actual data. In the case of spatio-temporal data, the account information includes geodesic coordinates of the most recent geolocation, timestamp and the nonce. The *nonce* is the total number of stored transactions for the account.

To efficiently support account state, the MPT has the following properties: (i) the authenticated data-structure that is able to quickly recalculate a tree root after an insert or update operation, for example, create an account or update the last geolocation and timestamp, respectively, (ii) a tree root depends only on the data, omitting the order in which updates are made, (iii) support for a fast roll-back operation to construct specific global states for each block reflecting only the fresh account information presented in the block and its past, and (iv) provide rapid answers to account specific questions, for example, 'What is the most recent position of account x?' or 'Does account with id xxx-xx exists?'.

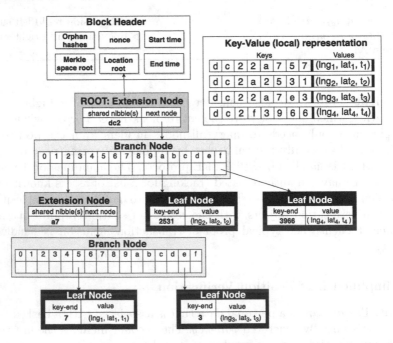

Fig. 1. The last positions of objects related to accounts in the system are represented by key-value pairs and encoded in Merkle Patricia-trie.

Every block header has a value of the MPT root, 'Location root', that reflects a distinct version of the global state per block. The MPT implementation introduces the value driven data-structure through referencing each node by its hash, therefore, key-value is stored in a levelDB database where the value is the string representation of a node and the key is its hash. Thus, redundant storage of the historical state of each node allows fast roll-back. MPT consist of three types of nodes, (i) a leaf node that stores key-value, (ii) an extension node stores a hash of another node, and (iii) branch nodes which are lists of a fixed length, typically 17. The first sixteen elements correspond to the sixteen possible hex characters in a key, and the final element holds a value if there is a [key, value] pair where the key ends at the branch node. A sample part of MPT version for a block is shown in Fig. 1.

Note that in a blockchain, the state of MPT changes just from one block to another, but the block-DAG allows several references from a block to other blocks that challenge the construction of MPT. In Algorithm 2, we introduce steps for a MPT snapshot formation per block.

Algorithm 2. A sub-routine for per block MPT formation

Require: B_{header}, B_{body}
Ensure: ROLLBACK(MPT,B_{header}.orphanHashes)   ▷ prepare MPT version
1: $S \leftarrow \emptyset$
2: $S \leftarrow \{T_i : \forall T_i, T_j \in B_{body} \text{ s.t. } \nexists(T_i.uid = T_j.uid \wedge T_i.t < T_j.t)\};$
3: $S \leftarrow \{T_i : \forall T_i, T_j \in \text{getAllBlocksTransactions}(B_{header}.\text{orphanHashes}),$
      $T_b \in B_{body} \text{ s.t. } \exists(T_i.uid = T_j.uid \wedge T_i.uid \neq T_b.uid \wedge T_i.t > T_j.t)\}$
4: MPT $\leftarrow S$ ▷ Extend MPT
5: $B_{header}.$"Location root" $\leftarrow \mathcal{H}_R(\textbf{MPT})$ ▷ Save fingerprint of MPT version

---

Theorem 1. *The query of an account's last position from MPT has the completeness of query answer (Property 4) for each block from GetRobustAccepted(G).*

Proof. Property 1 guarantees absence of conflicting transactions. The properties of MPT ((ii)–(iii) above) lead to a distinct version of tree root value among peer holder. Property 3 guarantees a monotonically decreasing time from a block to referenced blocks.

Account Location Tracking. We use Merkle-Patricia-trie and *location root* from B_{header} for verification of the most recent location of a particular account. We assume that the lightweight client has a fresh state of the block-DAG. For verification of its last location we consider the following steps:

(1) The client requests associated *account state* that stores the most recent position and a time-stamp.

(2) The request is handled on the peer side by MPT authenticated data structure. The request-results include the account state and additional information for authorization. The authorization information consists of the array of VOs (verification objects) formed by MPT in strict order and the block identifier of a B_{header} that includes *location root* hash value corresponding to a fresh version of the MPT on the peer side.

(3) The client locally verifies results on its block-DAG image w.r.t. block identifier and computes the 'Location root' by consequent hashing of $\mathcal{H}$ (obtained account state) with the VOs.

(4) If the previous step is positive then we have the definite most recent location and time of the client.

5   Spatio-Temporal Queries on Block-DAG

The spatio-temporal query is the combination of the two consequent procedure calls, first is the temporal range search procedure that retrieves a set of B_{header}; and the second is the spatial query over Mkd-tree executed for each B_{header}.

5.1   Temporal Queries on Block-DAG

The temporal part is regulated by considering the fact that every miner is expected to maintain the SPECTRE protocol. To enable a fast temporal search over the block-DAG, we introduce a temporal meta-information included in each block header. The temporal range query with a given time range β over blocDAG is the deterministic search procedure P over block-DAG topology $G(V, E)$. G has multiple source nodes s for each $v \in V$, by the source node, we consider recent nodes that reference the previous points. The temporal range search (*searchtime*) is implemented in a breadth-first (BFS) manner, the steps of which are listed below.

(1) Given a time range $\beta = (start\_time, end\_time)$, start BFS from the tips of the $GetRobustAccepted(G)$.

(2) As soon as BFS faces with B_{header} s.t. $[B_{header}.start\_time, B_{header}.end\_time] \cap [\beta.start\_time, \beta.end\_time] \neq \emptyset$, the B_{header} goes to the result set.

(3) The BFS runs until no new B_{header} occurs in range β and for all the next $B_{header}.end\_time < \beta.start\_time$.

Theorem 2. *The temporal range search over block-DAG topology has the completeness of query answer (Property 4) but does not satisfy the soundness property.*

Proof. Property 3 guarantees a monotonically decreasing $B_{header}.start\_time$ and $B_{header}.end\_time$ for each step of procedure σ_β. The σ_β accepts blocks in the way of a partial intersection of time ranges, therefore, the algorithm has Property 4 nonetheless a tiny number of outline transactions will be in resulting set that leads to an unsound answer.

5.2   Spatial Queries per Block

Definition 3 (Range Query). *Given a dataset of points D, and a range $q = \{x, y\}$ where $x, y \in \mathcal{R}^d$, range query seeks each $s \in D$ that is located inside the hyper-rectangle constituted by q.*

The range query complexity in kd-tree is $O(\sqrt{B_{size}} + k)$ where k is the number of answer points. An example of a spatio-temporal range query on the block-DAG is shown in Fig. 2. We show a two-dimensional Mkd-tree for ease of illustration, however, in the experiments we consider 3D Mkd-tree.

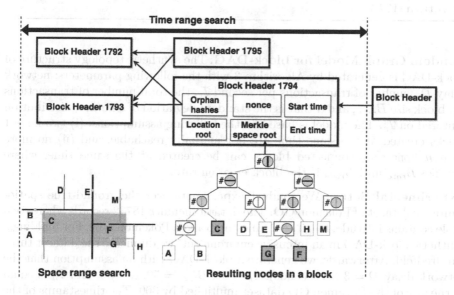

Fig. 2. Spatio-temporal range search on block-DAG: firstly filter by temporal range procedure σ_β and given time range β then space range query on every Mkd-tree by the hyper-rectangle $q = \{x, y\}$.

Definition 4 (k-nearest neighbors). *Given a dataset of points D, a scalar value k, and a query point $q = \{x\}$ where $x \in \mathcal{R}^d$, $\sigma_{k,q}(D)$ returns a subset of k points from D that are closest to q.*

The k-nearest neighbor algorithm maintains a priority queue to keep k closest points. The first k points are en-queued, then the algorithm traverses down the tree skipping bounding boxes where there is no chance to get a closer point than any of the k points found so far. The final stage is k points aggregation. Finally, TOP-k of all the k-NN results is computed (in case of several blocks).

6   Experimental Evaluation

In the section, we first present the random graph model for block-DAG. Then we briefly describe the experimental setup which is followed by the evaluation.

Algorithm 3. GENBLOCKDAG sub-routine for random graph generation

---

Require: $\mathcal{T}_n$, $\mathcal{T}_r$, α, $\mathcal{D}$, B_{size}
1: $\mathcal{T}_{rlist} \leftarrow \text{gen}(\mathcal{N}\left(\mathcal{T}_r, \alpha^2\right), \mathcal{T}_n)$ ▷ list of various transaction rates per second
2: $(V, E) \leftarrow (\text{genesis block}, \emptyset)$ ▷ initialize vertices and block references
3: $\omega \leftarrow 0$ ▷ remainder of transactions from previous chunk
4: **for** each chunk c of size $\mathcal{D}$ from $\mathcal{T}_{rlist}$ **do**
5:     $\mathcal{B} \leftarrow \text{round}((\text{sum}(c) + \omega)/B_{size})$ ▷ number of unconnected blocks while $\mathcal{D}$
6:     $E \leftarrow$ (connect all $\mathcal{B}$ to orphans of V)
7:     $V \leftarrow \mathcal{B}$
8:     $\omega \leftarrow (\text{sum}(c) + \omega) \bmod B_{size}$
9: **return** (V, E)

Random Graph Model for block-DAG. The synthetic topology structure of block-DAG is generated by Algorithm 3 with the following parameters: network delay $\mathcal{D}$, number of transactions per second $\mathcal{T}_r$, the total number of transactions $\mathcal{T}_n$, block size B_{size}, and α that is the standard deviation of a normal distribution centered on $\mathcal{T}_r$. The model is based on the following assumptions: (i) two honest blocks created at the same time are not mutually reachable, and (ii) no more than n honest un-connected blocks can be created at the same time, where $n = \mathcal{D} * B_{rate}$ and B_{rate} is the block creation rate.

Experimental Setup. We conduct experiments over the ground-true spatio-temporal dataset of Pokemon GO. The dataset includes 18732 records where each tuple contains latitude, longitude, timestamp and a Pokemon type. For building a synthetic block-DAG in an intensive environment, we duplicate the dataset three hundredfold. Afterwards, we generate a block-DAG with an assumption that the network delay $\mathcal{D} = 3$ (in seconds), $\mathcal{T}_r = 60$, $B_{size} = 50$, $\alpha^2 = 3$, and $\mathcal{T}_n$ is equal to the size of the Pokemon GO dataset multiplied by 300. The timestamps of the dataset are overwritten with the generated timestamps to maintain consistency. For query performance measurement, we repeat each experiment thirty times at each testing point to get descriptive and robust median points. All experiments are coded in Python[1] and performed on a Linux machine with Intel Core i7-6700 CPU 3.40GHz processor and 16 GB RAM.

Spatio-temporal Query Performance Analysis. For each query type described in section 5, we conduct three experiments to evaluate the performance of spatiotemporal queries ($\beta = 2$ weeks) with respect to (a) growing B_{size} from 30 to 110, (b) growing β from 0.5 to 11 hours, and (c) growing $\mathcal{T}_n$. We compare the query performance of the system with **search-scan** operation on a typical blockchain which includes two versions of it: **scan-time** and **scan-space**. The search-scan operation over a blockchain is a brute-force iteration among every block, therefore the scan-time must go through the total number of entries in the history since the block-headers in a typical blockchain has no any information about timestamps of transactions included in.

---

[1] https://github.com/ILDAR9/spatiotemporal_blockdag.

We performed extensive experiments for the queries that we propose. However, due to space limitation, we only report representative results. Hence, we report performance evaluation for *range query* and *k-NN query*. It can be observed that the block size is an important parameter for query performance. As the block size increases, the pruning is more effective and the total number of observations to be considered are reduced. Clearly, the temporal search on block-DAG in contrast with the scan-search is significantly faster. The results are shown in Fig. 3.

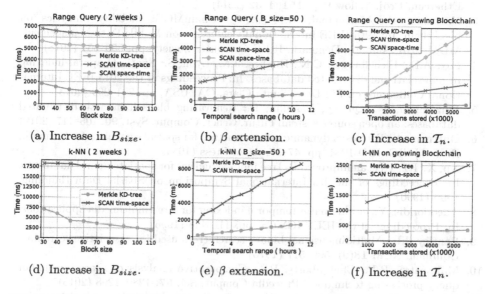

(a) Increase in B_{size}.   (b) β extension.   (c) Increase in $\mathcal{T}_n$.

(d) Increase in B_{size}.   (e) β extension.   (f) Increase in $\mathcal{T}_n$.

Fig. 3. Performance evaluation for range and *k-NN* queries. In b, c, e, f subfigures $B_{size} = 50$. Query parameters: $q = (\phi, \lambda) = (22.6, 114)$, $k = 15$.

7 Conclusion and Future Work

Ours is the first study on the topic and we have presented an efficient spatio-temporal data storage in public decentralized ledgers that maintains integrity through cryptographically signed history in block-DAG and maintains the efficient spatio-temporal queries without additional local indexing. We performed experiments to evaluate the proposed combination of Merkle kd-tree with block headers. The experiments show that we are able to handle the range query and the k-NN query efficiently. A number of future directions remain to be explored. For example, a focus on client-peer communication as it is crucial to enable lightweight clients. Also, due to the presence of malicious peers, *query authentication* is a mandatory point to consider. Additionally, more sophisticated queries similar to 'find all pairs of points whose distance is at most k within a time bounds', skyline k-NN queries, reverse k-NN, etc. are yet to be explored.

Acknowledgments. The work was partially supported by the CAS Pioneer Hundred Talents Program, China [grant number Y84402, 2017], and CAS President's International Fellowship Initiative, China [grant number 2018VTB0005, 2018].

References

1. Nakamoto, S.: Bitcoin: A Peer-to-Peer Electronic Cash System (2008)
2. Wood, G.: Ethereum: a secure decentralised generalised transaction ledger. Ethereum Proj. Yellow Pap. **151**, 1–32 (2014)
3. Nastrulin, B., Muzammal, M., Qu, Q.: ChainMOB: mobility analytics on blockchain. In: 19th IEEE International Conference on Mobile Data Management, MDM 2018, Aalborg, Denmark, IEEE Computer Society, pp. 556–557 (2018)
4. Fox, A.D., Eichelberger, C.N., Hughes, J.N., Lyon, S.: Spatio-temporal indexing in non-relational distributed databases. In: Proceedings of the 2013 IEEE International Conference on Big Data, Santa Clara, CA, USA, pp. 291–299 (2013)
5. Muzammal, M., Qu, Q., Nasrulin, B.: Renovating blockchain with distributed databases: an open source system. Futur. Gener. Comput. Syst. **90**, 105–117 (2019)
6. Guttman, A.: R-trees: a dynamic index structure for spatial searching. In: Yormark, B., (ed.) SIGMOD 1984, pp. 47–57. ACM Press (1984)
7. Xu, X.: RT-Tree: an improved R-Tree index structure for spatiotemporal databases. In: Proceedings of the 4th International Symposium on Spatial Data Handling, 1999 (1990)
8. Theodoridis, Y., et al.: Spatio-temporal indexing for large multimedia applications. In: Proceedings of the IEEE ICMCS, pp. 441–448 (1996)
9. Bentley, J.L.: Multidimensional binary search trees used for associative searching. Commun. ACM **18**(9), 509–517 (1975)
10. Mahapatra, R.P., Chakraborty, P.S.: Comparative analysis of nearest neighbor query processing techniques. Procedia Comput. Sci. **57**, 1289–1298 (2015)
11. Papadias, D., Tao, Y., Fu, G., Seeger, B.: Progressive skyline computation in database systems. ACM Trans. Database Syst. **30**(1), 41–82 (2005)
12. Li, F., et al.: Proof-infused streams: enabling authentication of sliding window queries on streams. In: Proceedings of the 33rd VLDB, pp. 147–158 (2007)
13. Mouratidis, K., et al.: Partially materialized digest scheme: an efficient verification method for outsourced databases. VLDB J. **18**(1), 363–381 (2009)
14. Hu, L., Ku, W., Bakiras, S., Shahabi, C.: Spatial query integrity with voronoi neighbors. IEEE Trans. Knowl. Data Eng. **25**(4), 863–876 (2013)
15. Komargodski, I., Naor, M., Yogev, E.: Collision resistant hashing for paranoids: dealing with multiple collisions. In: Nielsen, J.B., Rijmen, V. (eds.) EUROCRYPT 2018. LNCS, vol. 10821, pp. 162–194. Springer, Cham (2018). https://doi.org/10.1007/978-3-319-78375-8_6
16. Martel, C.U., et al.: A general model for authenticated data structures. Algorithmica **39**(1), 21–41 (2004)
17. Becker, G.: Merkle signature schemes, merkle trees and their cryptanalysis. Ruhr-University Bochum, Technical report (2008)
18. Xu, J., Wei, L., Zhang, Y., Wang, A., Zhou, F., Gao, C.: Dynamic fully homomorphic encryption-based merkle tree for lightweight streaming authenticated data structures. J. Netw. Comput. Appl. **107**, 113–124 (2018)

19. Lewenberg, Y., Sompolinsky, Y., Zohar, A.: Inclusive block chain protocols. In: Böhme, R., Okamoto, T. (eds.) FC 2015. LNCS, vol. 8975, pp. 528–547. Springer, Heidelberg (2015). https://doi.org/10.1007/978-3-662-47854-7_33
20. Mackey, G.E.: Efficient nearest neighbor searches in n-able tm. Technical report, Sandia National Laboratories (2010)

A Robust Spatio-Temporal Verification
Protocol for Blockchain

Bulat Nasrulin[1], Muhammad Muzammal[2,3], and Qiang Qu[2(✉)]

[1] Shenzhen College of Advanced Technology,
University of Chinese Academy of Sciences, Shenzhen, China
bulat@siat.ac.cn
[2] Shenzhen Institutes of Advanced Technology,
Chinese Academy of Sciences, Shenzhen, China
{muzammal,qiang}@siat.ac.cn
[3] Department of Computer Science, Bahria University, Islamabad, Pakistan
muzammal@bui.edu.pk

Abstract. Massive Spatio-temporal data is increasingly collected in a variety of domains including supply chain. The authenticity as well as the security of such data is usually a concern due to the requirement of trust in centralised systems. Blockchain technology has come to the forth recently and offers ways for trustless and reliable storage and processing of data. However, current blockchain proposals either do not support spatial data or make simplifying assumptions such as 'trusted' servers to process spatio-temporal data. We presume that the notion of 'trust' in a blockchain is too strong an assumption and propose a robust spatio-temporal verification protocol for the blockchain. In this work, we present a novel practical proof-of-location protocol on top of a permissioned blockchain. The protocol is instrumented by the implementation of an access control model and utilises a set of verification rules to create and verify spatio-temporal data points. We also propose a threat-to-validity model to evaluate the robustness of the verification protocol. The applicability and practicality of the protocol is demonstrated by the implementation of a supply chain case study as a proof-of-concept.

Keywords: Proof of location · Spatio-temporal consensus

1 Introduction

Many real-life applications collect spatio-temporal data for providing location-based services [8], for example, location-based web search, real-time traffic updates, nearest fuel station, and popular food points. Some applications offer incentives to the users [14] in the form of discounts or e-coupons for sharing location. One of the limitations of such applications is the assumption of 'trust' that the location shared by a client, for instance, an online check-in [13], is correct.

In some situations, it is important that the user data is accessible only if the presence of the user at a specific location is verified. For example, user data

© Springer Nature Switzerland AG 2018
H. Hacid et al. (Eds.): WISE 2018, LNCS 11233, pp. 52–67, 2018.
https://doi.org/10.1007/978-3-030-02922-7_4

is accessible to the physician only if the user is present at that time [19], or special kinds of certificates can be used only in limited locations. A number of application scenarios exist where a proof-of-location (PoL) is a requirement, for instance, to verify the authenticity of a financial transaction, to prevent impersonation attacks, and to reduce the possibility of spam and DOS attacks.

However, the implementation of a reliable location-based service is not obvious due to the requirements of (i) privacy and security, and (ii) integrity of data. An inherent limitation in today's centralised business solutions is the notion of 'trust' and the encryption mechanisms in centralised business solutions do not always guarantee data security as when such a system is compromised, sensitive user data is revealed. Similarly, business processes such as supply chain have a complex eco-system [17], for instance, in a minimalist settings, the stakeholders in a supply chain include sets of producers, logistic operators, distributors, and consumers. However, current supply chain solutions [2] lack transaction transparency due to trust-based centralised solutions.

Recently, the interest in the Blockchain systems and the advancements in Distributed Ledger Technology (DLT) have enabled 'trustless' reliable decentralised applications [18]. In short, a blockchain is a decentralised linked data structure with a state transition machine replication. Current blockchain systems lack the support for location-based services as the locations reported by the users through trusted servers are considered legitimate. However, the incentives associated with the locations [21] encourage false location reporting and issues such as GPS or Wifi spoofing [7] and colluding location attacks [22] are common.

1.1 Motivation

Consider a supply chain scenario where an item is transferred from a producer to a consumer by way of a transportation agent. The transportation of the product is of interest to the end-user, for example, the transportation time. However, the agent may wish to delay the shipment due to reasons such as putting multiple long-distance shipments together and thus, may attempt to report a location which is not true. Typically, special digital certificates are used for validating PoL. Many digital certificate-based solutions to attest the spatio-temporal location of an object have been proposed [6, 20]. However, current PoL systems lack robustness as they require trusted centralised servers for operation and are vulnerable to various privacy and security attacks. Therefore, a robust and reliable trustless PoL protocol is desired that ensures the integrity and security of the location data.

1.2 Our Contribution

In this paper, we present a robust and reliable proof-of-location protocol that is implemented using a permissioned blockchain and ensures the data security and integrity. The proposed system enables a reliable supply chain without the involvement of trusted third parties. Our contributions are as follows:

(1) We formalise the spatio-temporal and security requirements for a decentralised PoL protocol using digital certificates.
(2) We propose a practical robust PoL protocol that is trustless, decentralised and ensures the creation and validation of proof-of-location efficiently.
(3) We propose and evaluate the threats-to-validity for the protocol.
(4) We implement a proof-of-concept supply chain business case based on Hyperledger Iroha to demonstrate the applicability of the protocol.

Significance of the PoL Protocol. Many PoL protocols have been proposed in literature. This study formalises the problem (Sect. 3) by proposing the necessary and sufficient conditions that should be satisfied for a PoL protocol. A threats-to-validity model (Sect. 3.2) lists the possible threats to the protocol and the robustness of the protocol is demonstrated for the possible threats (Sect. 5).

The rest of this paper is organised as follows. In Sect. 2, we discuss related concepts including centralised approaches for achieving location proofs, previous attempts for spatio-temporal blockchains and location attacks. We formalise the problem in Sect. 3 and present the PoL protocol in Sect. 4. Threats-to-validity are evaluated in Sect. 5 and a proof-of-concept supply chain case study is presented in Sect. 6. Section 7 concludes this work.

2   Related Work

Large volumes of spatio-temporal data are being collected by Smart devices using location-based services. However, a device may not report the correct location due to reasons such as privacy concerns. The authenticity of location is established by techniques similar to location proofs [21]. A number of studies [8,20] have considered secure privacy-aware location proofs. Studies in literature can broadly be distinguished into infrastructure-based and infrastructure-independent location-proof studies. Infrastructure-based location verification assumes the presence of trusted access points, typically Wifi points, or other short-range communication media [9,15]. The location reported by such mechanisms is although correct, it has some limitations. For example, access points are hard to scale due to limited coverage. The study [8] considers wireless proofs using spatio-temporal properties of wireless channels whereas SecureRun [14] utilises wifi access points to get PoL.

Solutions for infrastructure-independent location verification are by way of a short-range communication between collocated smart devices. This approach mitigates the cost of deploying infrastructure but the proof generation is harder to create and is less secure. Moreover, users need to manually maintain a connection with nearby devices. Many studies [13,20] have considered similar ideas for the PoL. It should be clear that both classes of solutions assume a centralised trusted server to store PoL. A distinction is a protocol *PROPS* [6] that stores location proofs on a personal smart device. But this approach hardens spatio-temporal analytics, as user is required to be online. Another interesting proposal is the use of internet [1] for long-range location verification with the assumption that the protocol is resistant to proxy or relay attacks.

Some initial considerations for reliable blockchain PoL include Ethereum smart contracts *Sikorka*[1] and location-on-blockchain [3]. However, these proposals do not elaborate on practical details and security discussion. Further, public blockchains that use Nakamoto consensus [11] lack a transaction finality which is crucial for industrial applications such as supply chain. Currently, two projects, FOAM[2] and XYO[3], are under development that provide Ethereum smart contract-based reliable spatio-temporal information. The main idea of FOAM protocol is to create a spatial-index with crypto-spatial coordinates by the assignment of an unforgeable index to each object. However, the requirement of special hardware deployment is a major drawback. XYO Network, in contrast, is based on the assumption of a shared network where each participant has a clear role and gets incentives for forwarding, collecting or storing location data [10,12].

3   Proof-of-Location Problem

The objective of a PoL protocol is to attain failure-tolerance in a decentralised environment. Proof-of-location is generated via the interaction of multiple untrusted or semi-trusted parties including validators, provers, witnesses and network peers. In order to achieve security for a given region, a set of witnesses must be deployed. It has been shown that a total of $3f + 1$ participants can only tolerate f Byzantine failures [4]. This is impractical as it requires massive hardware deployment. We now formally define the problem.

Definition 1 (Proof-of-Location). *A proof-of-location is a verifiable digital certificate that attests the presence of a prover σ at location l and time t and is signed by an authorised witness ω.*

Following set of properties are desirable for a reliable PoL protocol, assuming the presence of malicious actors [6]:

Definition 2 (Completness). *A PoL is considered complete if σ is attested at l, at time t, by $\omega \in W$, such that ω is registered at location l.*

Definition 3 (Spatio-Temporal Soundness). *A PoL is spatio-temporal sound if σ is not able to obtain a proof when not physically present at $\{l, t\}$.*

Definition 4 (Non-Transferability). *A non-transferable PoL attesting the location of a σ is valid only if it was provided by a prover $\sigma \in S$.*

---

[1] Online at http://sikorka.io.
[2] Online at https://foam.space.
[3] Online at https://xyo.network/whitepaper.

Definition 5 (Tamper-Evidence). *A PoL is tamper-evident if tampering can be detected.*

Definition 6 (Proof-of-Location Problem). *A PoL is secure if it is complete, spatio-temporally sound, non-transferable and tamper-evident.*

Table 1. Table with useful notations

Notation	Meaning
G	Block transaction set
$P = [\gamma_1 \dots \gamma_n]$	Network peers set
$S = [\sigma_1 \dots \sigma_k]$	Prover set
$W = [\omega_1 \dots \omega_m]$	Witness set
Z	Proposal transaction set
f	Number of Byzantine failures of system entities
l	Reported location
t	Transaction timestamp
$\delta; \Delta$	Minimum network delay; network imprecision
θ	Proximity threshold
τ	Signed transaction
ζ	Maximum proposal threshold
ρ	Dynamic entity
κ_s, κ_p	Cryptographic keypair: secret key, public key

In addition to PoL, a peer-to-peer system must provide (i) security, (ii) liveness, and (iii) fault-tolerance. Note that (i)–(iii) are not simultaneously achieved [5], as (i) security requires all results to be valid for the network, (ii) liveness guarantees termination, and (iii) fault-tolerance requires a result regardless of the network failure (Table 1).

3.1   Protocol Entities

The entities can either be static or dynamic. Dynamic entities do not have a fixed location recorded in the blockchain and move in a limited geographical region and communicate with nearby witnesses. It is usually assumed that the movement is initiated by way of a smart-contract that defines the starting and ending points, e.g. an object transfer from entity A to entity B. Static entities, on the other hand, are associated with some specific location that is verified and recorded in a blockchain. Examples of such entities include producers, like farmers and regular customers, waiting for delivery at predefined locations. We consider the following static entities for the proposed protocol.

Witness. A witness is a static entity and is associated with a particular location which is agreed and stored by network peers. Communications are performed using a small communication delay that allows preventing proxy or relay attacks using delay-based challenge-response scheme. We assume that witness is able to perform cryptographic operations such as checking the validity of signatures and sign messages. Thus, the correctness of data is ensured by the witnesses as they provide data security and protection against the location attacks. We assume that there exist multiple witnesses such that the number of colluding malicious witnesses does not exceed a pre-defined threshold f_ω.

Prover. A prover communicates with nearby witnesses only if they are at a distance lower than a proximity threshold θ. We assume that a registered prover has a public/private keypair and the public key of the prover is known to all network peers as a pseudonym identifier. Prover can be either dynamic or static. When a dynamic entity is moving near a witness it performs a challenge-response authentication to collect location proofs. Similar authentication is performed for a static entity that desires to join the protocol.

Network peer. Transaction history is maintained by network peers in a blockchain. Blockchain supports identity mapping to a location, and the authenticity of the data in presence of adversaries. The network peers not only maintain system integrity but also provide PoL. Each peer is open for queries from witnesses, clients and other peers. Smart contracts are generated and executed as system transactions, for example, a transfer of goods from one owner to another. These smart contracts serve as a proof of an intent or an interaction between parties and are used during the validation process.

Client. A dynamic entity that moves in a restricted area is known to the client and the client makes queries to get the location or path of a dynamic entity at a specific time.

3.2   Threat Model

In order to create a PoL, a witness must be registered and verified at location (x, y). A witness performs proximity tests at a location within a maximum radius θ around a point (x, y). To tolerate f Byzantine failures $3f + 1$ nodes must be deployed [4]. This is a known lowest upper bound to achieve security in an asynchronous peer-to-peer network with possible Byzantine failures. We say that the protocol can tolerate up-to f_γ peer failures and f_ω witness failures. Users are registered with the network peers, and network peers maintain a list of users, witnesses and associated locations. The users and the witnesses are recognised by their public keys by the network peers. We assume that the public-private key pair is unique and is stored securely on personal devices.

For the threat model, we consider the following[4]: (i) a malicious prover submits a false location claim without being physically present at a location, (ii) a prover or network peer attempts to perform a colluding attack in order to manipulate an old or new location proof, (iii) a malicious witness either reports about a prover that is not present in close proximity, or does not report a prover

---

[4] We do not consider user privacy issues in this study and consider it as a future work.

presence that is in close proximity, and (iv) an adversary attempts a Sybil attack by way of creating multiple malicious users, peers or witnesses.

4   Proof-of-Location Protocol Design

We now present a protocol design that satisfies the conditions (Sect. 3) for a secure and reliable location tracking. We consider a peer-to-peer network where location proofs are generated by witnesses and are sent to peers for verification and storage. Peers exchange transactions and agree on the validity of a location proof by way of a consensus routine. An illustration of the PoL protocol design is presented in Fig. 1. We segregate between commands and queries to improve the performance of the system. A query is sent only to one peer which returns query results and a verification object that validates the query result. A command is packaged as a transaction and is put into the execution pipeline of the blockchain. Transaction validation is both via stateless and stateful validation. Stateless validation is for signature and time validation. Stateful validation is by way of a 'is_valid' predicate which validates transactions according to the system rules relative to the current local state. For example, the requirement of balance validation for an asset transfer.

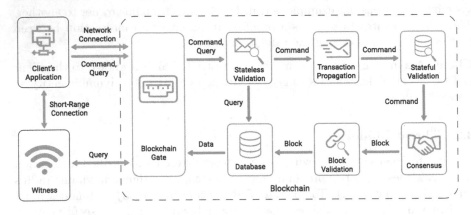

Fig. 1. An illustration of PoL protocol design

4.1   Proof-of-Location Consensus Algorithm

An outline of the consensus algorithm is given in Algorithm 1. The connection with a blockchain peer is established through a blockchain gate which is a network connection between a client and blockchain peer. Transactions are included in a proposal and the most popular proposal is decided by the consensus algorithm as a new block. Next, the block is validated and applied to the local database. The pipeline of the consensus algorithm is as follows: transaction is sent to the ordering peers (line 1), transactions are packaged into a proposal (line 3–7), proposal is shared with validating peers (line 8–9), the most popular

---

Algorithm 1. An outline of the spatio-temporal consensus algorithm

---

1 **function** RECIEVETRANSACTION(τ: Transaction; $P = [\gamma_1 \ldots \gamma_n]$: Network peers;
   ζ: Maxmimum proposal size; t_Z: Proposal time delay)
2    $Z \leftarrow \emptyset$ ▷ Proposal set
3   **for all** unprocessed transaction τ **do**   ▷ Process transaction queue
4   **if** STATELESSVALIDATION(τ, SENDER(τ), SIGNATURE(τ)) **then**
5          $Z \leftarrow \tau \cup Z$
6   **if** $|Z| < \zeta$ **then**
7   DELAY(t_Z)
8   Send proposal Z to peers $[\gamma_1 \ldots \gamma_n] \in P$

9 **function** PROCESSPROPOSAL(Z: proposal; $P = [\gamma_1 \ldots \gamma_n];$)
10   $Z_v \leftarrow$ STATEFULVALIDATION(STATE(γ_i), Z)   ▷ Verified proposal
11   $\Lambda \leftarrow$ LEADER(Z_v, STATE(γ_i))   ▷ Decide leader on proposal and state
12   Send VOTE(VP), $sig_{\gamma_i}(VP)$ to leader Λ

13 **function** PROCESSVOTE($[h_k, sig_{\gamma_j}]$: peer vote; $P = [\gamma_1 \ldots \gamma_n]$: Network peers;)
14   Vote h_k, sig_{γ_j} received from γ_j
15   **if** $\gamma_j \in P \wedge sig_{\gamma_j} \in$ STATE(γ_i) **then**   ▷ Peer γ_j is known to peer γ_i
16   votes$[h_k]$= votes$[h_k] \cup \gamma_i$
17   **for all** hash $h \in$ votes **do**
18   **if** $|votes[h]| \geq 2f_\gamma + 1$ **then**
19   Send block with hash h to peers $[\gamma_1 \ldots \gamma_n] \in P$

20 **function** PROCESSBLOCK(G: block; $P = [\gamma_1 \ldots \gamma_n]$: Network peers; $peer_i$: Current
   peer)
21   **if** BLOCKVALIDATION(G, textscState(γ_i)) **then**
22      $NewState =$APPLY(textscState(γ_i), G)
23   **return** $NewState$
24   $NewState \leftarrow$ SYNCHRONISE(P)   ▷ Synchronise state with peers
25   **return** $NewState$

---

proposal is decided with a leader-based consensus protocol (line 9–20) and is formed as a new block (line 20–21). A consensus process produces a validated block with signatures of peers agreeing on the validity of the block. The validity of the received blocks is decided with a block validation routine (line 23). Finally, the accepted block is applied to the local datastore (line 24).

Assume that ρ moves from A (i.e. l_A) to B (i.e. l_B). A transaction τ_s is issued that defines the movement, i.e. source, target, path, system and business rules.

When passing through some route, the dynamic entity ρ communicates with the witnesses $[\omega_1 \ldots \omega_n]$ with a location proof generation protocol. Each witness is bonded with at least one network peer that stores and transmits location proofs to other network peers. Peers ratify location proofs on their soundness and completeness. The final validity of a location proof is decided by the consensus mechanism (Algorithm 1). When ρ arrives at the target location, an ending transaction τ_e is generated by ρ which is a confirmation of the execution of the terms as defined in τ_s, for example, an asset transfer between A and B.

Network Initialisation. The first block generated during network initialisation is the genesis block G_0 which lists ledger rules including validation rules, participants, roles, permissions and threat model. Thus, parameters such as the number of permitted peers and witness failures are set during initialisation.

New Entity Join. When a new system entity, such as a network node or a client, wants to join the network, the following is performed: entity generates a cryptographic key-pair (κ_s, κ_p) followed by the creation of a join transaction which is sent to the blockchain nodes. If the role requires fixing a static location, location proofs are collected from the nearby witnesses. The signed join transaction is updated to the datastore after consensus.

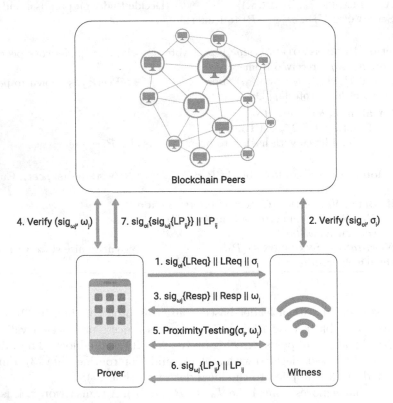

Fig. 2. An illustration of PoL generation

4.2   Proof-of-Location Generation

When a prover desires to obtain a proof of current location, the proof generation procedure is initiated in collaboration with the nearby witnesses. A short-range location proof is considered more reliable than a long-range location proof [16].

The proof generation protocol is illustrated in Fig. 2. The protocol consists of following phases:

(0) The prover σ_i is known to the blockchain peers due to the new entity join procedure. Next, σ_i prepares a list of witnesses in near proximity.

(1) Prover σ_i selects a witness ω_j and sends a location proof request. A message, $LReq||\sigma_i||sig_{\sigma_i}\{LReq\}$, is sent to witness ω_i. The message components include: $LReq$ to initialise the proximity testing, declared position $[x_i, y_i]$ at time t. Signature sig_{σ_i} and σ_i are used to authenticate the request.

(2) Witness processes the location proof request by validating the signature sig_{σ_i}, position $[x_i, y_i]$, timestamp t and the pseudonym σ_i. Position $[x_i, y_i]$ is validated relative to the proximity of the witness ω_j. Pseudonym σ_i must be known to the network peers and timestamp t must be within a valid range, i.e. $now - \delta \leq t \leq now + \delta + \Delta$, where δ is a minimum network delay, Δ is the network imprecision, now is the local clock time of the witness.

(3) Witness ω_j responses with a message $Resp$ whereas pseudonym ω_j and signature for $Resp$ message are used for the acknowledgement authentication.

(4) $Resp$ message from witness ω_j is validated with the help of network peers that includes the following checks: pseudonym ω_j is known to the peers, entity has a 'witness' role, the signature is valid and consistent with the pseudonym ω_j.

(5) γ_i and ω_j start a proximity testing protocol. Typically, witness and prover perform a fast bit exchange where witness sends some generated secret and waits for a response from the prover with the same secret. Each message is signed to protect against proxy attacks.

(6) After a set of rounds of proximity testing, witness is convinced about the declared location of γ_i. Witness sends signed location proof lp_{ij}, where lp contains declared location of entity γ_i, location of witness ω_j and current timestamp t_{ω_j}. lp_{ij} is sent to γ_i.

(7) Prover σ_i signs received location proof lp_{ij} and sends it to the network peers. Network peers share with each other the location proof and run consensus algorithm on the validity of the location proof.

Location Proof Verification. Location proof is considered valid if (i) it contains a valid timestamp, (ii) it is signed with at least $2f_\omega + 1$ witnesses and the prover, (iii) the witnesses and the prover are known, (iv) each witness is an authorised 'witness', and (v) all witnesses can validate the proximity of the prover in location $[x_i, y_i]$.

5 Threats-to-Validity Check

We now evaluate the threats-to-validity as we first discuss the adversary model.

5.1 Adversary Model

We distinguish among following adversary[5] types:

---

[5] We assume that an adversary can not violate cryptographic assumptions.

1. *Malicious prover.* An adversary tries to modify current real-time or position and prove the wrong location, or lie to the verifier about its position. A malicious prover attempts to (i) create a fake location proof, (ii) lie to the witness or the verifier, or (iii) steal a location proof of another user.
2. *Malicious verifier.* An adversary can approve fake location proof, or blacklist and censor transactions from a legitimate user.
3. *Malicious witness.* An adversary can try to foul a legitimate prover or verifier by a fake location proof. For example, proof of proximity of a physically-not-present or non-existing entity.
4. *Malicious network peer.* Malicious peer attempts to attack consensus process to include transactions that are not valid. Sybil attack and double spending are examples of such attacks.
5. *Colluding users.* A colluding attack with malicious parties is performed to accept and include a fake location proof. Colluding adversary attacks are considered hard to be detected and prevented [6].

5.2   Adversary Resistance

We now present the adversary resistance of the protocol according to the criteria discussed in Sect. 3.

Correctness. Assume that prover σ submits a proof of presence at location l and time t. The validity of the proof is decided by a consensus process. With the assumption that number of prover failures is less than f_γ and number of witness failures is less than f_ω, the protocol achieves completeness. Spatio-temporal soundness is achieved by validating the signatures and location information. A malicious prover may attempt a distance fraud. The proximity testing described in section 4.2 prevents a distance fraud.

Proof of Ownership. A proof of ownership for the location of prover σ requires the signature of σ which is not possible without the secret key κ_s^σ.

Authenticity. A change in the location proof invalidates the signatures which in turn makes location proof invalid. A location proof is stored in a blockchain which is maintained by network peers. As long as the number of malicious peers is less than f_σ, authenticity of the datastore is ensured.

Colluding Attacks. A fake location proof generated by a malicious witness with the help of a malicious prover is rejected if the number of witness signatures is less than $2f_\omega + 1$. Due to the assumption that the number of malicious witnesses is less than f_ω, it is not feasible to create a valid fake location proof.

6   Case Study: Supply Chain

A supply chain is a system of organisations, people, activities, information, and resources involved in moving a product or service from a set of suppliers to a set of customers. Typically, a supply chain can be represented as a sequence of transitions from a producer to the consumer with multiple intermediate points. We consider the transportation chain as the communication chain between two parties. Let's call this communication between party A and party B. Party A is a service provider for some goods, party B is a customer.

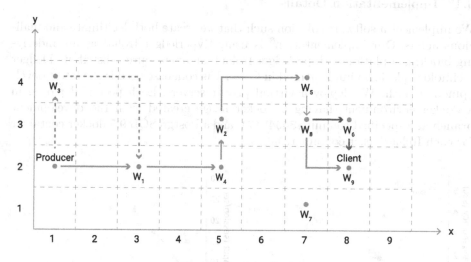

Fig. 3. Example Grid

Consider a 4×9 grid as shown in Fig. 3 with many stakeholders in the chain. We say that, Party A is a producer and party B is a customer who is intending to buy some amount of goods in exchange of some assets. The blockchain records the fact that party A is the owner of the commodity and party B is the liquidity bearer. Party A and Party B are associated with a static location. For example, party A is a producer in location $[1, 2]$, and party B is a consumer in location $[8, 2]$. To deliver goods from one location to another, a transport is used which is a dynamic entity that delivers goods in exchange of some fee. Witnesses $\omega_1, \ldots, \omega_9$ are present in different locations to generate location proofs. The proof pipeline is as follows:

1. Party A and party B agree on terms of exchange via a starting transaction: product P is put to the transport T in exchange to the payment from party B at location $[1, 2]$ to party A at time t_1 in order to deliver it to the B at location $[8, 2]$.
2. The product is transported from location $[1, 2]$ to location $[8, 2]$.
3. Through the road, witnesses are tracking all bypassing transports, perform a proximity test and report to the blockchain peers.
4. The proof of location transaction is sent to the blockchain nodes, where each node is validating it and writing the results to the underlying data-store.
5. When transport arrives to $[8, 2]$ client confirms the delivery with an ending transaction: a product is delivered to location $[8, 2]$ at time t_2 and is verified

by party B. Ending transaction closes the delivery as it transfers the required number of assets to the producer, fixes the ownership of the product with B and pays transportation fee for the transport T.

6.1   Implementation Details

We implement a software solution such that we create both legitimate and malicious actors. Our implementation[6] is using Hyperledger Iroha[7] as an underlying blockchain platform. Hyperledger Iroha is an open-source, distributed ledger technology platform implemented in C++. The codebase used is the latest development branch. We deploy 4 virtual private servers. Each peer is deployed in a docker environment, having a docker image created from the development branch at a specified commit (`3404b17`), and a PostgreSQL 9.5 docker container for each Iroha docker container.

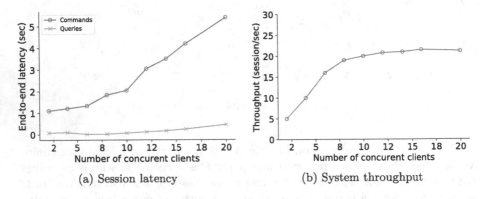

(a) Session latency   (b) System throughput

Fig. 4. System latency and throughput analysis

A transaction is implemented as an asset transfer to an agreed transport. The consumer sends a transaction and locks the funds of the transport by turning transport into a multi-signature account with quorum 2. Producer transfers the ownership of a product, adds new signatory and increases the quorum of transport account. In the end, transport is a multi-signature account that requires at least three signatures for any further transactions. This ensures the safety of the funds before reaching the consumer. Upon arriving at the desired location, transport creates an ending transaction that transfers coins to the producer, product ownership to the consumer and transportation fees to the transporter wallets. The transaction is checked by three parties for agreed number of coins and products.

We simulate a supply chain case and deploy transportation units in a test area and show how it affects the performance of the system. A witness is capable

---

[6] https://github.com/grimadas/iroha-supply-chain.
[7] https://github.com/hyperledger/iroha.

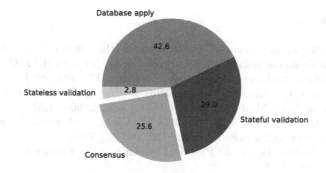

Fig. 5. Time distribution of a transaction at different execution stages

of attesting an area within one cell. An end-to-end transaction latency for the full pipeline for location proof transaction is shown in Fig. 4. The results show that with an increase in the number of concurrent clients generating and submitting a location proof, latency increases linearly. The system is capable of creating and validating 30 location proofs per second. The end-to-end latency increases linearly with the number of concurrent clients, while the latency of read-only queries changes insignificantly. As Iroha is still in development, a number of optimisations can be applied to improve the system performance. Figure 5 shows time distribution of different components. It can be seen that database apply operations are the most time-consuming.

7 Conclusion

In this paper, we propose a practical robust proof-of-location protocol on top of the blockchain. We implement a proof-of-concept supply chain and evaluate the security and the performance of the protocol in the presence of malicious actors. We illustrate in practice that system scales linearly with the increase of concurrent actors. To the best of our knowledge, ours is one of the first practical systems that achieves tamper-resistance for the accepted Proof-of-Location transactions. We show the utility of our solution in the supply chain domain. A dis-intermediation of actors with defined roles allows using both public and private domains.

A number of challenges still remain. For example, flexible privacy settings with zero-knowledge proofs and incentive design with game-theoretic guarantees. As a future work, we consider an extension of the protocol to work with dynamic witnesses and the deployment of the solution in real-world settings.

Acknowledgments. The work was partially supported by the CAS Pioneer Hundred Talents Program, China [grant number Y84402, 2017], and CAS President's International Fellowship Initiative, China [grant number 2018VTB0005, 2018].

References

1. Abdou, A.M., et al.: Location verification on the internet: towards enforcing location-aware access policies over internet clients. In: 2014 IEEE Conference on Communications and Network Security (CNS), pp. 175–183 (2014)
2. Ahi, P., Searcy, C.: Assessing sustainability in the supply chain: a triple bottom line approach. Appl. Math. Model. **39**(10–11), 2882–2896 (2015)
3. Brambilla, G., Amoretti, M., Zanichelli, F.: Using block chain for peer-to-peer proof-of-location. arXiv preprint arXiv:1607.00174 (2016)
4. Castro, M., Liskov, B., et al.: Practical byzantine fault tolerance. In: OSDI, vol. 99, pp. 173–186 (1999)
5. Fischer, M.J., Lynch, N.A., Paterson, M.S.: Impossibility of distributed consensus with one faulty process. J. ACM (JACM) **32**(2), 374–382 (1985)
6. Gambs, S., Killijian, M.O., Roy, M., Traoré, M.: PROPS: a PRivacy-preserving location proof system. In: IEEE 33rd International Symposium on SRDS, pp. 1–10 (2014)
7. Jafarnia-Jahromi, A., Broumandan, A., Nielsen, J., Lachapelle, G.: GPS vulnerability to spoofing threats and a review of antispoofing techniques. Int. J. Navig. Obs. (2012)
8. Javali, C., Revadigar, G., Rasmussen, K.B., Hu, W., Jha, S.: I am Alice, i was in wonderland: secure location proof generation and verification protocol. In: 2016 IEEE 41st Conference on Local Computer Networks (LCN), pp. 477–485 (2016)
9. Muzammal, M., Gohar, M., Rahman, A.U., Qu, Q., et al.: Trajectory mining using uncertain sensor data. IEEE Access **6**, 4895–4903 (2018)
10. Muzammal, M.: Renovating blockchain with distributed databases: an open source system. Futur. Gener. Comput. Syst. **90**, 105–117 (2019)
11. Nakamoto, S.: Bitcoin: a peer-to-peer electronic cash system (2009). https://bitcoin.org/bitcoin.pdf
12. Nasrulin, B., Muzammal, M., Qu, Q.: ChainMOB: mobility analytics on blockchain. In: 19th IEEE International Conference on Mobile Data Management, MDM, pp. 292–293 (2018)
13. Ni, X., et al.: A mobile phone-based physical-social location proof system for mobile social network service. Secur. Commun. Netw. **9**(13), 1890–1904 (2016)
14. Pham, A., Huguenin, K., Bilogrevic, I., Dacosta, I., Hubaux, J.P.: SecureRun: cheat-proof and private summaries for location-based activities. IEEE Trans. Mob. Comput. **15**(8), 2109–2123 (2016)
15. Qu, Q., Liu, S., et al.: Efficient online summarization of large-scale dynamic networks. IEEE Trans. Knowl. Data Eng. **28**(12), 3231–3245 (2016)
16. Ranganathan, A., Capkun, S.: Are we really close? Verifying proximity in wireless systems. IEEE Secur. Priv. (2017)
17. Stadtler, H.: Supply chain management - an overview. In: Stadtler, H., Kilger, C. (eds.) Supply Chain Management and Advanced Planning, pp. 9–36. Springer, Heidelberg (2008). https://doi.org/10.1007/978-3-540-74512-9_2
18. Underwood, S.: Blockchain beyond bitcoin. Commun. ACM **59**(11), 15–17 (2016)
19. Wan, J., Zou, C., Ullah, S., Lai, C., Zhou, M., Wang, X.: Cloud-enabled wireless body area networks for pervasive healthcare. IEEE Network **27**(5), 56–61 (2013)
20. Wang, X., Pande, A.: STAMP: enabling privacy-preserving location proofs for mobile users. IEEE/ACM Trans. Netw. **24**(6), 3276–3289 (2016)

21. Waters, B., Felten, E.: Secure, private proofs of location. Department of Computer Science, Princeton University, Princeton, NJ, USA, Technical report (2003)
22. Yang, J., Chen, Y., Trappe, W., Cheng, J.: Detection and localization of multiple spoofing attackers in wireless networks. IEEE Trans. Parallel Distrib. Syst. **24**(1), 44–58 (2013)

Towards an End-to-End IoT Data Privacy-Preserving Framework Using Blockchain Technology

Faiza Loukil[1](✉), Chirine Ghedira-Guegan[2], Khouloud Boukadi[3], and Aïcha Nabila Benharkat[4]

[1] University of Lyon, University Jean Moulin Lyon 3, CNRS, LIRIS, Lyon, France
faiza.loukil@liris.cnrs.fr
[2] University of Lyon, University Jean Moulin Lyon 3,
iaelyon school of Management, CNRS, LIRIS, Lyon, France
chirine.ghedira-guegan@liris.cnrs.fr
[3] Mir@cl Laboratory, Sfax University, Sfax, Tunisia
khouloud.boukadi@fsegs.usf.tn
[4] University of Lyon, INSA Lyon, CNRS, LIRIS, Lyon, France
nabila.benharkat@liris.cnrs.fr

Abstract. Internet of Things-based environments collect and generate huge amounts of data about users, their activities, and their surroundings, which can disclose some sensitive information and threaten their privacy. Hence, the user's collected and handled data by IoT-based applications need to be exploited and secured in an appropriate way to protect personal data and user's privacy. Therefore, we aim at improving the data ownership, transparency, and auditability for users. To this end, we propose an end-to-end privacy-preserving framework for the IoT data using blockchain technology. The smart contract use in our framework will hence enforce the privacy requirement compliance according to the user's (i.e., data owner) privacy preferences and end-user's (i.e., data consumer) requests. To do so, we detail the design of the system architecture by introducing its core components and functionalities and highlight through an example of how it operates in a real-world use-case.

1 Introduction

Internet of Things (IoT) consists of devices that collect, exchange, store, and process large amount of fine-granularity and high-frequency data in every aspect of life. Such detailed data improve delivering advanced services in a wide range of application domains. Indeed, service providers gather the IoT data and use them to personalize services, optimize decision-making process, and predict future trends. However, the IoT data raise security and privacy concerns. In fact, the users have a little or no control over the collected data about themselves [10].

Furthermore, due to the distributed nature of the IoT networks, security and privacy are recognized to be among the major challenges of the IoT domain.

© Springer Nature Switzerland AG 2018
H. Hacid et al. (Eds.): WISE 2018, LNCS 11233, pp. 68–78, 2018.
https://doi.org/10.1007/978-3-030-02922-7_5

Well-known security and privacy techniques, such as encryption, authentication, and role-based access control which are used in the context of conventional information systems failed to protect IoT data due to the variety of hardware platforms and limited computing resources [1]. For instance, well-known encryption protocols and privacy-preserving methods, such as RSA, fully homomorphic encryption, and differential privacy proved to be very expensive when running on devices with limited computing capabilities in the IoT domain [16].

In recent years, the blockchain emerged as a new technology. The first system based on this technology was Bitcoin [12], which enables users to securely transfer the currency (bitcoins) while eliminating the need to trust a centralized regulator. Ethereum [3] is another blockchain-based system that can also be used for the cryptocurrency. Unlike Bitcoin, Ethereum has the ability to use a smart contract, which is a common agreement between two or more parties. It stores information, processes inputs and writes outputs thanks to its predefined functions [3]. Ethereum requires paying currency to run smart contract to prevent infinitely runs. Since then, other projects demonstrated how the blockchain technology could be used to address other domains, like the IoT data privacy-preserving. Although several blockchain-based solutions [2,5,7,13] address the privacy issue in the IoT domain, they assumed that the IoT resources had sufficient resources to solve the Proof-Of-Work, which may not always be true as well as the others, did not address the whole IoT data lifecycle. Moreover, existing solutions did not consider all the privacy requirements, such as the purpose, retention duration, disclosure limitation, etc. that are defined by the privacy standard [8] and legislation [15] to preserve user privacy. In our work, the privacy requirements should cover the obligations that must be fulfilled by all the involved parties to preserve the privacy during the whole IoT data lifecycle.

Motivated by the legal rights imposed by the European General Data Protection Regulation (GDPR) [15], we focus on the privacy requirement enforcement to preserve privacy during the whole IoT data lifecycle. The objective of our work is to guarantee that the privacy requirements will be enforced while handling the shared IoT data that are collected by IoT resources. To this end, we propose an end-to-end privacy-preserving framework for the IoT based on the blockchain technology and more specifically on smart contract. The main purpose of the smart contract use is to enable our framework to express privacy-preserving policies. A policy is a set of conditions that the consumer needs to fulfill in order to handle a specific shared IoT data. Thus, the use of smart contract will prevent any privacy violation attempts. By protecting the shared data, these data can be only handled by invoking functions defined on the hosted smart contract in the blockchain. This implies that there are no parties that can get hold of the smart contract once hosted in the blockchain. Thus, smart contract enforces the data owner's privacy preferences, then the shared data will be handled as expected.

This paper is organized as follows. Section 2 discusses the existing researchers who studied blockchain and IoT technology integration to preserve privacy. Section 3 presents an overview of our proposed framework. Section 4 identifies its core components. Section 5 explains the framework's main functionalities.

Section 6 validates our solution in a healthcare scenario. Section 7 concludes the paper and presents some future endeavors.

2   Related Work

IoT and blockchain combination generates resilient, peer-to-peer systems, and the ability to interact with peers in a trustless and auditable manner [4]. Many researchers have studied the integration of blockchain and IoT technology.

Biswas and Muthukkumarasamy [2] for example proposed a blockchain based security framework to enable secure data communication in a smart city. However, their proposal is at a high level of abstraction and they do not provide any system design to prove the feasibility of their framework. On the other hand, Dorri et al. [5] proposed a lightweight and optimized blockchain for resource-constrained devices and applied it in a smart house scenario. This work focused on data store and access use cases by IoT resources. However, the system design included a centralized control node, which can be considered as a single point of failure that could damage availability. For their part, Hashemi et al. [7] proposed a decentralized solution for the sharing of data in the IoT environment, which consists of a distributed data storage system. However, the authors assumed that the IoT have sufficient resources to solve the Proof-Of-Work which may not always be true. In fact, solving the POW for Bitcoin requires a very sophisticated hardware. Ouaddah et al. [13] proposed FairAccess, a blockchain-based framework for access control. The authors relied on Bitcoin system and introduced new transaction types. The transactions are used to provide access control, and the blockchain is used to store and read the permissions. However, this work addressed the access control issue but did not address the whole data lifecycle.

To sum up, it can be said that most existing blockchain-based solutions concentrate on addressing the access control issue in the IoT field. Moreover, they are only concerned with one phase, and do not address the whole data lifecycle.

3   Proposed Solution Overview

Considering the legal rights imposed by the GDPR [15], it is necessary to ensure the privacy requirement compliance to preserve privacy during the whole data lifecycle, covering the collection, transmission, storage and processing phases. In our work, we focus on how to enforce these privacy requirements and obligations for the IoT environment. To this end, we propose PrivBlockchain, an end-to-end privacy-preserving framework for the IoT data. PrivBlockchain is based on these principles, namely (i) user-driven and transparency, (ii) distributed architecture and central authority lack, and (iii) fine-granularity privacy policies.

Figure 1 depicts the proposed architecture, which includes two types of network, namely private IoT network, which can be a smart home, smart building, etc. This network includes the IoT resources owned by a data owner, which can be an individual or an organization. The second network is the public IoT network, which represents the external domain of the private IoT network. Moreover, we distinguish three types of IoT network nodes, namely private, public, and storage nodes. Both public and storage IoT network nodes belong to the public network. The private node (i.e., gateway node or private IoT resource) is an IoT node that belongs to both the private and public IoT networks.

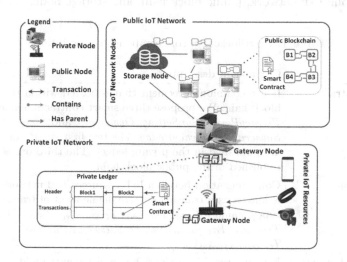

Fig. 1. PrivBlockchain architecture

In the private IoT network, each data owner has one or more high resource device, known as "gateway node", which is responsible for the other owned IoT resources. The communication between the owned IoT resources by the data owner (i.e., the private IoT resources) is stored in a private blockchain, which is called "private ledger". The communication between the private IoT nodes and the other nodes of the public IoT network is stored in a "public blockchain".

The reason behind using the blockchain technology to preserve privacy in the IoT domain is that the blockchain is an immutable public record of data secured by a network of peer-to-peer participants that use addresses as pseudonyms. Such user identity management improves anonymity and pseudonymity in an IoT network. In our case, we use a pair of public and private keys to ensure the anonymity and pseudonymity privacy properties. Moreover, using a different key pair in each transaction enforces unlinkability. For instance, the gateway node can use a different key pair in each transaction with the external nodes that belong to the public IoT network. Besides the privacy properties, namely anonymity, pseudonymity, and unlinkability [14] that are ensured by the blockchain, this final has the potential to enforce privacy requirement compliance. To this end, a set of privacy requirements is chosen that we agree critical in the area of IoT based on an extensive literature review, the ISO standard [8], and the GDPR [15].

We outline the proposed framework core components in the following section.

4 PrivBlockchain Core Components

Table 1 shows the main blockchain-based solution components. Indeed, the PrivBlockchain framework consists of nine core components, to know: smart contract, transaction, private IoT network, private ledger, gateway node, local storage, public IoT network, public blockchain, and storage node.

Table 1. PrivBlockchain core components description

Component	Component description
Smart contract	It is a common agreement that is hosted within the blockchain. We propose three smart contracts, namely *PrivacyPermissionSetting*, *Ownership*, and *SubscriptionPrivacyPolicy*. The two first smart contracts are published in the private ledger. The third one is published in the public blockchain
Transaction	Communication between IoT resources and network nodes is known as a transaction. We define a set of transaction types, namely T_{Add}, T_{Remove}, $T_{LocalStore}$, T_{Store}, T_{Access}, $T_{Monitor}$, $T_{GetPermission}$, $T_{GrantPermission}$, and $T_{GetSharedResource}$
Private IoT network	It is an area, like a smart home or a smart building, where its owner can control a set of private IoT resources
Private ledger	It is a local private blockchain that enables the data owner to control his own IoT resources
Gateway node	It is a device with a high memory and storage capabilities. Each gateway node is responsible for a set of private IoT resources, generates there keys and adds them to the IoT network
Local storage	It is a device used to store data locally. It saves the collected data by IoT resources for long-term storage before sending them to the external storage center, which is the storage node
Public IoT network	It is a peer-to-peer network that contains several nodes with different memory and storage capabilities
Public blockchain	It can be seen as the history of all the transactions that are sent by the public nodes to access or share IoT data in the public IoT network. In fact, it can ensure auditing functions
Storage node	It is a public IoT network node that offers a storing service for both public blockchain and data collected by the IoT resources

5   PrivBlockchain Functionalities

Based on the proposed smart contracts, our end-to-end privacy-preserving framework includes the following functionalities: (i) adding a new IoT resource to the *Ownership* smart contract, (ii) storing the collected data by a private IoT resource, and (iii) sharing IoT resource output with data consumers. The dynamic aspect of PrivBlockchain relies on those functionalities that we detail hereafter:

5.1   IoT Resource Add Transaction Protocol

Each gateway node publishes an *Ownership* smart contract that includes its own IoT resource addresses in the private ledger. For each IoT resource, a set of outputs can be added. A *PrivacyPermissionSetting* smart contract is associated with each IoT resource output. This smart contract enforces the data owner's privacy preferences about how the owned IoT resources must behave.

Figure 2 depicts the business process of adding a new IoT resource. We assume that the gateway node has created its *Ownership* smart contract and generated a pair of keys for an IoT resource.

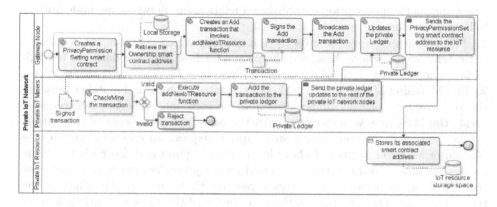

Fig. 2. Adding a new IoT resource business process notated in BPMN

5.2   IoT Resource Output Store Transaction Protocol

IoT resource periodically sends a $T_{LocalStore}$ transaction to the gateway node in order to locally store the collected data.

We assume that the IoT resource has an address and it knows the address of its associated smart contract and the gateway node address.

1. The IoT resource calculates the data hash, encrypts the collected data using the lightweight AES Encryption [9], and sends the encrypted data to the gateway node through a secure channel.

2. The gateway node decrypts the received data and calculates the data hash.

3. The IoT resource creates a new $T_{LocalStore}$ transaction that invokes the Local-Store function of its smart contract. The transaction includes the data type and hash. The $T_{LocalStore}$ transaction is sent to the appropriate gateway node because only the parent has the appropriate private key to sign the transaction.
4. When the gateway receives the transaction, it verifies the data integrity by comparing the data hash stored on the transaction and the calculated data hash in Step 2. In case of a match, the gateway signs the transaction and broadcasts it to the private IoT network. Otherwise, the transaction and data are rejected.
5. The miners validate and then execute the appropriate LocalStore function, which enables the IoT resource permission verification. The included data type on the transaction must belong to the allowed IoT resource outputs. If it belongs, the associated *PrivacyPermissionSetting* smart contract address of the IoT resource output is retrieved and compared with the included smart contract address on the transaction. If both addresses are the same and the storage permission is Permit, then the transaction is added to the ledger and the gateway node stores the data on the local storage. Otherwise, the received data are rejected.

5.3   Process of Sharing IoT Resource Output

The consumer's subscription establishes a relationship between one node and an IoT resource output.
1. The consumer creates a subscription request, which contains the requested data, why, when, where, how, to whom, and for how long the data are needed. This request is sent to the data owner address in a $T_{GetPermission}$ transaction.
2. The Matching Manager is included in the gateway node and matches between the data owner's privacy preferences and the consumer's subscription request. First, the Matching Manager evaluates the privacy requirements of the output privacy rule and the consumer's subscription request. In case of a match, the Matching Manager verifies if there is an already published file with the same requested data to retrieve the associated *SubscriptionPrivacyPolicy* smart contract address. Otherwise, it creates a new file that contains the result of the requested data. Then, it invokes the constructor of the *SubscriptionPrivacyPolicy* smart contract to create a new one. It specifies the hash of the file content and the address of the node that stores the shared file. After that, it sends the transaction to the gateway node to be signed and propagated to the network. Once the smart contract is created, the Matching Manager receives its address.
3. Once the Matching Manager gets the appropriate *SubscriptionPrivacyPolicy* smart contract address, it creates a $T_{GrantPermission}$ transaction that invokes the addConsumer function of this contract with the appropriate privacy permissions. It sends the transaction to the gateway node to be signed and propagated to the network. This transaction enables to add a new consumer to the list of the allowed consumers of the shared file.
4. When the data consumer receives the *SubscriptionPrivacyPolicy* smart contract address, it can send a $T_{GetSharedResource}$ transaction that invokes a set of the smart contract functions to handle the data. Before executing each function,

the set of the consumer's permissions is verified to enforce the data owner's privacy preferences. For instance, if the consumer has a permission to disclose the file content and the retention duration is not finished yet, it can invoke the **add-Consumer** function in order to add a new consumer to the file but with read-only permission and limited retention duration. We refer the reader to [1] for further details about the Business Process Models of the defined protocols in Sects. 5.1, 5.2, and 5.3.

6   Prototype and Validation

We implemented our proposed smart contracts using the Solidity language [6] and deployed it to the Ethereum test network, which is based on the go ethereum (gcth) implementation of the protocol released by the Ethereum foundation. The test network is identical to the production network except that the Ether has no real-world value. Given that our system does not rely on the currency transfer, the test network works like the real Ethereum network. MyEtherWallet [11] is used to access the network node information and control the public IoT network. Because synchronizing with the entire Ethereum test network would take too much time and resources, we decided to mount a local private Ethereum blockchain.

Insofar as PrivBlockchain is generic and could be used for a variety of IoT application domains, such as smart home, smart grid, etc., we applied it to the following scenario from the healthcare domain to validate our solution:

A patient named Bob needs to follow a healthcare protocol, which consists in practicing some sport activities and eating healthy meals. Bob owns a private IoT resource, which is a wearable device that collects the user's heart rate. This IoT resource can be connected to several IoT resources, such as smart equipment. Bob goes regularly to a modern gym, which uses fitness smart equipment. This smart equipment collects data from the different users' wearable devices and sends them to the gym data center to be stored. These stored data are analyzed to propose personalized recommendations for users, detect the most used equipment, propose group training programs, etc. Moreover, the gym offers a "healthy eat application" that proposes a set of healthy meals according to the needed calories for each specific user. Bob wants to connect his wearable device to a smart treadmill in the gym to monitor his performance by consistently measuring his heart rate and walking or running power. Hence, a need for a break or the water notifications could be sent to Bob when necessary. However, Bob is afraid that the gym center uses the collected data by his wearable device to monitor his activities or to disclosure them to third parties without his consent.

In order to reassure its clients, the gym center needs to opt for a solution that preserves the IoT data privacy and improves transparency. Indeed, the gym center relies on the PrivBlockchain solution in order to (i) gain the users' trust, (ii) guarantee the users' right to control their personal data while benefiting from personalized services, and (iii) be compliant with the privacy legislation [15].

---

[1] https://sites.google.com/view/privblockchain-protocols.

Figure 3 shows the major components and the interactions between them. It is worth noting that all the components are identified by Ethereum addresses and the interactions are shown as transactions in Fig. 4.

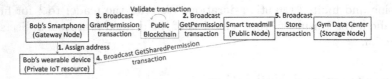

Fig. 3. Implementation components

Figure 4 depicts both adding an IoT resource and sharing its output protocols. Transactions shown in the middle succeed to add Bob's wearable device and the smart treadmill as a heart rate consumer. The figure left side details the addNewIoTResource function that is invoked by the gateway node to add Bob's wearable device and associate a *PrivacyPermissionSetting* smart contract to the output. The right side details the addConsumer function, which is defined in the *SubscriptionPrivacyPolicy* and invoked by a transaction sent by the smart treadmill. This transaction is validated and added to the public blockchain because the smart treadmill has the disclosure permission, otherwise, it is rejected.

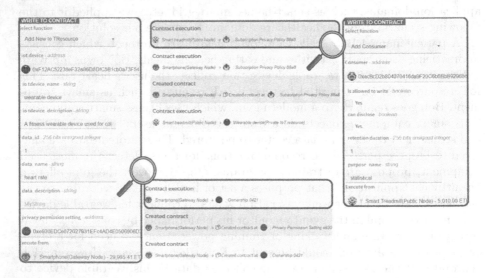

Fig. 4. Illustration of addNewIoTResource and addConsumer function invocations

7   Conclusion

In recent years, several researchers have agreed that the combination of blockchain and IoT generates resilient, truly distributed peer-to-peer systems and the ability to interact with peers in a trustless and auditable manner. However, few proposed solutions have dealt with the privacy issue in the IoT domain. Despite the increasing legislation pressure, the privacy requirements, such as consent and choice, purpose specification, and collection limitation, have been less addressed in the IoT domain. For these reasons, we have focused on the privacy requirement enforcement to preserve privacy during the whole IoT data lifecycle using the blockchain technology. To this end, we have proposed an end-to-end privacy-preserving framework for the IoT data based on smart contracts to enforce preserving privacy in the IoT domain. As 'proof of concept', we have established an initial implementation of PrivBlockchain with limited number of nodes. In our future work, we intend to experiment our framework in a large IoT network with multiple involved parties. We plain also to broadcast a large amount of transactions that ask and grant permissions, then analyze the PrivBlockchain's behavior over time. Moreover, we intend to perform comparative performance analysis by comparing the estimated computational costs of PrivBlockchain and some of the existing works using the *gas* as a unit of measure.

Blockchain analysis can possibly reveal the frequency of visiting a place or practicing an activity by a specific node. To overcome this problem, our framework enables the several addresses use for the same IoT resource. Besides, we intend to incorporate the use of differential privacy, a rigorous privacy model that preserves data privacy while maintaining utility in our framework. In fact, by adding some noise to the transactions, we can prevent blockchain analysis.

References

1. Bertino, E.: Data security and privacy in the IoT. In: EDBT, vol. 2016, pp. 1–3 (2016)
2. Biswas, K., Muthukkumarasamy, V.: Securing smart cities using blockchain technology. In: 2016 IEEE 18th International Conference on High Performance Computing and Communications; IEEE 14th International Conference on Smart City; IEEE 2nd International Conference on Data Science and Systems (HPCC/SmartCity/DSS), pp. 1392–1393. IEEE (2016)
3. Buterin, V., et al.: A next-generation smart contract and decentralized application platform. White paper (2014)
4. Christidis, K., Devetsikiotis, M.: Blockchains and smart contracts for the Internet of Things. IEEE Access 4, 2292–2303 (2016)
5. Dorri, A., Kanhere, S.S., Jurdak, R.: Towards an optimized blockchain for IoT. In: Proceedings of the Second International Conference on Internet-of-Things Design and Implementation, pp. 173–178. ACM (2017)
6. Ethereum project's Solidity Team: Solidity language (2016). http://solidity.readthedocs.io/en/develop/. Accessed 10 Aug 2018
7. Hashemi, S.H., Faghri, F., Rausch, P., Campbell, R.H.: World of empowered IoT users. In: 2016 IEEE First International Conference on Internet-of-Things Design and Implementation (IoTDI), pp. 13–24. IEEE (2016)
8. International Organization for Standardization: Information technology security techniques privacy framework, ISO/IEC 29100 (2011)
9. Landman, D.: Arduino Library for AES Encryption (2017). https://github.com/DavyLandman/AESLib. Accessed 10 Aug 2018

10. Maddox, T.: The dark side of wearables: how they're secretly jeopardizing your security and privacy (2015). https://www.techrepublic.com/article/the-dark-side-of-wearables-how-theyre-secretly-jeopardizing-your-security-and-privacy/. Accessed 10 Aug 2018
11. MyEtherWallet Team: Myetherwallet (2015). https://www.myetherwallet.com/. Accessed 10 Aug 2018
12. Nakamoto, S.: Bitcoin: a peer-to-peer electronic cash system (2008)
13. Ouaddah, A., Abou Elkalam, A., Ait Ouahman, A.: Fairaccess: a new blockchain-based access control framework for the Internet of Things. Secur. Commun. Netw. **9**(18), 5943–5964 (2016)
14. Pfitzmann, A., Hansen, M.: Anonymity, unlinkability, unobservability, pseudonymity, and identity management - a consolidated proposal for terminology (2005)
15. Regulation, General Data Protection: Regulation (EU) 2016/679 of the European Parliament and of the Council of 27 April 2016 on the protection of natural persons with regard to the processing of personal data and on the free movement of such data, and repealing Directive 95/46. Off. J. Eur. Union (OJ) **59**, 1–88 (2016)
16. Singla, A., Mudgerikar, A., Papapanagiotou, I., Yavuz, A.A.: HAA: hardware-accelerated authentication for Internet of Things in mission critical vehicular networks. In: 2015 IEEE Military Communications Conference, MILCOM 2015, pp. 1298–1304. IEEE (2015)

Security

i2kit: A Deployment Tool with the Simplicity of Containers and the Security of Virtual Machines

Pablo Chico de Guzmán(✉), Felipe Gorostiaga(✉),
and César Sánchez(✉)

IMDEA Software Institute, Madrid, Spain
{pablo.chico,felipe.gorostiaga,cesar.sanchez}@imdea.org

Abstract. Container virtualization technologies, like Docker, are becoming increasingly popular. Containers provide exceptional developer experience because containers offer lightweight isolation and ease of software distribution. Containers also solve a fundamental code portability problem.

In contrast, container virtualization is basically insecure when compared to virtualization based on hypervisors. Virtual machines are also better integrated with the rest of the cloud ecosystem. Sum it all, virtual machines are more suitable for production environments. However, virtual machines impose a non-negligible memory footprint and suffer longer boot times, which is impractical for local development. So far, there is no deployment infrastructure that allows both the developer experience of containers and the maturity and isolation capabilities of virtual machines.

We solve this problem in this paper by introducing *i2kit*, an orchestration tool that enjoys the best of both worlds: (1) the development workflow is untouched, containers can be used as usual; (2) at time of deployment, containers are transformed into virtual machines, keeping code portability, but providing better security and better integration with other cloud services. The tool *i2kit* creates virtual machines using Linuxkit. Linuxkit alleviates the drawback in size that using virtual machines would otherwise entail because the footprint of our Linuxkit distributions is only about 60 MB. The attack surface of the application is reduced since Linuxkit only installs the minimum set of OS dependencies to run containers. Finally, we report an empirical study using *i2kit* that allows us to conclude that *i2kit* is a promising technology for VM deployment of applications developed using containers.

Keywords: Virtualization · Orchestration · Security
Resource utilization

This research has been partially supported by: the EU H2020 project Elastest (num. 731535), by the Spanish MINECO Project "RISCO (TIN2015-71819-P)" and by the EU ICT COST Action IC1402 ARVI (*Runtime Verification beyond Monitoring*).

H. Hacid et al. (Eds.): WISE 2018, LNCS 11233, pp. 81–95, 2018.
https://doi.org/10.1007/978-3-030-02922-7_6

1 Introduction

Docker containers [1] have popularized the use of lightweight virtualization technologies such as LXC [2]. Some large companies report running all of their services in containers (e.g. [3]), and Container as a Service (CaaS) products are available from the main cloud players including Amazon EC2 Container Service, Azure Container Service, and Google Container Engine Service.

There are good reasons for the popularity of containers: containers provide extremely fast instantiation times, small per-instance memory footprints, high density on a single host and ease of software distribution. Figure 1 illustrates the differences between virtual machines and containers. Containers are lightweight because the operating system layer is not replicated for every application running on the same server. Developers are able to run third-party dependencies such as databases, message brokers, proxies, ... each in its own container. Additionally, everything is easily integrated with the application under development with enough isolation and density of containers to run many small services in the developer's local machine. In fact, containers have popularized the so-called micro-service architectures [4, 5].

At deployment time, containers solve a fundamental code portability problem. Containers are packaged with all the dependencies and libraries they need to run, making them portable between distributions. However, although the high density of containers is of great value in a *local environment* or for continuous integration (CI) jobs, it introduces new challenges in *production environments*. First, containers are poorly integrated with the rest of the cloud offering, such as auto-scalability, fault tolerance, load balancing, service discovery or networking. Container cluster management tools—like Kubernetes [6], Docker Swarm [7] and Mesos [8]—provide similar services at the cost of adding a new control plane layer, which requires additional setup steps, adds redundancy and might become hard to debug. [9] extends on the complexity of Kubernetes.

But the main challenge introduced by containers is security [10]. Container isolation is based on concepts like namespaces, cgroups, seccomp technologies,

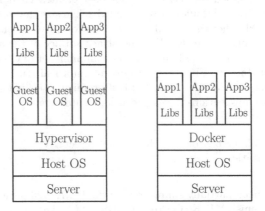

Fig. 1. Virtual machines vs. containers.

the user core Linux permission model or root user capabilities. These mechanisms provide an additional defense on top of application security, but it only takes a single kernel bug to bypass all these mechanisms and escape the container isolation model (see [11] for some vulnerabilities). Some use cases require a higher level of isolation, like sandboxes for running vulnerable or untrusted code, or multi-tenant environments in the case of hosted services. Note that trusted but vulnerable applications running on the same server might become an entry point for malicious agents.

Hypervisors, such as KVM [12], VMware ESXi [13], or Microsoft Hyper-V [14] are proven and mature technology that solved this problem years ago. Following the Linux philosophy of *Do one thing and do it well*, we propose a separation of concerns to provide the security of virtual machines, but the portability and simplicity of containers. In our approach, each container[1] is deployed in its own virtual machine. This container virtual machine (CM) is only meant to provide isolation between containers, and can be reduced to a minimum footprint. The current implementation of *i2kit* is based on Linuxkit [15], which is able to generate Linux distributions specialized to run containers with a memory footprint of approximately 60 MB. Smaller distributions also entail:

(a) the reduced attack surface of the system by having less software pre-installed;
(b) faster booting times since booting times are roughly linear in the size of the distribution.

Note that since CMs run containers, code portability and ease of software distribution is maintained.

CMs is the main concept behind *i2kit*, a container orchestrator introduced in this paper which uses the CM as the unit of deployment. The name *i2kit* stands for *immutable infrastructure kit*. Immutable infrastructure [16], also known as *i2*, is an approach to managing software deployments wherein the servers (where components run) are replaced rather than changed or modified in every software update. The tool *i2kit* recreates every CM on every deployment, following a pure *immutable infrastructure* approach. The *i2kit* orchestrator is inspired by Kubernetes, where applications are defined in a declarative way using a YAML Manifest File. The tool *i2kit* provides out of the box solutions for auto scalability, fault tolerance, load balancers, service discovery, rolling upgrades or networking, but instead of reimplementing these services (as Kubernetes does), *i2kit* reuses proven, mature and efficient cloud technology like auto scalability groups, load balancers or DNS services. In a nutshell, compared to containers, CMs provide better security and integration with the rest of the cloud offering.

The rest of the paper is organized as follows: Sect. 2 introduces the *i2kit* principles and explains how to integrate CMs with the rest of the cloud offering to provide common features available in other orchestrators like Kubernetes. Section 3 describes how CMs are generated from containers. Section 4 describes

---

[1] In this paper, we refer to containers or pods indistinctly. A pod is a group of strongly related containers that get deployed as a unit.

the implementation of the *i2kit* orchestrator. Section 5 measures the impact of *i2kit* on different metrics such as booting times, networking and cluster memory consumption. Finally, Sect. 6 concludes and describes some research lines for future work.

2   The Design of *i2kit*

The *i2kit* tool is open source, and it is actively under development at the IMDEA Software Institute[2]. The *i2kit* tool is a container orchestrator whose unit of deployment is the container virtual machine (CM). CMs are built using Linuxkit (see Sect. 3 for a detailed description). *i2kit* is inspired by Kubernetes and follows the best principles in the container ecosystem. Kubernetes is an evolution of the Borg [17] and Omega [18] cluster manager tools, adapted for containers, where applications are defined using a declarative model.

For example, Fig. 2 shows a simplified Kubernetes Deployment Manifest input (which we borrow as the input format for *i2kit*), and Fig. 3 represents the deployment of this Deployment Manifest in Kubernetes. Pods are the unit of deployment in Kubernetes, which are essentially a sandbox that allows running containers inside. Informally, a Pod fences an area of the host OS, builds a network stack, creates the necessary kernel name-spaces, and runs one or more containers. A Replica Set builds on top of a set of Pods. A Replica Set takes a Deployment Manifest and instantiates the desired number of replicas of the Pod. Replica Sets also instantiate a background reconciliation loop that ensures that the right number of replicas are always

```
name: myapp
replicas: 3
containers:
  nginx:
    image: nginx:1.7.9
    ports:
    - 80
  api:
    image: myapp:1.0
```

Fig. 2. *i2kit* manifest file.

running, forcing the reconciliation between the desired state and the current state. The K-Proxy is responsible for forwarding traffic between Pods, providing load balancing capabilities, based on the ports defined in the Deployment Manifest. Finally, the Service creates a reliable endpoint based on the Deployment Manifest **name** field resolving to the running Pods.

The architecture of *i2kit* is similar than Kubernetes, but *i2kit* replaces the Pod by the CM as the unit of deployment. The current implementation of *i2kit* is integrated with Amazon Web Services, although support for other cloud vendors is under development, and provides similar concepts than replica sets, k-proxy, and service by integrating the deployment of CMs directly with the rest of the cloud vendor offering. Figure 4 shows how the Deployment Manifest presented in Fig. 2 looks like in *i2kit*. The Autoscalabity Group [19] is the equivalent to Replica Sets. It ensures that three CMs are always running, recreating CMs if they become unreachable. The Elastic Load Balancer (ELB) [20] is responsible for forwarding traffic between CMs, providing load balancing capabilities. The DNS Record creates a reliable endpoint based on the Deployment Manifest **name**

---

[2] *i2kit* is available at www.github.com/pchico83/i2kit.

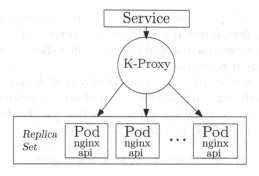

Fig. 3. A high-level view of a Kubernetes Deployment.

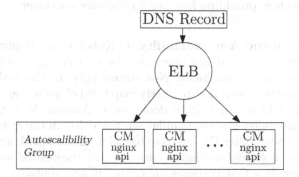

Fig. 4. A high-level view of a *i2kit* Deployment.

field resolving to the running CMs for service discovery capabilities. In a nutshell, *i2kit* mimics the Kubernetes architecture but using proven cloud vendor technology thanks to the better integration that virtual machines have compared to containers. We explain these concepts in the next subsections.

Declarative Model. A declarative model of *i2kit* works as follows:

- The user declares the desired state of his deployment in, for example, a YAML Deployment Manifest that includes a description of which container images to run, the ports exposed by each container, the number of replicas, the commands to execute, or how to instantiate environment variables.
- The orchestrator issues workloads to run the deployment in the cloud, in a manner that is completely transparent to the user.
- The orchestrator also watches the deployment state in order to restore the desired state in the event of a failure. For example, if a CM becomes unreachable for whatever reason, it is replaced by a new CM running the same containers.

There is a significant difference between the declarative approach described above and an imperative language to describe control planes. In an imperative model, the user issues a procedure with specific commands to reach the desired state. A declarative description is usually much shorter and simpler than a long sequence of imperative commands and describes the details of how to create and coordinate the different resources.

Figure 2 shows an example of an *i2kit* Deployment Manifest. Note that the *i2kit* declarative model supports the execution of several containers on the same CM, as Pods do. There are advanced use-cases that justify the run of multiple containers inside a single CM, for example:

- a log scraper tailing the output of the user container to a centralized logging service.
- a stats collector sending metrics to perform analytics.
- a sidecar container providing features for the user container.

Failure Tolerance and Autoscalability. In Kubernetes a Replica Set ensures that a fixed number of Pod instances are always running. Replica Sets also replace Pods that get unreachable. Not surprisingly, in the realm of virtual machines, there are solutions that perfectly map this behavior under the assumption of using the CM as the unit of deployment. Amazon Web Services offers Amazon Auto Scalability Groups (similar services exists in the rest of cloud vendors). AWS Auto Scalability Groups help to maintain the health and availability of a fleet of Amazon virtual machines, ensuring that the desired number of CMs is always running. If a CM becomes unhealthy, it gets replaced by a new CM running the same container versions. In addition, Auto Scalability Groups can be used to scale the application up and down under certain situations, such as high CPU or memory usage, or an increment in incoming requests. This allows very flexible deployments only consuming resources on demand, without the need of guessing infrastructure capacity in advance.

Note that in the event of a rolling update, new CMs are generated, and the Auto Scalability Group replaces every existing virtual machine by the new ones. The tool *i2kit* does follow a pure immutable infrastructure approach by design. In contrast, Kubernetes reuses the same cluster between deployments. Kubernetes Nodes might leak memory and become unreachable after a number of deployments. For example, it is very common that Kubernetes Nodes become unhealthy due to the lack of storage resources, for example by the garbage accumulated by old docker images from previous deployments. In contrast, following a pure immutable approach like *i2kit* brings important advantages. For example, in the event of a full system failure, applications can be recreated in a different availability zone with a single deploy command on the Deployment Manifest. Guessing the state of a Kubernetes cluster or recreating this state is more difficult and time-consuming.

Load Balancing. CMs are mortal and, in practice, it is not unusual that a given CM becomes unreachable due to networking issues or other software or hardware failures. On failure, Autoscalability Group replaces the dying CM with a new one, which probably is assigned a different IP. Moreover, when performing rolling updates the new CMs have different IPs than the old ones. Therefore, the application logic cannot rely on CM IPs.

The solution to this problem is the use of load balancers, which provide a reliable networking endpoint for a set of CMs. Amazon Web Services offers Elastic Load Balancers (similar services exists in the rest of cloud vendors). Elastic Load Balancers, not only provides a reliable networking endpoint for a set of CMs but also, as their name suggests, balance the incoming traffic between these CMs. Complex policies can be defined to customize how the traffic load is balanced. The port configuration of the AWS Load Balancer is created based on the information contained in the *i2kit* Deployment Manifest.

Service Discovery. The reliable networking endpoint provided by Elastic Load Balancers is not configurable, and it is randomly created at deployment time. This is not compatible with the idea of providing a Service Discovery mechanism based on the **name** property of Deployment Manifests.

Our solution is to use Amazon Web Services Route 53, which provides DNS as a service (similar services exists in the rest of cloud vendors). The approach of *i2kit* is to create a Route 53 Domain CNAME entry that resolves the *name* field of the Deployment Manifest File to the AWS Load Balancer endpoint that is proxying incoming traffic between the different CMs. In this manner, *i2kit* can provide the same service discovery mechanism—based on names—as Kubernetes.

3   Container Machine Generation

The main drawback of container virtual machines (CM) is a loss of performance because in principle a virtual machine imposes a non-negligible overhead on infrastructure resources compared to a container or Pod. However, there are tools that allow creating minimal Linux distributions specifically crafted to run containers. The current footprint of these distributions can get as small as 60 MB, a size comparable to container technology. The tool *i2kit* is built on the assumption that the overhead of running a container per virtual machine is acceptable. Moreover, it is to be expected that this figure will keep improving as leaner Linux distributions are developed (this is discussed in detail in Sect. 6).

The current implementation of *i2kit* uses Linuxkit [15], a toolkit for building custom minimal, immutable Linux distributions.[3] Linuxkit reads YAML templates that describe how to build a Linux distribution. The container information presented in an *i2kit* Deployment Manifest is transformed by *i2kit* into a Linuxkit template, in order to generate a minimal Linux distribution specialized in running these deployment containers. The result is shown in Fig. 5. From every container in the *i2kit* Deployment Manifest, *i2kit* extracts the container relevant information (such as container image, run command, environment variables) and adds an entry in the **services** section of the Linuxkit template.

---

[3] We are also exploring how to support alternative technologies to Linuxkit.

```
            kernel:
              image: linuxkit/kernel:4.9.63
              cmdline: "console=tty0"
            init:
              - linuxkit/init
              - linuxkit/runc
              - linuxkit/containerd
              - linuxkit/ca-certificates
            onboot:
              - name: sysctl
                image: linuxkit/sysctl
              - name: rngd1
                image: linuxkit/rngd
                command: ["/sbin/rngd", "-1"]
              - name: dhcpcd
                image: linuxkit/dhcpcd
              - name: metadata
                image: linuxkit/metadata
            services:
              - name: getty
                image: linuxkit/getty
                env: [INSECURE=true]
              - name: sshd
                image: linuxkit/sshd
              - name: nginx
                image: nginx:1.7.9
                capabilities: [all]
              - name: api
                image: myapp:1.0
                capabilities: [all]
            trust:
              org: [linuxkit, library]
```

Fig. 5. Linuxkit template.

In our example, this information is:

```
            services:
              - name: nginx
                image: nginx:1.7.9
                capabilities: [all]
              - name: api
                image: myapp:1.0
                capabilities: [all]
```

Note that the value `all` is used for the capabilities of the user containers, which is a limitation of the current *i2kit* implementation. Future work includes equipping *i2kit* with an analysis that limits the capabilities associated with every container.

The remaining fields in the Linuxkit template are pre-generated and are identical for every deployment. The filesystem of every custom distribution is currently initialized from the docker image *linuxkit/kernel:4.9.63*. Also, every custom distribution installs the *init* process, *runc*, and *containerd* to be able to run containers, and *ca-certificates* to be able to manage certificates. At boot-time, the following containers are executed in sequence order: *sysctl*, *rngb*, *dhcpcd* and *metadata*. These are basic services required by any software application. Note that *metadata* is installed to be able to manage Amazon Metadata from the CM itself (in this case, supporting other cloud vendors would require to changes). Then, the containers in the **services** section run as daemons in parallel, in particular *getty*, *sshd* and the containers defined in the *i2kit* Deployment Manifest. Finally, *i2kit* uses content-trust-delivery for images coming from the *linuxkit* and the *library* organizations.

Once the Linuxkit template has been generated, *i2kit* builds the minimal Linux distribution and uploads it as an Amazon Machine Image, which is then available to be consumed by the *i2kit* orchestrator. The next section explains the implementation of the *i2kit* orchestrator.

4   The *i2kit* Orchestrator

The current *i2kit* implementation supports deployments in Amazon Web Services, but support for other cloud vendor is work in progress. The Amazon Web Services driver makes use of the AWS Cloud Formation Service [21] to create the different resources. Cloud Formation receives JSON manifest files to create and manage a collection of related AWS resources, provisioning and updating them in an ordered and predictable fashion. Cloud Formation templates can specify rolling updates policies to be applied when the template is modified, allowing the simulation of Kubernetes rolling updates.

The *i2kit* orchestrator transforms *i2kit* Deployment Manifest into Cloud Formation templates once the Container Machine Image has been generated using LinuxKit. For example, assume the Amazon Image *ami-XXXXX* has been generated from the Deployment Manifest in Fig. 2 following the process explained in Sect. 3. Then, a simplified version of the Cloud Formation generated by *i2kit* is shown in Fig. 6.

The Cloud Formation template defines four different resources: **LaunchConfig**, **ASG**, **ELB**, and **DNSRecord**. The resource **LaunchConfig** defines how virtual machines will be created. In our case, each CM will run the AMI created by the process explained in Sect. 3. The next resource is **ASG**, an Auto Scalability Group which use the **LaunchConfigurationName** created above in order to create CMs. The minimum and the maximum number of CMs matches the number of replicas in the *i2kit* Deployment Manifest. Every CM generated by the Auto Scalability Group is associated with an Elastic Load Balancer defined also in the Cloud Formation template. **ELB** stands for the Elastic Load Balancer that takes the name from the Deployment Manifest **name** field. The **Listeners** information matches the **ports** section of the *i2kit* Deployment Manifest, where

```
AWSTemplateFormatVersion: 2010-09-09
Resources:
  LaunchConfig:
    Type: AWS::AutoScaling::LaunchConfiguration
    Properties: { ImageId: ami-XXXXX }
  ASG:
    Type: AWS::AutoScaling::AutoScalingGroup
    Properties:
      LaunchConfigurationName:
        Ref: LaunchConfig
      MaxSize: 3
      MinSize: 3
      LoadBalancerNames: { Ref: ELB }
  ELB:
    Type: AWS::ElasticLoadBalancing::LoadBalancer
    Properties:
      LoadBalancerName: myapp
      Listeners:
        LoadBalancerPort: 80
        InstancePort: 80
        Protocol: HTTP
DNSRecord:
  Type: AWS::Route53::RecordSet
  Properties:
    HostedZoneName: i2kit.com
    Name: myapp.i2kit.com
    ResourceRecords:
    - Fn::GetAtt:("ELB", "DNSName")
    Type: CNAME
```

Fig. 6. Cloud Formation template for the *i2kit* Manifest File.

the `Protocol` is inferred from the port number. Finally, the `DNSRecord` resource is a CNAME entry for the Route 53 Domain *i2kit.com*. This domain is received as a parameter of the *i2kit* tool. The CNAME entry is created based on the *i2kit* Deployment Manifest `name` field. It resolves to the `ELB` endpoint, providing service discovery for other deployments. Note that every CM deployed by *i2kit* adds a `DNS_SEARCH` entry in every container pointing to *i2kit.com*. This way, containers will resolve the name of any *i2kit* deployment without the need of specifying the full domain name (ending in *i2kit.com*).

5   Empirical Evaluation

This section compares *i2kit* versus the native Kubernetes implementation based on three different metrics: booting times, memory consumption and network performance.

Booting Times. The creation of a virtual machine running *i2kit* Linuxkit distributions in AWS takes about one minute while creating a Pod in Kubernetes takes only seconds (depending on the size and the local availability of docker images). Even though this booting time difference can be very relevant in local environments, it is less relevant in production environments. For example, it is a common practice to introduce a delay of at least 30 s between Pod creations during a rolling update, in order for load balancers to have enough time to be updated. Using this common practice induces a comparable delay to the time required to create *i2kit* CMs. Therefore, even though *i2kit* is slower than Kubernetes in terms of booting times, we argue that difference is not very relevant in production environments.

Note also that though we use AWS for deploying the Virtual Machines, these are not the "usual" virtual machines but much smaller specialized machines generated using Linuxkit.

Memory Consumption. The overhead that *i2kit* imposes for every Pod creation is a consequence of the overhead of the CM running the Linuxkit distribution. It is linear on the number of Pod instances. In contrast, the Kubernetes overhead for running a Pod is due to the overhead of running the Worker Node components, which are shared by several Pods. It is constant on the number of Pod instances.

Table 1. Total memory footprint of *i2kit* vs Kubernetes running the example in Fig. 2.

Pods	1	10	20	30	40
i2kit	78 MB	0.78 GB	1.56 GB	2.34 GB	3.12 GB
K8	1.94 GB	2.09 GB	2.27 GB	2.44 GB	2.62 GB

Table 1 displays the memory comparison between *i2kit* and Kubernetes for running the Deployment Manifest shown in Fig. 2 using different numbers of replicas. Table 1 shows that for a low number of replicas, *i2kit* requires significantly less memory than Kubernetes. Even though the growth in the memory required is faster for *i2kit* than for Kubernetes, *i2kit* is more memory efficient when running less than (approx.) 30 containers in the same Worker Node. Note that the Kubernetes web page [22] does not recommend running more than 30 Pods per Worker Node. Therefore, we can conclude that the memory consumption of *i2kit* behaves better than that of Kubernetes for standard workloads. In fact, we were not able to create with Kubernetes more than 42 Pods on the same Worker Node running on a *t2.xlarge* AWS EC2 Instance. Moreover, the data reported in Table 1 does not take into account the memory consumption of Master Nodes or the Kubernetes Distribute Storage Layer, which would report an even more favorable comparison to *i2kit*.

Finally, sharing a host between several containers imposes performance side effects on the containers running on the same host. Although some research has been done in this area, [23,24], it is a more mature approach to have resources affecting performance isolated at the hypervisor level.

Table 2. Network of *i2kit* vs Kubernetes.

Pods	1	5	25
i2kit	129.86 Mbps	128.191 Mbps	128.58 Mbps
K8-1	129.17 Mbps	25.92 Mbps	-
K8-5	108.44 Mbps	108.36 Mbps	21.73 Mbps
K8-25	97.95 Mbps	98.11 Mbps	97.84 Mbps

Network Performance. Table 2 shows the network performance comparison between *i2kit* and different Kubernetes cluster sizes. The experiment uses *iperf2* to measure the average network bandwidth consumed by each replica, where each replica runs an *iperf2* server. On the other hand, the *i2kit* configuration runs every *iperf2* server in its own CM using a *t2.large* AWS EC2 Instance. In the table, *K8-N* stands for a Kubernetes cluster with N Worker Nodes, where every Worker Node runs on a *t2.large* AWS EC2 Instance. In order to be able to measure the consumed bandwidth, every experiment runs a large amount of *iperf2* clients, where each client runs on its own VM. These clients first send traffic to warm the load balancers up and then synchronize to sending traffic at the same time for 3 min.

Table 2 indicates that *i2kit* scales linearly on the number of replicas, as expected. The network overhead of using an AWS Load Balancer is negligible. Note that the limit of the virtual machine incoming traffic is 130 Mbps. The row *K8-1* in Table 2 shows that the overhead imposed by Kubernetes when running on a single node is not very relevant (less than 2%). Since *K8-1* runs all Pod replicas on the same server, running more than one Pod replica quickly hits the VM incoming bandwidth limit, distorting the experiment for the case of 5 replicas. Moreover, we were not able to successfully run 25 Pods on a single Worker Node. The row *K8-5* shows that Kubernetes imposes an overhead of about 20% when the Kube-Proxy needs to forward traffic between five different Worker Nodes. As expected, the overhead grows with the cluster size, as we can see in the *K8-25* row, which accounts for a 30% network overhead. *K8-5* also shows how the traffic is dramatically affected by the virtual machine incoming bandwidth limit when running 25 Pod replicas.

This experiment exposes some side effects of running the additional control plane of Kubernetes on top of cloud vendor technology. In this case, the functionality provided by the K-Proxy is redundant as it is already provided by the AWS Load Balancers, and consequently this additional control plan imposes unneeded performance overheads. This experiment also illustrates that containers are poorly integrated with the rest of the cloud offering. For example, running more than one Pod per VM hits the VM incoming traffic limit. This issue does not happen in *i2kit*—which runs each replica in its own virtual machine—because virtual machines are better integrated with the networking capabilities of the cloud vendor.

6   Conclusions and Future Work

This paper has presented *i2kit*, a deployment tool that pursues the following main goals: (1) to preserve the docker development workflow untouched, as containers are a great fit for local environments; (2) to transform containers into lightweight virtual machines upon deployment for better isolation; (3) to provide fault tolerance, load balancing or service discovery without reimplementing these features in a new layer thanks to the better integration of virtual machines with the rest of the cloud offering.

The tool *i2kit* follows the Linux principle of *"Do one thing and Do it well."* The responsibility of virtual machines is to provide workload isolation and security. The responsibility of containers is to offer portability and ease of container image distribution. The drawback of using virtual machines for container isolation is higher resource utilization, but *i2kit* is specifically designed to exploit Linxkit to generate virtual machines with low memory footprint. The results in Sect. 5 suggest that the memory consumption of *i2kit* is better than the one of Kubernetes for standard workloads.

Note also that there is a very active research effort targeting VM optimization which *i2kit* can potentially leverage in terms of memory usage to get even better results. Kata Containers [25] fulfills similar goals than *i2kit* in terms of security, building lightweight virtual machines that feel like containers but provide the isolation level of virtual machines. The main difference is that Kata Containers are conceived to be a container runtime [26] instead of integrating with the rest of the cloud offering. LightVM [27] is a new virtualization solution based on Xen that is optimized to offer fast boot-times regardless of the number of active VMs. LightVM features a complete redesign of Xen's control plane reducing the hypervisor to a minimum. LightVM can boot a VM in 2.3 ms, comparable to fork/exec on Linux (1 ms), and two orders of magnitude faster than Docker. LightVM can pack thousands of LightVM guests on modest hardware with memory and CPU usage comparable to that of processes. The current *i2kit* implementation uses Linuxkit instead of LightVM because it is easily integrable with the AWS cloud offering. LightVM is based on Unikernels [28], which are also very promising on this area. Exploiting VM optimizations and Unikernels to improve *i2kit* further is a line of current and future work. Finally, another research line is to analyze synergies between *i2kit* and serverless architectures [29] provided by cloud vendors.

Section 5 also shows that the Kubernetes control plane introduces redundancies that can affect network performance (and probably other metrics). As we show with *i2kit* in this paper, containers can be integrated with the rest of the cloud offering without the need of adding a complex control plane for container orchestration. Also, Kubernetes is not a cloud native technology.

First, Kubernetes requires a Distributed and Reliable Store Cluster. The most common solution to this end is *etcd* [30], a Key-Value Store based on the Raft [31] protocol. Kubernetes also requires a cluster of Master Nodes. Master Nodes execute three different components: Api, Scheduler, and Replication Controller. Also, every Worker Node requires the Kubelet (responsible of executing the tasks assigned by the Scheduler) and the Kube-Proxy (responsible for service discovery and load balancing in a high-density container environment). Even further, users need to take into account that the components in the Kubernetes Control Plane are a runtime dependency for the applications. An error in the Kubernetes Control Plane is not only difficult to debug, but it also disturbs running applications by affecting, for example, service discovery. Managing a large cluster infrastructure and optimizing the scheduling of containers all backed by a complex distributed state store is counter to the premise of the cloud. Cloud vendors let users utilize resources as they go, without guessing capacity, and providing deep operational control without operational burden. The tool *i2kit* allows developers to write their code and have it run, without having to worry about configuring complex management tools. As a result, *i2kit* turns containers into a secure and cloud native technology.

Cloud native containers is also the goal of AWS Fargate [32]. Some differences are: (1) the *i2kit* declarative model is cleaner and allows the execution of several containers per VM, which is very valuable in advanced uses cases like sidecars or log/stats collectors. (2) *i2kit* future work conceives the option to install OS dependencies on the Linuxkit distribution. For example, tools like Sysdig [33] needs to be installed as a kernel module. (3) *i2kit* is open source and more flexible than AWS Fargate on controlling the runtime technology. Multi cloud vendor support is currently implemented for *i2kit*.

References

1. Merkel, D.: Docker: lightweight Linux containers for consistent development and deployment. Linux J. **2014**(239) (2014)
2. Wang, C.: LXC and Docker explained. http://www.infoworld.com/article/3072929/linux/containers-101-linux-containers-and-docker-explained.html
3. Clark, J.: EVERYTHING at Google runs in a container. http://www.theregister.co.uk/2014/05/23/google_containerization_two_billion/
4. Lewis, J., Fowler, M.: Microservices: a definition of this new architectural term. http://martinfowler.com/articles/microservices.html
5. Thönes, J.: Microservices. IEEE Softw. **32**(1), 113–116 (2015)
6. Burns, B., Grant, B., Oppenheimer, D., Brewer, E., Wilkes, J.: Borg, omega, and kubernetes. Commun. ACM **59**(5), 50–57 (2016)
7. Docker Swarm. https://github.com/docker/swarm
8. Hindman, B., et al. : Mesos: a platform for fine-grained resource sharing in the data center. In: Proceedings of NSDI 2011, pp. 295–308. USENIX Association (2011)
9. Moiron, J.: IsK8s Too Complicated? http://jmoiron.net/blog/is-k8s-too-complicated/
10. Mouat, A.: Five security concers when using Docker. https://www.oreilly.com/ideas/five-security-concerns-when-using-docker

11. Linux Kernel Security Vulnerabilities. https://www.cvedetails.com/vulnerability-list.php
12. Habib, I.: Virtualization with KVM. Linux J. **2008**(166) (2008). http://dl.acm.org/citation.cfm?id=1344209.1344217
13. Mishchenko, D.: VMware ESXi: Planning, Implementation, and Security, 1st edn. Course Technology Press, Boston (2010)
14. Velte, A., Velte, T.: Microsoft Virtualization with Hyper-V, 1st edn. McGraw-Hill, Inc., New York (2010)
15. LinuxKit. https://github.com/linuxkit/linuxkit
16. Fowler, C.: Trash Your Servers and Burn Your Code: Immutable Infrastructure and Disposable Components. http://chadfowler.com/2013/06/23/immutable-deployments.html
17. Verma, A., Pedrosa, L., Korupolu, M.R., Oppenheimer, D., Tune, E., Wilkes, J.: Large-scale cluster management at Google with Borg. In: Proceedings of EuroSys 2015. ACM (2015)
18. Schwarzkopf, M., Konwinski, A., Abd-El-Malek, M., Wilkes, J.: Omega: flexible, scalable schedulers for large compute clusters. In: Proceedings of EuroSys 2013, pp. 351–364. ACM (2013)
19. Auto Scalability Groups. https://aws.amazon.com/autoscaling/
20. Elastic Load Balancing. https://aws.amazon.com/elasticloadbalancing/
21. Cloud Formation. https://aws.amazon.com/cloudformation/
22. Building Large Kubernetes Clusters. https://kubernetes.io/docs/admin/cluster-large/
23. Delimitrou, C., Kozyrakis, C.: Quasar: resource-efficient and QoS-aware cluster management. SIGARCH Comput. Archit. News **42**(1), 127–144 (2014)
24. Mars, J., Tang, L., Hundt, R., Skadron, K., Souffa, M.L.: Bubble-up: increasing utilization in modern warehouse scale computers via sensible co-locations. In: Proceedings of MICRO 2011. ACM (2011)
25. Kata Containers. https://katacontainers.io
26. Ernst, E.: Kata containers doesnt replace kubernetes (2018). https://katacontainers.io/posts/why-kata-containers-doesnt-replace-kubernetes/
27. Manco, F., et al.: My VM is lighter (and safer) than your container. In: Proceedings of SOSP 2017, pp. 218–233. ACM (2017)
28. Madhavapeddy, A., Scott, D.J.: Unikernels: rise of the virtual library operating system. Queue **11**(11), 30:30–30:44 (2013)
29. Serverless Architectures. https://martinfowler.com/articles/serverless.html
30. etcd. https://github.com/coreos/etcd
31. Ongaro, D., Ousterhout, J.: In search of an understandable consensus algorithm. In: Proceedings of USENIX ATC 2014. USENIX Association, pp. 305–320 (2014)
32. AWS Fargate. https://aws.amazon.com/fargate/
33. Borello, G.: System and application monitoring and troubleshooting with Sysdig. USENIX Association, Washington, D.C. (2015)

Gradient Correlation: Are Ensemble Classifiers More Robust Against Evasion Attacks in Practical Settings?

Fuyong Zhang[1], Yi Wang[1(⊠)], and Hua Wang[2]

[1] Dongguang University of Technology, Dongguan, Guangdong, China
{zhangfy,wangyi}@dgut.edu.cn
[2] Institute for Sustainable Industries and Liveable Cities,
VU Research, Victoria University, Melbourne, Australia
hua.wang@vu.edu.au

Abstract. Pattern recognition is an essential part of modern security systems for malware detection, intrusion detection, and spam filtering. Conventional classifiers widely used in these applications are found vulnerable themselves to adversarial machine learning attacks. Existing studies argued that ensemble classifiers are more robust than a single classifier under evasion attacks due to more uniform weights produced on the basis of training data. In this paper, we investigate the problem in a more practical setting where attackers do not know the classifier details. Instead, attackers may acquire only a portion of the labeled data or a replacement dataset for learning the target decision boundary. In this case, we show that ensemble classifiers are not necessarily more robust under a least effort attack based on gradient descent. Our experiments are conducted with both linear and kernel SVMs on real datasets for spam filtering and malware detection.

Keywords: Adversarial machine learning · Ensemble classifiers Evasion attacks

1 Introduction

Learning-based classifiers are increasingly accepted as a versatile tool for data-intensive security tasks [7,13,14,23–26]. They have been successfully deployed in many cyber security applications such as biometric authentication, intrusion detection, malware detection, spam filtering, detection of malicious Web page and so on [16,28,29,32,34,36]. In these applications, binary classifiers are essential for the task of discriminating a malicious instance from a legitimate one. To boost the accuracy performance, an ensemble approach may be adopted by combining multiple classifiers together to form an integrated output [9,12,17].

This work was supported in part by National Natural Science Foundation of China (61403324) and Dongguan University of Technology (KCYKYQD2017003).

© Springer Nature Switzerland AG 2018
H. Hacid et al. (Eds.): WISE 2018, LNCS 11233, pp. 96–110, 2018.
https://doi.org/10.1007/978-3-030-02922-7_7

Unlike in other applications where the operating environment is static, these security-related tasks involve intelligent adversaries who are able to analyse vulnerabilities of learning-based models and adapt their attacks in response to system outputs. In such an adversarial setting, conventional learning-based classifiers are found to be susceptible to evasion attacks among other security issues [2,33]. In evasion attacks, the attacker is able to manipulate samples carefully to circumvent system detections. For example, in spam filtering, attackers can disguise their email behavior by misspelling bad words or adding normal words [35]. The PDFrate[1], a real-world deployed, well-known PDF malware detection system, can suffer substantial drops of detection accuracy when exposed to simple attacks [20].

The growing evidence of adversarial learning in different application domains has drawn significant attention of the research community in related fields [6,8,31]. There are several theoretical attempts to understand the rationale of inherent vulnerabilities in machine learning systems [6,22]. It was pointed out that the success of attacks against learning algorithms crucially depends on the amount and type of knowledge exposed to an attacker [6]. Regarding the targeted system, there are four level of knowledge [6]: (1) the training data $\mathcal{D}$; (2) the feature set $\mathcal{X}$; (3) the learning algorithm f, along with the objective function $\mathcal{L}$ minimized during training; and, possibly, (4) the targeted model parameters $\mathbf{w}$. Thus, the attacker's knowledge can be characterized in terms of $\theta = (\mathcal{D}, \mathcal{X}, f, \mathbf{w})$.

Most of previously reported successful attacks assume that the attacker has full knowledge of the targeted model, known as the "white-box" attack [6] or the worst-case attack [5]. Recently, there are studies discussing evasion attacks with limited knowledge of θ, mainly focusing on improving the robustness of a *single* classifier in specific application domains [8,35]. These methods argued that reducing the amount of knowledge available to the attacker or a proactive response to potential exploitation of such knowledge should provide adequate protection against adversarial data manipulation. Accordingly, several security evaluation measures were proposed to indicate the robustness of a learning-based classifier against evasion attacks. For example, hardness of evasion measure was defined as the average minimum number of features that have to be modified in a malicious sample to evade detection [35]. Another measure called *weight evenness* was proposed in [5] based on the observation that some features are highly discriminant than the others and if the adversary can identify them, e.g., by the associated weight values, it is not difficult to modify and get the malicious sample misclassified as a legitimate one. Under these security measures, it was shown that multiple classifier systems by averaging simpler classifiers such as classic SVMs can be exploited to improve the robustness against evasion attacks because more evenly distributed feature weights should require the adversary to manipulate a higher number of features [4,5,27].

In this paper, we re-investigate the security evaluation problem from another perspective. Our intuition is that, with small subsets or even zero knowledge of the target training data $\mathcal{D}$, ensemble learning may lead to a surrogate classifier

---

[1] http://pdfrate.com/.

with less variation thus more accurate estimation of gradients when approximating the target decision boundary. Accordingly, we introduce a new security measure called *Gradient Correlation* to evaluate the similarity of gradient estimation between the surrogate and the targeted systems. We build the ensemble on linear and kernel SVMs with averaging and voting strategies, respectively. Our experimental results on real-world datasets indicate that, unlike expected previously, ensembling base classifiers such as linear SVMs do *not* necessarily improve the robustness of classifiers against evasion attacks under all circumstances.

2   Related Work

The problem of evasion attack at test time has been considered in the literature [6,22]. Most of the studies are focusing on individual classifiers, either convex-inducing classifiers including SVM with simple decision functions [2,35] or more complex neural networks [10,33], and defence methods in specific application domains. There are relatively fewer discussions on the security of ensemble classifiers. Ensemble classifiers were originally proposed to improve the classification accuracy by combining multiple weak classifiers to cope with more complex hypothesis and nonlinear decision boundaries [11,30]. Because the feature weights can be more evenness through the combination of multiple single classifiers and the decision boundary is hard to find [5,27]. From this perspective, previous studies showed that ensemble classifiers are more robust than a single classifier under white-box attacks with full knowledge of the targeted system [4,5].

The white-box attack is extended in [5] to more general attacks on multiple classifiers. The paper proposed two limited knowledge attacks by assuming the feature set in an attacker's hand is not the same as the original one $\mathcal{X}$. To simulate the limited knowledge scenarios, the feature set $\hat{\mathcal{X}}$ assumed available to the attacker was generated by shuffling half or all features in $\mathcal{X}$ at random. However, it is not clear how attackers can obtain shuffled features in practical classifier systems. The security evaluation therein is based on the weight evenness which considers a classifier is more robust if the weights in $\mathbf{w}$ are more evenly distributed so that the attacker cannot easily discover the most salient features.

In more realistic settings, the attacker's knowledge is limited by restricting the training data $\mathcal{D}$ available to the attacker [2,3,35]. It was assumed that the available $\hat{\mathcal{D}}$ is either subsets of the original $\mathcal{D}$ or a surrogate collected from alternative sources with the same data distribution as the target. Gradient descent attacks were proposed to increase the probability of successful evasion by exploiting knowledge of the (estimated) decision boundary gained from the discriminant function of the target classifier. However, again, these methods are evaluated on a single SVM rather than ensemble classifiers and the security measures are in general based on the hardness of evasion and the weight evenness.

Gradient information was exploited to attack tree ensemble classifiers in [15,27]. It was shown that both gradient boosted trees and random forests are extremely susceptible to evasions under white-box attacks with full knowledge of the original training dataset $\mathcal{D}$ [15]. On the other hand, the study in [27] shows that an ensemble of classifiers, either decision trees or SVMs, can be used to detect evasion attacks by checking diversity in the ensembles themselves. Due to diversity and adjusting the voting threshold accordingly, ensemble trees are considered more robust than a single classifier against evasion attacks [27].

3 Background

Before proceeding to the proposed approach, here we briefly review SVM-based classifiers and introduce relevant notations used in this paper followed by the general formulation of evasion attacks.

3.1 Support Vector Machines

Given a training dataset $\mathbf{x}_1, \mathbf{x}_2, ..., \mathbf{x}_N$, the primal problem of linear SVM is to solve the following quadratic program:

$$\min_{\mathbf{w}, b, \xi} \frac{1}{2} ||\mathbf{w}||^2 + C \sum_{i=1}^{N} \xi_i \tag{1}$$

$$s.t. \quad y_i(\mathbf{w}^T \mathbf{x}_i + b) \geq 1 - \xi_i \text{ and } \xi_i \geq 0,$$

where $\mathbf{w}$ is the weight vector, b is the displacement, $\{\xi_i\}$ are slack variables defining the soft margin [11], and the regularization term C tunes the trade-off between the classification error and margin maximization. Once the parameters are solved, the discriminant function defining the decision boundary is given by

$$g_{linear}(\mathbf{x}) = \mathbf{w}^T \mathbf{x} + b. \tag{2}$$

The linear SVM can be extended to a more complex feature space by introducing some kernel function on $\mathbf{x}$. The discriminant function written in its dual form is

$$g_{kernel}(\mathbf{x}) = \sum_{i=1}^{N} a_i y_i K(\mathbf{x}, \mathbf{x}_i) + b. \tag{3}$$

where $\{\alpha_i\}$ are Lagrange multipliers in KKT conditions. We consider a kernel SVM with radial-basis function (RBF) where $K(\mathbf{x}, \mathbf{x}_i) = exp(-\gamma ||\mathbf{x} - \mathbf{x}_i||^2)$.

To build ensemble SVMs, we follow previous studies [5,27] by bagging sub-space features as it was reported to be more effective than bagging training subsets [27]. We adopt two aggregation methods, namely *averaging* and *voting*, to make the final decision over classification results of the independently trained SVMs. The voting method aggregates the results of individual base classifiers by majority votes. The averaging method has discriminant function as

$$g_{ensem}(\mathbf{x}) = \frac{1}{M} \sum_{m=1}^{M} \left[\sum_{i=1}^{N} a_i^m y_i^m K(\mathbf{x}^m, \mathbf{x}_i^m) + b^m \right]$$

$$= \frac{1}{M} \sum_{m=1}^{M} \sum_{i=1}^{N} a_i^m y_i^m K(\mathbf{x}^m, \mathbf{x}_i^m) + b^{avg} \tag{4}$$

where M is the number of base classifiers and b^{avg} is the averaged displacement. Note that the linear kernel is given by $K(\mathbf{x}^m, \mathbf{x}_i^m) = <\mathbf{x}^m, \mathbf{x}_i^m>$ in this case.

3.2   Evasion Attacks

In this attack mode, the attacker's goal is to have a malicious sample misclassified as benign at test time. To this end, the attacker needs to know the decision boundary or similar boundary of the targeted system. Without loss of generality, suppose that the discriminant function $g(\mathbf{x}) > 0$ for detecting a malicious sample $\mathbf{x}$ otherwise passing a legitimate one. The attack rationale is to find a sample $\mathbf{x}'$ that yields $g(\mathbf{x}) < 0$ by minimally manipulating the initial malicious sample $\mathbf{x}$, where the amount of manipulations is characterized by some distance function $d(\mathbf{x}, \mathbf{x}')$ in feature space. This general formula can be written as [35]

$$A(\mathbf{x}) = \operatorname{argmin}_{\mathbf{x}'} \ d(\mathbf{x}, \mathbf{x}'), \quad s.t. \ \ g(\mathbf{x}') < 0. \tag{5}$$

In the case when the features are Boolean as used in the paper, $d(\cdot, \cdot)$ corresponds to the Hamming distance which indicates the number of features that must be added (i.e., flipped from 0 to 1) or deleted (i.e., flipped from 1 to 0) from the initial attack sample $\mathbf{x}$.

Recall that the attacker's knowledge regarding the target classifier can be characterized in terms of $\theta = (\mathcal{D}, \mathcal{X}, f, \mathbf{w})$ as introduced in Sect. 1. It should be noted that only in white-box attacks the target $g(\mathbf{x})$ will be known to the attacker and the required $d(\mathbf{x}, \mathbf{x}')$ is minimal. In more realistic settings, the attacker can only obtain an estimated $\hat{g}(\mathbf{x})$ by constructing an approximated learner $\hat{f}$ with estimated parameters $\mathbf{w}$ trained on a surrogate training set $\hat{\mathcal{D}} = \{(\hat{\mathbf{x}}_i, \hat{y}_i)\}_{i=1}^{N_s}$ of N_s samples. The surrogate training data may be collected by the adversary via sniffing network traffic or augmenting from other sources. Sometimes, the attacker may obtain some true labels and/or subsets of the original training samples. In any case, there must be bias in the estimation should there be knowledge discrepancy about the target. It is intuitive that a better approximation of the surrogate will make the attacker manipulate fewer features to evade the detection and thus less secure/robust of the target classifier against the attack.

4   The Proposed Attack Approach

In this paper, we evaluate the robustness of ensemble SVMs under gradient attacks by assuming limited knowledge of the target's training data available to an attacker. As the surrogate dataset $\hat{\mathcal{D}}$ differs from $\mathcal{D}$, more or less there must be a distribution drift in attacker's learning the surrogate classifier $\hat{f}$ for simulating attacks. It is anticipated that the more drift from $\mathcal{D}$ the more bias in the learner estimate $\hat{f}$ and parameter estimates in $\hat{\mathbf{w}}$ that are trained on the surrogate dataset $\hat{\mathcal{D}}$. In any case, if the attacker can obtain from θ a better discriminant function $\hat{g}(\mathbf{x})$ that closely approximates the target $g(\mathbf{x})$ he is able to manipulate an evasion sample more effectively with fewer efforts.

Accordingly, we consider two attack scenarios with respect to the attacker's knowledge on $\mathcal{D}$. The first scenario is called *the subset scenario* which assumes the attacker knows a subset of training data, i.e., $\hat{\mathcal{D}} \subset \mathcal{D}$. We gradually vary the size of $\hat{\mathcal{D}}$ to evaluate the classifier's robustness under evasion attacks with respect to the distribution drift between the surrogate and the target datasets. In particular, when $\hat{\mathcal{D}} = \mathcal{D}$ it is equivalent to the "white-box" attack in which the surrogate classifier can be regarded as a reproduction of the target and $\hat{g}(\mathbf{x}) = g(\mathbf{x})$ yields the worst-case scenario for evasion with least efforts. The second scenario is called *the surrogate data scenario* which assumes the attacker does not know any instance of the original training data but is able to collect a surrogate dataset $\hat{\mathcal{D}}$ resemble the data distribution of $\mathcal{D}$.

To solve the optimization problem in (5), we assume $\hat{g}(\mathbf{x})$ to be differentiable almost everywhere and adopt the gradient descent attacks which were shown to be effective against single SVMs [2,8]. For classifiers with binary features, the procedure of gradient descent attacks can be found as follows. Firstly, the gradients in $\mathbf{g}$ have to be sorted in descending order of their absolute values, and feature values $\mathbf{x}$ of the malicious sample have to be sorted accordingly. We denote the sorted gradients as $g_1, g_2, ..., g_n$, and the features as $x_1, x_2, ..., x_n$, where $|g_1| \geq |g_2| \geq ... \geq |g_n|$. Then, for $i = 1, 2, ..., d$:

- If $g_i > 0$ and $x_i = 1$, set x_i to 0;
- If $g_i < 0$ and $x_i = 0$, set x_i to 1 (if it is possible);
- otherwise, x_i is left unmodified.

The following subsections explicitly give the gradient formular of discriminant functions for the comparing classifiers. We also propose a new evaluation measure called *gradient correlation* to indicate the similarity of gradients between the surrogate and the targeted systems for constructing evasion samples offline with limited knowledge on θ.

4.1   Gradients of Single SVMs

The most important thing in gradient decent attack is to know the gradient of a classifier. In this section, we give the gradient of single SVMs. The gradient of ensemble SVMs is given in Sect. 4.2.

The gradient of linear-SVM is quite simple which is

$$\nabla g(\mathbf{x}) = \mathbf{w} \tag{6}$$

For kernel-SVM, the gradient is

$$\nabla g(\mathbf{x}) = \sum_{i=1}^{N} a_i y_i \nabla K(\mathbf{x}, \mathbf{x}_i) \tag{7}$$

For RBF kernel, $\nabla K(\mathbf{x}, \mathbf{x}_i) = -2\gamma exp(-\gamma\|\mathbf{x} - \mathbf{x}_i\|^2)(\mathbf{x} - \mathbf{x}_i)$, so the gradient of RBF-SVM is

$$\nabla g(\mathbf{x}) = \sum_{i=1}^{N} a_i y_i [-2\gamma exp(-\gamma\|\mathbf{x} - \mathbf{x}_i\|^2)(\mathbf{x} - \mathbf{x}_i)] \tag{8}$$

4.2   Gradients of Ensemble SVMs

For averaging linear-SVMs, we can see that its discriminant function is still a linear function. Its gradient is just like linear-SVM which is the averaged weight vector $\mathbf{w}^{avg}$. We use the same gradient in voting linear-SVMs.

For averaging RBF-SVMs, the gradient is

$$\nabla g(\mathbf{x}) = \frac{1}{M} \sum_{m=1}^{M} \sum_{i=1}^{N} a_i^m y_i^m \nabla K(\mathbf{x}^m, \mathbf{x}_i^m) \tag{9}$$

where

$$\nabla K(\mathbf{x}, \mathbf{x}_i) = -2\gamma exp(-\gamma \|\mathbf{x}^m - \mathbf{x}_i^m\|^2)(\mathbf{x}^m - \mathbf{x}_i^m) \tag{10}$$

This gradient is also used by voting RBF-SVMs.

4.3   The Gradient Correlation Measure

Kolcz and Teo [18] proposed a measure to evaluate the weight evenness of a classifier which is

$$F(k) = \frac{\sum_{i=1}^{k} w_i}{\sum_{j=1}^{n} w_j} \tag{11}$$

where $k = 1, 2, ..., n$, w_i is the absolute value of its original weight, and $w_1, w_2, ..., w_n$, denote the weights sorted in descending order of their absolute value.

However, this measure is not a scalar. The weight distribution is most even when every weight is identical, which corresponds to $F(k) = k/n$. The most uneven distribution is when only one weight is not zero where $F(k) = 1$ for each k. Accordingly, Biggio et al. [5] proposed a normalized measure $(E) \in [0, 1]$, called weight evenness, based on $F(k)$.

The weight evenness measure was used in addition to hardness of evasion to indicate the robustness of a linear classifier. It is worth noting that in more practical settings the weight eveness measured on a surrogate classifier is not necessarily the same as that on the target model. To address this problem, we propose a more universal measure to evaluate the similarity of gradient estimation between the surrogate and the targeted systems. The gradient correlation (GC) measure is given by:

$$GC = \frac{\sum_{k=1}^{n} C(k)}{n} \tag{12}$$

where

$$C(k) = \frac{\sum_{i=1}^{k} g_i'}{\sum_{i=1}^{k} g_i} \tag{13}$$

Let $\mathbf{g}^+$ denotes the original gradient vector of the targeted system, $\mathbf{g}$ is the vector which sorted $|\mathbf{g}^+|$ in descending order, i.e., $g_1 \geq g_2 \geq ... \geq g_n$. $\mathbf{g}'$ is the gradient vector of surrogate system with the absolute gradient value of targeted system

---

Algorithm 1. Gradient Correlation

Input: $[\mathbf{g}^+, \mathbf{f}^+]$, $\mathbf{g}^+$: the original gradient vector of targeted system, $\mathbf{f}^+$: the features used in the targeted system; $[\mathbf{g}^-, \mathbf{f}^-]$, $\mathbf{g}^-$: the original gradient vector of surrogate system, $\mathbf{f}^-$: the features used in the surrogate system, $\mathbf{f}^- \subseteq \mathbf{f}^+$; n: the number of features used in the targeted system; m: the number of features used in the surrogate system.

Output: GC

1: $[\mathbf{g}, \mathbf{f}] \leftarrow$ sort $|\mathbf{g}^+|$ in descending order;
2: $[\mathbf{g}^*, \mathbf{f}^*] \leftarrow$ sort $|\mathbf{g}^-|$ in descending order;
3: $j \leftarrow 1$;
4: **while** $j \leq m$ **do**
5:      $p \leftarrow$ find the position of f_j^* in $\mathbf{f}$ if exist, otherwise $p \leftarrow 0$;
6:      **if** $p > 0$ **then**
7:          $g_j' \leftarrow g_p$
8:      **else**
9:          $g_j' \leftarrow 0$
10:     **end if**
11:     $j \leftarrow j + 1$;
12: **end while**
13: **if** $m < n$ **then**
14:     $g_j' \leftarrow 0, j = m, m+1, ..., n$;
15: **end if**
16: $C(k) = \frac{\sum_{i=1}^{k} g_i'}{\sum_{i=1}^{k} g_i}, k = 1, 2, ..., n$;
17: $GC = \frac{\sum_{k=1}^{n} C(k)}{n}$.

---

for the same features between the targeted and the surrogate systems. n is the number of features used in the targeted system. The detailed procedure is given by Algorithm 1.

Form Algorithm 1, we can see that $GC \in [0, 1]$, $GC = 0$ and $GC = 1$ correspond respectively to the most uncorrelated and the most correlated gradient distribution. Larger GC means attacker knows more about the gradient of targeted system and the attacks are more effective.

5   Evaluation

In this section, we evaluate the robustness of ensemble SVMs and a single SVM trained on the same dataset for spam email filtering and malware detection tasks. In the subset attack scenario, we gradually increase the size of $\hat{\mathcal{D}}$ by 10%, 20%, ..., 100% of the original training dataset. The samples in $\hat{\mathcal{D}}$ are randomly selected from $\mathcal{D}$. In the second attack scenario of surrogate datasets, we also vary the amount of attacker's knowledge by portions but the training data is from an alternative source rather than the targeted system. Each experiment was run 30 times and the results were averaged to produce the figures.

It is worth noting that for ensemble learning, the surrogate and the targeted systems are different in each run for the features were selected randomly, albeit the target was trained on exactly the same dataset. Each ensemble classifier contains 100 independent base classifiers. As suggested by Ho [12], using half of the features resulting in the best or very close to the best accuracies. We set the feature bagging ratio to 50% for all ensemble classifiers.

In our evaluation tasks, we do not restrict the attack ability which means the attacker can manipulate a malicious sample using whatever computing and time resources needed based on the available knowledge. For security evaluation, we adopt both the conventional hardness of evasion [35] and the proposed gradient correlation measures. The following shows our results performed on two real application datasets.

5.1   Spam Email Filtering

Experimental Setup: The PU3 dataset was considered in spam email classification task [1,21]. There are 11 subdirectories in PU3 (part1, ..., part10, unused) and the first 10 subdirectories, which consists of 1820 spam and 3310 legitimate emails, were used in our experiments. In PU3 dataset, the messages are "encoded" with digital numbers. For the evaluation task, we first extracted words (i.e., features) from emails in the first 5 subdirectories and more than 30,000 features were extracted. Then we reconstructed every email with binary features, which is 1 or 0 represent a feature presence or absence respectively in an email. For keeping computational complexity manageable, we used a feature selection approach, information gain [19], to reduce the feature space to 200 features without loss the classification accuracy significantly.

After turned every email in PU3 to the new feature space, we split the 4130 emails into 3 different subsets. Subset 1 included 608 spam and 769 legitimate emails, which was used as the training dataset. Subset 2 included 604 spam and 773 legitimate emails, which was used as the surrogate dataset. Subset 3 included 608 spam and 768 legitimate emails, which was used as the test dataset. Each email in 3 subsets was different. The training data was used by the targeted systems to train classifiers. The evaluation was carried out on the test data. For linear-SVM and linear-SVM ensemble, the SVM regularization parameter C was set to $C = 1$. For RBF-SVM and RBF-SVM ensemble, we set the SVM regularization parameter $C = 100$ and the kernel parameter $\gamma = 0.01$.

Experimental Results: Table 1 shows classification accuracies achieved by the targeted systems. We observe that ensemble classifiers improved the classification accuracy as expected and that the base classifier using RBF-SVMs slightly outperforms that using linear-SVMs. Figure 1 shows that, under the subset attack scenario, ensemble SVMs are more robust than a single SVM when there is a significant amount (more than 30%) of the original training data are exposed to an attacker. When the available data is reduced to less than 30% ensemble classifiers become more susceptible to evasion for both linear-based and RBF-based SVMs. Under the surrogate data scenario, which is shown on the right side of

Fig. 1, the amount of surrogate data is not as critical as that in the subset scenario. In this case, the ensemble classifiers are always easier to be compromised by manipulating fewer features on average for evading the target classifier.

Table 1. Classification accuracy

Single Linear-SVM	Averaging Linear-SVMs	Voting Linear-SVMs	Single RBF-SVM	Averaging RBF-SVMs	Voting RBF-SVMs
0.9440	0.9565	0.9573	0.9448	0.9599	0.9592

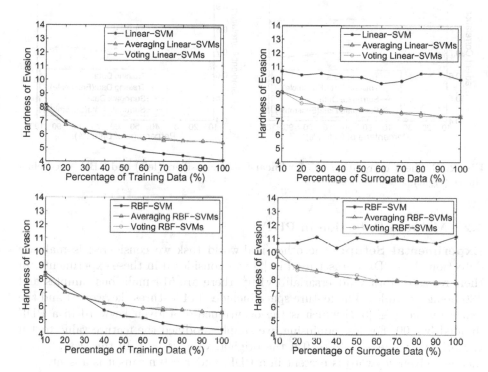

Fig. 1. Hardness of evasion (i.e. average minimum number of modified words to let all spam emails classified as legitimate) in the subset scenario (left) and the surrogate data scenario (right).

Figure 2 plots gradient correlation measures for the two attack scenarios. It can be seen that ensemble SVMs always have higher gradient correlation scores than single SVMs, which supports the observation in Fig. 1. A higher correlation score indicates a higher similarity level between gradient estimates of the surrogate and the targeted systems, and thus more prone to be compromised by evasion attack. In the subset data scenario, the gradient correlation score rises with an increasing percentage of training data exposed to the attacker for both single and ensemble classifiers. In the surrogate data scenario, however, the change is rather flat as the distribution drift is determined by the nature of the two data sources rather than the surrogate data size.

Another interesting observation in Fig. 2 is that an attacker can use less than 50% of the original training dataset to build a surrogate system of ensemble classifiers that closely mimics the targeted system performance and acquire a proactive response to a malicious input. Whereas for a single SVM system it will require more than 80% of the target training dataset to build a resembled surrogate system.

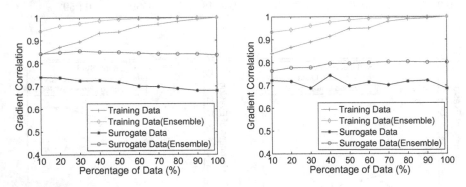

Fig. 2. Gradient correlation GC for linear-based classifiers (left) and RBF-based classifiers (right).

5.2   Malware Detection in PDF

Experimental Setup: The other real-world task we considered is malware detection. The PDF dataset used in [2] was considered in these experiments. In their released library adversarialib v1.0[2], there are 514 malicious samples and 486 benign samples. The feature space includes 114 features (keywords) and the feature value $x \in [0, 1]$ which is the occurrence of a given keyword in a PDF divided by 100. For less confusing, we simply modified the feature value to 1 if the original value $x > 0$, or 0 if the original value $x = 0$. In this case, $x = 1$ means a given keyword is present in a PDF, and $x = 0$ means it is absent.

As discussed in [2], it is hard to remove an embedded object (keywords) from a PDF file without corrupting its structure. But inserting new objects (keywords) through adding a new version to a PDF file is quite easy [2,35]. In our experiments, keywords only can be added cannot be removed which means a feature value only can be modified from 0 to 1, cannot be modified from 1 to 0. For this reason, only the original gradient value $g_i < 0$ need to be considered when calculating GC.

Following the experimental setup used in spam email filtering task, we also split the 1000 samples into 3 different subsets. Subset 1 included 162 malicious and 171 benign samples, which was used as the training dataset.

[2] http://pralab.diee.unica.it/.

Subset 2 included 182 malicious and 152 benign samples, which was used as the surrogate dataset. Subset 3 included 170 malicious and 164 benign samples, which was used as the test dataset. Also, each sample in 3 subsets was different. For linear-SVM and linear-SVM ensembles, the SVM regularization parameter C was set to $C = 1$. For RBF-SVM and RBF-SVM ensembles, we set the SVM regularization parameter $C = 100$ and the kernel parameter $\gamma = 0.01$.

Experimental Results: For this dataset, the security measures in terms of hardness of evasion are very close between the single and the ensemble classifiers in Fig. 3. Nevertheless, it can still be observed that ensemble SVMs tend to require fewer features modified on average for evasion when an attacker has less knowledge on $\mathcal{D}$. This is more obvious by the gradient correlation scores shown in Fig. 4 where gradient estimations are more accurate by ensemble classifiers for both linear- and kernel-based SVMs. This indicates that ensemble SVMs are more susceptible to gradient descent attacks with limited knowledge. In all cases, there is no much difference in security performance between the two aggregation methods of averaging and voting.

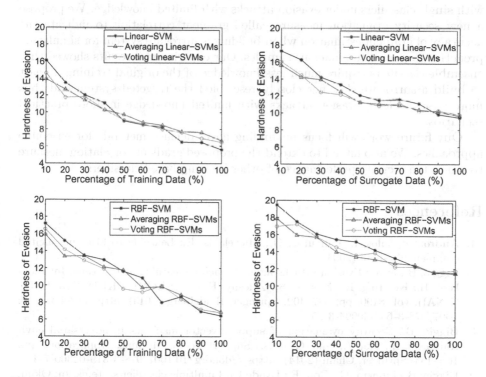

Fig. 3. Hardness of evasion (i.e. average minimum number of added keywords to make every malicious PDF classified as benign) in the subset scenario (left) and the surrogate scenario (right).

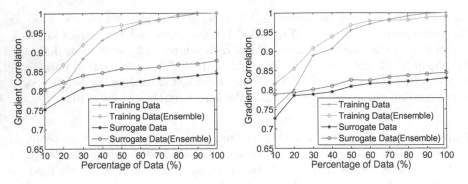

Fig. 4. Gradient correlation GC for linear-based classifiers (left) and RBF-based classifiers (right).

6   Conclusion and Future Work

In this paper, we investigated the robustness of ensemble classifiers comparing with single classifiers under evasion attacks with limited knowledge. We propose a new security evaluation measure called gradient correlation to indicate the accuracy of gradient estimation when building a surrogate system for simulating proactive responses to malicious samples. Our experimental results showed that ensemble classifiers require much less knowledge of the original training dataset to build a surrogate classifier closely resembled the targeted system and thus more susceptible to evasion attacks with limited knowledge in more practical scenarios.

Our future work will focus on finding novel defence methods for ensemble approaches. We also intend to extend the proposed gradient correlation measure to study the security performance of other learning-based classifiers.

References

1. Androutsopoulos, I., Paliouras, G., Michelakis, E.: Learning to filter unsolicited commercial e-mail (2004)
2. Biggio, B., et al.: Evasion attacks against machine learning at test time. In: Blockeel, H., Kersting, K., Nijssen, S., Železný, F. (eds.) ECML PKDD 2013. LNCS (LNAI), vol. 8190, pp. 387–402. Springer, Heidelberg (2013). https://doi.org/10.1007/978-3-642-40994-3_25
3. Biggio, B.: Security evaluation of support vector machines in adversarial environments. In: Ma, Y., Guo, G. (eds.) Support Vector Machines Applications, pp. 105–153. Springer, Cham (2014). https://doi.org/10.1007/978-3-319-02300-7_4
4. Biggio, B., Fumera, G., Roli, F.: Evade hard multiple classifier systems. In: Okun, O., Valentini, G. (eds.) Applications of Supervised and Unsupervised Ensemble Methods, pp. 15–38. Springer, Heidelberg (2009). https://doi.org/10.1007/978-3-642-03999-7_2

5. Biggio, B., Fumera, G., Roli, F.: Multiple classifier systems for robust classifier design in adversarial environments. Int. J. Mach. Learn. Cybern. **1**(1–4), 27–41 (2010)
6. Biggio, B., Roli, F.: Wild patterns: ten years after the rise of adversarial machine learning. arXiv Preprint (2017). http://arxiv.org/abs/1712.03141
7. Cheng, K., et al.: Secure k-NN query on encrypted cloud data with multiple keys. IEEE Trans. Big Data **1**, 1–1 (2015)
8. Demontis, A., et al.: Yes, machine learning can be more secure! A case study on android malware detection. IEEE Trans. Dependable Secur. Comput. (2017, in press). https://ieeexplore.ieee.org/document/7917369
9. Dong, Y.S., Han, K.S.: Boosting SVM classifiers by ensemble. In: The 14th International Conference on World Wide Web, pp. 1072–1073, WWW 2005. ACM (2005)
10. Goodfellow, I.J., Shlens, J., Szegedy, C.: Explaining and harnessing adversarial examples. In: The International Conference on Learning Representations, ICLR 2015 (2015)
11. Hastie, T., Tibshirani, R., Friedman, J.: The Elements of Statistical Learning, 2nd edn. Springer, New York (2009)
12. Ho, T.K.: The random subspace method for constructing decision forests. IEEE Trans. Pattern Anal. Mach. Intell. **20**(8), 832–844 (1998)
13. Kabir, E., Mahmood, A., Wang, H., Mustafa, A.: Microaggregation sorting framework for k-anonymity statistical disclosure control in cloud computing. IEEE Trans. Cloud Comput. (2015, in press). https://ieeexplore.ieee.org/document/7208829
14. Kabir, M.E., Wang, H., Bertino, E.: A role-involved purpose-based access control model. Inf. Syst. Front. **14**(3), 809–822 (2012)
15. Kantchelian, A., Tygar, J., Joseph, A.: Evasion and hardening of tree ensemble classifiers. In: International Conference on Machine Learning, pp. 2387–2396 (2016)
16. Khalil, F., Li, J., Wang, H.: An integrated model for next page access prediction. Int. J. Knowl. Web Intell. **1**(1–2), 48–80 (2009)
17. Kim, H.C., Pang, S., Je, H.M., Kim, D., Bang, S.Y.: Constructing support vector machine ensemble. Pattern Recogn. **36**(12), 2757–2767 (2003)
18. Kołcz, A., Teo, C.H.: Feature weighting for improved classifier robustness. In: Sixth Conference On Email and Anti-spam, CEAS 2009 (2009)
19. Kolter, J.Z., Maloof, M.A.: Learning to detect malicious executables in the wild. In: Proceedings of the Tenth ACM SIGKDD International Conference on Knowledge Discovery and Data Mining, pp. 470–478. ACM (2004)
20. Laskov, P., et al.: Practical evasion of a learning-based classifier: a case study. In: 2014 IEEE Symposium on Security and Privacy (SP), pp. 197–211. IEEE (2014)
21. Mujtaba, G., Shuib, L., Raj, R.G., Majeed, N., Al-Garadi, M.A.: Email classification research trends: review and open issues. IEEE Access **5**, 9044–9064 (2017)
22. Papernot, N., Mcdaniel, P., Sinha, A., Wellman, M.: SoK: Towards the science of security and privacy in machine learning. arXiv Preprint, pp. 1–19 (2016). http://arxiv.org/abs/1611.03814
23. Peng, M., Zeng, G., Sun, Z., Huang, J., Wang, H., Tian, G.: Personalized app recommendation based on app permissions. World Wide Web **21**(1), 89–104 (2018)
24. Shah, Z., Mahmood, A.N., Barlow, M., Tari, Z., Yi, X., Zomaya, A.Y.: Computing hierarchical summary from two-dimensional big data streams. IEEE Trans. Parallel Distrib. Syst. **29**(4), 803–818 (2018)
25. Shen, Y., Zhang, T., Wang, Y., Wang, H., Jiang, X.: Microthings: a generic iot architecture for flexible data aggregation and scalable service cooperation. IEEE Commun. Mag. **55**(9), 86–93 (2017)

26. Shu, J., Jia, X., Yang, K., Wang, H.: Privacy-preserving task recommendation services for crowdsourcing. IEEE Trans. Serv. Comput. (2018, in press). https://ieeexplore.ieee.org/document/8253516
27. Smutz, C., Stavrou, A.: When a tree falls: using diversity in ensemble classifiers to identify evasion in malware detectors. In: NDSS (2016)
28. Sun, X., Li, M., Wang, H., Plank, A.: An efficient hash-based algorithm for minimal k-anonymity. In: Proceedings of the Thirty-First Australasian Conference on Computer Science, vol. 74, pp. 101–107. Australian Computer Society, Inc. (2008)
29. Sun, X., Wang, H., Li, J., Zhang, Y.: Injecting purpose and trust into data anonymisation. Comput. Secur. **30**(5), 332–345 (2011)
30. Vapnik, V.: The Nature of Statistical Learning, 1st edn. Springer, New York (1999)
31. Wang, G., Wang, T., Zheng, H., Zhao, B.Y.: Man vs. machine: practical adversarial detection of malicious crowdsourcing workers. In: USENIX Security Symposium, pp. 239–254 (2014)
32. Wang, H., Cao, J., Zhang, Y.: Ticket-based service access scheme for mobile users. In: Australian Computer Science Communications, vol. 24, pp. 285–292. Australian Computer Society, Inc. (2002)
33. Xu, W., Qi, Y., Evans, D.: Automatically evading classifiers. In: Proceedings of the 2016 Network and Distributed Systems Symposium (2016)
34. Yi, X., Sun, H., Jafar, S.A., Gesbert, D.: Tdma is optimal for all-unicast dof region of tim if and only if topology is chordal bipartite. IEEE Trans. Inf. Theory **64**(3), 2065–2076 (2018)
35. Zhang, F., Chan, P.P., Biggio, B., Yeung, D.S., Roli, F.: Adversarial feature selection against evasion attacks. IEEE Trans. Cybern. **46**(3), 766–777 (2016)
36. Zhang, Y., Shen, Y., Wang, H., Zhang, Y., Jiang, X.: On secure wireless communications for service oriented computing. IEEE Trans. Serv. Comput. **11**(2), 318–328 (2018)

An Improved Lightweight RFID Authentication Protocol for Internet of Things

Xu Yang[1]([✉]), Xun Yi[1], Yali Zeng[2], Ibrahim Khalil[1], Xinyi Huang[2], and Surya Nepal[3]

[1] School of Science, RMIT University, Melbourne, VIC 3000, Australia
xu.yang@rmit.edu.au
[2] College of Mathematics and Informatics, Fujian Normal University, Fuzhou, China
[3] CSIRO Data61, Sydney, NSW 2122, Australia

Abstract. With the widely development and deployment of Radio Frequency Identification (RFID) technology for nowadays' object automatic identification, it has became one of the core technologies of the Internet of Things (IoT). RFID authentication is a primary approach to secure a RFID system and make it privacy-friendly. There are many RFID authentication protocols proposed to tackle the RFID security, privacy and efficiency concerns. However, with the increasingly stringent security and privacy requirements and limited computation capacity of tags, most of these protocols have suffered with serious security weaknesses and inefficient performance. In this paper, we firstly give an overview on Kaur et al.'s protocol and point out the security deficiencies of their protocol. Then, we propose an improved lightweight anonymous authentication protocol for RFID systems using elliptic curve cryptography (ECC) algorithm. The security analysis shows that the proposed protocol achieves mutual authentication, confidentiality, anonymity as well as resistance to various attacks, such as replay, impersonation and modification attacks, etc. Furthermore, performance evaluation indicates that the proposed protocol significantly reduces the computation cost by at least 3 times and decreases the communication cost by at least 50% compared to previous RFID authentication protocols.

Keywords: Internet of Things · RFID · Authentication · Privacy Security · Efficiency

1 Introduction

In the past decade, the Internet of Things (IoT) has drawn significant attention. IoT has emerged as one of the most powerful information and communication technology (ICT) paradigms of the 21st century. It has been used to provide services and resources in different domains, including government, education, commerce, healthcare, and so on [1]. Obviously, nowadays IoT is a technological

© Springer Nature Switzerland AG 2018
H. Hacid et al. (Eds.): WISE 2018, LNCS 11233, pp. 111–126, 2018.
https://doi.org/10.1007/978-3-030-02922-7_8

revolution that represents the future of computing and communications, and its development needs the support from some innovational technologies [24]. Radio frequency identification (RFID) is seen as one of the core technologies and the pivotal enablers of IoT.

IoT comprises things that have unique identities and are connected to the Internet. While many existing devices, such as networked computers or 4G-enabled mobile phones, already have some forms of unique identities and are connected to the Internet, the focus on IoT is in the configuration, control and networking via the Internet of devices or "things" that are traditionally not associated with the Internet. These include devices such as RFID tags.

In order to connect the objects in IoT, they should be firstly identified by IoT. So RFID, which uses radio waves to automatic identify items, becomes the role to provide this function. A mature RFID technology provides a strong support for IoT. The tags contain electronically-stored information. Passive tags collect energy from a nearby RFID reader's interrogating radio waves. Active tags have a local power source (such as a battery) and may operate hundreds of meters from the RFID reader. RFID not only can be labeled as a replacement of bar code, but also can track items in real-time to get important information about their location and status. It has already applied in some valuable applications, such like retail, health-care, facilities management, supply chains, passports, et al. [16,25].

RFID authentication is a primary approach to secure an RFID system and make it privacy-friendly. As is shown in Fig. 1, a typical RFID system is composed of three main components: the RFID tag, the RFID reader and the backend server. In the system, the communication between the reader and backend server is assumed to be secure. However, the communication between the tag and the reader is insecure. Tags are attached to the objects to identify them uniquely and store the identification information of the objects. The reader could activate the tag by sending the radio frequency (RF) signals to communicate and exchange information with it and finally submit the relevant data to the backend server.

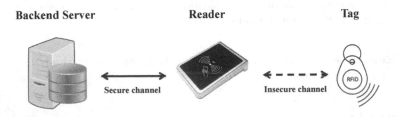

Backend Server   **Reader**   **Tag**

Secure channel   Insecure channel

Fig. 1. A typical RFID system

However, because of the usage of the wireless communication and signal broadcasting technique, the RFID systems are vulnerable to many security attacks and privacy disclosure threats. Additionally, due to strictly limited calculation resources, small storage capacity and faint power supply of low-cost tags, it is difficult to apply complicated operations to an RFID system and these factors are hindering the rapid spread of this technology [12]. Therefore, it is important for an RFID system to provide secure, anonymous, and efficient authentication process for the server and the tags in the system.

Over the past few years, many researchers are committed to design the secure, anonymous and efficient authentication methods for RFID systems. Generally, they can be broadly classified into two categories: non-public key cryptosystem (NPKC) based protocols and public key cryptosystem (PKC) based protocols.

The NPKC-based RFID authentication protocols have better performance because no complex operations are needed. Therefore, many NPKC-based RFID authentication protocols have been proposed for various practical applications. For example, cyclic redundancy code (CRC) checksum-based protocols [5,26], simple bit-wise operations (such as OR, AND, and XOR)-based protocols [21,22], one-way hash functions-based protocols [6,10] and symmetric encryption algorithms-based protocols [8,13]. The aforementioned NPKC-based RFID authentication protocols are suitable for RFID systems in terms of reduced computational and storage costs.

However, in terms of security, these approaches cannot match the PKC-based protocols, which tend to provide higher security [4]. Due to the development of the microelectronic technique, some complex PKC algorithms have been directly used for the RFID systems [9,15]. Compared to these PKC algorithms, the elliptic curve cryptography (ECC) algorithm, which can provide similar security level but with a shorter key size and lower computation cost, is more practical and suitable for RFID systems [17]. Lee et al. [17] proposed a provably secure RFID authentication protocol using the ECC technique and claimed their scheme could provably secure. However, their protocol could not resist the impersonation attack and the tracking attack [3]. In order to enhance security, Lee et al. [18] proposed another ECC-based RFID authentication protocol. However, their work failed again [7]. Afterward, many ECC-based RFID authentication protocols was successively proposed [11,19,20,23]. Also, these protocols cannot reach the security and efficiency requirements. Most recently, Kaur et al. [14] propose a lightweight authentication protocol for RFID-enabled systems based on ECC. They claim that their protocol can achieve mutual authentication, anonymity, as well as resistance to various security attacks. However, as we will show in this study, Kaur et al.'s protocol [14] cannot provide these important security requirements and goals.

In order to address the weaknesses in recent authentication protocols for RFID systems, in this paper, we propose an improved lightweight authentication protocol for RFID systems based on ECC. In a nutshell, our main contributions can be summarized as follows.

(1) We give an overview and cryptanalysis on Kaur et al.'s protocol [14] which show that Kaur et al.'s protocol [14] cannot achieve some important security goals, including vulnerable to impersonation attack, difficult to distinguish the tags, and so on.

(2) We propose an improved ECC-based lightweight authentication protocol which can overcome the deficiencies in Kaur et al.'s protocol [14]. The security analysis shows that our protocol can provide mutual authentication, confidentiality, anonymity, as well as attack resistance.

(3) We evaluate the performance of the proposed ECC-based authentication protocol to show that our protocol significantly reduces the computation delay by at least 3 times and decreases the communication cost by at least 50% compared to the existing ECC-based authentication protocols in RFID systems.

The remainder of this paper is organized as follows. Section 2 discusses some preliminaries, including the elliptic curve group, the mathematical problems and the security requirements. In Sect. 3, we review Kaur et al.'s protocol [14], and describe the weaknesses in their protocol. In Sect. 4, we present an improved RFID authentication protocol. Section 5 and Sect. 6 provide the security analysis and performance evaluation of our protocol, respectively. At last, we conclude the paper and look into potential future works in Sect. 7.

2   Preliminaries

In this section, we briefly introduce the elliptic curve cryptography, the corresponding mathematical problems over it, and the security requirements for an RFID authentication protocol.

2.1   Elliptic Curve Cryptography

Let F_q be a prime finite field, $E/F_q : y^2 = x^3 + a \cdot x + b$ is an elliptic curve defined over F_q, where $a, b \in F_q$ and $\Delta = 4a^3 + 27b^2 \neq 0 \bmod q$. Let P be an element of a large prime order q in E/F_q. The points on E/F_q together with an extra point Θ, called the point at infinity, form a group $G = \{(x, y) : x, y \in F_q; (x, y) \in E/F_q\} \cup \{\Theta\}$. G is a cyclic additive group of composite order q. Besides, a scalar multiplication over E/F_q can be computed as follows: $tP = \underbrace{P + P + \cdots + P}_{t\ times}$, where t is an integer.

2.2   Mathematical Problems

There exist the following problems over the elliptic curve group which have been widely used in the design of authentication protocols.

Discrete Logarithm (DL) Problem: For a random chosen value $a \in \mathbb{Z}_q^*$ and the generator P of G, given aP, it is computationally intractable to compute the value a.

Computational Diffie-Hellman (CDH) Problem: For random chosen values $a, b \in \mathbb{Z}_q^*$ and the generator P of G, given aP and bP, it is computationally intractable to compute the value abP.

2.3   Security Requirements

RFID authentication is one of the most important steps to ensure secure communication in the RFID systems. According to previous research efforts on RFID security [10,11,19,23], to guarantee a secure communication, the design of a RFID authentication protocol should satisfy the following basic requirements:

(1) **Mutual authentication:** It is essential for the authentication protocol to provide the mutual authentication between the server and the RFID tag in order to ensure the legitimacy.
(2) **Confidentiality:** The secret information stored in the RFID tag, such as identity and secret key, should be protected. That is, these secret information cannot be retrieved by the adversary through analyzing the messages in the communication channels.
(3) **Anonymity:** To protect RFID tag's privacy, an RFID authentication protocol should provide anonymity. The adversary is unable to extract any tag's real identity.
(4) **Attack resistance:** The RFID authentication protocol should be able to withstand various common attacks such as replay attack, impersonation attack, modification attack, cloning attack, and DoS attack.

3   Review and Cryptanalysis of Kaur et al.'s Protocol

In this section, we will give a brief review on Kaur et al.'s protocol [14] and describe the deficiencies in their protocol. For more detailed information, please refer to the original paper in [14]. To facilitate the understanding, we list some related notations which use in this paper in Table 1.

3.1   Review of Kaur et al.'s Protocol

There are three phases in their protocol, including initial setup phase, server authentication phase, and tag authentication phase.

Initial Setup: In this phase, both the server and the tag are equipped with public and private keys, and the public parameters of elliptic curve domain $params = \{F_q, E/F_q, P, q, n\}$. For the server side in this phase, it chooses a random number $x_S \in Z_n^*$ as private key and computes $X_S = x_S \cdot P$ as public key. The server also keeps all the valid RFID tags' public keys X_T and the corresponding identities id. For the tag side, it owns the private key $x_T \in Z_n^*$ and the public key $X_T = x_T \cdot P$.

Server Authentication: In this phase, the server selects a random number $r_1 \in Z_n^*$ and computes $R_1 = r_1 \cdot P$. Then, it also computes an authentication token $Auth_S = x_S \cdot R_1$ for tag to verify. Finally, it sends a message $M_1 = (id, r_1, Auth_S)$ to the tags in its vicinity, where the id is the target tag's identity who needs to be authenticated by the server. Once the tags receive M_1, they firstly check whether the id in M_1 is the same as their own identity (id_T). If not match, ignores

Table 1. Notations in the protocol

Notation	Description
n, q	Two large prime numbers
F_q	A prime finite field
E/F_q	An elliptic curve E over F_q
G	A cyclic additive group, $G = \{(x, y) : x, y \in E/F_q\} \cup \{\Theta\}$
P	A generator point with n order of the elliptic curve E
x_S	The private key of the server
X_S	The public key of the server
x_T	The private key of the tag
X_T	The public key of the tag
id	Real identity of the tag
M_1, M_2	The messages send by the server or the tag
ts	A time stamp
$H()$	A secure general hash function
pid	A pseudo identity
Ver_S, Ver_T	A verification token generated by the server or the tag

this message. Otherwise, the tag with id computes another authentication token $Auth'_S = r_1 \cdot X_S$ to compare with the received token $Auth_S$ in M_1. If they are equal, the server is successfully authenticated by the tag. Otherwise, the tag also drops this message.

Tag Authentication: In this phase, the tag selects a random number $r_2 \in Z_n^*$ and computes $R_2 = r_2 \cdot P$. Then, it also computes an authentication token $Auth_T = x_T \cdot R_2$ for the server to verify. Finally, it sends a message $M_2 = (r_2, Auth_T)$ to the server. Once the server receive M_2, it checks authentication taken by computing another authentication token $Auth'_T = r_2 \cdot X_T$ to compare with $Auth_T$ in M_2. If they are equal, the tag is successfully authenticated by the server. Otherwise, the server also drops this message and declares this tag as malicious.

3.2   Cryptanalysis of Kaur et al.'s Protocol

Through security and performance analysis, Kaur et al. [14] claims that their protocol can fulfill mutual authentication and anonymity, and can also resist various kinds of attacks. However, in this subsection, we point out that their protocol fails to serve its purposes and has several deficiencies.

(1) **Impersonation attack**

Adversary $\mathscr{A}$ can impersonate the server to pass the authentication. Firstly, according to the protocol, $\mathscr{A}$ chooses a random number $a_1 \in Z_n^*$ and then directly computes an authentication token $Auth_{A_1} = a_1 \cdot X_S$. Finally, $\mathscr{A}$ generates a message $M_{A_1} = (id, a_1, Auth_{A_1})$. This message can be sent to any tags as long as changing the identity id. Correspondingly, once the target tag receives this message, it checks the token by comparing whether $Auth_{A_1}$ is equal to $a_1 \cdot X_S$ or not. Obviously, the verification can be easily passed. Also, adversary $\mathscr{A}$ can impersonate the tag to pass the authentication. When the server sends out a message $M_1 = (id, r_1, Auth_S)$, $\mathscr{A}$ can monitor and intercept this message. Then, $\mathscr{A}$ randomly chooses a random number $a_2 \in Z_n^*$ and computes $Auth_{A_2} = a_2 \cdot X_T$. Finally, $\mathscr{A}$ generates a message $M_{A_2} = (a_2, Auth_{A_2})$ and sends to the server as a response message. Similarly, once the server receives this message, it can authenticate this tag by checking $Auth_{A_2} = a_2 \cdot X_T$. Therefore, this protocol fails to resist impersonation attack.

(2) **Difficult to distinguish the tags**

When the server receives many messages from different tags, such like $M_2' = (r_2', Auth_T')$ and $M_2'' = (r_2'', Auth_T'')$, it is difficult for the server to confirm which exactly tag the message belongs to. Because there are only two parameters without any relevant to tag's identity in the received message. Thus, the server can only use the violence analysis method to confirm. The detailed process is that the server tries all the public keys of valid tags to check whether the received token $Auth_T$ equals to $r_2 \cdot X_{T_i}$ (where $i = 1, 2, \ldots, n$, n is the number of valid tags). Obviously, this method results in a very expensive computation overhead and high verification time delay because of up to n times elliptic curve scalar multiplication (ECSM) operations.

(3) **Failure of other security requirements**

According to the above analysis, we can easily conclude that this protocol cannot provide the **mutual authentication** because it is unable to authenticate each other (between server and tags) to ensure the legitimacy. Also, their protocol cannot provide **anonymity** for the tag since the real identity of all tags are transferred in plaintext in the messages. Besides, their protocol cannot resist the **modification attack**. For example, the adversary $\mathscr{A}$ can modify id to anyone's identity in M_1 while the tags cannot detect the modification. Furthermore, $\mathscr{A}$ can use any previous message $M_1 = (id, r_1, Auth_S)$ and replay to the tags. The tags cannot find out it is a replay message because there is no detection mechanism for the **replay attack**.

4 Improved Protocol

In this section, we propose a new ECC-based authentication protocol in RFID systems, which can overcome the deficiencies in Kaur et al.'s protocol [14]. Our protocol consists of two phases: setup phase and authentication phase.

4.1   Setup Phase

In this phase, both the server and the tag are equipped with public and private keys, and the public parameters of elliptic curve domain $params = \{E, F_q, P, q, n\}$. The detail of this phase is shown as follows.

(1) **Server side:** The server chooses a random number $x_S \in Z_n^*$ and computes $X_S = x_S \cdot P$. The server also keeps all the valid RFID tags' public keys X_T and their corresponding identities id (these identities of tags are only known by the server and each corresponding tag).

(2) **Tag side:** All tags owns their unique real identity id, the private key $x_T \in Z_n^*$ and the public key $X_T = x_T \cdot P$ (note that different tags have different identity, private key and public key).

Overall, the public information contains $\{params, X_S, X_T\}$. The server keeps private key x_S and all tags' identities id as the secret information and the tag keeps private key x_T and their own identity id as the secret information.

4.2   Authentication Phase

The mutual authentication between the server and the tag happens in this phase. The detailed process is shown in Fig. 2 and can be described as follows:

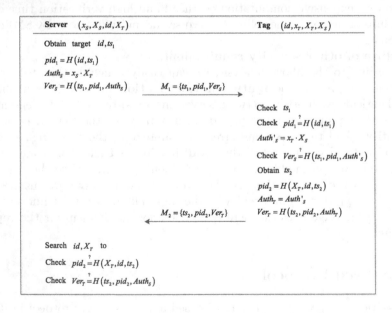

Fig. 2. The proposed RFID authentication protocol

(1) The server firstly obtains the current time stamp ts_1 and determines the identity id of target tag which it wants to communicate. Then it computes $pid_1 = H(id, ts_1)$, $Auth_S = x_S \cdot X_T$, and $Ver_S = H(ts_1, pid_1, Auth_S)$. Then the server broadcasts the message $M_1 = \{ts_1, pid_1, Ver_S\}$ to the tags in its vicinity.

(2) **Server $\rightarrow$ Tag:** $M_1 = \{ts_1, pid_1, Ver_S\}$

(3) When receiving the message M_1 from the server, the tags firstly check the freshness of time stamp ts_1 in M_1 to prevent the replay attack. If ts_1 is beyond the service expiration time, the tags drop this message. Otherwise, they use their identity id to check whether the equation $pid_1 = H(id, ts_1)$ holds. If it does not hold, M_1 will be dropped. Otherwise, the tag continues the authentication phase. (Note that only the target tag can successfully check the equation, since the id is specially owned by the target tag.)

(4) Afterward, the target tag computes $Auth'_S = x_T \cdot X_S$ and checks whether the equation $Ver_S = H(ts_1, pid_1, Auth'_S)$ holds. The server is successfully authenticated by the tag if the equation holds. Otherwise, this tag regards M_1 as a modified message and drops it.

(5) The target tag then obtains the current time stamp ts_2, and computes $pid_2 = H(X_T, id, ts_2)$, $Auth_T = Auth'_S$, and $Ver_T = H(ts_2, pid_2, Auth_T)$. Finally, it sends the response message $M_2 = \{ts_2, pid_2, Ver_T\}$ to the server.

(6) **Tag $\rightarrow$ Server:** $M_2 = \{ts_2, pid_2, Ver_T\}$

(7) Once the server receives the message M_2. It firstly checks the freshness of ts_2 to prevent the replay attack. Then it determines the corresponding public key X_T by the recently used identity id of tags for checking whether $pid_2 = H(X_T, id, ts_2)$ holds (here we use both X_T and id to resist the replay attack of reusing M_2 as M_1). Once successfully checked, it further checks whether the equation $Ver_T = H(ts_2, pid_2, Auth_S)$ holds. The tag is successfully authenticated by the server if the equation holds. Otherwise, the server regards M_2 as a modified message and drops it.

5   Security Analysis

In this section, we discuss the security of our improved ECC-based RFID authentication protocol. We demonstrate that the proposed protocol can satisfy the security requirements presented in Sect. 2.3.

Mutual authentication: The mutual authentication between the server and the tag in our protocol can be divided into two aspects, one is server to tag authentication and another is tag to server authentication.

- Server to tag authentication: In the step 2 of the proposed authentication phase, the server sends the message $M_1 = \{ts_1, pid_1, Ver_S\}$ to the tag, where $pid_1 = H(id, ts_1)$, $Auth_S = x_S \cdot X_T$, and $Ver_S = H(ts_1, pid_1, Auth_S)$. Only the target tag has the real identity id to successfully pass the verification of pid_1. After the successfully verification by the target tag, only this tag (except the server) can compute $Auth'_S = x_T \cdot X_S$ and further check the equation

$Ver_S = H(ts_1, pid_1, Auth'_S)$. Unless any adversary can get the knowledge of the real identity id of this tag and use $X_T = x_T \cdot P$ and $X_S = x_S \cdot P$ to obtain the value of $Auth_S = x_S \cdot x_T \cdot P$. However, it contradicts with the assumption of the CDH problem. Therefore, our proposed protocol can provide the server to tag authentication.

- Tag to server authentication: In the step 6 of the proposed authentication phase, the tag sends the message $M_2 = \{ts_2, pid_2, Ver_T\}$ to the server, where $pid_2 = H(X_T, id, ts_2)$, $Auth_T = x_T \cdot X_S$, and $Ver_T = H(ts_2, pid_2, Auth_T)$. While receiving the message, the server can determine the exact tag's public key X_T by checking $pid_2 = H(X_T, id, ts_2)$ of the recently used identity list id of tags. Then it further checks the verification value Ver_T by using the authentication token $Auth_S$. Similarly, only the server and the target tag have the knowledge of this authentication token $Auth_S/Auth_T$. Due to the assumption of the CDH problem, it is impossible for any adversary to use $X_T = x_T \cdot P$ and $X_S = x_S \cdot P$ to obtain the value of $Auth_T = x_S \cdot x_T \cdot P$. Therefore, our proposed protocol can provide the tag to server authentication. Overall, the mutual authentication between the server and the tag is provided in the proposed protocol.

Confidentiality: In the proposed protocol, the tag's real identity id and private key x_T are included in the message $M_1 = \{ts_1, pid_1, Ver_S\}$ and $M_2 = \{ts_2, pid_2, Ver_T\}$, where $pid_1 = H(id, ts_1)$, $pid_2 = H(X_T, id, ts_2)$ and $Auth_T = x_T \cdot X_S$. Since only the server and the tag itself know about id, the adversary cannot get id by analyzing pid_1. Besides, because of the assumption of the DL problem, it is impossible for the adversary to compute x_T by giving X_T and P. Similarly, the private key x_S of the server is securely protected in the messages under the assumption of the DL problem. Therefore, our proposed protocol can provide confidentiality for both the server and the tag.

Anonymity: In the proposed protocol, the tag's real identity id is included in the message $M_1 = \{ts_1, pid_1, Ver_S\}$ and $M_2 = \{ts_2, pid_2, Ver_T\}$, where $pid_1 = H(id, ts_1)$ and $pid_2 = H(X_T, id, ts_2)$. Due to the confidentiality of id and the protection of one-way hash function, it is difficult for the adversary to know the real identity of the tag. Besides, due to the variability of the time stamp ts_2 in message $M_2 = \{ts_2, pid_2, Ver_T\}$, it is difficult for the adversary to trace the location of the tag by collecting messages or to distinguish any two messages which were sent out by the same tag. Therefore, our proposed protocol could provide anonymity for the tag.

Attack resistance: Our protocol can resist the following attacks.

- Replay attack: A replay attack is infeasible in our protocol since we use a time stamp ts to prevent the replay attacks; that is, any replay message goes beyond the service expiration time. Even if ts can be updated by an attacker in the replay message, the attacker cannot generate a valid pseudo identity pid and the verification value Ver_S/Ver_T related to this new ts. So it is impossible to successfully pass the verification due to the freshness of ts.

- Impersonation attack: Assume that the adversary wants to impersonate the sever to communicate with the tag. It firstly needs to generate a valid message $M_1 = \{ts_1, pid_1, Ver_S\}$, where $pid_1 = H(id, ts_1)$, $Auth_S = x_S \cdot X_T$, and $Ver_S = H(td_1, pid_1, Auth_S)$. Due to the confidentiality of id and x_S, it is impossible for the adversary to generate valid pid_1 and Ver_S. Similarly, assume that the adversary wants to impersonate the tag to communicate with the server. It firstly needs to generate a valid message $M_2 = \{ts_2, pid_2, Ver_T\}$, where $pid_2 = H(X_T, id, ts_2)$, $Auth_T = x_T \cdot X_S$, and $Ver_T = H(ts_2, pid_2, Auth_T)$. Due to the confidentiality of id and x_T, it is also impossible for the adversary to generate valid pid_2 and Ver_T. Therefore, the proposed protocol is able to resist the impersonation attack.

- Modification attack: As is analyzed above, due to the confidentiality of id and private key x_S/x_T and the difficulty of DL problem and CDH problem, it is unable for the adversary to successfully modify the parameters, such as pid_1, Ver_S, pid_2 and Ver_T, in the messages. Also, the one-way hash function in the authentication message can ensure the data integrity. Therefore, it is impossible to modify a valid message during authentication.

- Cloning attack: In setup phase of the proposed protocol, every tag has its own identity id and private key x_T, which are randomly chosen numbers. Assume that the adversary could obtain some tags' identity and private key. However, it cannot get other tags' identity and private key by using the known tags' identity and private key since there are no relationship between these tags. Therefore, our proposed protocol could resist cloning attack.

- DoS attack: Our protocol employs only hash operations for the verification in both the tag side (the verification on $pid_1 = H(id, ts_1)$) and the server side (the verification on $pid_2 = H(X_T, id, ts_2)$). Due to extremely low computational consumption of the general hash operation, it makes our proposed authentication protocol could overcome the DoS attack.

6   Performance Evaluation

In this section, we discuss the performance of our improved ECC-based RFID authentication protocol and compare with several most recent ECC-based RFID authentication protocols [11,14,19,23]. We analyze the performance of our protocol in terms of computation and communication costs. All tests are performed on a laptop with the following specifications: CPU: 1.6 GHz Intel Core i5, Memory: 4 GB 1600 MHz DDR3, the macOS High Sierra operation system, and we also used a high performing implementation from libgmp via the gmpy2 python module (https://gmpy2.readthedocs.io/en/latest/) and pairing-based library (version 0.5.14, https://crypto.stanford.edu/pbc/). For the convenience of the evaluation, we assume that $E : y^2 = x^3 + ax + b \bmod q$ is an elliptic curve defined over a prime finite field, where exists an additive group G generated by a point P with the order q and $a, b \in Z_n^*$ (n is a 160-bit prime number, the size of q relevant to NIST recommended key size [2]).

6.1   Computation Cost

The computation overhead represents the processing delays of the cryptography operations at the side of the server and the tag. We denote the time cost of some cryptographic related operations as follows:

- T_M : denotes the execution time for one elliptic curve scalar multiplication operation.
- T_A : denotes the execution time for one elliptic curve point addition operation.
- T_H : denotes the execution time for one general hash function operation.

According to NIST recommended key size [2], we test the computation time and communication size in 5 rounds based on the key sizes of ECC: 160 bits, 224 bits, 256 bits, 384 bits and 512 bits. As is known, with the increasing of the key size, the security of ECC operations will be enhanced, but it also will result in the low efficiency of computation and communication costs. The time cost of executing ECC operations on different key sizes is shown in Table 2. Besides, the time cost of T_H is 0.002 ms. (We run 100 times for all the implementation to finally get the average operation time.)

Table 2. The time cost of executing ECC operations on different key sizes

Key size (bits)	T_M (ms)	T_A (ms)
160	0.966	0.006
224	1.166	0.009
256	1.326	0.011
384	1.619	0.014
512	2.090	0.016

Based on the above implementation, we compare the computation cost in the side of both the server and the tag for our protocol and other four protocols proposed by Kaur et al. [14], Liao et al. [19], He et al. [11], and Shen et al. [23], respectively. The comparison results of computation cost are listed in Table 3. According to Table 3, we also simulate the implementation of the computation cost on different key sizes of ECC in Fig. 3.

Table 3. Comparison of computation cost

	Kaur et al.'s protocol [14]	Liao et al.'s protocol [19]	He et al.'s protocol [11]	Shen et al.'s protocol [23]	Our protocol
The server	$3T_M$	$5T_M + 3T_A$	$3T_M + 8T_A$	$3T_M + 3T_H$	$T_M + 3T_H$
The tag	$3T_M$	$5T_M + 3T_A$	$3T_M + 8T_A$	$3T_M + T_A + 4T_H$	$T_M + 4T_H$
Total	$6T_M$	$10T_M + 6T_A$	$6T_M + 16T_A$	$6T_M + T_A + 7T_H$	$2T_M + 7T_H$

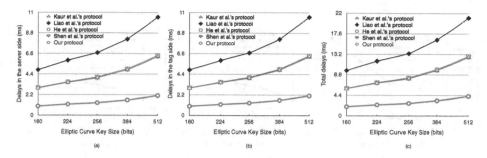

Fig. 3. Comparison of computation costs on different key sizes of ECC: (a) delays in the server side; (b) delays in the tag side; (c) total delays

As is the results shown in Table 3 and Fig. 3, our protocol only uses one time elliptic curve scalar multiplication operation and several times hash function operations for both the server and the tag. By comparing with the other four protocols, we can see that our protocol is quite lightweight, since our protocol significantly reduces the computation cost on both the server side and the tag side by at least 3 times.

6.2   Communication Cost

The communication cost of relevant ECC-based RFID authentication protocols is analyzed in this subsection. The communicational cost denotes the length of the messages transmitted between the server and the tag while executing the authentication process. We assume that the sizes of an identity, a timestamp, and a general hash function output are 32 bits, 32 bits, and 160 bits, respectively. Also, the length of an elliptic curve point is twice the key size of an elliptic curve. (For example, an elliptic curve with length of 160 bits, then the length of an elliptic curve point is 320 bits.) Based on these assumption, we analyze the communication costs on our protocol and the other four protocols. As is listed in Table 4, it shows the comparison of communication consumption results on different key sizes of ECC. According to Table 4, we also do a simulation of the communication cost on different key sizes of ECC in Fig. 4.

Table 4. Comparison of communication cost

	Kaur et al.'s protocol [14]	Liao et al.'s protocol [19]	He et al.'s protocol [11]	Shen et al.'s protocol [23]	Our protocol
The server	$192 + 2k$ bits	$4k$ bits	$4k$ bits	$160 + 4k$ bits	$32 + 32 + 160 = 224$ bits
The tag	$160 + 2k$ bits	$4k$ bits	$4k$ bits	$320 + 2k$ bits	$32 + 32 + 160 = 224$ bits
Total	$352 + 4k$ bits	$8k$ bits	$8k$ bits	$480 + 6k$ bits	448 bits

*: k denotes the key size of ECC, where $k = 160, 224, 256, 384, 512$ bits according to NIST recommended key size [2].

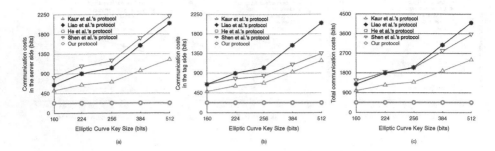

Fig. 4. Comparison of communication costs on different key sizes of ECC: (a) in the server side; (b) in the tag side; (c) total communication costs

As is the results shown in Table 4 and Fig. 4, it shows that our protocol has a lower communication cost by comparing with the other four protocols (our protocol decreases the communication cost by at least 50%). Besides, the communication cost of our protocol is not influenced by the changes on key sizes of ECC. That is, our protocol keeps a constant communication cost.

Overall, based on the aforementioned evaluation results, we can conclude that our ECC-based RFID authentication protocol achieves outstanding performance in terms of computation and communication costs in comparison with other recently proposed works.

7   Conclusion

In this paper, we firstly give an overview and cryptanalysis on Kaur et al.'s protocol [14]. Then, we propose an improved lightweight RFID authentication protocol for IoT systems based on ECC algorithm, which overcomes the deficiencies in Kaur et al.'s protocol [14]. Through the security analysis, we show that the proposed protocol achieves mutual authentication, confidentiality, anonymity as well as resistance to various attacks. Furthermore, by comparing with previous RFID authentication protocols, the performance evaluation indicates that the proposed protocol is quite efficient because of the lower computation and communication costs. In our future work, we will try to enhance the security level by using the formal security proof and provide more detailed performance analysis.

Acknowledgement. This work is partial supported by Australian Research Council Discovery Project (DP160100913: Security and Privacy of Individual Data Used to Extract Public Information), Data61 Research Collaborative Project (Enhancing Security and Privacy in IoT), the Distinguished Young Scholars Fund of Fujian, China (2016J06013) and Fujian Provincial Department of Education Project, China (JOPX15066).

References

1. Atzori, L., Iera, A., Morabito, G.: The Internet of Things: a survey. Comput. Netw. **54**(15), 2787–2805 (2010)
2. BlueKrypt: Nist key length recommended (2016). https://www.keylength.com/en/4/
3. Bringer, J., Chabanne, H., Icart, T.: Cryptanalysis of EC-RAC, a RFID identification protocol. In: Franklin, M.K., Hui, L.C.K., Wong, D.S. (eds.) CANS 2008. LNCS, vol. 5339, pp. 149–161. Springer, Heidelberg (2008). https://doi.org/10.1007/978-3-540-89641-8_11
4. Burmester, M., De Medeiros, B., Motta, R.: Robust, anonymous RFID authentication with constant key-lookup. In: Proceedings of the 2008 ACM Symposium on Information, Computer and Communications Security, pp. 283–291. ACM (2008)
5. Chien, H.Y., Chen, C.H.: Mutual authentication protocol for RFID conforming to EPC class 1 generation 2 standards. Comput. Stand. Interfaces **29**(2), 254–259 (2007)
6. Cho, J.S., Yeo, S.S., Kim, S.K.: Securing against brute-force attack: a hash-based RFID mutual authentication protocol using a secret value. Comput. Commun. **34**(3), 391–397 (2011)
7. van Deursen, T., Radomirovic, S.: Untraceable RFID protocols are not trivially composable: attacks on the revision of ec-rac. IACR Cryptology ePrint Archive 2009, 332 (2009)
8. Feldhofer, M., Dominikus, S., Wolkerstorfer, J.: Strong authentication for RFID systems using the AES algorithm. In: Joye, M., Quisquater, J.-J. (eds.) CHES 2004. LNCS, vol. 3156, pp. 357–370. Springer, Heidelberg (2004). https://doi.org/10.1007/978-3-540-28632-5_26
9. Gaubatz, G., Kaps, J.P., Ozturk, E., Sunar, B.: State of the art in ultra-low power public key cryptography for wireless sensor networks. In: Third IEEE International Conference on Pervasive Computing and Communications Workshops, PerCom 2005 Workshops, pp. 146–150. IEEE (2005)
10. Gope, P., Hwang, T.: A realistic lightweight authentication protocol preserving strong anonymity for securing RFID system. Comput. Secur. **55**, 271–280 (2015)
11. He, D., Kumar, N., Chilamkurti, N., Lee, J.H.: Lightweight ECC based RFID authentication integrated with an ID verifier transfer protocol. J. Med. Syst. **38**(10), 116 (2014)
12. Juels, A.: RFID security and privacy: a research survey. IEEE J. Sel. Areas Commun. **24**(2), 381–394 (2006)
13. Juels, A., Molnar, D., Wagner, D.: Security and privacy issues in e-passports. In: First International Conference on Security and Privacy for Emerging Areas in Communications Networks, SecureComm 2005, pp. 74–88. IEEE (2005)
14. Kaur, K., Kumar, N., Singh, M., Obaidat, M.S.: Lightweight authentication protocol for RFID-enabled systems based on ECC. In: 2016 IEEE Global Communications Conference (GLOBECOM), pp. 1–6. IEEE (2016)
15. Kaya, S.V., Savaş, E., Levi, A., Erçetin, Ö.: Public key cryptography based privacy preserving multi-context RFID infrastructure. Ad Hoc Netw. **7**(1), 136–152 (2009)
16. Lee, I., Lee, K.: The Internet of Things (IoT): applications, investments, and challenges for enterprises. Bus. Horiz. **58**(4), 431–440 (2015)
17. Lee, Y.K., Batina, L., Verbauwhede, I.: EC-RAC (ECDLP based randomized access control): provably secure RFID authentication protocol. In: 2008 IEEE International Conference on RFID, pp. 97–104. IEEE (2008)

18. Lee, Y.K., Batina, L., Verbauwhede, I.: Untraceable RFID authentication protocols: revision of EC-RAC. In: 2009 IEEE International Conference on RFID, pp. 178–185. IEEE (2009)
19. Liao, Y.P., Hsiao, C.M.: A secure ECC-based RFID authentication scheme integrated with ID-verifier transfer protocol. Ad Hoc Netw. **18**, 133–146 (2014)
20. Lv, C., Li, H., Ma, J., Zhang, Y.: Vulnerability analysis of elliptic curve cryptography-based RFID authentication protocols. Trans. Emerg. Telecommun. Technol. **23**(7), 618–624 (2012)
21. Peris-Lopez, P., Hernandez-Castro, J.C., Estevez-Tapiador, J.M., Ribagorda, A.: EMAP: an efficient mutual-authentication protocol for low-cost RFID tags. In: Meersman, R., Tari, Z., Herrero, P. (eds.) OTM 2006. LNCS, vol. 4277, pp. 352–361. Springer, Heidelberg (2006). https://doi.org/10.1007/11915034_59
22. Peris-Lopez, P., Hernandez-Castro, J.C., Estévez-Tapiador, J.M., Ribagorda, A.: LMAP: a real lightweight mutual authentication protocol for low-cost RFID tags. In: Proceedings of 2nd Workshop on RFID Security, p. 06 (2006)
23. Shen, H., Shen, J., Khan, M.K., Lee, J.H.: Efficient RFID authentication using elliptic curve cryptography for the Internet of Things. Wirel. Pers. Commun. **96**(4), 5253–5266 (2017)
24. Tan, L., Wang, N.: Future internet: the Internet of Things. In: 2010 3rd International Conference on Advanced Computer Theory and Engineering (ICACTE), vol. 5, p. V5-376. IEEE (2010)
25. Xu, L.D., He, W., Li, S.: Internet of Things in industries: a survey. IEEE Trans. Ind. Inform. **10**(4), 2233–2243 (2014)
26. Yeh, T.C., Wang, Y.J., Kuo, T.C., Wang, S.S.: Securing RFID systems conforming to EPC class 1 generation 2 standard. Expert. Syst. Appl. **37**(12), 7678–7683 (2010)

Dynamic Transitions of States
for Context-Sensitive Access
Control Decision

A. S. M. Kayes[1]([⊠]), Wenny Rahayu[1], Tharam Dillon[1], Syed Mahbub[1],
Eric Pardede[1], and Elizabeth Chang[2]

[1] La Trobe University, Melbourne, Australia
{a.kayes,w.rahayu,t.dillon,s.mahbub,e.pardede}@latrobe.edu.au
[2] University of New South Wales, Canberra, Australia
e.chang@adfa.edu.au

Abstract. Due to the proliferation of data and services in everyday life, we face challenges to ascertain all the necessary contexts and associated contextual conditions and enable applications to utilize relevant information about the contexts. The ability to control context-sensitive access to data resources has become ever more important as the form of the data varies and evolves rapidly, particularly with the development of smart Internet of Things (IoTs). This frequently results in dynamically evolving contexts. An effective way of addressing these issues is to model the dynamically changing nature of the contextual conditions and the transitions between these different dynamically evolving contexts. These contexts can be considered as different states and the transitions represented as state transitions. In this paper, we present a new framework for context-sensitive access control, to represent the dynamic changes to the contexts in real time. We introduce a state transition mechanism to model context changes that lead the transitions from initial states to target states. The mechanism is used to decide whether an access control decision is granted or denied according to the associated contextual conditions and controls data access accordingly. We introduce a Petri net model to specify the control flows for the transitions of states according to the contextual changes. A software prototype has been implemented employing our Petri net model for detection of such changes and making access control decisions accordingly. The advantages of our context-sensitive access control framework along with a Petri net model have been evaluated through two sets of experiments, especially by looking for re-evaluation of access control decisions when context changes. The experimental results show that having a state transition mechanism alongside the context-sensitive access control increases the efficiency of decision making capabilities compared to earlier approaches.

Keywords: Context-sensitive access control
Dynamic changes to the contextual conditions · States
Transitions of states · Petri net model

© Springer Nature Switzerland AG 2018
H. Hacid et al. (Eds.): WISE 2018, LNCS 11233, pp. 127–142, 2018.
https://doi.org/10.1007/978-3-030-02922-7_9

1 Introduction

Access control is a cornerstone of the treatment of security of stored data that has been widely investigated in today's dynamic environments [1]. Among the available access control mechanisms in the literature, the traditional Role-Based Access Control (RBAC) [2,3] and Attribute-Based Access Control (ABAC) [4] solutions are two representative and reliable security models for many practical applications to safeguard data and information resources. Due to the flexibility in administration when dealing with large number of users in connection with the embodiments of the user-role and role-permission associations, the traditional RBAC model [2] and the spatial and temporal RBAC models [5,6] have been widely accepted by security practitioners and scientific communities. On the other hand, the ABAC models [4] differ from the RBAC models, replacing the roles and other relevant authorities by a set of attributes and grants users' accesses to data based on the relevant attributes that are possessed by the users. However, these traditional role-based and attribute-based access control models are not adequate to incorporate the dynamic contexts into the policies.

The computing technologies have been changed over the last several years and this has created the need for the solution to the problem of controlling data access in today's dynamic environments. Many organizations nowadays have been seeking appropriate context-sensitive access control mechanisms for utilizing data resources. For example, a nurse can access a patient's daily medical records when they both are co-located in the general ward of the hospital (which is a positive policy). On the other hand, a negative policy states that a nurse cannot perform a given action (e.g., read or write) to a patient's records from the outside of the hospital. The existing context-sensitive access control approaches [7–9] have used such positive and negative policies [10] to grant access to data according to the contexts. In recent times, we have been moving towards the Internet of Things (IoTs) and the number of IoT sensors are growing at a rapid rate. When different types of IoT conditions which characterize different contexts are collected from these enormous numbers of sensors, which continuously generate a massive amount of data, the traditional context-sensitive access control mechanisms [11,12] (i.e., manually specifying the full set of policies) become infeasible. For example, a doctor should have access to a patient's emergency medical records to save his life from a critical heart attack while he is on the move from his office to the emergency ward.

How can we be sure that the access control decisions can be re-evaluated when there are dynamic changes to the relevant contexts? It is often easy to manually check the relevant contextual conditions that are associated with the applicable access control policies. However, building a required context-sensitive access control system for a dynamic environment is too complicated for the large number of policies. On the one hand, it is really a big challenge to specify the full set of positive and negative policies, due to the presence of the dynamic contextual conditions. On the other hand, there is a need to deal with the issue of dynamicity of contexts in real-time when context changes (e.g., making an access control decision while on the move from general ward to ICU).

In this research, our aim is to introduce a state transition model to evaluate context-sensitive access control decisions when there are dynamic changes to the contextual conditions. The significant contributions are listed as follows. We first introduce the formalization of the state transition model in the treatment of context-aware access control (CAAC) issues. In particular, the state transition mechanism facilitates access control decision making when the context changes. Using our state transition mechanism, we build a Petri net model in a way that specifies the control flows for the transition of states according to the context changes. The fine-grained access control decisions along with a Petri net model is one of the main contributions in this paper. We implement an android studio-based software prototype along with a colored Petri net for detecting contextual changes and making required access control decisions accordingly. Through two sets of experiments, we evaluate access control decisions including the dynamic changes to the contextual conditions. The experimental results demonstrate the efficiency of decision making capabilities of our proposed approach compared to relevant earlier access control solutions.

The organization of our paper is as follows. Section 2 presents the motivation and hypotheses of this study. Section 3 discusses the formalization and methodology of our proposed state transition model. Section 4 examines the state transition mechanism using Petri net, including a prototype testing. Section 5 evaluates our proposed approach with respect to a relevant earlier approach. Section 6 briefly discusses the related work. Finally, Sect. 7 concludes the paper and outlines future challenges.

2   Research Motivation

In this section, we present a roadside emergency assistance scenario to motivate our work.

– *The scenario begins with patient James who has experienced a heart attack while driving in a remote area. John, who is a paramedic, has been called to investigate the patient's condition and give him necessary treatments to reduce the risk of a heart attack. John visits the roadside heart attack spot with an ambulance that consists of a body sensor network supporting a wide variety of IoT devices and monitoring applications. He needs to access James' regular medications and all of his previous medical history to treat him properly and save his life from such an emergency heart attack situation.*

The contextual conditions such as locations, request times, relationships and situations are involved in this roadside assistance scenario and consequently the access control decisions are based on the dynamically changing values of such contexts. The relevant user and resource-centric data should be used to extract relevant information about the contexts. In particular, these contexts can be extracted from people entering data like user profiles and from measurement data like IoT data. However, collecting and analyzing IoT contexts from all the sources or sensors are not feasible as there are millions of sensors connected to the Internet. In addition, an access control mechanism or application should have an awareness of relevant information about such contexts and also about dynamic changes to the contexts. For example, John, who is a paramedic and has necessary medical training, should be allowed to access a patient's regular medication and provide necessary pre-hospital treatments to the patient on roadside environments. The relevant contextual conditions, such as John and James are co-located, need to be checked and satisfied when making an access control decision. However, someone on the roadside is not permitted to provide the required treatments to the patients without necessary medical training.

In general, we need to consider the necessary access control policies with the roles of the person and the relevant contextual conditions as policy constraints. On the one hand, we need to specify all the positive and negative policies [10]. For a given access control request, if there is no relevant policy, the default access control decision is usually taken as *"denied the access"*. In absence of such negative policies, the default "denied" decisions may lead to an adverse effect like policy rule conflicts [13]. For example, John, who is normal paramedic, should be permitted to access James' medical history and provide him emergency treatments when there is a potential life-threatening situation. However, the associated contextual conditions need to be assured for granting access to the medical history of James, such as a "life-threatening" situation has occurred. As such, we need to specify corresponding context-sensitive access control policies, including all the possible contextual conditions. In addition, we need to consider the dynamically changing nature of such contextual conditions and consequently re-evaluate the access control decisions when there are dynamic changes to such contexts. For example, when the context changes (e.g., James' situation becomes normal), John should not be allowed to access James' medical history and provide him emergency treatments as roadside assistance.

In the light of the above hypothesis, we can use the basic state transition mechanism [14] to model all the possible "granted" and "denied" states, including all the values of the contextual conditions.

3   Formalization of Our State Transition Mechanism

In this section, we discuss the formalization of state transition mechanism in building our context-sensitive access control framework.

In the literature, the term "context" has been defined by many researchers and the entity-centric concept of context, that has been claimed by Dey [15] provides a general characterization of context in pervasive (context-aware) computing environments. However, Dey's and other earlier definitions of context are not enough to cover dynamic changes to the contexts. Thus, we define the following definition of context that can be used to specify the access control-specific entities and to identify the relevant contextual conditions (i.e., how an access control decision can be made by satisfying a relevant context.).

Definition 1 *(Definition of Context). Context description or simply context is the set of dynamic contextual conditions that are used to make a particular access* *control decision. These conditions are used to characterize the state of the access control-specific entities. Let us consider "c_{acd}" is a context that is composed of a set of contextual conditions "$cc_1, cc_2, cc_3, ..., cc_i$" , then we can specify the following expression.*

$$c_{acd} = \{cc_1, cc_2, cc_3, ..., cc_i \mid c_{acd} \in C_{ACD}\} \tag{1}$$

Example 1. *Let us consider a positive policy from our scenario, John, who is a paramedic, is granted to access necessary medical records (MR) of James to save his life from critical heart attack when they are co-located at the scene of the accident. The context (c_{acd_1}) that is associated with this positive policy* < **user(John), access(granted), data(MR)** > *can be specified as follows.*

$$cc_1 = healthStatus(Patient) = \text{``critical''}$$
$$cc_2 = isColocatedWith(Paramedic, Patient) = \text{``yes''} \tag{2}$$
$$c_{acd_1} \triangleq cc_1 \wedge cc_2$$

Example 2. *Let us consider a negative authorization policy from the same application scenario, John, who is a paramedic, is not permitted to access James' medical history (MH) from anywhere of the roadside location where they are not co-located. The context (c_{acd_2}) that is associated with this negative policy* < **user(John), access(denied), data(MH)** > *can be specified as follows.*

$$c_{acd_2} \triangleq cc_2 \tag{3}$$

The same co-located contextual condition (cc_2) has been associated with both the contexts of c_{acd_1} and c_{acd_2}. There are two different types of contextual conditions, sensed and inferred contextual conditions.

Definition 2 *(Definition of Contextual Conditions). A sensed contextual condition is captured independently from context sources and an inferred contextual condition is derived from available conditions.*

Example 3. *Let us say the location co-ordinates or access request times can be captured directly without using any other conditions. However, the current health status of a patient can be derived from the body sensor network data. Considering an earlier research on fuzzy context information system [9], a patient's current health status is "66% normal with criticality level 34%" that is derived based on the raw contextual facts (e.g., pulse rate).*

3.1   Analysis of Dynamic Changes to the Contexts

Dynamic change to the context is typically driven by a distance measure between the new context and the old context. When there are dynamic changes to any of the contextual conditions that are associated with a context, the distance value indicates how similar the new and old contexts are.

Definition 3 *(Definition of Distance Function). The distance function is defined by the pairwise distances between the contextual conditions that are associated with the new and old contexts. This function is expressed by Eq. (4).*

$$distance(c_{acd}^{new}, c_{acd}^{old}) \triangleq distance(cc_i^{new}, cc_i^{old}) \qquad (4)$$

In the above expression, $distance(c_{acd}^{new}, c_{acd}^{old})$ is the distance between new and old contexts, $distance(cc_i^{new}, cc_i^{old})$ is the pairwise distances between new and old contextual conditions, c^{new} is a new context, c^{old} is an old context, cc_i^{new} is the new contextual condition and cc_i^{old} is an old contextual condition.

The absolute values of the pairwise distances between new and old contextual conditions is used to measure the distance between the new and old contexts. The pairwise distance between the new and old contextual condition is defined as follows (see the Eq. 5), where $w_{cc_i}^{new}$ is the weight of the new contextual condition and $w_{cc_i}^{old}$ is the weight of the old contextual condition.

$$distance(cc_i^{new}, cc_i^{old}) \triangleq |w_{cc_i}^{new} - w_{cc_i}^{old}| \qquad (5)$$

We consider the corresponding distance vectors W_{cc_i} of the weights of the new $(w_{cc_i}^{new})$ and old $(w_{cc_i}^{old})$ contextual conditions. The vectors W_{cc_i} are designed to maximize dynamic range of the distances between the contextual conditions.

$$|w_{cc_i}^{new} - w_{cc_i}^{old}| \triangleq \begin{pmatrix} 0 \\ 1 \\ 0 <> 1 \end{pmatrix} \qquad (6)$$

In the above equation, the pairwise distance between the new and old contextual conditions is measured based on the range from 0 to 1.

Example 4. *Based on the policy specified in Example 1, the contextual conditions "healthStatus (cc_1)" and "co-located (cc_2)" that are associated with the context. The distance between the new and old contexts is "0", for the first time when the access request is originated from the user and there are no dynamic changes to the contextual conditions at that particular situation (i.e., both the pairwise distances between the new and old contextual conditions are "0"). Based on the Eqs. (4), (5) and (6), we can write the following expressions.*

$$distance(c_{acd_1}^{new}, c_{acd_1}^{old}) \triangleq |w_{cc_i}^{new} - w_{cc_i}^{old}|$$
$$|w_{cc_1}^{new} - w_{cc_1}^{old}| \triangleq 0 \qquad (7)$$
$$|w_{cc_2}^{new} - w_{cc_2}^{old}| \triangleq 0$$

3.2   Definitions of States and Analysis of the Transitions of States

We consider all the relevant access control decisions and their associated contexts to define the possible states. We represent the transition of states in terms of contexts involved in making access control decisions and also when there are dynamic changes to such contexts.

Definition 4 *(Definition of State). A state is composed of the role, data, resource, context and decision (e.g., "granted" decision). It can be formally described using the following four-tuple notation.*

$$state \triangleq < role, data, context, decision > \tag{8}$$

According to the decision values, we categorize granted, denied and intermediary states and we define these three states as follows.

Definition 5 *(Definition of Granted, Denied and Intermediary States). A granted state means a user who can play the required role can have the complete access to the data, by satisfying the associated context. A denied state means a user cannot have any access to associated data. An intermediary state means a user by playing a role and satisfying the context can have partial access to data.*

The transition of states can be determined using the possible states with the corresponding distance values of the associated contextual conditions. In particular, a state (that we call an initial state) changes to the next state based on such a distance value. The transition of states can be formalized as follows.

$$State(s_t) \xrightarrow{d} State(s_{t+1})$$
$$State(s_t) \xrightarrow{d} State(s_{t-1}) \tag{9}$$

In the above expression, "s_t" is an initial state, "d" is the distance variable according to the Eq. (6) and "s_{t+1}" or "s_{t-1}" is the next state.

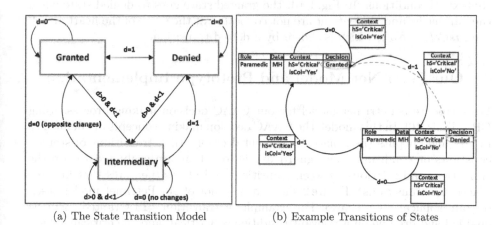

(a) The State Transition Model   (b) Example Transitions of States

Fig. 1. Proposed model for transitions of states

The state transitions diagram based on the all three basic states (i.e., granted, denied and intermediary) is illustrated in Fig. 1(a), where rectangles represent states, arrows connecting states represent transitions and arc labels represent transition conditions. In general, the transition of states occurs as follows, depending on the values of the distance variable "d".

(i) For "*d is 1*" when there are dynamic changes to the contextual conditions, the granted state goes to denied, the denied state goes to granted or an intermediary state goes to denied state.

(ii) For "*d is 0*" when there are no changes, the granted, denied or intermediary state goes to granted, denied or intermediary state respectively.

(iii) For "*d is 0<> 1*" (i.e., d is greater than 0 but less than 1), the granted, denied or intermediary state goes to intermediary state.

(iv) For "*d is 0*" when there are opposite changes, an intermediary state goes to granted state.

Based on the above-specified conditions in Steps (i) to (iv), we can formalize the following expressions.

$$
\begin{aligned}
s_{granted} &\xrightarrow{d=0} s_{granted} \xrightarrow{d=1} s_{denied} \xrightarrow{d>0\,\&\,d<1} s_{inter} \\
s_{denied} &\xrightarrow{d=0} s_{denied} \xrightarrow{d=1} s_{granted} \xrightarrow{d>0\,\&\,d<1} s_{inter} \\
s_{inter} &\xrightarrow{d=0\ (no\ changes)} s_{inter} \xrightarrow{d=0\ (same\ opposite\ changes)} s_{granted} \\
&\quad s_{inter} \xrightarrow{d>0\,\&\,d<1} s_{inter} \xrightarrow{d=1} s_{denied}
\end{aligned}
\tag{10}
$$

Considering our application scenario and based on the positive and negative authorization policies specified in Examples (1) and (2), Fig. 1(b) shows several specific transitions between "*Granted*" and "*Denied*" states and their corresponding distance values, according to the dynamic changes to the associated contextual conditions. In Fig. 1(b), the granted state goes to denied state when the paramedic and the patient are not co-located at the scene of the heart attack (i.e., *isCol = 'No'*), which is shown by a dotted transition.

4   Our Petri Net Model and Prototype Implementation

We introduce a Petri net model for our CAAC decision making process, using CPN Tools [16]. In this model, the CAAC decision making concepts are defined as inputs and outputs (i.e., places), different decisions (i.e., transitions of states), constants or conditions (i.e., guard expressions that are evaluated to fire the transition) and relations between transitions and places (i.e., arcs that link the places and transitions). Figure 2 shows a snapshot of our Petri net model based on our application scenario. For example, a transition "T1:Granted" will be enabled and fired when a relevant condition is 'Critical' and the criticality level 50% to 75% (which is a guard condition). In particular, a transition is fired when (i) the relevant tokens in the input places are equal to the weights of the arcs and these tokens are traveled from the places to the relevant transitions and (ii) the transition conditions according to the guard expressions are satisfied. After firing the transitions, the tokens are distributed to the corresponding output places. In a relevant earlier work, we have introduced a fuzzy context information system to derive fuzzy contextual conditions [9]. In this research, we adapt this fuzzy context model to implement and measure the distance vector for health status

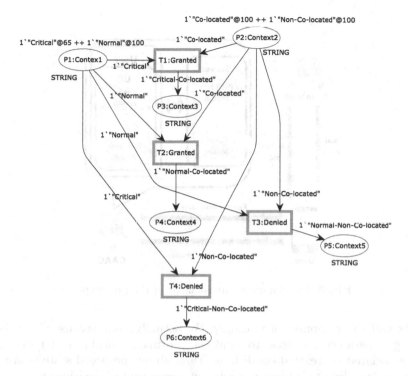

Fig. 2. A snapshot of our petri net model

context. For example, in Fig. 2, we use a temporal condition "Critical@65", which refers to a patient's current health status is "65% Critical". When a patient's current health status is transformed from a 100% normal state to 65% critical state, we measure the distance value (i.e., d) is 0.65. This Petri net model is linked to a context-sensitive access controller to make possible decisions for the users to access the requested data resources.

We have developed our prototype on the Java platform along with Android Studio IDE [17] and other widely supported open source tools. Figure 3 gives a pictorial overview of the tools and technologies that are used to implement different software components of our prototype. We have used SQLite relational database [18] to implement a data store of the patients' medical health records, including their previous health history. We have used different XML-based technologies to build the mobile interfaces for our prototype, including the Petri net markup language for our Petri net model [19]. Other than the health status context, we also consider different locations (e.g., the current locations of the patient and paramedic) as contextual conditions in our prototype. We have stored different location coordinates in the relational database and simulated such locations in our laboratory setup. The development environment of our mobile prototype (see Fig. 3) has mainly client-side (e.g., user interfaces) and server-side (e.g., Petri-net, policies) parts. We have developed a mobile application and used our implemented user interfaces to deal with different users'

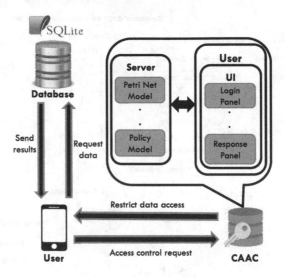

Fig. 3. Development environment of the prototype

requests and the responses accordingly. We actually limit the users' access to data (e.g., a paramedic's access to a patient's normal medical records) according to the associated contextual conditions. Through our proposed state transition mechanism, we also re-evaluate the relevant access control decisions when there are dynamic changes to the contextual conditions.

Figure 4 demonstrates the relevant access control decisions for a paramedic John's request, including the results of a patient James' different medical records.

Fig. 4. John's access request for James' health records

The left part of the Fig. 4 shows John is allowed to access James' health records by satisfying the following contextual conditions: when John is co-located with James at the scene of the accident and his health status is critical. Another access request result is shown in the right part of the same Fig. 4, where John has only limited access to James' medical records as his health status becomes normal. Overall, we have tested our proposed context-aware access control (CAAC) approach with regards to context changes and the prototype implementation can provide an infrastructure support for the practitioners to build relevant CAAC applications in today's dynamic environments.

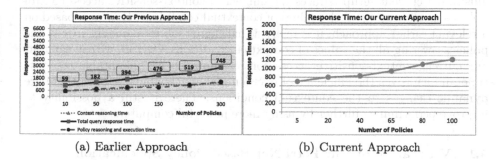

(a) Earlier Approach   (b) Current Approach

Fig. 5. Performance with respect to number of policies

5   Experimental Evaluation and Verification

We conduct two sets of experiments to evaluate our proposed approach compared to a relevant earlier approach [7]. In particular, we evaluate the access control decisions when there are dynamic changes to the contextual conditions.

In our first set of experiments, we measure the query response time (i.e., performance overheads) with respect to different number of CAAC policies along with relevant contextual conditions. We specify the separate access control policies for the different values of the relevant contextual conditions. As such, according to a relevant earlier approach [7], we have a larger size of ontology knowledge-base, including role, data resource, context and policy ontologies. The experimental results are illustrated in Fig. 5(a). In this set of experiments, we can see that the query response time is linearly increasing and for the 300 policies it measures 2.9 s (and at that point the ontology size is 748 kilobytes). For any large policy-base, what we see that the applied approach is very expensive. This is due to the large number of policies and the complex reasoning task (context and policy reasoning) behind the data access query.

In our second set of experiments, we use our current approach to measure the query response time. In particular, we quantify the performance with respect to different number of context-sensitive access control policies in conjunction with our state transition mechanism (i.e., Petri net model). In this current setup, we don't have a larger size of the policy-base, no complex reasoning task for policy selection and reasoning is involved. Particularly, this is due to using a state transition model with Petri net. The evaluation results are illustrated in Fig. 5(b), where we can see that the query response time measures 1.2 s with respect to 100 policies. This size of the policy-base has covered the 300 policies using an earlier approach. Having a state transition mechanism, the performance can be improved using our current approach and we can achieve fine-grained access control decisions, detecting dynamic changes to the contexts.

In the above experiments that were conducted, our proposed access control approach can detect context changes and consequently facilitates the fine-grained access control decisions when context changes. Having a state transition mechanism, our current setup reduces the number of context-sensitive access control policies incorporating all the values of contextual conditions. However, based on a relevant earlier approach, it is really a challenging job to manually specify the policies covering all the values of contextual conditions. Overall, we can say that our current context-sensitive access control approach has acceptable response time and improves the efficiency of decision making capabilities when there are dynamic changes to the contextual conditions. There is still a possibility to improve performance further by using more powerful computers.

5.1   Verification of Our Petri Net-Based Policy Specification

In this section, we verify the Petri net model that is associated with our context-sensitive access controller by extracting necessary contextual conditions for its correctness. In particular, we verify the correctness of the model through the execution of the Petri net and firing each of the transition. In order to check the correctness of the specification, we apply the top-down approach that starts from the first level of the Petri net model and advances level by level. We know that if the model objects (i.e., places in the Petri net model) are not suitable or available, the relevant transitions cannot be fired. Conversely, if the objects are available and suitable, then the relevant transitions can be enabled for firing.

Figure 6 shows our initial simulation results in which one rule is fired based on the two tokens ("Normal" and "Co-located") and the transition "T2:Granted" is enabled accordingly. We then extract the relevant contextual conditions, changes these conditions values and passes such values as tokens. Then, we check all the transitions again in our Petri net model as well as the relevant status. The transition "T4:Denied" can be enabled later when there are dynamic changes to the contextual conditions, i.e., when John and James are not co-located with each other and James' health condition is critical. Based on our simulation results in Fig. 6, the transition "T4:Denied" is fired later when other two tokens ("Critical" and "Non-Co-located") are traveled from P1 and P2 places to T4, i.e., when there are dynamic changes to the contextual conditions.

Overall, we have checked the status of the relevant transitions (enabled or disabled) when their are the dynamic changes to the contextual conditions. We have changed some contextual conditions through passing tokens and checked the transitions accordingly to verify the correctness of the specification. Other than

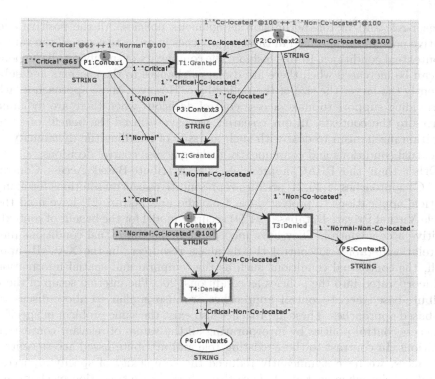

Fig. 6. Simulation results using proposed petri net model

the above-mentioned correctness approach, we have used CPN Tools [16]) to verify the correctness of our Petri net model and specification.

6   Related Work

In this section, we include a brief review of relevant access control solutions, including the context-sensitive role-based and attribute-based approaches. In particular, we briefly analyze the contributions of our proposed solution compared to existing state-of-the-art context-sensitive access control solutions. Our analysis focuses mainly on the key aspect of our proposed approach, that is, to deal with the dynamic changes to the contextual conditions and consequently re-evaluates access control decisions.

The traditional Role-Based Access Control (RBAC) approach [2] has considered the users' identities and roles as conditions to make access control decisions. The temporal and spatial RBAC approaches [5, 6] have considered further conditions like temporal and spatial information for making access control decisions. These approaches are not truly context-aware and do not provide adequate functionalities to integrate dynamic contexts into the access control policies.

On the other hand, the existing Context-Aware role-based Access Control (CAAC) approaches [7–9, 11, 12] incorporate different contextual conditions into the user-role and/or role-permission assignments policies. All the positive and negative policies are specified in these approaches in order to cover all the values of contextual conditions. In this fashion, the complexity of specifying all the policies can be increased when there are dynamic changes to the contexts and subsequently the larger the size of the policy-base. However, these approaches lack in providing support to make access control decisions when there are dynamic changes to the contexts. In our research, we have utilized the benefit of state transition mechanism to deal with such an issue of handling the dynamicity of contextual conditions and consequently making access control decisions.

Other than the RBAC approaches, the Attribute-Based Access Control (ABAC) approaches [4,20,21] are also representative security solutions for many practical applications to protect data. Colombo and Ferrari [21] have used the Oracle Virtual Private Database (VPD) security model for the benefit of context-sensitive access control. They have incorporated the contextual conditions into the relational tables to control the database resources. In the XACML approach, the contextual conditions such as the temporal and spatial information are incorporated into the policies as conditions [20]. The internal setup of these attribute-based access control approaches are also similar to above-discussed role-based approaches. These approaches also exist the same problem in specifying access control policies by incorporating all the values of relevant contextual conditions. In contrast to these existing role and attribute-based access control approaches, we have significantly reduced the complexity of specifying access control policies when context changes through the state transition mechanism.

Overall, the computing technologies have been changed over the last few years and we need a flexible policy-based solution to access required data resources in today's dynamic environments. As such, there is a need to consider the dynamically changing contextual conditions into the access control policies. The existing access control solutions are not adequate to deal with such dynamicity of contexts. However, based on the dynamic nature of the context information, it is really an important issue to detect dynamic changes to the contextual conditions and to make access control decisions accordingly. Having a state transition mechanism, our proposed context-sensitive access control solution is able to detect dynamic changes to the contextual conditions and consequently re-evaluates access control decisions when context changes.

7   Conclusion and Future Research Directions

We have addressed an important research issue with regards to accessing data when there are dynamic changes to the contexts. The existing context-sensitive access control solutions can lead to a much larger policy-base when incorporating the dynamic values of such contextual conditions into the policies. We have introduced a new context-aware access control (CAAC) solution along with the state transition mechanism to deal with the dynamicity of contexts. First, we have presented a formal model for specifying the state transition mechanism in building our CAAC approach. Then, we have implemented a mobile-based prototype including a colored Petri net model for detecting dynamic changes to the contexts and making access control decisions for the users to access required data resources. We have demonstrated the applicability of our proposed access control approach by evaluating the performance and the correctness of the specification. The experimental results show that our approach along with the state transition model has better performance than a relevant earlier approach.

In this paper, we have considered several contextual conditions and built corresponding distance vectors based on the dynamic changes to such conditions. However, while it is beyond the scope of this paper, it may require special modeling to build such distances between initial and current contextual conditions, which are domain dependent, and thus, further investigation to effectively model such distance vectors is required in the future. We have used the CPN tools to verify the correctness of the specification of our state transition mechanism. In this aspect, further investigation is also required to justify the feasibility of the underlying formalization of the specification. The main purposes of the specification of such state transition mechanism are to (i) identify the dynamic changes to the contextual conditions at runtime and (ii) relate such dynamic conditions to the applicable policies in such a manner that the context-specific access control decisions can be evaluated.

References

1. Weiser, M.: Some computer science issues in ubiquitous computing. Commun. ACM **36**(7), 75–84 (1993)
2. Sandhu, R.S., Coyne, E.J., Feinstein, H.L., Youman, C.E.: Role-based access control models. IEEE. Computer **29**, 38–47 (1996)
3. Wang, H., Cao, J., Zhang, Y.: A flexible payment scheme and its role-based access control. IEEE TKDE **17**(3), 425–436 (2005)
4. Servos, D., Osborn, S.L.: Current research and open problems in attribute-based access control. ACM Comput. Surv. **49**(4), 65:1–65:45 (2017)
5. Joshi, J.B., Bertino, E., Latif, U., Ghafoor, A.: A generalized temporal role-based access control model. IEEE TKDE **17**(1), 4–23 (2005)
6. Damiani, M.L., Bertino, E., Catania, B., Perlasca, P.: GEO-RBAC: a spatially aware RBAC. ACM TISSEC **10**(1), 2 (2007)
7. Kayes, A.S.M., Han, J., Colman, A.: OntCAAC: an ontology-based approach to context-aware access control for software services. Comput. J. **58**(11), 3000–3034 (2015)
8. Hosseinzadeh, S., Virtanen, S., Rodríguez, N.D., Lilius, J.: A semantic security framework and context-aware role-based access control ontology for smart spaces. In: SBD@SIGMOD, pp. 1–6 (2016)
9. Kayes, A., Rahayu, W., Dillon, T., Chang, E., Han, J.: Context-aware access control with imprecise context characterization through a combined fuzzy logic and ontology-based approach. In: CoopIS 2017, vol. 10573. LNCS, pp. 132–153. Springer, Cham (2017). https://doi.org/10.1007/978-3-319-69462-7_10
10. Damianou, N., Dulay, N., Lupu, E., Sloman, M.: The ponder policy specification language. In: Sloman, M., Lupu, E.C., Lobo, J. (eds.) POLICY 2001. LNCS, vol. 1995, pp. 18–38. Springer, Heidelberg (2001). https://doi.org/10.1007/3-540-44569-2_2

11. Kulkarni, D., Tripathi, A.: Context-aware role-based access control in pervasive computing systems. In: SACMAT, pp. 113–122 (2008)
12. Schefer-Wenzl, S., Strembeck, M.: Modelling context-aware rbac models for mobile business processes. IJWMC **6**(5), 448–462 (2013)
13. Sloman, M.: Policy driven management for distributed systems. Journal of network and Systems Management **2**(4), 333–360 (1994)
14. Chang, E., Gautama, E., Dillon, T.S.: Extended activity diagrams for adaptive workflow modelling. In: IEEE ISORC-2001, pp. 413–419 (2001)
15. Dey, A.K.: Understanding and using context. Pers. Ubiquitous Comput. **5**(1), 4–7 (2001)
16. CPNTools: A tool for editing, simulating, and analyzing colored petri nets (2018). http://cpntools.org/
17. Android-Studio-IDE: Android studio for building apps (2018). https://developer.android.com/studio/
18. SQLite: It is a self-contained and mostly used SQL database engine in the world (2018). https://www.sqlite.org/index.html
19. PNML: The petri net markup language (PNML) is a proposal of an XML-based interchange format for petri nets (2018). http://www.pnml.org/
20. Rissanen, E.: XACML v3.0 core and hierarchical role based access control (RBAC) profile version 1.0. In: OASIS Standard (2014). http://docs.oasis-open.org/xacml/3.0/rbac/v1.0/xacml-3.0-rbac-v1.0.html
21. Colombo, P., Ferrari, E.: Towards virtual private NoSQL datastores. In: ICDE, pp. 193–204. IEEE (2016)

Social Network and Security

CoRank: A Coupled Dual Networks Approach to Trust Evaluation on Twitter

Peiyao Li[✉], Weiliang Zhao, and Jian Yang

Department of Computing, Macquarie University, Sydney, Australia
peiyao.li@students.mq.edu.au, {weiliang.zhao,jian.yang}@mq.edu.au

Abstract. The trust evaluation of information and people is crucial for maintaining an open and healthy Online Social Networks (OSN) platform for society. In this work, we develop a Coupled Dual Networks Trust Ranking (CoRank) method to measure the trustworthiness of users and tweets by analysing user/tweet behaviour on Twitter. This method goes beyond the existing solutions that use a single network to represent both users and tweets. A set of experiments have been conducted against the real data collected from Twitter. The experimental results show the effectiveness and robustness of the proposed method. We compare our solution with two baseline methods PageRank and TURank, and analyse how our approach outperforms the existing ones.

Keywords: Trust · Online social networks · Twitter

1 Introduction

Nowadays people can easily get in touch with others via various online social networks (OSN) platforms, such as Twitter, Facebook, Snapchat. These platforms allow people to communicate in various ways, typically, spreading news, posting personal thoughts and feelings, sharing photos. Today there are around 6,000 tweets, on average, generated on Twitter every second, and the number of monthly active users has reached more than 330 million worldwide in the first quarter of 2018 [2]. The rapid growth of Twitter has brought significant changes to people's life: people rely more and more on OSN such as Twitter to get news and information. As a result, trust has become a crucial issue for people to effectively obtain information and to make decisions. It is highly desirable to have mechanisms to evaluate trustworthiness of users and information, and build up a healthy social and business environment on OSNs.

In order to evaluate the trustworthiness of posts and people on OSN, we can utilise the available information around them. For instance on Twitter, for users, we can get the followers, the tweets they post, and the posts that mention the users; for tweets, we can get their retweet, replies, and contents. Intuitively we can use these information as the basics to build a network in which hopefully trust on users and information can be evaluated. Questions are raised: (1) how

© Springer Nature Switzerland AG 2018
H. Hacid et al. (Eds.): WISE 2018, LNCS 11233, pp. 145–160, 2018.
https://doi.org/10.1007/978-3-030-02922-7_10

this network shall be established; (2) how the values of the nodes and edges are determined and calculated to get the trust values of the people and tweets; (3) how the trust of people and the trust of their posts impact each other's trust value.

Trustworthiness of users and the information produced by users has attracted significant attentions in recent years [12]. Existing mechanisms of trust evaluation can be classified into two main categories: sentiment-based trust evaluation [8,10,18] and social graph-based trust evaluation [7,15]. The former focuses on evaluating the credibility of tweets by analysing their contents. This type of work generally extracts features from tweet contents and train various classifiers to identify the credible and incredible tweets. Some of the content based approaches also consider social relationships to assist in trust evaluation [18]. The social graph based methods generally focus on trust rankings of users and tweets by modeling social network structure through the interactions between/among users and tweets. The graph-based methods have also investigated contextual information in trust ranking to build a hybrid model for inferring user and tweet trustworthiness [16]. Twitter network consists of complex user and tweet relationships that makes it challenge to evaluate the trustworthiness of users and tweets as well as to discover the trustworthy links between them. The effectiveness of both types of trust evaluation is affected by the sufficiency and comprehensiveness of the Twitter network model. Therefore building up a sophisticated network model that captures the characteristics and relationships of both users and tweets behaviours is the first step towards effective trust calculation, which paves the way for developing any advanced trust evaluation mechanisms including contents and context analysis.

In this work, we are inspired by the collaborative trust concept in PageRank [14]: "*if a page was mentioned by a trustworthy or authoritative source, it is more likely to be trustworthy or authoritative.*". We use the basic assumptions: (a) a user connected with more trustworthy users and tweets is more trustworthy; (b) a tweet posted by a more trustworthy user and connected with more trustworthy tweets is more trustworthy. With these assumptions, we develop a Coupled Dual Networks Trust Ranking (CoRank) method. We construct a user network and a tweet network which are coupled with each other. The user network is built up with follow relationships. The tweet network is built up with retweet and reply. The user network and tweet network are coupled by mention and post. With this CoRank setting, we will be able to calculate the trust value of users and tweets separately and incorporate and adjust each other's trust value explicitly. We believe this separation and coupling of user network and tweet network is important for clarity and accuracy which is demonstrated by the experimental results. We also propose functions to map statistical numbers of interactions and actions to corresponding trust values to overcome the saturation effect. The trust ranking algorithm is developed to evaluate the trust values of involved users and tweets. We conduct the experiments on real-world Twitter dataset collected within the Tasmania state of Australia to validate the effectiveness of the proposed method, and compare it with the well-known method PageRank and the highly cited TURank.

The rest of the paper is organised as follows. Section 2 proposes a coupled dual networks model. Section 3.1 describes the details of the CoRank trust evaluation method. Section 4 analyses and discusses the experimental results. Section 5 reviews some related work. Section 6 concludes the paper.

2   Coupled Dual Networks Model

In order to evaluate the trustworthiness of users and tweets, we will take account of a rich set of actions and interactions on Twitter platform including *follow, post, mention, retweet,* and *reply.* In order to represent the complex relations between users and tweets, we construct a user network and a tweet network. The user network is built up with users as nodes and follow relations as edges. The tweet network is built up with tweets as nodes and interactions including retweets and replies as edges. These two networks are coupled with each other by actions including post and mention. This work will develop an approach for trust evaluation of users and tweets based on the analysis of the constructed coupled user network and tweet network.

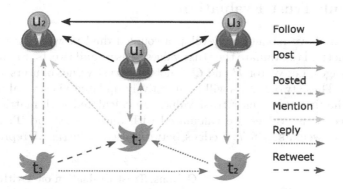

Fig. 1. A coupled user network and tweet network model.

Figure 1 illustrates the coupled dual networks model. The user network is represented as a directed graph $G_U = (U, E_u)$, where U are nodes representing users, and E_u are edges representing follow relations among users. A directed edge $e_u \langle u_i, u_j \rangle \in E_u$ represents that user u_i follows user u_j, where $u_i, u_j \in U$. A tweet network has tweets as nodes and interactions between them as edges. The interactions include *retweet* and *reply.* The tweet network is represented as a directed graph $G_T = (T, E_t)$, where T are nodes representing tweets and E_t are edges representing interactions between tweets. A directed edge $e_t \langle t_i, t_j \rangle \in E_t$ stands for tweet t_i retweets or replies to t_j, where $t_i, t_j \in T$. These two networks are coupled by *mention* and *post.* Let E_c be the set with all edges that couple these two networks. A directed edge $e_p \langle u_i, t_k \rangle$ stands for that user u_i posts a tweet t_k; a directed edge $e_{p'} \langle t_k, u_i \rangle$ stands for that a tweet t_k is posted by user u_i; a directed edge $e_m \langle t_k, u_i \rangle$ stands for that a tweet t_k mentions user u_i, where $u_i \in U, t_k \in T, \{e_p \langle u_i, t_k \rangle, e_{p'} \langle t_k, u_i \rangle, e_m \langle t_k, u_i \rangle\} \in E_c$.

It is worth noting that any user can follow, mention, retweet and reply to many trustworthy users, but the number of users he/she interacts with will not reflect how trustworthy he/she is. Therefore, we only consider the interactions in the direction from the action initiator to the ones he/she interacts with. But for the post actions, we consider the edges in both directions of user to tweet and tweet to user. This is because when a user posts a tweet, this tweet inherits the trustworthiness from the user. If this tweet is retweeted or replied by other trustworthy users, it also propagates corresponding trustworthiness back to its owner. These two edge types with reverse directions are known as *post* and *posted* edge. In the coupled user network and tweet network, the in-degree edges of each node are the incoming directed edges of this node. The trust value of this node will be affected by the trust values of nodes on the other side of these edges. The out-degree edges of each node are the outgoing directed edges of this node. The trust value of this node will affect the trust values of nodes on the other side of these edges.

3   CoRank Trust Evaluation

Based on the above model, we develop a coupled dual networks trust ranking (CoRank) method to evaluate the trust values of users and tweets. Figure 2 shows the major stages. Let vector P and Q denote the trust values of users and tweets respectively. These two vectors will be iteratively updated via several processing stages. At the first stage, user trust values are calculated with matrices U and M, and tweet trust values are calculated with matrices N and T. Matrix U represents the weights of *follow* edges between users. Matrix M represents the

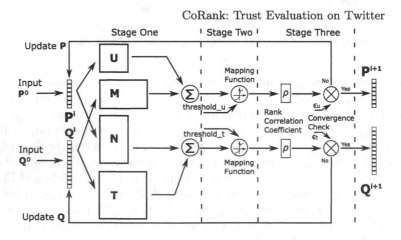

Fig. 2. Process of CoRank

weights of *mention* and *posted* edges between tweets and users. Matrix $\mathbf{N}$ represents the weights of *post* edges between users and tweets. Matrix $\mathbf{T}$ represents the weights of *retweet* and *reply* edges between tweets. Details of these matrices are described in Sect. 3.1. At stage two, with the outputs of stage one as input, mapping functions are used to transform the calculated statistical numbers to corresponding trust values of users and tweets. At stage three, the convergence is checked for completing the calculation or entering next round calculation.

3.1 Trust Values Calculation

We use $\boldsymbol{P}$ to represent the trust values of users, and $\boldsymbol{Q}$ to represent the trust values of tweets. We specify a set of matrices to quantify relations involved in the trust evaluation as follows:

- Matrix $\mathbf{U}$ with size $|U| \times |U|$ represents the trust values of users inherited from other users who follow them. If user u_i is followed by user u_j, $u_{ij} = w_f$ where w_f is the weight of the *follow* edge $e_u \langle u_j, u_i \rangle$ between the two users in the user network.
- Matrix $\mathbf{T}$ with size $|T| \times |T|$ represents trust values of tweets obtained according to retweet and reply relations. If t_i is retweeted by t_j, $t_{ij} = w_{rt}$, where w_{rt} is the weight of the *retweet* edge $e_t \langle t_j, t_i \rangle$ between t_i and t_j. If t_i is replied by t_j, $t_{ij} = w_{rp}$, where w_{rp} is the weight of the *reply* edge $e_t \langle t_j, t_i \rangle$ between t_i and t_j.
- Matrix $\mathbf{M}$ with size $|U| \times |T|$ represents trust values of users obtained according to tweets that mention them or posted by them. If u_i is mentioned by t_j, $m_{ij} = w_m$, where w_m is the weight of the *mention* edge $e_m \langle t_j, u_i \rangle$ between u_i and t_j. If u_i posts t_j, $m_{ij} = w_p$, where w_p is the weight of the *posted* edge $e_{p'} \langle t_j, u_i \rangle$ between u_i and t_j.
- Matrix $\mathbf{N}$ with size $|T| \times |U|$ represents trust values of tweets obtained according to users who post them. If t_i is posted by u_j, $n_{ij} = w_u$, where w_u is the weight of the *post* edge $e_p \langle u_j, t_i \rangle$ between t_i and u_j.

The weights of edges are evaluated based on the characteristics of different relations between/among users and tweets. We will provide detailed discussions in Sect. 3.2. Based on the above matrices, we build up the following equations to calculate $\boldsymbol{P}$ and $\boldsymbol{Q}$:

$$\boldsymbol{P} = \alpha_t \mathbf{M} \boldsymbol{Q} + \alpha_u \mathbf{U} \boldsymbol{P} \tag{1}$$

$$\boldsymbol{Q} = \beta_u \mathbf{N} \boldsymbol{P} + \beta_t \mathbf{T} \boldsymbol{Q} \tag{2}$$

where $\alpha_t, \alpha_u, \beta_u, \beta_t$ are weight parameters, and will be discussed in Sect. 4. In these equations, $\boldsymbol{P}$ and $\boldsymbol{Q}$ are coupled with each other and they will be calculated iteratively with multiple rounds.

3.2 Weights Evaluation

In the coupled user network and tweet network, trust values of nodes are evaluated according to the trust values of nodes linked with them. It is necessary to judge the relative contributions of different edges when evaluating the trust values of users and tweets in the coupled network.

Weights of Edges Between Users. The user network only contains *follow* edges. To evaluate the weight of each *follow* edge, we firstly analyse the follow relationships between users. Let O_{u_i} be the number of users following user u_i, and R_{u_i} be the number of users followed by user u_i. Users with a high degree of trustworthiness in real-world normally have big values of O_{u_i}, such as celebrities, politicians, and etc. However, these users with a large number of followers may have small R_{u_i}. Intuitively, the more followers a user has, the more trustworthy he/she is. However, it is arbitrary to evaluate one's trust ranking just based on follower numbers. A spammer can easily attract a large number of followers by following a huge number of users. This phenomenon is known as an unwritten etiquette on Twitter. It is believed that to follow someone who is following you shows a good manner [7]. Therefore, spammers and marketers will take advantage of this general practice and manipulate their trust rankings. To deal with this, we calculate the weights of in-degree *follow* edges for a user u_i as follows:

$$w_{f_i} = \frac{O_{u_i}}{O_{u_i} + R_{u_i} - C_{u_i}} \tag{3}$$

Here, C_{u_i} is the number of u_i's followers who are also followed by u_i. This weight represents the proportion of one's followers among all the users linked with this user.

Weights of Edges between Tweets. The tweet nodes in the tweet network are linked by edges including *reply* and *retweet* edges. We use the examples in Fig. 1 to evaluate weights of these edges in different scenarios. In the diagram, a tweet t_3 retweets t_1, and t_2 replies to t_1. The trust value of t_1 is affected by trust values of t_2 and t_3 and the weights of the edges linking t_1 with t_2 and t_3. These actions both demonstrate acceptance and attention towards t_1 from the other two tweets. The degree of acceptance by retweeting and replying is different. Weights of these two types of edges should be evaluated based on principles: (a) the retweet action shows fully acceptance of the original tweet; (b) the reply action cannot reflect the level of acceptance towards the content of the original tweet. In this work, we set $w_{rt} = 1.0$ for all the *retweet* edges, and $w_{rp} = 0.5$ for all the *reply* edges. We also deal with situations that one user u_i receives multiple retweet or reply actions from the same user u_j. In such cases, the weight w_{rt} or w_{rp} will be divided by the corresponding number of retweets or replies from the user u_j.

Weights of Edges Between Users and Tweets. There are three types of edges coupling the user network and the tweet network. They include *mention*, *post*, and *posted* edges. In Twitter, one can post a tweet that mentions other users in the purpose of bringing the attentions from the mentioned ones. If user u_i has been mentioned by more trustworthy tweets than user u_j, then we consider u_i to be more trustworthy than u_j. However, in reality, one could intentionally mention a large number of users expecting that the mentioned users will in return mentioning back. This type of mention interactions can be often seen in conversations. The weight w_m of *mention* edges is bigger if a user receives more mentions than the user mentions others. We calculate the weight as follows:

$$w_{m_i} = \frac{M_{u_i}}{M_{u_i} + Y_{u_i}} \qquad (4)$$

where M_{u_i} is the number of tweets mentioning u_i, and Y_{u_i} is the number of mention tweets posted by u_i. If u_i is mentioned by a tweet posted by u_j, the weight of the *mention* edge between u_i and this mention tweet is divided by the number of tweets mentioning u_i that are posted by u_j.

A tweet is more trustworthy when it is posted by a more trustworthy user. We set $w_u = 1.0$, which means the trust values of tweets will inherit the trust values of users who post them. Tweets posted by one user can be categorised into three groups: tweets with in-degree edges, tweets with out-degree edges, and isolated tweets. An isolated tweet is the one with no in-degree or out-degree edges in the tweet network. The trust values of isolated tweets and tweets with only out-degree edges in the tweet network are not affected by other tweets. Therefore, they are insignificant in the trust evaluation for users who post them. Thus the weights of *posted* edges linked with these two types of tweets are zeros. Only the tweets with in-degree edges in the tweet network will affect the trust values of their posting users. We set $w_p = 1.0$ for *posted* edges that link users with tweets which are retweeted or replied.

3.3 Mapping Functions

The trust values of users will be evaluated according to users who follow them, tweets that mention them, and tweets that are posted by them. The trust values of tweets will be evaluated according to users who post them, tweets that retweet or reply to them. After one round calculation with Eq. 1, we will apply mapping functions to process the calculation results for P and Q. We believe that the linearly summing over related values in Eq. 1 can not satisfy the trust evaluation purpose. It is necessary to apply appropriate mapping functions. The design principles of these mapping functions should reflect the following intuitions:

- if one node has a high degree of trustworthiness, then increasing the number of trustworthy nodes linked to it will not significantly increase its trust value (saturation effect);
- if one node has a low degree of trustworthiness, then decreasing the number of trustworthy nodes linked to it will not significantly decrease its trust value;

- if one node has a medium degree of trustworthiness, then increasing or decreasing the number of trustworthy nodes linked to it will significantly change its trust value.

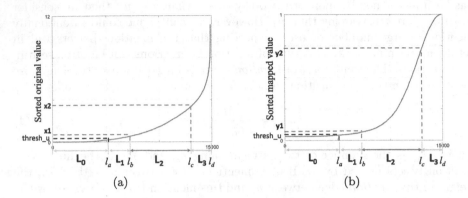

(a)   (b)

Fig. 3. (a) Trust values before mapping; (b) Trust values after mapping.

Figure 3 shows the user trust values before and after mapping. In Fig. 3a, the y-axis represents the original trust values of P generated by Eq. 1 in one round of iteration with all values sorted in the ascent order. Let S denote the sorted trust values of U. The x-axis are the corresponding position indices of U ranging from 1 to N, $N = |U|$. All the position indices are stored in a vector L, and $L = [1, 2, ..., N]$. Let $l_1 = 1$ be the index of user with the smallest trust value $S[1]$; while $l_N = N$ be the index of the user with the maximum trust value $S[N]$. We use a threshold to select a subset of users that are insignificant for the system. Users with trust values smaller than or equal to the threshold are stored in an index vector L_0. These users are the ones either isolating from or having very limited links to other nodes. Therefore, they have trust values close to zero. If we consider trust values of these users when calculating the parameters of the mapping functions, the results might be dominated by the amount of small values. By given a threshold value $threshold\_u$, we can select the index set L_0 based on the below condition:

$$\forall S[l_m] \le S[l_a] = threshold\_u, \quad \text{where } l_m \in L_0, \ S[l_m], S[l_a] \in S \quad (5)$$

Here, l_a is the index of user who has trust value $S[l_a]$, and we have $L_0 = \{l_1, l_2, ..., l_a\}$. Let L^* denote the rest of indices after eliminating L_0 from L. Then we select the second set of indices L_1 that belongs to the lower $y1 \times 100$ percent of L^*. Similarly, we select another set L_3 from the top $(100 - y2 \times 100)$ percent of L^*. The rest indices are put into L_2. Let l_b be index of the user with the maximum trust value $S[l_b]$ among the users within index set L_1. So we have

$$\forall S[l_n] \le S[l_b], \quad \text{where } l_n, l_b \in L_1, \ S[l_n] \in \{S[l_{a+1}], S[l_{a+2}], ..., S[l_b]\} \quad (6)$$

With the same method, we can obtain index l_c and l_d with $S[l_c]$ and $S[l_d]$ being the maximum trust values of corresponding users within index sets L_2 and L_3, respectively. The mapping function makes the trust values of users within index set L_2 have significant difference and the trust values of users within index set L_1 and L_3 have small difference. This is similar as the well-known saturation effect. In this work, we employ the sigmoid function Eq. (7) to achieve the mapping purpose.

$$f(x) = \frac{1}{1 + e^{-k(x-x_0)}} \qquad (7)$$

The parameters k and x_0 of the sigmoid function are calculated based on $(S[l_b], y1)$ and $(S[l_c], y2)$. We then map P to the range of $[0 \sim 1]$ by Eq. (7) with each element in P as x. The same processes are performed to calculate parameters of the sigmoid function used to map Q to the corresponding values in each round of the iteration. The recursion stops when values P and Q are converged.

3.4 Convergence

The Spearman's rank correlation coefficient (SRCC), as shown in Eq. (8), has been widely used to measure the difference between two rank sets [5].

$$\rho = 1 - \frac{6 \sum (x_i - y_i)^2}{N^3 - N} \qquad (8)$$

The x_i and y_i are the ranks of user u_i among two ranking sets. This rank correlation coefficient assesses monotonic relationships between the two variables without being impacted by any other statistical nature of the variables. In our proposed method, we recruit the same measurement scheme to determine the value of ρ, which indicates the convergence of the trust ranks. Intuitively, if the corresponding variables in $\mathbf{P}^{i+1}$ and $\mathbf{P}^i$ have more similar ranks, the SRCC value between the two sets is close to 1; otherwise, if the corresponding ranks are dissimilar, the value is close to -1.

4 Experiments and Results

4.1 Dataset

We build up a dataset based on data collected from Twitter platform. This dataset includes data from Twitter associated with users from an island state, Tasmania, of Australia. Tasmania is geographically isolated and it may provide a good foundation for us to only consider the tweets and users in such a community. A total of 14,894 users are collected as the Tasmania User Group (TUG). We also collect a total of 412,475 pairs of follow relationships among the TUG user set. We retrieve 8,401,895 tweets in total, and 1,200,365 of them are with relations with other tweets or users in the TUG dataset.

4.2   Effectiveness of CoRank

We evaluate CoRank on the TUG dataset to calculate trust values of users and tweets. In Eq. 1, α_u, β_u and α_t, β_t are the weight parameters when considering the contributions of trust values of users and tweets in the calculation. Values of these parameters are determined based on the importance of the user and tweet related interactions. According to [5], the more important or reputable users are the ones who receive more mentions or retweets than the ones with more followers. Thus, we set $\alpha_u = \beta_u = 0.4$ and $\alpha_t = \beta_t = 0.6$ to emphasis on the measurement of tweet related activities. The trust ranking converges when the Spearman's correlation coefficient ρ between ranks generated in two consecutive iterations reaches ϵ. In the experiments, we set $\epsilon_u = \epsilon_t = 0.999$. Table 1 shows the top 10 users according to their trust values calculated by the proposed method. A smaller rank value indicates a higher rank. Majority of these users are the official accounts of some organisations, such as news and media, Tourism, educational institute, and etc. These ten top users have all been followed by a large amount of users and posted tweets that have been frequently retweeted or replied to. They also have been mentioned by a large number of tweets. Not only the quantity of the related users and tweets of these ten users is big, but also most of these related nodes have high ranks. For example, the 10th user *Cricket Tasmania* has got 51.87% of his followers ranked at the top 30% of all the users. There are 83.13% of the users who retweeted or replied to *Cricket Tasmania* that are ranked in the same top range. Furthermore, most of the users who post tweets that mention *Cricket Tasmania* are the ones with high trust values. Approximately 88% of them are ranked at the top one-third of all the users.

Table 1. Statistics of top 10 users calculated by CoRank. Fo: number of followers; Po: number of tweets posted by the user; RTRP: number of tweets that retweet or reply to the user; Men: number of tweets that mention the user.

Rank	Name	Type	Fo	Po	RTRP	Men
1	ABC News Tasmania	News Media	3545	6188	8544	6890
2	Discover Tasmania	Tourism	2848	8824	8387	11087
3	The Mercury	News Media	2892	3037	2799	9314
4	University Of Tasmania	Education	1771	3431	3343	8174
5	Alex Johnston	Sports Star	1247	3443	2304	6007
6	Will Hodgman MP	Politician	1581	1854	1513	3651
7	Tasmania Police	Government	2134	2833	2758	2446
8	Brand Tasmania	Tourism	1995	2811	2604	1509
9	Lara Giddings	Politician	1485	1655	1170	2040
10	Cricket Tasmania	Sports	1416	3314	3062	3774

Table 2. Statistics of three users with similar numbers of in-degree links.

	Followers	Posts	Mention	Retweets	Replies	Rank
User A	487	1092	867	2015	93	36
User B	523	584	148	2311	80	85
User C	224	366	2714	246	404	191

Table 2 shows three users with similar numbers of in-degree links in the coupled dual networks. With the privacy concern, we use User A, User B, and User C to distinguish these users. User A and User B have both received a large number of retweets. User C has been mentioned by a large number of tweets. Apparently, the total numbers of nodes related to these three users are similar. But they have received quite different ranks. We analyse all users and tweets that are related to the trust values of these three users. Figure 4 shows the numbers of linked users/tweets at different rank ranges. We use black coloured bars to represent data related to user A, gray coloured bars for user B, and white coloured bars for User C. Figure 4a shows the numbers of the three users' followers ranked in five rank ranges (1–3000, 3000–6000, 6000–9000, 9000–12000, and 12000–15000). It shows that user A and B have been followed by more high ranked users compared to that of user C. Figure 4b shows the numbers of user posted tweets in different rank ranges. User A has posted more tweets that have been ranked at the top 20 percent of the total 1,200,365 tweets (tweets that have interactions with other tweets or users) than that posted by the other two users. The distribution of the mention tweet ranks in Fig. 4c shows that although User C has been mentioned by more tweets, but most of these mention tweets (approximately 75%) have been ranked in the bottom twenty percent of all the tweets. Figure 4d shows that User A have received more retweets and replies that are ranked in higher ranges compared to that of the other two users. Although User B has a larger number of retweets and replies, but these retweets and replies are not as trustworthy as those related to User A.

We evaluate the users who post tweets that mention the three users in Fig. 5a. Although there is a large number of tweets which mention User C, but the number of users who post those tweets is smaller than the number of users who mention User A. This is a common phenomenon on Twitter that one user often receives multiple mention tweets from another user. We have paid attention to this phenomenon and have handled the issue (see details in Sect. 3.2). In Fig. 5b, the users who post tweets that retweet or reply to User B seem to have higher ranks than those users who retweet or reply to User A and User C. But the retweets and replies posted by these users may not have high trust values (see Fig. 4d). This demonstrates that the trust values of tweets are not only determined by the trust values of the tweet owners, but also are affected by trust values of other tweets that related to them.

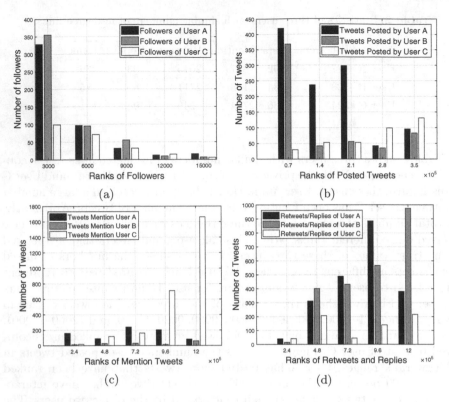

Fig. 4. Distribution of ranks of (a) users' followers, (b) tweets posted by users, (c) tweets mention users, (d) tweets retweet or reply to users.

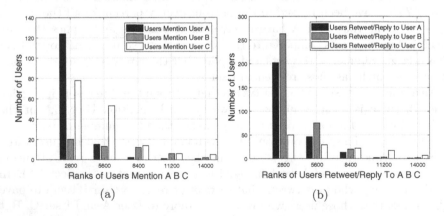

Fig. 5. Distribution of ranks of (a) users who post tweets that mention the three users, (b) users who post tweets that retweet or reply to the three users.

4.3   Comparing with PageRank and TURank

This section provides a comparison between CoRank and two baseline methods, PageRank and TURank [17]. The PageRank only considers follow relationships. TURank considers *follow, post, posted* and *retweet* (RT) relationships. Both PageRank and TURank construct one network with user nodes or user nodes and tweet nodes, and perform a normalisation over all nodes. CoRank constructs a user network and a tweet network which are coupled by *post, posted* and *mention* relations between users and tweets. All of PageRank, TURank and CoRank are effective in evaluating trustworthiness of users when they have large numbers of interactions with other users in the TUG dataset. Most of the top 100 users ranked by these three methods are the same. Some users in the TUG dataset have been abnormally ranked by PageRank and TURank. With CoRank, these users have received reasonable ranks. Some anomalies shown in Table 3 are discussed as follows:

Table 3. Statistics and ranks of users by PageRank, TURank and CoRank.

User	Follower	Posts	Mention	Retweets	PageRank	TURank	CoRank
User 1	3545	6188	6890	8544	1	1	1
User 2	1057	2159	1355	1344	33	36	16
User 3	1222	2893	815	707	22	41	37
User 4	70	2140	420	38	1118	1127	497
User 5	55	219	11	38	1405	2822	1354
User 6	1	378	1	1	9221	1478	10019
User 7	1	3224	0	1	9261	1301	10374

PageRank is Dominated by Follow Relations. Table 3 shows User 3 with a higher rank than that of User 2 when only considering follow relations in PageRank. TURank and CoRank have assigned User 3 with a lower rank than User 2 because User 3 has been mentioned and retweeted by less tweets.

PageRank and TURank Have not Considered *Mention*. *Mention* is an important feature on Twitter which should be considered in trust evaluation. None of PageRank or TURank has considered *mention* relations. Users with large numbers of mentions from trustworthy users are under-ranked in PageRank and TURank, in particular when these users have small numbers of followers and retweets. For example, User 4 has 38 retweets and more than 400 mentions. Both TURank and PageRank rank this user out of the top 1000 most trustworthy users. With CoRank, this user is ranked within the top 500 users.

TURank Uses an Inappropriate Weighting Method. With TURank, the weight of each follow link of a user has been divided by the total number of users followed by this user. The trust value of a user followed by a trustworthy user becomes smaller when more users are followed by this user. The weight of each post link of a user has been divided by the total number of tweets posted by this user. The trust value of a tweet posted by a trustworthy user becomes smaller when more tweets are posted by this user. The corresponding trust values of linked users and tweets are under evaluated. For example, User 5 has 55 followers including User 1, 57.41% of these followers fall in the top 1000 users ranked by TURank and CoRank. User 5 also has 38 retweets and most of these retweets are ranked high. User 6 and User 7 only follow and interact with each other but both of them have been ranked quite high with TURank. Comparing with them, User 5 is ranked much lower. This is due to the inappropriate weighting method that benefits users who have limited interactions with a small number of users. Overall, with TURank, approximately 10% of the top 5000 users only have limited interactions with a single user. The users with limited interactions with small number of users are normally assigned low trust values with our proposed CoRank method.

5 Related Work

In the last decade, various studies have been done in evaluating trust of users or information (such as web page in PageRank) [8,18]. Trustworthiness and influence of users or information are often studied together [6,9]. A few techniques have been developed to maximise the propagation of useful information and prevent the spread of information from unreliable resources [1]. The work in [13,16] focuses on the *follow* relationship with an assumption that users followed by more people are more trustworthy. [5] proves the insufficiency of only considering follow relationships and shows how both *retweet* and *mention* are significant for evaluating users' influence and/or trustworthiness. There is a range of methods that consider different types of interactions in their models [17]. In all these methods, users and tweets have not been considered in separate networks and it is difficult to deal with the coupling effect between them.

PageRank [14] has been widely used as the core concepts in *PageRank-like algorithms* [3,4,17]. ObjectRank [3] applies PageRank algorithm to a more complexed problem by considering different types of nodes and edges to specify the network structure. TURank is proposed in [17] by extending ObjectRank to the OSN domain to cover relations between user and user, user and tweet, and tweet and tweet. TURank can achieve better results comparing with PageRank and HITS [11]. FadeRank [4] improves the TURank by considering the temporal features of interactions. With the same network structure of TURank, FadeRank can not implicitly represent the complicated relations between/among users and tweets.

In our work, we propose a duel networks method CoRank for trust evaluation. The characteristics and contributions of the CoRank method can be summarised as follows:

- Trust on users and tweets are different, but reciprocally affect each other, therefore they should be represented separately and linked together. The proposed CoRank model has the capability to explicitly capture the contributions from user-user, tweet-tweet and user-tweet interactions into trust evaluation for users and tweets. The coupled networks allow us to update the trust values: using trust values of tweets to update the user trust values and vice versa till they converge.
- This work also develops mapping functions to overcome the saturation effect when transforming the statistical numbers of interactions/actions of users/tweets to trust values. These mapping functions consider the distribution of the statistical numbers of interactions/actions of users/tweets and assign trust values in a more reasonable way.

6 Conclusion and Future Work

We have developed a coupled dual networks trust ranking approach with a dual networks model, trust mapping functions, and weight assignments for various interactions. The trustworthiness of users and tweets can be calculated with the proposed approach. The overall algorithm with detailed functions and parameter evaluation has been developed and implemented. A set of experiments has been conducted against the real data of Tasmania users and tweets from Twitter platform (Tasmania is an island state of Australia). We have compared our results with that of the well-known ranking methods as PageRank and TURank. We have also provided detailed discussion on a set of scenarios to show the effectiveness of our proposed approach, in particular when handling cases which are imperceptible with existing methods. In the future, we will integrate interactions and tweet content to evaluate the trustworthiness of users and tweets. We will also investigate the effect of temporal features of different interactions in the trust evaluation.

References

1. Al-Garadi, M.A., Varathan, K.D., Ravana, S.D., Ahmed, E.: Analysis of online social network connections for identification of influential users: survey and open research issues. ACM Comput. Surv. **51**(1), article 16 (2018)
2. Aslam, S.: Twitter by the numbers: stats, demographics & fun facts. https://www.omnicoreagency.com/twitter-statistics/. Accessed 3 Apr 2018
3. Balmin, A., Hristidis, V., Papakonstantinou, Y.: Objectrank: authority-based keyword search in databases. In: VLDB, vol. 30, pp. 564–575 (2004)
4. Bartoletti, M., Lande, S., Massa, A.: Faderank: an incremental algorithm for ranking Twitter users. In: Cellary, W., Mokbel, M.F., Wang, J., Wang, H., Zhou, R., Zhang, Y. (eds.) WISE 2016, Part II. LNCS, vol. 10042, pp. 55–69. Springer, Cham (2016). https://doi.org/10.1007/978-3-319-48743-4_5

5. Cha, M., Haddai, H., Benevenuto, F., Gummadi, K.: Measuring user influence in Twitter: the million follower fallacy. In: AAAI on Weblogs and Social Media, pp. 10–17 (2010)
6. Eliacik, A.B., Erdogan, N.: Sentiment analysis, influential user, social network analysis, microblogging service. Expert. Syst. Appl. **92**, 403–418 (2018)
7. Gayo-Avello, D.: Nepotistic relationships in Twitter and their impact on rank prestige algorithms. Inf. Process. Manag. **49**(6), 1250–1280 (2013)
8. Gupta, A., Kumaraguru, P., Castillo, C., Meier, P.: TweetCred: real-time credibility assessment of content on Twitter. In: Aiello, L.M., McFarland, D. (eds.) SocInfo 2014. LNCS, vol. 8851, pp. 228–243. Springer, Cham (2014). https://doi.org/10.1007/978-3-319-13734-6_16
9. Hong, R., He, C., Ge, Y., Wang, M., Wu, X.: User vitality ranking and prediction in social networking services: a dynamic network perspective. IEEE Trans. Knowl. Data Eng. **29**(6), 1343–1356 (2017)
10. Huang, H., et al.: Tweet ranking based on heterogeneous networks. In: Proceedings of COLING, pp. 1239–1256 (2012)
11. Kleinberg, J.M.: Authoritative sources in a hyperlinked environment. J. ACM **46**, 604–632 (1999)
12. Li, M., Sun, X., Wang, H., Zhang, Y., Zhang, J.: Privacy-aware access control with trust management in web service. World Wide Web **14**, 407–430 (2011)
13. Nagmoti, R., Teredesai, A., Cock, D.M.: Ranking approaches for microblog search. In: International Conference on Web Intelligence and Intelligent Agent Technology, pp. 153–157 (2010)
14. Page, L., Brin, S., Motwani, R., Winograd., T.: The pagerank citation ranking: bringing order to the web. Technical report, Stanford Digital Library Technologies Project (1998)
15. Wang, G., Wang, H., Tao, X., Zhang, J.: A self-stabilizing algorithm for finding a minimal positive influence dominating set in social networks. In: Proceedings of ADC, pp. 93–99 (2013)
16. Weng, J., Lim, E.P., Jiang, J., He, Q.: TwitterRank: finding topic-sensitive influential twitterers. In: ACM WSDM, pp. 261–270 (2010)
17. Yamaguchi, Y., Takahashi, T., Amagasa, T., Kitagawa, H.: TURank: Twitter user ranking based on user-tweet graph analysis. In: Chen, L., Triantafillou, P., Suel, T. (eds.) WISE 2010. LNCS, vol. 6488, pp. 240–253. Springer, Heidelberg (2010). https://doi.org/10.1007/978-3-642-17616-6_22
18. Zhao, L., Hua, T., Lu, C.T., Chen, I.R.: A topic-focused trust model for Twitter. Comput. Commun. **76**, 1–11 (2016)

Social Context-Aware Trust Prediction: Methods for Identifying Fake News

Seyed Mohssen Ghafari[✉], Shahpar Yakhchi, Amin Beheshti,
and Mehmet Orgun

Macquarie University, Sydney, Australia
{seyed-mohssen.ghafari,Shahpar.Yakhchi}@hdr.mq.edu.au,
{amin.beheshti,mehmet.orgun}@mq.edu.au

Abstract. Fake news, a type of yellow journalism or propaganda, consist of false or incorrect information and have the potential to spread very fast on online social networks. This false information is mainly distributed by social actors who has influence in a specific context (e.g., politics or industry) with the intent to mislead in order to damage an entity (e.g., a politician or a product). Identifying fake news is a challenging task and requires analyzing the reliability, truth, or ability (i.e., trust) of social actors in a certain context and to a certain extent. To address this challenge, in this paper, we present a context-aware trust prediction approach which considers the notion of a context (which conceptually refers to any knowledge to specify the condition of an entity) as well as the social actor's behavior (supported by theories from social psychology) as first class citizens. We present novel algorithms that employ social context factors inspired by social phycology theories and mathematically model our approach based on Tensor Decomposition. We perform an extensive empirical study and present evaluation results on the effectiveness and the quality of the results using real-world datasets.

Keywords: Trust prediction · Fake news detection
Social networks analytics

1 Introduction

Fake news, a type of yellow journalism or propaganda, has become a major worldwide political and media theme [9]. For example, an analysis by Buzzfeed (buzzfeed.com) found that false or incorrect stories about the 2016 U.S. presidential election received more engagement on Facebook than the top news stories on the election itself. Fake news consist of false or incorrect information and have the potential to spread very fast on online social networks (OSNs). For governments and enterprises, this acceleration of content on OSNs presents a real problem and requires to not only monitor closely what is being said in the news but also in what contexts, to ensure that they are not associated with content that is detrimental to their policies/brand values [5,9]. This false information is

© Springer Nature Switzerland AG 2018
H. Hacid et al. (Eds.): WISE 2018, LNCS 11233, pp. 161–177, 2018.
https://doi.org/10.1007/978-3-030-02922-7_11

mainly distributed by social actors - individuals or collectives such as political parties, trade unions, social movements - who have influence in a specific context (e.g., politics) with the intent to mislead in order to damage an entity (e.g., politician). Identifying fake news is a challenging task and requires analyzing the reliability, truth, and ability (i.e. *Trust*) of the social actor in a certain context and to a certain extent.

In this context, the notion of trust can play a vital role in identifying a reliable source of information, which we can accept and/or share [11,28]. It is important to recognize the context-dependent nature of trust. For example, if Peter (the patient) trusts Diana (the doctor) in the health context, it may not be the case that Peter trusts Diana in politics. Therefore, it is important to estimate a trust relation between a pair of users who usually have not interacted with each other before. This process, known as *trust prediction* [37] can be roughly categorized into two main groups [25]: network-based and interaction-based trust models. The approaches in the first category are based on the concept of Web-of-trust or Friend-of-a-Friend (FOAF). In this category, which also includes the trust propagation-based [12] and inference-based [32] methods, each user assumes to have a trust network that contains friends (social network actors/users) as the nodes and relationships (value of their trust relations) among them as the edges [25]. This assumption can be invalid or too strong as in reality, in many online communities, either there is no way to know a web of trust or the connectivity can be very sparse [19]. In the second category, relevant features derived from user actions (user factors, i.e., features associated with a given user who can be a trustor or trustee) and interactions (interaction factors, i.e., features associated with the interaction that occurs between a pair of users in the trustor-trustee roles) are the key factors for social trust computation [19]. From the perspective of context, each of network-based and interaction-based trust models can be also divided into two subgroups: context-less [28] and context-aware [32] approaches.

While existing trust prediction solutions - which are mainly based on the network-based trust evaluation models as well as the concept of web-of-trust - do a great job in understanding the trust between a pair of users, they do not consider the notion of a context (which conceptually refers to any knowledge to specify the condition of an entity) as well as the social actor's behavior (supported by theories from social psychology) as first class citizens. To fill this gap, in this paper, we present a novel hybrid context-aware trust prediction approach. This approach employs both interaction-based and network-based trust models and does not depend on the web-of-trust concept. To the best of our knowledge, this is the first study of predicting pairwise user trust using a hybrid context-aware trust prediction model in the literature that does not depend on the web-of-trust concept. The unique contributions of this paper are as follows:

– We propose a context-aware approach, namely TDTrust, based on Tensor Decomposition (TD) for predicting trust values in OSNs. We mathematically model our approach based on TD and it directly considers the context of trust.

- We present algorithms, for predicting trust values in OSNs, employing social context factors inspired by phycology theories such as Social Penetration Theory.
- We propose a new mechanism to evaluate the level of expertise of the users on OSNs, by analyzing and mining the content of reviews and posts.
- We conduct a comprehensive experiment on two real-world datasets to demonstrate the superior TDTrust performance compared to the state-of-the-art approaches.

The rest of the paper is organized as follows: Sect. 2 presents the background and related work. We present the hybrid context-aware trust prediction approach in Sect. 3. In Sect. 4, we present the results of the evaluation of the proposed approach before concluding the paper with remarks for future directions in Sect. 5.

2   Background and Related Work

With more than one billion users on Facebook (facebook.com), and millions more active users on online social networks (OSNs) such as Google Plus (plus.google.com) and Twitter (twitter.com), it is clear that OSNs provide important channels through which many people access news stories and important information. For example, according to a recent study, in United States 62% of people get their news from social media [35]. Besides opportunities for governments and enterprises to share their thoughts and participate in ongoing conversations on OSNs, there is also a threat for the spread of false stories that appear to be news. For instance, an analysis by Buzzfeed (buzzfeed.com) found that false or incorrect stories about the 2016 U.S. presidential election received more engagement on Facebook than the top news stories on the election itself. As another example, based on a study [24] on 14 million tweets in Twitter, many social bots leveraged influence maximization techniques [8] to target influential users (social network users who have great influence on people) on Twitter and to spread misinformation, i.e., false or incorrect information. The threats of fake news can be more frustrating, if we know that "An individual user with no track record or reputation can in some cases reach as many readers as those of Fox News, CNN, or the New York Times" [1]. In such scenarios, trust can play an important rule to slow down the speed of spreading the fake news. In the trust context, the users need to not only monitor closely what is being said on a (social) news, but also in what contexts, to ensure they are not associated with content that is detrimental to their interest.

In the following, we present and classify the related work in trust evaluation approaches, based on the structure and context dimensions, and discuss the added value of our approach compare to the state of the art techniques.

2.1   Structure Dimension

Network-Based Trust Models. Golbeck et al. [12] proposed a trust inference approach based on the FOAF concept that can determine which two pairs of users trust each other and in which topic. In the same line of work, Zhang et al. [36] proposed an approach by which the source user accept the recommendation from similar neighbor nodes (other users who are directly connected to the target user). In addition, Kim et al. [14] proposed an approach to build a web-of-trust based on the implicit feedbacks of users in a certain context. The network-based trust evaluation approaches have shortcomings due to the data sparsity problem in OSNs, as in the real-world scenarios a user may not have a web-of-trust or the existing web-of-trust may be too sparse [19]. Besides, in some cases these approaches may fail to "capture actual interactions between members" [25].

Interaction-Based Trust Models. Liu et al. [19] proposed a classification approach for trust prediction in OSNs based on the action and interactions of the users. A similar approach presented by Nepal et al. [22], proposed a trust prediction model that considers two types of trust: trust of the other users to a target user, and the trust value that a user has to a network. These approaches only focus on the users' interactions and do not consider the social network structure, which may contain important information about the users and the type of relations among them.

Hybrid Trust Models. Hybrid trust models combine the network-based and interaction-based models. In particular, they simultaneously consider the previous interactions of users and the information about the social network structure into account [29].

2.2   Context

Contextual Trust Prediction. The notion of trust is context-dependent. Therefore, trusting someone under one type of context does not guarantee trusting her in other types [27]. Moreover, **context** (which influences in the building of a trust relationship between the trustor and the trustee [31]) is multi-faceted [37]. In social society, the context about the interaction between two participants can be considered as the *Social Context* and the *Interaction Context* can provide more information such as the type or time of services, location and so on [37]. In addition, context-aware approaches try to consider different social context factors to evaluate a potential trust relation.

Context-less Approaches. The context-less approaches do not distinguish between different types of trust relations in OSNs. They can be roughly divided into two groups: unsupervised [13] and supervised [19] methods. Supervised methods treat the trust prediction problem as a classification problem, where trustable users have been labeled after extracting some features from available sources [7,20]. Unsupervised approaches make a trust prediction with the help of some trust network properties, such as trust propagation [10]. The approach presented in hTrust [28], is based on one of the well-known social theories, Homophily [28], and shows that two similar users have a higher potential to establish a trust relationship. In addition, in sTrust [33], another social psychology based trust prediction approach, which is based on social status theory, users with higher statuses can have a higher chance to be considered as trust-worthy users. We observe that, all of the above approaches mostly focus on one type of trust evaluation methods and the majority of them are based on the network-based models. More importantly, these approaches neglect to consider the context of trust, which can have a significant impact on the performance of trust prediction.

Context-Aware Approaches. Liu et al. [18] highlighted the importance of the context of trust as an essential factor for trust prediction approaches. However, minor efforts have been conducted in the literature to consider the context of trust as a first class citizen. In this line of work, Zheng et al. [37] proposed a context-aware approach to take both the user's properties and features of contexts into account. Social trust proposed as a novel probabilistic social context-aware trust inference approach, exploits some textual information to deliver better results [32]. In this approach, trust is inferred along the paths connecting two users. Thus, if two users are not connected by any path, no trust between them can be predicted. Similar to this approach, Liu et al. [17] proposed a context-aware trust prediction approach based on the web of trust concept and considered some social context factors, such as users' location, previous interactions, social intimacy degree with other users, the existing trust relations and so on. Although these studies put the first step towards considering the context of trust as a key factor, their success is not guaranteed because of their high dependency to the concept of Web-of-Trust and due to the data sparsity problem.

Added Value of Our Approach. Most of the existing trust evaluation approaches either do not consider the context of trust, or they highly depend on the concept of Web-of-Trust. For instance, among the few attempts that consider the context of trust, although some of them [17,32,37] capture the social intimacy degree, their main focus is on trust inference and trust propagation. Furthermore, since trust relations follow a power law distribution, which means "a large proportion of users specify a few trust relations" [28], this causes performance degradation in such trust inference approaches. To fill this gap, in this paper, we present a novel hybrid context-aware trust prediction approach. This approach employs both interaction-based and network-based trust models and does not depend on the web-of-trust concept.

3 TDTrust: Social Context-Aware Trust Prediction

We propose a context-aware approach, namely TDTrust, based on Tensor Decomposition (TD) for predicting trust values in OSNs. We mathematically model our approach based on TD. We present algorithms, for predicting trust values in OSNs, employing social context factors inspired by phycology theories, e.g., Social Penetration Theory.

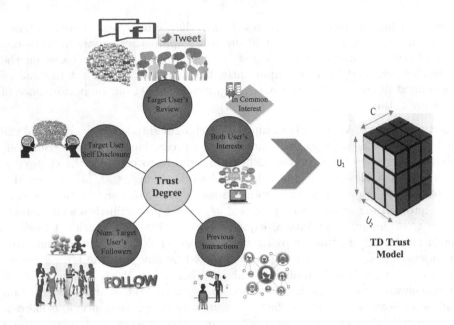

Fig. 1. The proposed TDTrust framework

3.1   Social Context

In this section, we describe the social context factors that we employ in our hybrid trust prediction approach.

Network-Based Social Context Factors. We first introduce *Level of Expertise*, *Interest*, and *Number of Followers* as important factors for capturing the network-based features.

A. *Level of Expertise.* A recommendation from an expert person in a certain domain is more acceptable compared to the less knowledgeable person [32]. In this paper, the Level of Expertise of a target user will be calculated by (i) evaluating her activeness in a certain context, where we assume that a target user is active in a certain context, if the number of her posts/reviews is equal or more than the average number of other users' posts/reviews; and (ii) considering other users' opinion about her posts/reviews, whether other users liked/highly rated those posts/reviews or not. For instance, consider that David (a user) has written many reviews/posts related to sport topics which are highly rated by other users. In contrast, Sarah only posted two reviews/posts on the same topic and no one rated them so far. Therefore, it can be argued that David can be trusted more in the context of sports from another user point of view.

Let $U = ...u_m$ denote the set of users, and $C = ...c_k$ the set of contexts of trust relations. Let n_i denote the total number of posts/reviews by user u_i, for i = 1, ..., m and n_{i,c_k} the total number of post/reviews by user u_i in context c_k

for $i = 1, ..., m$ and $k = 1, ..., z$. Let a_{i,c_k} denote the status of user u_i where $a_{i,c_k} = 0$ means u_i is inactive and $a_{i,c_k} = 1$ means u_i is active. Each review/post is evaluated by other users through giving a score s (it can be in scale of 1 to 3, on Likert scale, i.e., a psychometric scale commonly involved in research that employs questionnaires).

$$a_{i,c_k} = \begin{cases} 1, & \text{if } n_{i,c_k} >= \dfrac{\sum_{r=1}^{m} n_{r,c_k}}{m} \\ 0, & \text{otherwise} \end{cases} \tag{1}$$

we calculate the average score of posts/reviews that u_i gained from other users in c_k:

$$S_{i,c_k} = \frac{1}{n_{i,c_k}} \sum_{r=1}^{n_{i,c_k}} s_r^{i,c_k}, \tag{2}$$

where s_r^{i,c_k} is the score value of r_{th} review of u_i that was archived in c_k. At the end, the level of expertise u_i in the c_k can be calculated by:

$$v_{i,c_k} = S_{i,c_k} \times n_{i,c_k} \tag{3}$$

B. *Interest.* Preference could be conceived of as an individual's attitude towards a set of objects [32]. In this research, we consider p_{ic_k} as the interest of u_i in the context c_k, which means the topics/categories of items that a user's posts/reviews belong to. It is within the scope of c_k when $p_{ic_k} = 1$, and it is not when $p_{ic_k} = 0$.

C. *Number of Followers.* In this paper, NoF_A denotes the number of followers of user A, and is equal to the "number of followers" / "number of people who read and rated the user's reviews". Assume that David has a higher number of follower compared to Sarah. It has been validated in Social Science theories [17], that the recommendation from David (with a larger number of followers) is more credible.

Interaction-Based Social Context Factor. This term is related to the interactions between the participants: *Frequency and Quality of Previous Interactions* (FQPI). To have a hybrid trust prediction approach, we consider the interactions that the source user had with the target user. Therefore, if a source user highly rates/likes a target user's review/post, there is a high potential that she could be trusted. We calculate FQPI as follows:

$$FQPI_{u_i,u_j,c_k} = \frac{1}{n_{e,c_k}} \sum_{b=1}^{n_{e,c_k}} R_{b,c_k}^e, \tag{4}$$

where $FQPI_{u_i,u_j,c_k}$ is the average value of ratings (R_{b,c_k}) that u_i gave to the u_j in the context c_k and n_{e,c_k} is the total number of ratings that u_j received from u_i in c_k.

Social Context Factors Related to the User Behavior. This factor monitors the previous user's behavior in OSNs: *Self-disclosure*. The Social Penetration Theory (SPT) proposes that "as relationships develop, interpersonal communication moves from relatively shallow, non-intimate levels to deeper, more intimate ones" [2]. Based on this theory, self-disclosure (sometimes known by self-presentation), which can be defined as revealing personal information such as personal motives or desires, feelings, thoughts, and experiences to others [2], can be the reason of development of a relationship or the reason of an increase intimacy level. Moreover, since the level of trust has relation with the level of intimacy [32], we consider the fact that self-disclosure can influence the trust value between two users.

In this paper, we detect self-disclosure of a target user based on analyzing her textual materials in OSNs. For this reason, we employ one of the most famous text analyzer tools, called LIWC (Linguistic Inquiry Word Count) [23]. This tool can detect the worlds related to the personal feelings, emotions, thoughts, which are signs of self-disclosure in social presentation theory. We can evaluate the level of self-disclosure of the user u_i in the context c_k with the following formula.

$$Sd_{ic_k} = \frac{self_{ic_k}}{W_{i,c_k}}, \tag{5}$$

where $self_{ic_k}$ is the number of self-disclosure words that u_i wrote to express his feelings, thoughts and so on in the context c_k, and W_{i,c_k} is the total number of words that he wrote in his posts/reviews in the context c_k.

3.2   Trust Prediction Mechanism

We present algorithms, for predicting trust values in OSNs, employing social context factors inspired by phycology theories such as Social Penetration Theory. For trust prediction, we propose the following formula:

$$B_{ijc_k} = w_1(P_{ic_k} \times P_{jc_k}) + w_2 \times v_{jc_k} + w_3 \times NoF_j' + w_4 \times Sd_{jc_k} + w_5 \times FQPI_{ijc_k}, \tag{6}$$

where B_{ijc_k} is the trust degree that we employ to predict the trust value between u_i and u_j in the context c_k, and w_z, $z = \{1, \cdots, 5\}$, is the controlling parameter to control the impact of social contextual parameters. We use $(p_{ic_k} \times p_{jc_k})$ to make sure both users have the same interest. Since NoF_j could be a large number, we normalize it to the rage of 0 and 1 by feature scaling as (same normalization procedure for FQPI):

$$NoF_j' = \frac{NoF_j - min(NoF)}{max(NoF) - min(NoF)} \tag{7}$$

Problem Statement. Suppose we have n users $U = \{u_1, \cdots, u_n\}$, and $G \in R^{n \times n}$ be a square matrix that contains the trust relations between users, where $G(i,j) = 1$ means u_i trusts u_j, and $G(i,j) = 0$ indicates there is no trust relation between u_i and u_j. The G matrix is sparse in OSNs [28], and thus in order to deal with this data sparsity, the U low-ranked matrix as $U \in R^{n \times d}$, $d \ll n$ should be extracted. Tang et al. [28] propose a trust prediction approach based on Matrix Factorization (MF) as follows:

$$min_{U,H} \|G - UHU^T\|_F^2 + \alpha \times (\|U\|_F^2 + \|H\|_F^2), \ U > 0, H > 0, \qquad (8)$$

where U represents the users' interest, d represents the facets of these interests, and H matrix contains compact correlations among U [33]. In this paper, we assume that we have m users $U = \{u_1, \cdots, u_m\}$, k context of trust as $C = \{c_1, c_2, \cdots, c_k\}$, and $G \in R^{n \times n \times k}$ is a three way tensor that contains the trust relations between users, where $G(i, j, k) = 1$ means u_i trusts u_j in the context of c_k, and $G(i; j; k) = 0$ indicates there is no trust relation between u_i and u_j in c_k. We can have the following assumptions:

$$B_{ijc_k} \times (C \odot U_{2j})U_{1i}^T = 0$$
$$B_{ijc_k} \times (C \odot U_{2j})U_{1i}^T = 1, \qquad (9)$$

where U_1 and U_2 are the first and the second users dimensions of G and U_{1i} and U_{2j} represent the $i_t h$ user in the first dimension of G and the $j_t h$ user in the second dimension of G, respectively. Based on Wang et al. proposed approach [34], and also considering the CPD/Parafac model [26] for learning three f-dimensional matrices $U_1 \in R^{n \times f}$, $U_2 \in R^{n \times f}$, and $C \in R^{k \times f}$; then, the predicted tensor could be calculated by sum of the inner products:

$$\tilde{G} = \sum_{r=1}^{f} U_{1_{u_1 r}} U_{2_{u_2 r}} C_{cr} = <U_{1_{u_1}}, U_{2_{u_2}}, C_c> \qquad (10)$$

Modeling Proposed Trust Prediction. To solve any non-convex problem in tensors, we can fix some of the dimensions to change the problem to a linear problem. Based on mentioned assumptions, we propose the following formulas with fixing U_1:

$$\lambda \times (\sum_{i}^{n} \sum_{k=1}^{m} (min\{0, f(B_{ijc_k})((C \odot U_2)U_1^T)\})^2), \qquad (11)$$

where λ is a controlling parameter. There is also a same procedure with fixing C and U_2. We propose a tensor decomposition model based on CPD/Parafac [26,33], which is one of the ranking decomposition approaches for tensors [26]:

$$min_{U_1,C,U_2} \|G_1 - (C \odot U_{2j})U_{1i}^T\|_F^2 + \lambda \times (\sum_{i}^{n} \sum_{j!=i}^{n} (min\{0, f(B_{ijc_k})((C \odot$$

$$U_{2j})U_{1i}^T)\})^2) + \alpha \times (\|U_1\|_F^2 + \|C\|_F^2 + \|U_2\|_F^2)$$
$$U_1 \geq 0, U_2 \geq 0, C \geq 0, \qquad (12)$$

where α is a controller parameter to control U_1, C, and U_2. The procedures for G_2 and G_3 are the same. Moreover, the Lagrangian function of the above formula would be:

$$
\begin{aligned}
L(G_1; U_1, C, U_2) = &\, Tr((G_1 - (C \odot U_2)U_1^T)(G_1 - (C \odot U_2)U_1^T)^T) \\
&+ \lambda \times (Tr((B(C \odot U_2)U_1^T)(B(C \odot U_2)U_1^T)^T) + \alpha \times Tr(U_1 U_1^T)) \\
&+ \alpha \times Tr(CC^T) + \alpha \times Tr(U_2 U_2^T)
\end{aligned}
\tag{13}
$$

We follow the same procedure for calculating G_2 and G_3. Then, with the help of the Alternating Least Squares (ALS) algorithm, we can update U_1, C, and U_2. In this paper, we apply an updating approach presented in [15]:

$$
\Theta_i = \Theta_i \left(\frac{\dfrac{\partial C(\Theta)^-}{\partial \Theta_i}}{\dfrac{\partial C(\Theta)^+}{\partial \Theta_i}} \right)^a,
\tag{14}
$$

where $C(\Theta)$ is the cost function and Θ is the variable that is not negative. The operators $X \bullet Y$ and X/Y, which are used in the followings formula, are element-wise operations. Moreover, $(\partial C(\Theta)^-)/(\partial \Theta_i)$ and $(\partial C(\Theta)^+)/(\partial \Theta_i)$ are the negative and positive parts of the derivative, respectively. In addition, based on the partial derivative of L with respect to U_1, C, and U_2 and if we assume that $L(G_1; U_1, C, U_2)/\partial U_1 = 0$, $L(G_1; U_1, C, U_2)/\partial C = 0$, $L(G_1; U_1, C, U_2)/\partial U_2 = 0$, and since it is difficult to find the optimal solution for U_1 and U_2 and C at the same time [30] [16] with respect to the Karush Kuhn Tucker (KKT) complementary condition, we leverage the approach presented by Tang et al. [27] and propose three updating rules for U_1 (updating rule for C, and U_2 follows the same procedure):

$$
U_1 \leftarrow U_1 \bullet \left(\frac{2G_1^T(C \odot U_2)}{(C \odot U_2)U_1(C \odot U_2) + (C \odot U_2)^T U_1(C \odot U_2) + \lambda \times B(C \odot U_2)U_1 B(C \odot U_2) + \lambda \times B^T(C \odot U_2)^T U_1 B(C \odot U_2) + 2aU_1} \right)
\tag{15}
$$

Summary. We presented a context-aware trust prediction approach. We want to know how this approach can help a user to choose a trustable source of information among the available users in his social networks. First of all, since this approach is context-aware, it only considers the context that matters to the user (e.g. politics), it gives weight to the users who are interested in politics (based on their reviews/posts), weights the users who are experts in political context, weights the ones who more disclose themselves, weights the users who have more followers, and finally, gives weight to those who had frequent and positive interactions with the user. Based on this information, TDTrust establishes a trust degree between the user and the rest of users in his OSNs and with the help of the algorithm presented in Algorithm 1, it can determine whether user A is going to trust user B in a specific context and based on his expertise and behaviors or not. Detecting reliable (trustworthy) sources of information in a specific context, in turn, can slow down or stop the spreading of fake news in an OSN.

Algorithm 1. Algorithm 1: Trust Prediction with TDTrust

1: Establishing the trust degree matrix in each context
2: Randomly initializing U_1, U_2, C
3: **while** It is not the convergent state **do**
4:     $D = -2G_1^T (C \odot U_2)$
5:     $E = (C \odot U_2)U_1(C \odot U_2) + (C \odot U_2)^T U_1 (C \odot U_2) + B(C \odot U_2)U_1 B(C \odot U_2) + B^T (C \odot U_2)^T U_1 B(C \odot U_2) + 2\alpha U_1$
6:     for j = 1 to m do
7:         for r = 1 to k do
8:             $U_1 \longleftarrow U_1 \bullet (\dfrac{D_{ijr}}{E_{ijr}})$
9:         end for
10:    end for
11:    Do the same procedure for updating C and U_2
12:    end for
13: **end while**
14: return U_1, C, U_2

4   Experiments

4.1   Datasets and Baseline Characteristics

To evaluate our approach, we use two real-world datasets, Epinions (with 1050 users, network density of 0.0093, and 10264 trust relations) and Ciao (with 1000 users, network density of 0.0087, and 8726 trust relations) [27], that contain users' ratings and their reviews, and their trust relations. Unlike Ciao, the Epinions dataset does not include the information like who rated whom. Hence, we test our model on Ciao and Epinions datasets, while we ignore the users' previous interactions factor when we employing Epinions and discuss the importance of these interactions in trust prediction scenarios. We compare our approach against 4 other approaches: (i) hTrust [28]: it is based on the Homophily Theory and it predicts the trust relations between the users based on the rating similarities of the users: hTrust was reviewed in Sect. 2. (ii) sTrust [33]: it is based on the Social Status theory and it predicts the trust relation between users based on the status of each user in OSNs sTrust was reviewed in Sect. 2. (iii) Zheng [37]: it is a context-aware approach and since it is not completely based on trust inference and the web-of-trust concept, we could employ it in our experiments. (iv) Random Trust Prediction: it randomly predicts the trust value between the users.

4.2   Experimental Setup

We employ evaluation approaches presented in Tang and Liu [27]. We use Mean Absolute Error (MAE) and Root Mean Squared Error (RMSE) as follows:

$$MAE = \frac{1}{N} \sum_{i,j} | T_{R_{i,j}} - T_{P_{i,j}} | \tag{16}$$

$$RMSE = \sqrt{\frac{1}{N}\sum_{i,j}(T_{R_{i,j}} - T_{P_{i,j}})^2}, \tag{17}$$

where in both metrics, N is the number of predicted trust values, $T_{R_{i,j}}$ is the already expatiating trust values, and $T_{P_{i,j}}$ is the predicted trust values by the trust prediction approaches mentioned in this paper. Furthermore, the lower MAE and RMSE are more desirable for any approach. It is worth mentioning that, our approach can reach its highest performance when $\alpha = 0.1$. We investigate the effects of the different λ values on TDTrust in the next subsections.

4.3 Experimental Results

We did not compare our approaches with some of the existing context-aware approaches, like [17,18,32], as they are based on trust propagation and follow a different path to evaluate the trust values. In the other words, in their approaches, A trusts B if there is an indirect or direct trust path between them, so they calculate the trust values by "inferring trust" among the users, which is totally different from our approach that tries to "predict trust" among any pair of the users even with the lack of direct or indirect trust path in-between. The only context-aware approach that we can compare TDTrust with it, is Zheng [37] as it is not completely based on trust inference and proposes a trust prediction approach, as well. Moreover, parameters of approaches in this paper were defined by applying cross-validation and are set as $\lambda = 0.5$, $\alpha = 0.1$ and f=100.

Table 1. Experimental results on Ciao dataset

Training	Metrics	Random	sTrust	hTrust	Zheng	TDTrusr
60%	MAE	5.829	1.991	1.442	1.281	1.026
60%	RMSE	6.956	2.204	1.657	1.45	1.195
70%	MAE	5.757	1.815	1.238	1.153	0.928
70%	RMSE	6.811	2.098	1.435	1.298	1.138
80%	MAE	5.504	1.796	1.095	0.984	0.886
80%	RMSE	6.624	2.021	1.132	1.059	0.987
90%	MAE	5.453	1.641	0.956	0.884	0.768
90%	RMSE	6.233	1.911	0.978	0.96	0.876

Table 2. Experimental results on Epinions dataset

Training	Metrics	Random	sTrust	hTrust	Zheng	TDTrusr
60%	MAE	6.248	1.681	1.79	1.691	1.61
60%	RMSE	6.85	1.795	1.71	1.7	1.519
70%	MAE	6.029	1.515	1.443	1.398	1.356
70%	RMSE	6.463	1.505	1.624	1.556	1.412
80%	MAE	5.916	1.438	1.129	1.104	1.089
80%	RMSE	6.239	1.414	1.383	1.301	1.287
90%	MAE	5.753	1.314	1.089	1.019	0.969
90%	RMSE	6.042	1.378	1.11	1.029	0.991

Ciao Dataset. As Table 1 demonstrates, with the increase of the training data size, the accuracy of the models would be increased. Moreover, in the all experiences, TDTrust outperformed other approaches in both MAE and RMSE metrics. For example, when training size is 90%, it has 10%, 12%, seven times, and two times higher accuracy improvements regarding RMSE and 15%, 20%, two times, and seven times higher accuracy improvements in terms of MAE compared to the Zheng, hTrust, sTrust, and Random, respectively. After TDTrust, Zheng has the best performance. For instance, with 90% training data size, Zheng has 5%, 2 times, and 6 times higher accuracy improvement regarding RMSE and 8%,

2 times and 5 times higher accuracy improvements in terms of MAE compared to hTrust, sTrust, and Random, respectively. Next, hTrust has the best performance compared to the sTrust and Random models in both MAE and RMSE metrics. In the rest of the training data sizes also Zheng and hTrust keep these good performances. Finally, as it is clear in Table 1, random trust prediction model has the worst prediction accuracy.

Epinions Dataset. As Table 2 demonstrates, although TDTrust has higher prediction accuracy in Epinions dataset compared to other approaches, its effectiveness may be affected by ignoring the user's previous interactions. With the 90% training data size, TDTrust has around 5%, 10%, 15%, and more than three times higher accuracy regarding RMSE and 5%, 10%, 30%, and around six times higher accuracy improvement in terms of MAE compared to Zheng, hTrust, sTrust, and Random models. TDTrust also outperforms other approaches with the other training data size, according to Table 2. In addition, Random policy hast the worst prediction accuracy, and Zheng and hTrust are more accurate compared to sTrust.

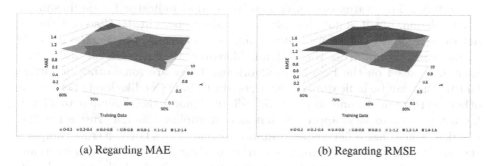

(a) Regarding MAE   (b) Regarding RMSE

Fig. 2. Effects TDTrust regularization

4.4   The TDTrust Regularization Effects

In this paper, we apply λ as a control variable to control our proposed trust prediction model effects. In this subsection, we investigate the effects of this variable on the accuracy of TDTrust. For this reason, in different training data size, we consider different values for $\lambda = \{0.1, 0.5, 0.8, 0.9, 10\}$. Our experiments demonstrate: (i) The best performance of TDTrust can be achieved, when the $\lambda = 0.5$; (ii) TDTrust's accuracy increases when λ increases from 0 to 0.5; and (iii) For $\lambda > 0.5$ we see a gradual performance degradation in TDTrust, especially when $\lambda > 1$. Figure 2 can illustrates the different performance of TDTrust when we apply different values for λ.

4.5   Discussion

The experimental results demonstrate that TDTrust has the best performance compared to four other approaches. In one hand, most of the existing trust prediction approaches do not consider the context of trust. Hence, TDTrust succeeded to improve its trust prediction accuracy by being a context-aware approach. On the other hand, among the rare attempts that consider the context of trust, most of them are mostly rely on the web-of-trust concept and they try to infer trust values between the users. Their main problem is in the cases that there is not any direct or indirect trust path between the users, which it may be in the most of the real-world scenarios, they may fail to accurately predict the trust values. However, TDTrust is not relying on web-of-trust concept.

The main reason for Random policy to have the worst performance is the fact that it does not follow any particular approach to measure the potential trust value. Moreover, sTrust has lower prediction accuracy compared to TDTrust, since it is a context-less model. However, a question arises why the accuracy of sTrust is lower than hTrust, which also is another context-less approach? This may because of the fact that sTrust only follows a simple procedure in which whenever a user has a higher status in OSNs, it is more likely to be trusted by other users. The status of a user may be a good indicator for establishing a trust relation, but it is not the only factor. Many users in the Ciao or Epinions datasets based on the sTrust have a high social status, however, the majority of the users do not trust them at all. Moreover, they extract the status of a user only based on the PageRank algorithm. There are some other important factors that can be indicators of the users' status in OSNs, like level of expertise, other users' opinion, and so on, which sTatus ignores them. Similar to sTrust, hTrust is a context-less approach, but since it applies different trust evaluation metrics, it may has a better performance compared to sTrust. hTrust compares the similarities between the users according to their mutual interested items and also the similarities between their rating styles; both of which are good metrics for calculating the similarity degree. Finally, according to our experimental results, although TDTrust and Zheng are both context-aware, TDTrust is more effective. The reason may be the fact that, in contrast to the Zheng, TDTrust directly apply the context in its mathematical model and it also monitors the users' previous behaviors.

After all, although SDTrust has a good performance, it totally ignores the time factor. Most of the existing trust prediction approaches assume that a trust value is a fixed value and do not change over time. However, this is not true in the real-world scenarios. If A trusts B in time τ, this trust value may be changed after a period of time ρ. This trust value may increase in the time $\tau + \rho$, or even decreases and becomes a distrust value. We further investigate the time-based relationships [3] and effects of time factor in our future works. Finally, we will also focus on Information extraction [4,6] and Natural language processing techniques to check the content of news, and then detect the fake news.

Summary. TDTrust is one of the few approaches that directly considers the context of trust in its model. TDTrust employs a three-dimensional tensor decomposition model for its trust prediction procedure. Furthermore, TDTrust employs factors that are available in most of the cases in OSNs, which may make it a feasible solution for OSNs. Our experimental results demonstrate the superior its performance.

5   Conclusion and Future Works

Slowing down and stopping the spread of fake news became a challenge in OSNs. In this paper, we have proposed a new context-aware trust prediction approach, TDTrust, to deal with the fake news by detecting the trustworthy source of information. TDTrust is a hybrid approach that combines the network-based and interaction-based trust prediction and at the same time monitor the social actor's behavior (supported by theories from social psychology). The experimental results demonstrate that the superior performance of TDTrust over the two state-of-the-art approaches. As an ongoing and future work, we are extending TDTrust to consider the time factor and model the evolution of trust over time among users and based on different topics and contexts. We will use the time-aware TDTrust to identify cases of Propaganda [21], a systematic form of purposeful persuasion that attempts to influence the emotions, attitudes, opinions, and actions of specified target audiences for ideological, political or commercial purposes through the controlled transmission of one-sided messages via mass and direct media channels.

References

1. Allcott, H., Gentzkow, M.: Social media and fake news in the 2016 election. J. Econ. Perspect. **31**(2), 211–236 (2017)
2. Altman, I., Taylor, D.: Social Penetration: The Development of Interpersonal Relationships. Holt, Rinehart and Winston, New York (1973)
3. Beheshti, A., Benatallah, B., Motahari-Nezhad, H.R.: ProcessAtlas: a scalable and extensible platform for business process analytics. Softw. Pract. Exp. **48**(4), 842–866 (2018)
4. Beheshti, S.-M.-R., Benatallah, B., Motahari-Nezhad, H.R., Sakr, S.: A query language for analyzing business processes execution. In: Rinderle-Ma, S., Toumani, F., Wolf, K. (eds.) BPM 2011. LNCS, vol. 6896, pp. 281–297. Springer, Heidelberg (2011). https://doi.org/10.1007/978-3-642-23059-2_22
5. Beheshti, S., et al.: Process Analytics - Concepts and Techniques for Querying and Analyzing Process Data. Springer, Cham (2016). https://doi.org/10.1007/978-3-319-25037-3
6. Beheshti, S., Benatallah, B., Venugopal, S., Ryu, S.H., Motahari-Nezhad, H.R., Wang, W.: A systematic review and comparative analysis of cross-document coreference resolution methods and tools. Computing **99**(4), 313–349 (2017)
7. Chen, S.D., Chen, Y., Han, J., Moulin, P.: A feature-enhanced ranking-based classifier for multimodal data and heterogeneous information networks. In: IEEE 13th International Conference on Data Mining, USA, pp. 997–1002 (2013)
8. Chen, W., Wang, Y., Yang, S.: Efficient influence maximization in social networks. In: SIGKDD, pp. 199–208. ACM (2009)

9. Gartner: Top Predictions for IT Organizations and Users in 2018 and Beyond (2017). https://www.gartner.com/newsroom/id/3811367

10. Golbeck, J.: Using trust and provenance for content filtering on the semantic web. In: Proceedings of the Workshop on Models of Trust on the Web, at the 15th WWW Conference (2006)

11. Golbeck, J.: Trust and nuanced profile similarity in online social networks. ACM Trans. Web **3**(4), 1–33 (2009)

12. Golbeck, J., Parsia, B., Hendler, J.: Trust networks on the semantic web. In: Klusch, M., Omicini, A., Ossowski, S., Laamanen, H. (eds.) CIA 2003. LNCS (LNAI), vol. 2782, pp. 238–249. Springer, Heidelberg (2003). https://doi.org/10.1007/978-3-540-45217-1_18

13. Guha, R.V., Kumar, R., Raghavan, P., Tomkins, A.: Propagation of trust and distrust. In: 13th International Conference on World Wide Web, WWW, USA, pp. 403–412 (2004)

14. Kim, Y.A., Le, M., Lauw, H.W., Lim, E., Liu, H., Srivastava, J.: Building a web of trust without explicit trust ratings. In: ICDE, pp. 531–536 (2008)

15. Krompaas, D., Nickel, M., Jiang, X., Tresp, V.: Nonnegative tensor factorization with RESCAL. In: Tensor Methods for Machine Learning, ECML workshop (2013)

16. Lenhart, A., Purcell, K., Smith, A., Zickuhr, K.: Social media mobile internet use among teens and young adults. In: Pew Internet American Life Project (2010)

17. Liu, G., et al.: Context-aware trust network extraction in large-scale trust-oriented social networks. World Wide Web **21**, 713–738 (2018)

18. Liu, G., Wang, Y., Orgun, M.A.: Social context-aware trust network discovery in complex contextual social networks. In: AAAI, pp. 101–107 (2012)

19. Liu, H., et al.: Predicting trusts among users of online communities: an epinions case study. In: 9th ACM Conference on Electronic Commerce (EC), USA, pp. 310–319 (2008)

20. Ma, N., Lim, E., Nguyen, V., Sun, A., Liu, H.: Trust relationship prediction using online product review data. In: CIKM-CNIKM, pp. 47–54 (2009)

21. Nelson, R.A.: A Chronology and Glossary of Propaganda in the United States, pp. 241–243. Greenwood Publishing Group, Westport (1996)

22. Nepal, S., Sherchan, W., Paris, C.: Strust: a trust model for social networks. In: TrustCom, pp. 841–846 (2011)

23. Pennebaker, J., Chung, C., Ireland, M.: Development and psychometric properties of LIWC (2007)

24. Shao, C., Ciampaglia, G.L., Varol, O., Flammini, A., Menczer, F.: The spread of fake news by social bots. CoRR abs/1707.07592 (2017)

25. Sherchan, W., Nepal, S., Paris, C.: A survey of trust in social networks. ACM Comput. Surv. **45**, 1–33 (2013)

26. Sidiropoulos, N.D., Lathauwer, L.D., Fu, X., Huang, K., Papalexakis, E.E., Faloutsos, C.: Tensor decomposition for signal processing and machine learning. IEEE Trans. Signal Process. **65**, 3551–3582 (2017)

27. Tang, J., Liu, H.: Trust in Social Media. Synthesis Lectures on Information Security, Privacy, and Trust. Morgan & Claypool Publishers, San Rafael (2015)

28. Tang, J., Gao, H., Hu, X., Liu, H.: Exploiting homophily effect for trust prediction. In: International Conference on Web Search and Data Mining, WSDM, Italy, pp. 53–62 (2013)

29. Trifunovic, S., Legendre, F., Anastasiades, C.: Social trust in opportunistic networks. In: INFOCOM IEEE Conference on Computer Communications, pp. 1–6 (2010)

30. Trifunovic, S., Legendre, F., Anastasiades, C.: Social trust in opportunistic networks. In: INFOCOM IEEE Conference on Computer Communications Workshops, pp. 1–6 (2010)
31. Uddin, M.G., Zulkernine, M., Ahamed, S.I.: CAT: a context-aware trust model for open and dynamic systems. In: Symposium on Applied Computing, Brazil, pp. 2024–2029 (2008)
32. Wang, Y., Li, L., Liu, G.: Social context-aware trust inference for trust enhancement in social network based recommendations on service providers. World Wide Web **18**, 159–184 (2015)
33. Wang, Y., Wang, X., Tang, J., Zuo, W., Cai, G.: Modeling status theory in trust prediction. In: 21th AAAI Conference on Artificial Intelligence, USA, pp. 1875–1881 (2015)
34. Wang, K., Zhang, R., Liu, X., Guo, X., Sun, H., Huai, J.: Time-aware travel attraction recommendation. In: Lin, X., Manolopoulos, Y., Srivastava, D., Huang, G. (eds.) WISE 2013. LNCS, vol. 8180, pp. 175–188. Springer, Heidelberg (2013). https://doi.org/10.1007/978-3-642-41230-1_15
35. Wu, L., Liu, H.: Tracing fake-news footprints: characterizing social media messages by how they propagate. In: WSDM, pp. 637–645 (2018)
36. Zhang, Y., Chen, H., Wu, Z.: A social network-based trust model for the semantic web. In: Yang, L.T., Jin, H., Ma, J., Ungerer, T. (eds.) ATC 2006. LNCS, vol. 4158, pp. 183–192. Springer, Heidelberg (2006). https://doi.org/10.1007/11839569_18
37. Zheng, X., Wang, Y., Orgun, M.A., Liu, G., Zhang, H.: Social context-aware trust prediction in social networks. In: Franch, X., Ghose, A.K., Lewis, G.A., Bhiri, S. (eds.) ICSOC 2014. LNCS, vol. 8831, pp. 527–534. Springer, Heidelberg (2014). https://doi.org/10.1007/978-3-662-45391-9_45

Privacy Preserving Social Network
Against Dopv Attacks

Yumeng Fu[1,2(✉)], Wei Wang[1,2], Hao Fu[1,2], Wu Yang[1,2], and Dan Yin[1,2]

[1] Harbin Engineering University, Harbin 150001, China
{2012201203,w_wei,061505,yangwu,yindan}@hrbeu.edu.cn
[2] Information Security Research Center, Harbin, China

Abstract. The published multi-social network graphs contain numerous private information. To protect these information, researchers try to simulate attack models and design protection schemes. In this paper, we propose a heuristic attack model based on Dopv (Degree of paired vertices) attack. The attacker by defrauding trust or browse homepage to acquires the victim's degrees (number of friends) from two published social network graphs and combine them into Dopv. Based on Dopv attack, attacker locates target candidates then compare nodes similarity by the same attributes or labels to find out target. To avoid this attack and protect the individuals' privacy, we propose a new solution called Pvk-degree anonymity (Paired vertices k-degree anonymous). In Pvk-degree anonymity, the probability of a real user being re-identified is no more than 1/k. We devise algorithms to achieve the Pvk-degree anonymity that preserves the original vertex set in the sense that we allow the edge modified but no deletion of vertices. The experimental results show that our approach can preserve the privacy and guarantee the utility of social network graphs effectively against Dopv attacks.

Keywords: Multi-social network · Dopv attack · Pvk-degree anonymous

1 Introduction

The more social network users there are, the more useful data they contain. These data are modeled as a graph structure that each vertex represents an individual and the social relationship and activities are summarized by the edges. Preserving privacy of these published data has become a major direction for researchers [1, 2, 20]. Since graph contains abundant information in the vertices and edges, there are a variety of privacy attacks. Some attacks are based on background knowledge, including node existence, node properties [3, 5–7, 11], node label [13], node connection [8, 16], edge weights [15, 17], edge label and some auxiliary figure method [4] etc., some according to the subgraph structure [9, 10, 12, 18].

In this paper, we proposed Dopv attack model, based on the degree of vertex pair that a user from two different social networks. The rise of multiple social media poses new challenges for privacy. Some users share or forward the content of their own account on another platform on one social media for advertising, dating, etc. This lack of privacy

© Springer Nature Switzerland AG 2018
H. Hacid et al. (Eds.): WISE 2018, LNCS 11233, pp. 178–188, 2018.
https://doi.org/10.1007/978-3-030-02922-7_12

and security awareness provides attackers with a good way to attack. According to the account association information disclosed by the user, the attacker obtains the Dopv in two social networks formed by friend relationships and finds the user from two graphs. In the example of Fig. 1, Bob, who is attacked by the attacker, has the degree of pair vertex $\{d1 = 2; d2 = 2\}$ that means $\left(Bob_1, Bob_2\right)_{Dopv} = (2,2)$. According to $(2,2)$, the attacker found candidate $\{A\text{-}2, A\text{-}3\}$ from (a2), and $\{B\text{-}2, B\text{-}4\}$ from (b2). The attacker can easily find the nodes A-2 and B-2 are Bob from candidate sets by comparing the similar attributes carried by candidate nodes without knowing any other information before. It is easy to see that (b2) is 2-degree anonymity but did not protect the victim.

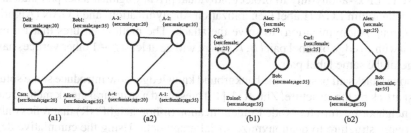

Fig. 1. An example of the Dopv attack. Original social networks (a1)(b1) and Naïve anonymized networks (a2)(b2)

To avoid Dopv attack, we introduce the concept of *Pvk*-degree anonymity, which according to the previous work obtain the associated attributes to generate Dopv and regard Dopv as anonymous object. Finally, for each Dopv, there are at least $k-1$ others equals to it. And at least 2k candidate points obtained by Dopv from two graphs can also represent at least k real users, this is to effectively cut off the attacker's further contrast identify by vertex similarity. The experimental results show that the social networks anonymized by our approaches preserve the effectiveness of the original social networks.

Challenges: In the *Pvk*-degree anonymity, the manipulated object Dopv consists of degree of the two nodes that make up the Dopv are unlinked on a single graph but require an overall anonymity to against multi-graphs degree attack. Another challenge is the attack model is established and feasible, but the data set relationship mapping of the attack model is difficult to establish. This issue involves the identification of multiple social networks and belongs to another research point, and relevant research members in our team are following up, so we have to experiment with artificial datasets to establish quasi-identifier associations.

2   Related Work

Liu et al. [3] studied a k-degree anonymity algorithm that based on dynamic programming to prevent vertex re-identify in graph, so that for any vertex n exist at least $k-1$ other vertices have the same degree in the public-shed graph. The article also combined with dynamic programming and greedy to optimize time complexity. Casas-Roma et al. [5] presented a k-degree anonymity algorithm on large, undirected and unlabeled simple networks. This paper implements UGMA (short for Univariate Micro-aggregation for

Graph Anonymization) algorithm to process datasets containing more than $2^q (q \geq 10)$ vertices. Tai et al. [8] introduced the privacy risk in public-shed social networks in terms of a new type of attack, called a friendship attack, and proposed k^2-degree anonymity to protect against such attacks. In this paper, they developed an Integer Programming formulation to find optimal solutions and designed an efficient heuristic approach for anonymizing large-scale social network. In addition, they discuss the extension of the heuristic approach to handle the general friendship attack of a degree sequence of length. Sun et al. [9] identified a k-NMF anonymity on the number of mutual friends, which protects against the mutual friend attack in social network publication. Liu C et al. [13] proposed a LP k^2-anonymity to protect individual privacy against label pair attack, and provide algorithm LGA (Label Generalization Anonymization) and LGAN (Algorithm Label Group Anonymization) to solve the attack. The method ensures that for every vertex v with an edge of label pair (l_1, l_2), there will be at least $k - 1$ other vertices having an edge of the same label pair.

Above attacks all based on the background knowledge, now introduce some solution against attack by the structure. Zhu et al. [18] presented a structural attack method called n-hop neighbor Feature for Node Reidentification (n-hop neighFNR) that relies only on the network structure to deanonymize social graph data. Using the cumulative degree of n-hop neighbors as the regional feature and combining with the simulated annealing-based graph matching method, with the aid of auxiliary graph, they can reidentify the nodes in anonymized social graphs. Tai et al. [19] proposed the structural diversity which ensures the existences of the existence of at least k communities containing vertices with the same degree for every vertex in the graph. They formulated the k-Structural Diversity Anonymization (k-SDA) to protects the community identify of each individual and an Integer Programming formulation to find the optimal solution to k-SDA, and three scalable heuristics to solve the large instances of k-SDA with different perspectives.

3   Problem Definition

In this paper, we model two social networks as two undirected simple graphs $G_1 = (V_1, E_1)$ and $G_2 = (V_2, E_2)$, where V_1 is a set of vertices representing the individuals from G_1, and V_2 from G_2. Besides, $E_1 \subseteq V_1 \times V_1$ is the set of edges representing the relationship of individuals from G_1 and $E_2 \subseteq V_2 \times V_2$ from G_2. $\tilde{G}_1$ and $\tilde{G}_2$ represent their published graphs form. First of all, let's see how we cause the attack on multi-graphs

Definition 1. Dopv (Degree of pair vertices) The victim's degree is d_1 in graph G_1, and d_2 in graph G_2, we called (d_1, d_2) is Dopv that degree of paired vertices. If a target named Alice whose degree is d_1 in G_1 and d_2 in G_2, then (d_1, d_2) means Alice's Dopv and express as $Alice_{Dopv} = (d_1, d_2)$.

Definition 2. Dopv attack In published anonymous graphs G_1 and G_2 and the attack aims Alice (a pseudonym). According to the background knowledge $Alice_{Dopv} - (d_1, d_2)$, attacker based on d_1 get a candidates vertices form set A from G_1 and based on d_2 get b candidates form set B from G_2.

A is candidates set from G_1 and B from G_2, and same vertices ID represent the same user. Figure 2 means that in the A and B sets, there are only a few or one pair of vertices representing a real user. We know that one user's nodes more similar so the attacked object Alice easy be found. Have to say, this attack still established even $a \geq k$ and $b \geq k$. Above is the Dopv attack. Contrary to this attack, we propose an anonymous solution.

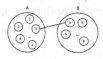

Fig. 2. Candidates sets and mapping relationship

Definition 3. Pvk-degree Anonymity Pvk–degree anonymity means put the Dopv that attacker' background knowledge as protected object, so that for each $Object_{Dopv} = (d_x, d_y)$, there are at least $k - 1$ other objects have same Dopv. Furthermore, if some points in G_1 and G_2 meet the conditions that in G_1 but no mapping in G_2 or reverse, we call them Npp, means unpaired and use $Object_{Npp_x} = d_x$, or $Object_{Npp_x} = d_y$ to said. For the Npp, we make sure that there are at least other $k - 1$ nodes have same degree in the graph to which it belongs. Especially, we ignored the user whose $d_x = 0$&&$d_y = 0$.

See the example in Fig. 3. Six nodes in (a1) and five nodes in (a2), First, we can see the $Node(1, 3, 4)_{Dopv} = (2, 2)$, $Node(2, 5)_{Dopv} = (3, 3)$, it means the Dopvs in (a1)(a2) have fit $Pv2$-degree anonymity. Then, only $Node(6)_{Npp_x} = 2$ in the unpaired set and degree 2 appear three times in all degrees. We said (a1)(a2) fit $Pv2$-degree anonymity. In (b1)(b2), we get $Node(1, 2, 3)_{Dopv} = (2, 2)$ fit $Pv3$-degree anonymity, then insert the value from $Node(4)_{Npp_x} = 2$, we call the (b1)(b2) reach $Pv3$-degree anonymity.

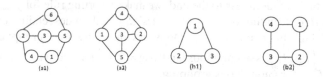

Fig. 3. Examples of Pvk-degree anonymity

Furthermore, Pvk-degree anonymity has the following properties of downward closure protection.

PROPOSITION 1. If graphs G_1 and G_2 are Pvk-degree anonymity, G_1 and G_2 are also k-degree anonymity.

PROPOSITION 2. If graphs G_1 and G_2 are Pvk_1-degree anonymity, then they also are Pvk_2-degree anonymity for every $k_2 \leq k_1$.

With the above properties, we can know the complete anonymous process, then we need introduce another concept for anonymous and compute modify cost.

Definition 4. Pvk-anonymous Sequence A sequence vector f is k-anonymous, if for any entry with value as v, there exist at least $k - 1$ other entries with value as v [3]. We use two sequences f_{Dopv}, f_{Npp_x} $\left(\text{or } f_{Npp_y} \right)$ to express the degrees satisfy the k-anonymous sequences in algorithm. For example, we get the sets of $f_{Dopv} = \{(2,2),(3,3),(2,2),(2,2),(3,3)\}$ and $f_{Npp_x} = \{2\}$ in Fig. 3 (a1)(a2).

Definition 5. Anonymous cost of Dopv We can vividly interpret Dopv as the point of the first quadrant in two-dimensional coordinates and the Dopv set as a set of points randomly distributed in the first quadrant. Anonymous cost of Dopv is:

$$C_{Dopv} = \Sigma \left(\left| d_{ori_x} - d_{targ_x} \right| + \left| d_{ori_y} - d_{targ_y} \right| \right)$$

where d_{ori} means original degree and d_{targ} is target degree.

4 Anonymity Approach

At the beginning of this section, we need to introduce some of the remarks below. We call the vertices from G_1 as x_i, $i = 1, 2, 3, \ldots, m$, m is the number of vertices in the G_1, and use y_j indicate the points from G_2, $j = 1, 2, 3, \ldots, n$, n is the number of vertices from G_2. $Node(q)_{Dopv} = \left(d_{x_i}, d_{y_j} \right)$, $q = 1, 2, \ldots, p, x_i \in G_1$ and $y_j \in G_2$, p is the number of Dopv so $p \le m$ and $p \le n$.

4.1 Order and Group Dopv

Due to the different density of Dopv distributions and the uncertainty of the point will bring unpredictable randomness to the end, we make a sorting rule for f_{Dopv}. According to the value of d_x descending order, if the nodes' d_x equal, in descending order of d_y. In addition, set "flag" bit mark whether Dopv has in anonymous group: flag $= 1$ said $Node(q)_{Dopv} = \left(d_{x_i}, d_{y_j} \right)$ is anonymous and flag $= 0$ is not anonymous. We have the sorted vector f'_{Dopv} for anonymous grouping.

Then, we should make f_{Dopv} be k-anonymous sequence. Select the first one *flag = 0* in the current f'_{Dopv}, and select at least $k - 1$ candidate Dopvs among the remaining members to form a new anonymous group, and select a center Dopv in the anonymous group to anonymize the group members make sure the same Dopv. During the selection process, involving group anonymous edits minimize the problem, we need compute formula below.

$$C_{sel(x_i,y_i)} = \left| d_{x_i} - d_{x_j} \right| + \left| d_{y_i} - d_{y_j} \right| \tag{1}$$

where $(x_i, y_i)_{Dopv} = (d_{x_i}, d_{y_i})$ from f_{Dopv} and is the first point *flag = 0* in current f'_{Dopv}, $\left(d_{x_j}, d_{y_j} \right)$ means all other Dopvs, $1 \le j \le p$ and $j \ne i$. For one point $(x_i, y_i)_{Dopv}$, there are an ascending sequence $f_s^{(i)}$ to save all $C_{sel(x_i,y_i)}$.

Thirdly, we focus on total anonymous cost. When we try to get final anonymous groups, we need choose at least $k - 1$ Dopvs with current first point $flag = 0$ make an anonymous group, then decision out who will be the goal Dopv in this group. When we have obtained the anonymous groups and their aims Dopv, we have succeeded in anonymizing all pairs of points. We use the $1, 2, ..., g$ to give the group members numbers, and in order to avoid single anonymous group is too large, we artificially rule if group member g is greater than $2 k$ we no longer consider adding new members for this group, so $g < 2k$.

$$C_{gro(x_u, y_u)} = \sum_{i=1}^{g} (\left| d_{x_i} - d_{x_u} \right| + \left| d_{y_i} - d_{y_u} \right|) \tag{2}$$

where Eq. (2) calculates the group anonymizing cost of (x_u, y_u) as the target point. Then we use the (3) choose the target point Dopv below.

$$(x_u, y_u) = min_{1 \le u \le g} \{ \sum_{i=1}^{g} (\left| d_{x_i} - d_{x_u} \right| + \left| d_{y_i} - d_{y_u} \right|) \} \tag{3}$$

The members in the group are respectively treated as the target Dopv to obtain the $C_{gro(x_u, y_u)}$ value, and the minimum C_{gro} value (x_u, y_u) is taken as the center of the anonymous group.

$$C_{merge} = C_{gro(x_{u'}, y_{u'})} + \left| d_{x_i} - d_{x_{u'}} \right| + \left| d_{y_i} - d_{y_{u'}} \right| \tag{4}$$

$$C_{new} = C_{gro(x_{u_j}, y_{u_j})} \tag{5}$$

We set $(x_i, y_i)_{Dopv}$ is the current Dopv in f_{Dopv}, $(x_j, y_j)_{Dopv}$ is the current selected Dopv in $l_s^{(i)}$. In current $l_s^{(i)}$, two situations are encountered for $(x_j, y_j)_{Dopv}$ when generating anonymous groups by selecting points in a forward-to-backward order.

Case1. If $flag = 0$, represent $(x_j, y_j)_{Dopv}$ no exists in any other anonymous group. Then we put it into the new group g_i that $(x_i, y_i)_{Dopv}$ formed, and calculated group anonymous costs C_{new} when the new group members equal to k.

Case2. If $flag = 1$, indicate $(x_j, y_j)_{Dopv}$ have belong to some anonymous group g_j, then we need to compute the value of C_{merge} in formula (4), the anonymous cost of the g_j after $(x_i, y_i)_{Dopv}$ put into the group.

In the resulting C_{merge} and C_{new}, select the group in which the minimum value exists and put $(x_i, y_i)_{Dopv}$ in it. If $minC_{merge} < C_{new}$, then release newly formed group members besides $(x_i, y_i)_{Dopv}$, and set the $(x_i, y_i)_{Dopv}$'s $flag = 1$. Otherwise, set all new group members' $flag = 1$.

Finally, the points in f_{Dopv} have find their own anonymous group, while the anonymous group's target Dopv has also been identified. We store the points from G_1 in f_{Dopv}

to f_{Dopv_x} and the points from G_2 into f_{Dopv_y}. We also design the design a structure to store the original and target degrees for each of points f_{Dopv_x} and f_{Dopv_y}.

4.2 Npp Anonymous

After processing *Algorithm1*, we handle the *Npp* next. First, we use a structure that stores f_{Dopv_x} and f_{Dopv_y} to store f_{Npp_x} and f_{Npp_y}, then fill the original and set the target $= 0$. Next, we set f_{G_1} and f_{G_2}, and

$$f_{G_1} = f_{Dopv_x} \cup f_{Npp_x} \ and f_{G_2} = f_{Dopv_y} \cup f_{Npp_y} \tag{6}$$

sort them in descending order of goal. Obviously, as the join of *Npp*, the f_{G_1} and f_{G_2} can not satisfied k anonymous.

Havel Theorem [21]. Set non-negative integer sequence $d = \{d_1, d_2, \ldots, d_n\}$ meet: $(d_1 + d_2 + \ldots d_n) \bmod 2 = 0$, $n - 1 \geq d_1 \geq d_2 \geq \ldots d_n \geq 0$, then d can be simple graphical if and only if

$$d' = (d_2 - 1, d_3 - 1, \cdots, d_{d1+1} - 1, d_{d1+2}, \ldots, d_n) \tag{7}$$

is simple.

By Havel theorem we can know, if the final sequences violation Havel Theorem, the anonymous graphs can not be generated. Therefore, in order to satisfy the Havel theorem, we have designed a double-target degrees candidate method as far as possible to ensure the success of anonymity.

Since $f_{G_1} \left(or f_{G_2} \right)$ is arranged in descending order of the target degree, *1* to *p* in the sequence is an anonymous node, and $p + 1$ to *m*(or *n*) are nodes with a target degree of *0* from $f_{Npp_x} \left(or f_{Npp_y} \right)$. Anonymous from the *m*th (or *n*th), set it to degree $d_m(ord_n)$, we need select two candidates as the d_m's $\left(or\ d'_n s \right)$ target degree and one of the candidate points is an odd number and the other is an even number. We use d_{odd} save the odd candidate, and d_{even} save the even candidate.

$$d_{odd} = min_{odd} \left\{ |d_m - d_x| \right\} \ and \ d_{even} = min_{even} \left\{ |d_m - d_x| \right\} \tag{8}$$

Where $d_x = d_1, d_2, d_3, \ldots, d_p$' target degree. And we use Δd_{m1} and Δd_{m2} record the differences.

$$\Delta d_{m1} = min\{|d_m - d_{even}|, |d_m - d_{odd}|\}$$
$$\Delta d_{m2} = max\{|d_m - d_{even}|, |d_m - d_{odd}|\}$$

After solving all the double candidates, set the target degree of all the points as the first candidate and check whether the generated degree sequence $f'_{G_1} \left(or f'_{G_2} \right)$ satisfies the Havel theorem. If satisfied, output anonymous sequence $f'_{G_1} \left(or f'_{G_2} \right)$, else selects the node with smallest Δd_2 among all second candidates and write the corresponding target degree.

4.3 Modify the Graphs

After getting $f'_{G_1} = \{d_1^{x'}, d_2^{x'}, d_3^{x'}, \ldots, d_m^{x'}\}$ and $f'_{G_2} = \{d_1^{y'}, d_2^{y'}, d_3^{y'}, \ldots, d_m^{y'}\}$, we sort the unprocessed nodes in the two k-anonymous sequences in non-ascending order, as $d'_1 \geq d'_2 \geq d'_3 \geq \ldots d'_t, d'_i$ is the current degree of processing graph, t is the number of unprocessed nodes and id (d'_i) said the degree of d'_i node label. If $d'_1 \geq 0$, following the order of d'_2 to d'_t, we add in the $\tilde{G}_1 = (V_1, E'_1)(or\tilde{G}_2 = (V_2, E'_2))$ the edges (id (d'_i), id (d'_i)) that exist in the original graph $G_1 = (V_1, E_1)(orG_2 = (V_2, E_2))$, where $d'_1 > 0$ and $d'_i > 0$ the value of d'_1 and d'_i subtract 1. During the process, there are possibilities encountered:

Case1. There are not enough d'_i matching edges in the original graph $G_1 = (V_1, E_1)(or\ G_2 = (V_2, E_2))$, we add the edges that do not appear in anonymous graph in ascending order.

Case2. In the original graph, there are more than d'_i matching edges, we add the edges in anonymous graph until $d'_i = 0$.

Mark the id (d'_1) as "processed" and repeat the above steps. When all id (d') is "processed", output the anonymous graphs $\tilde{G}_1 = (V_1, E'_1)(or\ \tilde{G}_2 = (V_2, E'_2))$.

5   Experimental Results

In this section, we conduct experiments on multiple datasets to evaluate the performance of the proposed graph anonymization algorithms.

5.1   Datasets

Compared with the general graph structure, social network needs to satisfy two characteristics that node degree accord with power law distribution [22] and "small-world phenomenon". We use the R-MAT Model [14] to generate artificial datasets to satisfy two special properties of social networks (Table 1).

Table 1. Detailed parameters of the dataset

	Graph	Vertices	Edges	AVL
G_A	G_{A1}	128	1280	1.787
	G_{A2}	128	1920	1.79
G_B	G_{B1}	1024	51200	1.961
	G_{B2}	1024	112000	1.787

5.2 Evaluating Pvk-Degree Anonymization Algorithm

We evaluate the performance of all process of the anonymity algorithms by measuring the running time, average clustering coefficient, average path length and the ratios of edges change and show the results in Figs. 4, 5, 6 and 7.

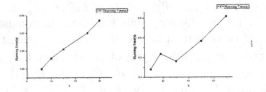

Fig. 4. G_A and G_B's running time

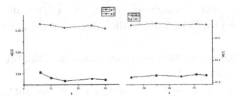

Fig. 5. Average clustering coefficient

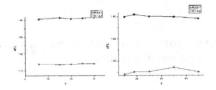

Fig. 6. Average path length

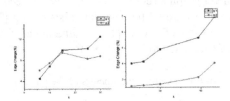

Fig. 7. Percentages of the edges add and deleted

Running Time: Figure 4(a) show the runtime on the network G_{A1} and G_{A2} with respect to different k values, and Fig. 4(b) shows the G_B's. We can know that in most cases, the running time increase when the k value increases from the figure, and the time increase as the vertices increase.

Average Clustering Coefficient (ACC): Figure 5 compares the average clustering coefficients of the original graphs and the anonymized graphs. the basic trend is that ACC have a slight fluctuation, but no significant increase and decrease.

Average Path Length (APL): The average path lengths on two datasets for the different value k, we can see the change in Fig. 6. The APL of the graph anonymized is very close to the APL of the original graphs, and the same number of vertices, the more edges of the original graph, the smaller the average length of the path after the anonymous.

Percentages of edges changed: We consider the edge changes in our algorithm. Figure 7 show the edge changes compare with original graphs. The changes include the radios of the edges added and edges deleted. In our algorithm, we try to change the fewest edges to realize pvk-degree anonymous, and we have different k to verification algorithm's efficient.

6   Conclusions

In this paper, we have identified a new problem of Pvk-degree anonymity on the double vertices of one person, which protects against Dopv attack in multi social networks publication. To solve this problem, we designed heuristic algorithm which consider the similarity of one person's double vertices and the utility of the graphs. By the algorithm, we also ensure the k-degree anonymity in each graph. The experimental results demonstrate that our approaches can ensure the Pvk-degree anonymity while preserve much of the utility in the original multi social networks.

FutureTask. We are studying the relationship between real data sets. After obtaining certain results, we will further verify the efficiency and feasibility of the algorithm on large real data sets. Improve the conclusion of the experiment.

Acknowledgements. This work is supported by National Natural Science Foundation of China under Grant 61572459, 61672180. The paper is funded by the International Exchange Program of Harbin Engineering University for Innovation-oriented Talents Cultivation.

References

1. Wu, X., Ying, X., Liu, K., et al.: A survey of algorithms for privacy-preservation of graphs and social networks. In: Managing & Mining Graph Data Chapter, pp. 421–453 (2009)
2. Casas-Roma, J., Herrera-Joancomartí, J., Torra, V.: A survey of graph-modification techniques for privacy-preserving on networks. Artif. Intell. Rev. **47**(3), 1–26 (2016)
3. Liu, K., Terzi, E.: Towards identity anonymization on graphs. In: ACM SIGMOD International Conference on Management of Data, pp. 93–106. ACM (2008)
4. Narayanan, A., Shmatikov, V:. De-anonymizing Social Networks. In: IEEE Symposium on Security and Privacy, 2009, pp. 173–187. IEEE (2009)
5. Casas-Roma, J.: An algorithm for k-degree anonymity on large networks. In: International Conference on Advances in Social Networks Analysis and Mining, pp. 671–675. IEEE (2013)
6. Casas-Roma, J., Herrera-Joancomartí, J.: k-Degree anonymity and edge selection: improving data utility in large networks. Knowl. Inf. Syst. **50**(2), 1–28 (2016)

7. Bredereck, R., Froese, V., Hartung, S., et al.: The complexity of degree anonymization by vertex addition. Theoret. Comput. Sci. **607**(P1), 16–34 (2015)
8. Tai, C.H., Yu, P.S., Yang, D.N., et al.: Privacy-preserving social network publication against friendship attacks. In: ACM SIGKDD International Conference on Knowledge Discovery and Data Mining, pp. 1262–1270. ACM (2011)
9. Sun, C., et al.: Privacy preserving social network publication against mutual friend attacks. In IEEE, International Conference on Data Mining Workshops, pp. 71–97. IEEE (2014)
10. Assam, R., Hassani, M., Brysch, M., et al.: (k, d)-Core anonymity: Structural anonymization of massive networks. ACM (2014)
11. Ma, T., Zhang, Y., Cao, J., et al.: KDVEM: a k-degree anonymity with vertex and edge modification algorithm. Computing **97**(12), 1165–1184 (2015)
12. Wu, H.: A Clustering Bipartite Graph Anonymous Method for Social Networks. J. Inf. Comput. Sci. **10**(18), 6031–6040 (2013)
13. Liu, C., Yin, D., Li, H., Wang, W., Yang, W.: Preserving privacy in social networks against label pair attacks. In: Ma, L., Khreishah, A., Zhang, Y., Yan, M. (eds.) WASA 2017. LNCS, vol. 10251, pp. 381–392. Springer, Cham (2017). https://doi.org/ 10.1007/978-3-319-60033-8_34
14. Chakrabarti, D., et al.: R-MAT: a recursive model for graph mining. In: Siam International Conference on Data Mining, Lake Buena Vista, Florida, USA. DBLP, April 2004
15. Li, Y., Shen, H.: Anonymizing graphs against weight-based attacks. In: IEEE International Conference on Data Mining Workshops, pp. 491–498. IEEE Computer Society (2010)
16. Bhagat, S., Cormode, G., Krishnamurthy, B., et al.: Class-based graph anonymization for social network data[J]. Proc. Vldb Endow. **2**(1), 766–777 (2009)
17. Liu, X., Yang, X.: A generalization based approach for anonymizing weighted social network graphs. In: International Conference on Web-Age Information Management, pp. 118–130. Springer-Verlag (2011)
18. Zhu, T., Wang, S., Li, X., et al.: Structural attack to anonymous graph of social networks. Math. Probl. Eng. **2013**(2), 1–8, 21 November 2013
19. Tai, C.H., Yu, P.S., Yang, D.N., et al.: Structural diversity for privacy in publishing social networks. In: Eleventh Siam International Conference on Data Mining, SDM 2011, April 28– 30, 2011, Mesa, Arizona, USA, pp. 35–46. DBLP (2012)
20. Hakimi, S.L.: On realizability of a set of integers as degrees of the vertices of a linear graph. I. J. Soc. Ind. **10**(3), 496–506 (1962)
21. Wasserman, S., Faust, K.: Social Network Analysis. Encycl. Soc. Netw. Anal. Min. **22**(Suppl 1), 109–127 (1994)

A Hybrid Approach for Detecting Spammers in Online Social Networks

Bandar Alghamdi[1,2](✉) 🆔, Yue Xu[1](✉) 🆔, and Jason Watson[1](✉) 🆔

[1] Faculty of Science and Engineering, Queensland University of Technology,
Brisbane 4000, Australia
bandar.alghamdi@hdr.qut.edu.au, alghamdib@ipa.edu.sa,
{yue.xu, ja.watson}@qut.edu.au
[2] Institute of Public Administration, Riyadh 11141, Saudi Arabia

Abstract. Evolving behaviours by spammers on online social networks continue to be a big challenge; this phenomenon has consistently received attention from researchers in terms of how it can be combated. On micro-blogging communities, such as Twitter, spammers intentionally change their behavioral patterns and message contents to avoid detection. Many existing approaches have been proposed but are limited due to the characterization of spammers' behaviour with unified features, without considering the fact that spammers behave differently, and this results in distinct patterns and features. In this study, we approach the challenge of spammer detection by utilizing the level of focused interest patterns of users. We propose quantity methods to measure the change in user's interest and determine whether the user has a focused-interest or a diverse-interest. Then we represent users by features based on the level of focused interest. We develop a framework by combining unsupervised and supervised learning to differentiate between spammers and legitimate users. The results of this experiment show that our proposed approach can effectively differentiate between spammers and legitimate users regarding the level of focused interest. To the best of our knowledge, our study is the first to provide a generic and efficient framework to represent user-focused interest level that can handle the problem of the evolving behaviour of spammers.

Keywords: Spammers · Behaviour · Detection · Online social networks

1 Introduction

Recent developments in the field of online social networks have led to the integration of Online Social Networks (OSNs) into nearly all aspects of everyday activity; however, spammers take advantage of these services for malicious purposes. With the increase in the influence of OSNs among users, a large platform has been established that spammers use to spread spam messages [1]. Spam tweets refer to unsolicited tweets containing malicious links that direct victims to external sites containing malware downloads, phishing scams, drug sales, etc. [2]. Spammers utilize different methods in spreading spam content, either using compromised accounts with already established reputations and exploiting the inherent trust of these accounts to spread

© Springer Nature Switzerland AG 2018
H. Hacid et al. (Eds.): WISE 2018, LNCS 11233, pp. 189–198, 2018.
https://doi.org/10.1007/978-3-030-02922-7_13

malicious messages [3] or creating fake accounts that appear to be legitimate to mimic legitimate user behaviour by posting spam content and normal content.

In this paper, we propose a novel two-stage approach to detecting spammers in online social networks. In the first stage, we inspect the user's interest to determine focused-interest and diverse-interest users and then represent each group with the most effective features. In the second stage, separate classifiers will be used to classify user as spammer or legitimate user. This work considered user interests as a key point since the engagement of users in any activity is driven by their interests. The main contribution of this study can be summarized as follows:

- We propose a novel approach for characterizing spammer behaviour based on two levels of user interest: focused and diverse.
- On the basis of focus level, we represent users with features, using existing features and our proposed new feature that describes the consistency between explicit user's interest in the profile and user's content.
- We implement our approach by combining unsupervised clustering and supervised classification to detect spammers in OSNs.

2 Related Work

Many studies have been conducted to investigate spammers' behaviour in online social networks, and researchers have shown an increased interest in this regard. A survey of potential solutions and challenges on spam detection in online social network has been proposed by [4]. Previous works have focused on characterizing spammers' behaviour using different features and approaches [5–8]. A considerable amount of literature has been published on spam detection using **content-based** features [1, 9]. The statistical analysis of language, such as linguistics evolution, self-similarity and vocabulary, are the primary features used for spam detection. Although they perform well in detecting spam tweets, their limitation relies in the fact that content features alone cannot be used to properly analyze spammer behaviour. Some works consider only **user's profile** [10] to detect spam users, without considering content features due to the idea that this is a fast and effective way. They capture users' behaviour and identify certain patterns from the profile to detect spammers and compromised accounts. However, these studies were limited to characterizing spammer behaviours in regard to a few aspects, and they showed a lack of classification accuracy as their approaches are not sophisticated.

It has conclusively been shown that combining content and profile features provides a comprehensive understanding of spammer behaviour [5, 6, 11]. They determine that there is a strong and consistent correlation between the profile and content for all suspicious accounts. Such a combination shows the fundamental characteristics of spammers from different views and provides a different level of detection rates. In our present study, we extend this combination by considering content features and user demographic data with a focus on the user's interests. The premise being that this more sophisticated way of differentiating user behavior will allow for enhanced detection of spammers.

Subsequent approaches have been proposed using **topics features** to detect spammers. [12] Performed an experiment using the standard Latent Ditichlet Allocation (LDA) approach to measure the degree of change in user's interest to detect human-like spammers. The results of this study indicate that spam users either concentrate on certain topics or have interests in some topics. Similarly, legitimate users mainly focus on limited topics. Alternatively, Nilizadeh et al. [13] identified different communities that share similar topics of interest and inspected the dissemination path to predict the pattern of posting within and outside of the community in order to detect malicious messages. The limitation of the above approaches is that they still characterize spammers with unified features, whereas spammers behave differently and should be represented by different features.

3 Proposed Method

We propose a two-stage approach to constructing a more accurate classifier for identifying spammers. The components of our approach are shown in Fig. 1. The first stage is to separate users into focused and diverse groups. The second stage is to construct a classifier which consists of two sub-classifiers for the focused and diverse users, respectively. In the following sections, we explain each component in detail.

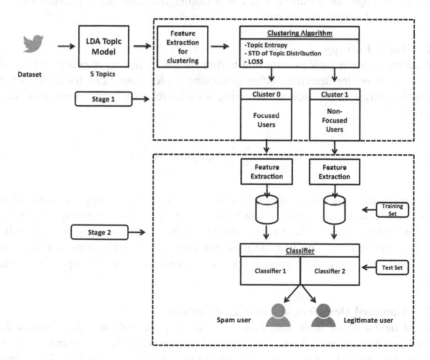

Fig. 1. A proposed framework with two stages to detect spam users.

3.1   Modelling User Information Interests Based on LDA Topic Models

Latent Dirichlet Allocation (LDA) was first introduced by Blei [14] as an example of a topic model. Each document d_i is represented as a bag of words $W = \{w_1, w_2 \ldots \ldots w_M\}$, and M is the number of words. Each word is attributable to one of the document's topics $Z = \{z_1, z_2 \ldots \ldots z_K\}$, and k is the number of topics. φ_j is a multinomial distribution over words for topic z_j, $\varphi_j = \langle p(w_1|z_j), \ldots \ldots, p(w_M|z_j) \rangle$, $\sum_{i=1}^{M} p(w_i|z_j) = 1$. φ_j is called the topic representation for topic z_j. θ_i is another multinomial distribution over topics for document d_i. $\theta_i = \langle p(z_1|d_i), p(z_2|d_i), \cdots, p(z_k|d_i) \rangle$, and $p(z_j|d_i)$) indicates the proportion of topic z_j in document d_i. θ_i is called the topic distribution for document d_i. We considered each user's tweets as one document. The user's information interest is reflected in the tweet content, and we apply the LDA Topic Model to generate 5 topics for each user and get the topic probabilities for each single user. From these topics distribution values, we extract the three topical features to measure the user's interest change as the following section explains.

3.2   Depicting Users' Interest as a Focused or Diverse Interest

The following section details three topic-based features: Topic Entropy, Standard Deviation of Topic Distribution, and Local Outlier Standard Score (LOSS), for clustering users as focused and diverse.

3.2.1   Topic Entropy
User's interest is a reliable feature that is difficult for spammers to evade and that can therefore be used for detection. After generating topics from LDA for each user, we used topic entropy to measure the diversity of topics for each user, using the following equation:

$$H(u) = - \sum_{i=1}^{k} p(z_i|u) \log_2 p(z_i|u) \tag{1}$$

$p(z_i|u)$ is the topic distribution for user u. $H(u)$, is the entropy value that shows different levels of focus interest. It can help us later in the clustering stage to get users with low topic entropy or high topic entropy. A user with low topic entropy will be more concentrated meaning that the user is interested on limited topics, while a user with higher entropy will be spread somewhat evenly over many topics as Table 1 shows.

3.2.2   Standard Deviation of Topic Distribution
Standard deviation of topic distribution is also a good indicator for differentiating focused and diverse users. Using the standard deviation, we have a 'standard' way of knowing how spread out the topics are from the mean of a given user. This demonstrates the degree of change in topics for a particular user as Table 1 shows.

Table 1. Users having different topic distributions, entropy and standard deviation

User	Std	Topic entropy	Topic 0	Topic 1	Topic 2	Topic 3	Topic 4
Focused user	0.4446	0.0355	0.0023	0.0007	0.0007	0.9953	0.00077
Diverse user	0.0216	1.6048	0.2035	0.1996	0.1871	0.1760	0.2336

3.2.3 Local Outlier Standard Score

This topic-based feature was first proposed by [12] to discriminate human-like spammers from legitimate users using topic distribution. Liu used this feature for classification purposes, but we use it for a clustering technique to separate users into focused and diverse groups. This feature measures the degree of interest of user in respect to a certain topic, using the following equation:

$$\mu(u_i) = \frac{\sum_{k=1}^{k} p(z_k|u_i)}{k} \qquad LOSS(u_{ik}) = \frac{x_{ik} - \mu(x_i)}{\sqrt{\sum_k (x_{ik} - \mu(x_i))^2}} \qquad (2)$$

Where, $\mu(u_i)$ is the average degree for all topics for a certain user. If we extract k topics for each user, we will end up with a vector of k features for each user, $LOSS(u_{i1}), \ldots\ldots, LOSS(u_{iK})$.

3.2.4 Features Vector for Clustering

In this paper, we propose to use topic entropy, standard deviation of topics distributions, and the vector of LOSS as the features to cluster users. Formally, for each user i, let k be the number of topics, we can calculate a total of $N = k + 2$ topical features which include k $LOSS$ features $LOSS(u_{i1}), \ldots\ldots, LOSS(u_{ik})$, topic entropy $H(u_i)$ and standard deviation $Std(u_i)$ Our goal is to construct a -N-dimensional feature vector $V_i = \langle v_{i,0}, \ldots v_{i,N} \rangle$ from the topic distribution values.

From the Honeypot dataset [15], we generate two clusters based on the proposed topical features. The result showed in Fig. 2 indicates that both clusters contain spammers and non-spammers as well. Table 2 shows the average feature vectors (i.e., centroids) of the two clusters. Based on the feature values in the centroids, we can decide that the users in Cluster 0 are more focused than the users in Cluster 1.

Table 2. Average feature vectors centroids

Attribute	Cluster 0	Cluster 1
Topic 0 LOSS	0.1126	0.1774
Topic 1 LOSS	0.7051	0.2255
Topic 2 LOSS	0.0617	0.1644
Topic 3 LOSS	0.0247	0.2092
Topic 4 LOSS	−0.2888	−0.3005
Std of topics dis	0.2987	0.2321
Topic Entropy	0.8468	1.0584

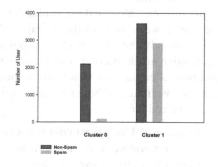

Fig. 2. Clusters based on topical features.

3.2.5   Focused and Diverse Interest

We have two clusters: one that are focused interest (Cluster 0) and the second one that has diverse interest (Cluster 1) as Fig. 2 shows. According to our observation of the clusters, we have 2251 users in the Cluster 0, with a total of 2263 legitimate users and 287 spam users. This cluster represents focused users and composes primarily of legitimate users, at 94% of the cluster. Even the number of spammers is low, it is interesting that a small number of spammers can be focused. On the hand, Cluster 1 is a mixed cluster, which composes of 2891 spam users and 3612 legitimate users, as Fig. 2 shows. After exploring this cluster, we have observed that legitimate users tend to have different interests and that this is common for normal users. Figures 3 and 4 show the cumulative distribution function (CDF) of standard deviation and topic entropy.

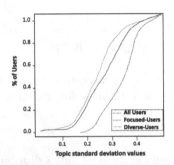 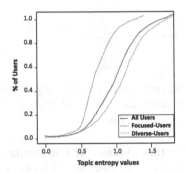

Fig. 3. CDF of std of topic distribution.   **Fig. 4.** CDF of topic entropy.

3.3   Supervised Learning Stage

Classification will be used to identify spammers in each of the clusters. We use our propose features based on users' self-description and other existing features to represent each group with more effective features.

3.3.1   Features for Classification

Interest consistency between the description and tweets [16]: Users provide their interest in the description statement in their profile. The users' descriptions provide explicit information about their interests, whereas the users' tweets reveal their interest implicitly. We calculated the similarities between each single user's description and all the users' tweets and got the average values. For this feature, legitimate users showed that averagely the content of their posts has relatively higher similarity to their self-description in comparison with that of spammers. This feature reveals the users interest and how often the user tweets in this interest. The primary hypothesis of this feature is that the integration of both post and profile features are essential to properly understand users' behaviour.

1. **Number of Unique words:** It has been proved that legitimate normal accounts are more innovative in their use of language, while spammers may repeat themselves more often [9]. However, this feature is not applicable to all spammers. In our

dataset, we have found that spammers and legitimate users in Cluster 0 (focused users) somehow exhibit similar use of the same word and do not post a large number of unique words. The number of unique words is more effective for diverse users.

2. **Average "@username" per tweet:** the insertion of @ is essentially used to deliver messages to another user. This is very common behaviour by spammers and has been examined in previous studies. The use of @ is found to be a good feature for diverse- users, to discriminate spam from non-spam users. The reason for this is that users with focused-interest (both spam and non-spam) generally exhibit similar behaviour of posting similar content and posting @username very often., whereas, it is more useful for diverse-users, as the normal user does not use @ significantly.

3. **Number of links:** This feature is very similar to the use of @ for both groups. Among focused users, spammers and non-spammers post a large number of links to target users. If we consider this feature as a unified feature for spam detection, we would have misclassification of legitimate users who have focused-interest. Diverse-interest users vary in term of using links, and spammers show a higher usage of links in their tweet than normal users do.

4. **Following and followers:** this feature can be used in both clusters. Following behaviour is abused by spammers to gain access to many targeted users. This behaviour is a common characteristic of spammers and has been extensively used for spammer detection by [9, 17]. It is worth mentioning that followers as a feature is not present in diverse-users, whereas this feature has a high contribution to the class attribute in focused user. Spam users in the diverse group are hidden as legitimate accounts that have a good reputation and they do not seek to have more followers in order to appear as legitimate accounts. Interestingly, this feature is significant in focused users, as spammers tend to appear as legitimate users, and they use third parties to get more followers [17].

3.3.2 Features Selection

We select features that best classify users using the correlation-based feature algorithm [18]. We utilize the correlation-based feature (CFS) algorithm to select the most effective features for each cluster. This algorithm evaluates the worth of a subset of attributes by considering the individual predictive ability of each feature along with the degree of redundancy between them. The subsets of features that are highly correlated with the class while having low inter-correlation with class are preferred and irrelevant features should be ignored. We apply this algorithm to all features, and result is presented on Table 3. The CFS's function works as follows:

$$Ms = \frac{k\overline{r_{cf}}}{\sqrt{k + k(k-1)\overline{r_{ff}}}} \tag{3}$$

Where Ms is the heuristic 'merit' of the feature subset s containing features, $\overline{r_{cf}}$ is the mean feature-class correlation over the feature in s (i.e., $f \in s$) and $\overline{r_{ff}}$ is the average feature-features inter-correlation.

Table 3. Each cluster has somewhat different features, where demographic features are very effective in focused users and the content features with demographic features are effective in diverse-interest users.

	Cluster 0 (Focused interest)	Cluster 1 (Diverse interest)
Demographic features	Number of followings Number of followers Std of following Ratio of following & followers Change rate of following	Number of followings Std of following Ratio of following & followers Change rate of following
Content features	No very effective	*Number of unique words* *Average at@ per tweet* *Number of links* *Average value of cosine similarity*

4 Experiment and Evaluation

In this section, we first describe the implementation of our detection approach, then the dataset, and the ground truth for evaluation. Finally, we evaluate the detection rate of our proposed method using the standard metrics. We use Weka machine learning [19] to test the classification algorithm. We use default values for parameters using 10-fold cross-validation, where the original sample (data) is divided into 10 sub-samples and 10 training and validation steps are performed. For the training, nine sub-samples are used, and the remaining sub-samples are used for validation. We chose the Honeypot dataset [15]. It is worth mentioning that the dataset was reduced because of the limited number of users who had descriptions and the limited tweets that were not enough to identify user's interest. The ultimate dataset contained 5875 spam users with a total of one million tweets, and 3178 non-spam users with 572,040 tweets.

For clustering, we calculated topical features based on the topic model as described in Sect. 3.2. We used the expectation maximization (EM) algorithm and clustered the users into two different groups.

4.1 Detection Results

We use the Random Forest algorithm to perform the detection mechanism. We calculated the average values for: precision, recall and F-measure; the results are 0.962, and the accuracy is 96.24%. The outcome of this detection method indicates that we have high accuracy, and this might be overconfidence, because we handled each group (focused and diverse) separately. To prevent this, we trained the data using the Random Forest as one group without clustering, and we got an accuracy of 94%. The results indicate that our proposed method performs well, which shows that spammers' behaviour cannot be characterized with unified features, and the technique of grouping users based on focused and diverse interest has confirmed that the selection of features is crucial to detecting spammers with higher accuracy. Our proposed method of measuring similarity between the user's description and tweets has uncovered smart

Table 4. Comparisons of different classification algorithms.

Method	Precision	Recall	F1-score	Accuracy
SVM	0.882	0.877	0.862	87.75%
J48	0.954	0.953	0.954	95.37%
Decision table	0.949	0.95	0.949	94.92%
Random forest	0.962	0.962	0.962	96.24%

behaviour with less effort than existing approaches do [3, 10]. We further tested different classification algorithms as Table 4 shows; Random Forest achieves the best performance.

5   Conclusion and Future Work

Due to the ability of spammers to use different strategies to evade detection, we conducted an extensive study of user interest evolution patterns. We propose a method to quantify changes in user interest and depict evolution patterns for user interest to understand the degree of focus interest. Based on the level of focus interest, we put forward a framework that combines the clustering algorithm with supervised machine learning to detect spammers in online social networks. Our experiment, based on a real-world dataset, reveals the differences between spammers and legitimate users in terms of focus level of interest and shows that user interest evolution patterns are indeed sufficient to represent and detect spammers with different features. There are many potential directions for future work on this research project. It would be interesting to explore user interest in a dynamic way through different activities to characterize user interest patterns comprehensively. In addition, our detection approach is offline, so it would be interesting to be online real-time detection system.

References

1. Martinez-Romo, J., Araujo, L.: Detecting malicious tweets in trending topics using a statistical analysis of language. Expert Syst. Appl. **40**(8), 2992–3000 (2013)
2. Benevenuto, F., et al.: Detecting spammers on twitter. In: Collaboration, Electronic Messaging, Anti-Abuse and Spam Conference (CEAS) (2010)
3. Egele, M., Stringhini, G., Kruegel, C., Vigna, G.: COMPA: detecting compromised accounts on social networks. In: NDSS, 2013. NDSS, San Diego (2013)
4. Kaur, R., Singh, S., Kumar, H.: Rise of spam and compromised accounts in online social networks: a state-of-the-art review of different combating approaches. J. Netw. Comput. Appl. **112**, 53–88 (2018)
5. Fu, Q., et al.: Combating the evolving spammers in online social networks. Comput. Secur. **72**, 60–73 (2018)
6. Sedhai, S., Sun, A.: Semi-Supervised Spam Detection in Twitter Stream. IEEE Trans. Comput. Soc. Syst. **5**(1), 169–175 (2018)

7. Almaatouq, A., et al.: If it looks like a spammer and behaves like a spammer, it must be a spammer: analysis and detection of microblogging spam accounts. Int. J. Inf. Secur. **15**(5), 475–491 (2016)

8. Sedhai, S., Sun, A.: Hspam14: a collection of 14 million tweets for hashtag-oriented spam research. In: Proceedings of the 38th International ACM SIGIR Conference on Research and Development in Information Retrieval. ACM, Santiago, Chile (2015)

9. Alfifi, M., Caverlee, J.: Badly evolved? exploring long-surviving suspicious users on twitter. In: International Conference on Social Informatics. Springer, Cham (2017)

10. Ruan, X., et al.: Profiling online social behaviors for compromised account detection. IEEE Trans. Inf. Forensics Secur. **11**(1), 176–187 (2016)

11. Shen, H., et al.: Discovering social spammers from multiple views. Neurocomputing **255**, 49–57 (2016)

12. Liu, L., et al.: Detecting "Smart" spammers on social network: a topic model approach. arXiv preprint arXiv:1604.08504 (2016)

13. Nilizadeh, S., et al.: POISED: spotting twitter spam off the beaten paths. In: Proceedings of the 2017 ACM SIGSAC Conference on Computer and Communications Security. ACM, Dallas (2017)

14. Blei, D.M., Ng, A.Y., Jordan, M.I.: Latent dirichlet allocation. J. Mach. Learn. Res. **3**, 993–1022, 2003

15. Lee, K., Caverlee, J., Webb, S.: Uncovering social spammers: social honeypots + machine learning. In: The 33rd International ACM SIGIR Conference on Research and Development in Information Retrieval. ACM, New York (2010)

16. Alghamdi, B., Xu, Y., Watson, J.: Malicious behaviour analysis on twitter through the lens of user interest. In: Boo, Y.L., Stirling, D., Chi, L., Liu, L., Ong, K.-L., Williams, G. (eds.) AusDM 2017. CCIS, vol. 845, pp. 233–249. Springer, Singapore (2018). https://doi.org/10.1007/978-981-13-0292-3_15

17. Lee, K., Eoff, B.D., Caverlee, J.: Seven months with the devils: A long-term study of content polluters on twitter. In: International Conference on Weblogs and Social Media ICWSM, AAAI (2011)

18. Hall, M.A.: Correlation-based feature selection for machine learning, in Computer Science, p. 171. Hamilton, Waikato (1999)

19. Witten, I.H., et al.: Data Mining: Practical Machine Learning Tools and Techniques, 4th edn. Morgan Kaufmann, United States (2016)

DUAL: A Deep Unified Attention Model with Latent Relation Representations for Fake News Detection

Manqing Dong[1(✉)], Lina Yao[1], Xianzhi Wang[1], Boualem Benatallah[1], Quan Z. Sheng[2], and Hao Huang[1]

[1] University of New South Wales, Sydney, NSW 2052, Australia
manqing.dong@student.unsw.edu.au, lina.yao@unsw.edu.au
[2] Macquarie University, Sydney, NSW 2109, Australia

Abstract. The prevalence of online social media has enabled news to spread wider and faster than traditional publication channels. The easiness of creating and spreading the news, however, has also facilitated the massive generation and dissemination of fake news. It, therefore, becomes especially important to detect fake news so as to minimize its adverse impact such as misleading people. Despite active efforts to address this issue, most existing works focus on mining news' content or context information from individuals but neglect the use of clues from multiple resources. In this paper, we consider clues from both news' content and side information and propose a hybrid attention model to leverage these clues. In particular, we use an attention-based bi-directional Gated Recurrent Units (GRU) to extract features from news content and a deep model to extract hidden representations of the side information. We combine the two hidden vectors resulted from the above extractions into an attention matrix and learn an attention distribution over the vectors. Finally, the distribution is used to facilitate better fake news detection. Our experimental results on two real-world benchmark datasets show our approach outperforms multiple baselines in the accuracy of detecting fake news.

1 Introduction

People are increasingly referring to social media as an alternative to traditional news channels for news seeking. Due to the easiness of spreading online, large volumes of fake news are generated on a daily basis for various purposes such as rumor spreading, new product promotion, or misleading people to believe some ideas [13]. The spread of fake news can adversely affect our lives in profound ways, making it one of the most significant issues in both the social and trust computing domains. First, fake news can be easily manipulated by advertisers with bad intentions to induce people to make unwise decisions. For example, some advertising news intentionally persuades consumers to accept biased or false beliefs [13]; many people tend to follow their like-minded peers and receiving news that promotes their favored existing narratives, resulting in an echo

H. Hacid et al. (Eds.): WISE 2018, LNCS 11233, pp. 199–209, 2018.
https://doi.org/10.1007/978-3-030-02922-7_14

chamber effect. Second, online news, including fake news, could spread fast and quick to make an impact. Third, fake news affects the way people interpret and respond to the real news in the long term and undermines the foundation of an honest society. A fake news not only makes it harder for people to distinguish the real news[1] but also causes confusion and triggers people's distrust in one another.

Considering the negative impact of fake news, many efforts have been contributed to the detection of fake news. The existing approaches generally belong to either news content-based or social context-based models. The former focuses on mining the lexical features, syntactic features of news' headlines, body text, and images, writing styles of news, or more recently, deep syntax and rhetorical structure [10]. The latter, in contrast, focuses on the side information of news, such as user-based (e.g., user's age, number of followers) [1], post-based (e.g., post time)[9], and relationship-based (e.g., user-posts network, friendship network) characteristics. Besides a single type of feature, some work leverages multiple aspects of features to detect fake news [2,4,8,11]. A limitation of such work is that most of them simply treat different types of features equally or combine them linearly before feeding them into classification models. This way, they neglect important clues such as the cross information between heterogeneous features and their structured associations.

We propose a unified framework that comprehensively incorporates the two types of features and their cross information for fake news detection. In a nutshell, we make the following contributions:

- We propose a hierarchical attention-based feature extraction scheme to distill the salient information and relationships from news content and social context, followed by a binary classification method that combines those features via an attention matrix for fake news detection.
- We develop a mechanism for preprocessing the two scopes of features which has suitability for diverse fake news datasets: an attention-based bidirectional Gated Recurrent Units (GRU) model to obtain robust representations of news content and a deep neural network to nonlinearly transform the related social context information into discriminating feature representations, as well as the corresponding learning methods.
- We evaluate our framework on two real-world benchmark datasets. The experimental results demonstrate its effectiveness in detecting fake news and its superior performance to the baselines and other competitive approaches. We have made the related source code[2] publicly available for the reproduction of our model by other researchers.

---

[1] https://nytimes.com/2016/11/28/opinion/fake-news-and-the-internet-shell-game.html?_r=0.

[2] https://github.com/dongmanqing/A-Deep-Unified-Attention-Model-with-Latent-Relation-Representations.

2   Related Work

2.1   Non-hybrid Models for Fake News Detection

For detecting fake news, we need to consider various domains of information, such as text, relationship, speaker/writer files. There have been plenty of work researching with single domain information. Generally, the works could be divided into two aspects: *news content based models* and *social context based models*. For news content based models, some researchers focus on knowledge-based techniques to detect the truthfulness of news. For example, in the work by Shi et al. [12], the authors used a knowledge graph to check whether the claims in news content can be inferred from existing facts in the knowledge graph. Some focused on mining the writing style of the news [10]. For social context based models, the side information is more concerned. For example, Ma et al. [9] use Recurrent Neural Network (RNN) to capture the changes in posts over time and to detect rumors.

2.2   Hybrid Models for Fake News Detection

To comprehensively learn from a dataset, it is better to combine the different type of features. A related problem is how to effectively combine those features. One simple way is extracting different kinds of features and feeds them into traditional classification models. For example, Janze et al. [4] extract different kinds of features for detecting fake Facebook news posted during the U.S. presidential election of 2016[3]. They consider cognitive cues (e.g., word counts, the polarity of opinion), visual cues (e.g., the brightness of the posted picture), affective cues (e.g., the number of votes for the post), and behavior cues (e.g., the number of comments). They combine those features as feature vectors and input them into models like Support Vector Machine (SVM) and Random Forest. These methods neglect the relationship between the features and normally one model could not well grasp all different kinds of features at once. The other way is finding the relationship between the features and dealing with them in a hierarchical way. In Ruchansky's work [11], they considered three aspects of features: the text of an article, the user response it receives, and the source users promoting it. A model called CSI (capture, score and integrate) is proposed in [6], where the 'capture' component uses Recurrent Neural Network (RNN) capture the temporal pattern of user activity on a given article, 'score' learns the source characteristic based on the behavior of users, and 'integrate' unifies these two hidden representations to do the classification.

3   The Methodology

Similar to [13], we define fake news as *a news article that is intentionally and verifiably false* and formulate fake news detection as a classification problem.

---

[3]   https://github.com/BuzzFeedNews/2016-10-facebook-fact-check.

Given a set of news content $X_c = \{x_{c_1}, x_{c_2}, \ldots, x_{c_N}\}$ and their side information (e.g., speaker's or writer's profiles, reviewers feedback), $X_s = \{x_{s_1}, x_{s_2}, \ldots, x_{s_N}\}$, the goal is to predict the labels for news, denoted by $Y = \{y_1, y_2, \ldots, y_N\}$, where $y_i = 1$ if news i is fake news and otherwise truthful news. The structure of our model is shown in Fig. 1.

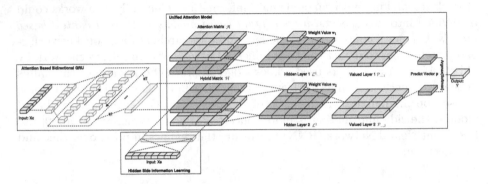

Fig. 1. Example of a binary classification with our model. First we have content information X_c go through with an attention based bidirectional GRU to get hidden representation $\mathcal{H}_c$, and then side information X_s go though with deep neural networks to get hidden representation $\mathcal{H}_s$. Hidden matrix $\mathcal{H}$ is then generated by unifying $\mathcal{H}_c$ and $\mathcal{H}_s$. For binary classification, we calculate two attention matrices $\mathcal{A}$, to get the hidden layer $\mathcal{L}^1$ and $\mathcal{L}^2$. The final prediction $\hat{y}$ is obtained by giving weights $\mathbf{w}$ to the two hidden layers.

3.1   Input Preprocessing

Side Information Preprocessing. We consider two types of the side information: speakers' profiles and reviewers' feedback. Speaker's profiles usually contain categorical features such as speaker's id (or name), speaker's job title, speaker's party, and the news release platform. For categorical variables that with a large number of factors, we use a clustering strategy to obtain the features and consider both counts and the accuracy for each category. For details, we first divide the features into a certain number of categories (here we used 5) according to the counts in training dataset and then set the value of according categories as the history trustworthy values in training dataset. For example, in the training dataset in Table 1, 'Jack' posted 30–39 posts and among those posts, the proportion for truthful news is 0.78. Thus, the posts by 'Jack' is designated the speaker ID feature of $[0, 0, 0, 0.78, 0]$. For every speaker ID that appears only in the test dataset, we set this feature in 0–9 and the credit history as 0.5.

Table 1. Example for Speaker ID preprocessing

Speaker ID	0–9	10–19	20–29	30–39	above 39
Jack	0	0	0	0.78	0

Reviewers' feedback contains features like support/oppose responses or like/-dislike/share toward the articles. By counting the number of features like supports #*supports* or oppositions #*oppositions*, we get numerical variables representing reviewers feedback information for later analysis.

Content Information Preprocessing. We use embedding methods for preprocessing the content information [3]. Word embedding is a set of techniques for natural language processing, which represents words or phrases by generating and using some lower dimensional vectors. It has recently gained success in many applications which further boost its application.

3.2 Latent Features Learning

Content Latent Features Learning. As mentioned before, the traditional RNN has the ability to better capture contextual information and cannot pay attention to the salient parts of text. Thus, it is inefficient in containing global information [6]. To deal with this inefficiency, researchers have proposed adding attention mechanisms into it, where attention mechanism could give each word/sequence an attention value [15]. Here, we use an attention-based bi-directional GRU model that follows the design of [15] to encode the news content sequences and learn the attention values for words in each sentence. We can get the latent content vector $\mathcal{H}_c$ by summing up the weighted word annotations.

Side Information Latent Features Learning. For learning the side information, we mainly used deep neural networks to obtain hidden representations of side information features: $\mathcal{H}_s = \sigma(X_s W_s + b_s)$, where $W_s \in \mathbb{R}^{M_1 \times D_1}$ are weights and $b_s \in \mathbb{R}^{D_1}$ are biases in neural networks. D_1 is the dimension of the latent side information vectors, it is a self defined parameters, and normally we set this dimension based on the dimension of the input vectors. σ is activate function such as sigmoid function [5].

3.3 Unified Attention Model

From above we get hidden representations for side information $\mathcal{H}_s \in \mathbb{R}^{D_1}$ and content information $\mathcal{H}_c \in \mathbb{R}^{D_2}$. To fully obtain the cross-domain information of content features and side information features, we take the cross product of these two hidden representation vectors as a hidden matrix:

$$\mathcal{H} = \mathcal{H}_s \otimes \mathcal{H}_c \tag{1}$$

To match this hybrid hidden representation $\mathcal{H}$ to the final prediction, we construct K attention matrices $\mathcal{A}^k$ to capture the relationship between $\mathcal{H}$ towards each label $k \subseteq K$, separately. Combined with those attention matrices, we obtained the target layer $\mathcal{L}^k$ for label k, by following the next steps:

$$\mathcal{T}^k = \sigma(U^{k^\top} \mathcal{H} V^k + B^k) \tag{2}$$

$$\mathcal{A}^k_{ij} = \frac{\exp \mathcal{T}^k_{ij}}{\Sigma_j \Sigma_i \exp \mathcal{T}^k_{ij}} \cdot D_1 D_2 \tag{3}$$

$$\mathcal{L}^k = \mathcal{A}^k \odot \mathcal{H} \tag{4}$$

where we first construct the transform matrix $\mathcal{T}$ for the hidden matrix $\mathcal{H}$ by multiplying two reversible weight matrix U^k and V^k, and then through an activate function. U^k and V^k are weight matrix and $U^k \in \mathbb{R}^{D_1 \times D_1}$, $V^k \in \mathbb{R}^{D_2 \times D_2}$, B is bias matrix and $B \in \mathbb{R}^{D_1 \times D_2}$, and σ is an activation function. The attention matrix $\mathcal{A}$ is calculated by softmax function, which could be regarded as normalized importance weights for each dots in the hidden matrix $\mathcal{H}$. When D_1 and D_2 are big (e.g. when D_1, D_2 equal to 50, then $D_1 \times D_2$ will be 2500), most of $\mathcal{A}_{ij}$ would be a quite small value that near zero. Thus the former softmax value then multiplies the dimension D_1 and D_2 as shown in Eq. 11. After all, we get a weighted version of a hidden matrix $\mathcal{H}$, which is layer $\mathcal{L}^k$, towards label k by taking the dot product of $\mathcal{A}^k$ and $\mathcal{H}$.

We also give each layer a weight towards the final prediction, where for a layer $\parallel$ be predicted to label k, we want to give more weight for this layer. We defined the weights as $\mathbf{w} = \{w_1, \ldots, w_K\}$, and they are calculate by the following,

$$\mathbf{h} = \sigma(\mathbf{u}^\top \mathcal{H} \mathbf{v}) \tag{5}$$

$$w_k = \frac{\exp h_k}{\Sigma_k \exp h_k} \tag{6}$$

where $\mathbf{u} \in \mathbb{R}^{D_1 \times K}$, $\mathbf{v} \in \mathbb{R}^{D_2 \times K}$ are transform vectors, σ is an activation function. Thus we first learn a hidden weight $\mathbf{h}$ for layer weights $\mathbf{w}$, and then use softmax as a normalization function to get each element in $\mathbf{w}$.

By multiply the weights $\mathbf{w}$ to $\mathcal{L}^k$ we got the final prediction cube $\mathcal{P} \in \mathbb{R}^{D_1 \times D_2 \times K}$, where

$$\mathcal{P}_{.,.,k} = w_k \cdot \mathcal{L}^k \tag{7}$$

For the final prediction, we sum up the elements in each layer in $\mathcal{P}$ by the axis i and j, which means we sum up the elements in $\mathcal{P}_{.,.,k}$. By doing so, we get the final predict vector $\mathbf{p}$ which consists of the probability that a sample to be label k, where is listed below:

$$\mathbf{p} = softmax(\Sigma_i \Sigma_j \mathcal{P}_{i,j,k=1}, \ldots, \Sigma_i \Sigma_j \mathcal{P}_{i,j,k=K}) \tag{8}$$

Then the final prediction is

$$\hat{y} = \operatorname{argmax}(\mathbf{p}) \tag{9}$$

We take cross entropy loss as the loss function for our prediction, where

$$L_p(X, Y, \Theta) = -\Sigma_k y \log \hat{y} \tag{10}$$

In the training process, we replace $\hat{y}$ to $\mathbf{p}$ for back-propagation. Also, we consider regularizing the parameters to prevent over-fitting. Here, we choose ℓ_2 norm for each parameter. Thus, the total loss is summing up the prediction loss and the regularization function. The optimization goal is to minimize the total loss:

$$\operatorname*{argmin}_{\Theta} L(X, Y, \Theta) = L_p(X, Y, \Theta) + \lambda \parallel \Theta \parallel_2, \qquad (11)$$

here, we use Θ represents all the parameters used in our algorithms, and λ is a hyper parameter. As for learning the parameters, they are mainly updated by back propagation via Adam optimization methods [5].

4   Experiments

4.1   Dataset Description

Evaluations are performed using two real-world manual labeled benchmark datasets for fake news detection task: LIAR dataset [14] and Buzzfeed News[4]. For each dataset, we consider a binary classification problem to distinguish truthful news from fake news. The LIAR Dataset contains 12,836 short statements from 3,341 speakers covering 141 topics in POLITIFACT[5]. Each news item includes text content, topic, and speaker profile (speaker name, title, party affiliation, current job, the location of the speech, and credit history). The labels take discrete values from 1 to 6 corresponding to pants-fire, false, mostly-false, half-true, mostly-true, and true. Data with the first three labels are labeled as fake news and the others are labeled as truthful news. We further deduct parts of the dataset for balancing the dataset, and we get around 4,000 truthful statements and 4,000 fake news. And Buzzfeed News Dataset contains 2,282 posts published on Facebook from 9 news agencies over a week close to the 2016 U.S. election. Every post and linked article were fact-checked claim-by-claim by 5 BuzzFeed journalists. Each article includes category (e.g., mainstream), public account, post type, spreading counts (number of shares, likes, and comments). We regard data with the label 'mostly true' as truthful news and others ('no factual content' and 'mixture of true and false') as fake news. Among the 2,282 posts, almost 75% of them are truthful news. We balanced the number of truthful and fake news with a 50-50 proportion and finally got 800 items for final prediction.

4.2   Comparison of the Results

Our default parameter settings are 2 fully connected layers for extracting side information, word embedding with dimension 200, and the attention matrix with 20×20. And we compare with a set of state-of-the-art methods and baselines. 1. Rubin et al. [10]: Rubin extracted news style-based features by combines the vector space model and rhetorical structure theory, and they used Support

---

[4] https://github.com/BuzzFeedNews/2016-10-facebook-fact-check.
[5] http://www.politifact.com/.

Vector Machine for the classification. 2. Castillo et al. [1]: This method predicts news veracity using social engagements. The features are extracted from user profiles and friendship network. 3. Janze et al. [4]: as mentioned in related work, Janze et al. considered various type of features for detecting the fake news and used SVM as the classifier. Since there are no visual cues in LIAR dataset, we didn't extract visual features from Buzzfeed News as well. 4. Yunfei et al. [8] propose to utilized a similar way as ours for predicting fake news. They used attention based LSTM model to learn context hidden representation and LSTM for learning the speaker's file. The difference is they concatenate the hidden representations and feed them into a softmax layer for prediction. 5. X_c only: where only content features are concerned and is processed with attention based bidirectional GRU for prediction. 6. X_s only: where only side information is concerned and is processed with deep neural networks. 7. $X_s + X_c + NN$: the side information and content features are simply concact and feed into deep neural networks. 8. $X_s + X_c + biNN$: first learn the hidden representations for both side information and content with the same methods as our model's, and then processed them with bilinear neural networks [7].

Table 2. Prediction comparison in two datasets

Methods	LIAR dataset				Buzzfeed news			
	Acc	Pre	Rec	F1	Acc	Pre	Rec	F1
Rubin et al.	0.5719	0.5953	0.5336	0.5445	0.6108	0.6022	0.5612	0.5551
Castillo et al.	0.7798	**0.7772**	0.7914	0.7837	0.7475	0.7357	0.7833	0.7562
Janze et al.	0.7319	0.6433	0.7029	0.6718	0.7900	0.8163	0.7692	0.7921
Yunfei et al.	0.8063	0.7361	0.7943	0.7641	0.8081	0.8125	0.7959	0.8041
X_c only	0.6225	0.5619	0.5191	0.5396	0.6531	0.5417	0.6842	0.6047
X_s only	0.7300	0.6686	0.6786	0.6722	0.7959	0.7500	**0.8182**	0.7826
$X_s + X_c + NN$	0.7838	0.7625	0.7386	0.7504	0.8081	0.8542	0.7736	0.8119
$X_s + X_c + biNN$	0.8113	0.7361	0.8045	0.7688	0.8182	0.8750	0.7778	0.8235
Ours	**0.8283**	0.7621	**0.8112**	**0.7859**	**0.8384**	**0.9167**	0.7857	**0.8462**

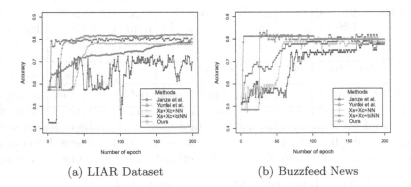

(a) LIAR Dataset   (b) Buzzfeed News

Fig. 2. Comparing with hybrid methods

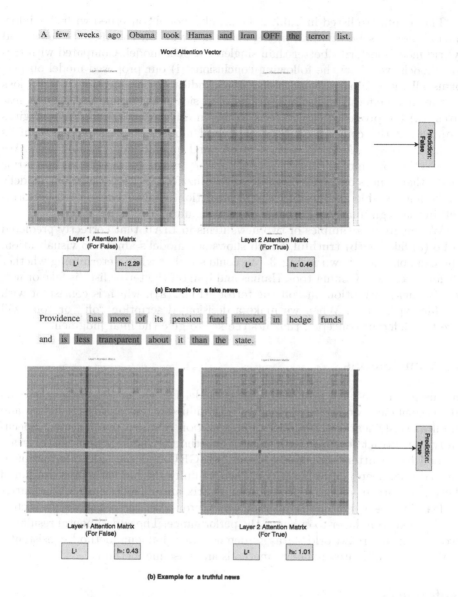

Fig. 3. Examples for news of LIAR that are correctly predicted to be (a) fake or (b) truthful. In each sub-figure, the first line shows the attention to the content information. The highlight color is changing from shallow yellow to dark orange, and the darker the highlight, the bigger the attention value. Two matrices show the attention matrices when doing the prediction. And the red rectangles below shows the weights for the attention matrices. Layer $\mathcal{L}^1$ is for fake and layer $\mathcal{L}^2$ is for true. For example, $\mathcal{L}^1$ presents the attention when we search for the cues for false information. And then followed with red rectangles, which show the weights for the attention matrices, to enhance the former judgment. (Color figure online)

The results are listed in Table 2. Generally, model concerned with side information performs better than model concerned with only content features, and hybrid model performs better than singleton aspect model. Compared with several models, we draw the following conclusions (i) our proposed model outperforms all the comparison methods, which indicate the attention matrix does capture more effective relationships between side information and content features; (ii) the proposed structure is fixable in different datasets. Figure 2 gives examples of the comparison between hybrid models (Janze's model, Yunfei's model, $X_s + X_c + NN$, $X_s + X_c + biNN$) and our model on two datasets. We can see that our model converges to an optimal accuracy rapidly and performs better than others on LIAR dataset. As for Buzzfeed News dataset, our model's prediction is stable and produces good prediction, while Yunfei's work performs well in the beginning but becomes over-fitting after several iterations.

We are giving examples of two news items in LIAR that correctly predicted to be (a) fake or (b) truthful, which shows our model's ability to visualization. The examples are shown in Fig. 3. We could see that when determining whether "A few weeks ago Obama took Hamas and Iran off the terror list" is fake or not, we give much attention on "off the terror" (Fig. 2(a)), which is consistent with our life experiences. When we make a decision, description "off" or "on" will give two different concepts, thus it's the key to give the final judgment.

5   Conclusions

In this paper, we have proposed a novel model DUAL for detecting fake news which combines two scope of features with unified attention for prediction and simultaneously learns the hidden representation of these two features. Specifically, we extract features with adaptive methods, where content features are learned by an attention-based bidirectional GRU and the side information is learned from deep neural networks. Then we unify the hidden representation of these two features as the representation matrix and introduce attention matrix for learning the attention distribution over vectors. We derive the weight vectors for the prediction layer to enhance the performance. The experimental results on two benchmark real-world datasets demonstrate that our method consistently beats a series of state-of-the-art methods and baseline methods.

References

1. Castillo, C., Mendoza, M., Poblete, B.: Information credibility on twitter. In: World Wide Web, pp. 675–684. ACM (2011)
2. Dong, M., Yao, L., Wang, X., Benatallah, B., Huang, C., Ning, X.: Opinion fraud detection via neural autoencoder decision forest. arXiv:1805.03379 (2018)
3. Huang, C., Yao, L., Wang, X., Benatallah, B., Sheng, Q.Z.: Expert as a service: Software expert recommendation via knowledge domain embeddings in stack overflow. In: ICWS, pp. 317–324. IEEE (2017)
4. Janze, C., Risius, M.: Automatic detection of fake news on social media platforms. In: Pacific Asia Conference on Information Systems (2017)

5. Kingma, D., Ba, J.: Adam: a method for stochastic optimization. arXiv preprint arXiv:1412.6980 (2014)
6. Lai, S., Xu, L., Liu, K., Zhao, J.: Recurrent convolutional neural networks for text classification. In: AAAI, vol. 333, pp. 2267–2273 (2015)
7. Lin, T.Y., RoyChowdhury, A., Maji, S.: Bilinear CNN models for fine-grained visual recognition. In: International Conference on Computer Vision, pp. 1449–1457 (2015)
8. Long, Y., Lu, Q., Xiang, R., Li, M., Huang, C.R.: Fake news detection through multi-perspective speaker profiles. In: International Joint Conference on Natural Language Processing, vol. 2, pp. 252–256 (2017)
9. Ma, J., Gao, W., Mitra, P., Kwon, S., Jansen, B.J., Wong, K.F., Cha, M.: Detecting rumors from microblogs with recurrent neural networks. In: IJCAI (2016)
10. Rubin, V.L., Lukoianova, T.: Truth and deception at the rhetorical structure level. J. Assoc. Inf. Sci. Technol. 66(5), 905–917 (2015)
11. Ruchansky, N., Seo, S., Liu, Y.: CSI: a hybrid deep model for fake news detection. In: CIKM, pp. 797–806 (2017)
12. Shi, B., Weninger, T.: Fact checking in heterogeneous information networks. In: WWW, pp. 101–102 (2016)
13. Shu, K., Sliva, A., Wang, S., Tang, J., Liu, H.: Fake news detection on social media: a data mining perspective. ACM SIGKDD Explor. Newslett. 19(1), 22–36 (2017)
14. Wang, W.Y.: "liar, liar pants on fire": a new benchmark dataset for fake news detection. arXiv preprint arXiv:1705.00648 (2017)
15. Yang, Z., Yang, D., Dyer, C., He, X., Smola, A.J., Hovy, E.H.: Hierarchical attention networks for document classification. In: HLT-NAACL, pp. 1480–1489 (2016)

Social Network

Extracting Representative User Subset of Social Networks Towards User Characteristics and Topological Features

Yiming Zhou[1], Yuehui Han[1], An Liu[1], Zhixu Li[1], Hongzhi Yin[2], and Lei Zhao[1(✉)]

[1] School of Computer Science and Technology, Soochow University, Suzhou, China
{ymzhou,yhhan}@stu.suda.edu.cn, {anliu,zhixuli,zhao1}@suda.edu.cn
[2] School of Information Technology and Electrical Engineering,
The University of Queensland, Brisbane, Australia
db.hongzhi@gmail.com

Abstract. Extracting a subset of representative users from the original set in social networks plays a critical role in Social Network Analysis. In existing studies, some researchers focus on preserving users' characteristics when sampling representative users, while others pay attention to preserving the topology structure. However, both users' characteristics and the network topology contain abundant information of users. Thus, it is critical to preserve both of them while extracting the representative user subset. To achieve the goal, we propose a novel approach in this study, and formulate the problem as RUS (Representative User Subset) problem that is proved to be NP-Hard. To solve RUS problem, we propose a method KS (K-Selected) that is consisted of a clustering algorithm and a sampling model, where a greedy heuristic algorithm is proposed to solve the sampling model. To validate the performance of the proposed approach, extensive experiments are conducted on two real-world datasets. Results demonstrate that our method outperforms state-of-the-art approaches.

Keywords: Representative · Social networks · Characteristics
Topological features

1 Introduction

In social networks such as Twitter, Facebook, and Sina Weibo, a large number of users post their tweets, spread viewpoints, and share ideas [26]. Due to the diversity of users' characteristics and the largeness of networks, it is intractable and impractical to analyze users and the network in its entirety [14]. Therefore, if a representative user subset containing the most information of the original dataset could be selected, the original dataset will be more "human-readable" and help us analyze users. For example, when conducting surveys and collecting feedbacks in Human-Computer Interaction, it is significant to select

© Springer Nature Switzerland AG 2018
H. Hacid et al. (Eds.): WISE 2018, LNCS 11233, pp. 213–229, 2018.
https://doi.org/10.1007/978-3-030-02922-7_15

representative users, as they have a high representative degree and the number of these users is much smaller than that of the original set [23]. Compared with directly analyzing all users, it will be more effective and efficient to conduct a study on a small subset of representative users that represent behaviors and preferences of the original dataset [8, 28].

Much effort has been devoted to extracting a small subset of users from social networks, mainly divided into three aspects: users' influence analysis [1, 3, 5, 19, 22], finding opinion leaders [11], and detecting communities [10, 12, 16]. They have proposed effective solutions to find a group of users who have great influence or have similar topics and preferences. However, such users are not enough to represent the original dataset. This is because, authors [1, 5, 10–12, 16] only focus on preserving the users' characteristics or topological features when extracting users. Recently, in order to sample representative users from social networks, some studies [14, 18, 21, 24, 27] utilized the graph structure to study the social network topology and solved sampling problems in the dynamic environment [21]. In [14], Maiya et al. proposed a sampling method to produce subgraph samples to represent the community structure of the original network. Obviously, the study is limited as it only considers the structure of the network when extracting representative users. Subsequently, Tang et al. [23] proposed two types of sampling models to sample representative users according to their attributes. Though relationships between users have been considered when computing the representative degree, the topology of the original network was ignored. Despite of the great contributions made by existing studies, none of them focus on preserving both users' characteristics and the network structure while extracting representative users.

To present the reason behind the extraction of representative users, we give the following example, where we also discuss which kind of users will be selected. When a mobile retail company hopes to conduct a survey of users' feedbacks on their products, it is inefficient to send questionnaires to all customers to collect surveys. This is because the number of customers is usually very large, and customers may come from different groups, such as students, office workers, and celebrities. People from different groups usually have different characteristics and there are many communities in each group like colleges, companies or some clubs [29]. In addition, people from the same group or community usually have some similar characteristics or opinions [23]. How to select representative users from these diverse customers is a great challenge. To achieve customers' feedbacks comprehensively, we need to collect feedbacks from each group and community. In other words, the representative users should have more coverage in users' characteristics and come from as more communities as possible. As a result, the survey will be more systemic and convincing. In general, our goal is to extract representative users who have more coverage both in users' characteristics and the topology of networks.

More clearly, we use the following example to present the details of extracting representative users. As shown in Fig. 1, there are eight users from two different communities. Each element of the representative degree matrix can be obtained by computing the distance between two users' characteristics. The details of computing the representative degree are presented in Sect. 3. Given the size of the representative subset ($k = 2$), if we extract the representative subset based on users' characteristics, $S = \{u_2, u_5\}$ will be returned. Because u_2 has high representative degree of characteristics to u_1, u_3, and u_4 while u_5 has high representative degree of characteristics to u_6, u_7, and u_8. Thus, the representative degree of users' characteristics of S is the highest. However, u_2 is a marginal user, which has a small number of neighbours and low representative degree on the topology in Community 1. On the other hand, if we extract the representative subset based on the network topology, $S = \{u_1, u_5\}$ will be selected, as $\{u_1, u_5\}$ has more neighbours such that they will carry more information of the network topology. However, u_1 has much less representative degree of users' characteristics than that of u_2 and u_4 in Community 1, which will lead to the loss of users' characteristics. If we consider users' characteristics and topological features simultaneously, the representative subset $S = \{u_4, u_5\}$ will be selected, since u_4 and u_5 can preserve high representative degree both of users' characteristics and the network topology. Based on the example, given the size of the representative subset k, we aim at extracting users that have more coverage in users' characteristics and topological features. In our method, we propose a sampling model to integrate these two factors. We formulate this problem as RUS problem and give the definition in Sect. 3.

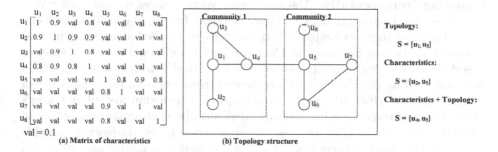

(a) Matrix of characteristics   (b) Topology structure

Fig. 1. Illustration of extracting representative subset from the perspective of both users' characteristics and topological features.

Intuitively, to solve RUS problem, a straightforward way is to extract two subsets respectively from the perspective of users' characteristics and topological features. Then, the intersection of these two subsets is the representative subset of users. However, the size of this final representative subset may not be equal to k. In the worst case, the subset would be empty. Therefore, we propose a sampling model to extract the subset with the consideration of both users' characteristics and topological features.

Specifically, our contributions in this paper are as follows:

- Different from previous work that extracts representative users only preserving users' characteristics or topological features, we preserve both of them. The new problem is formulated as RUS problem.

- We prove that RUS problem is NP-Hard and propose our method KS, which combines a clustering algorithm and a sampling model. For the sampling model, we propose a greedy heuristic algorithm.
- We conduct experiments on two real-world datasets to compare the performance of our method with that of other methods based on four metrics. The experimental results show that KS has the best performance on both datasets.

The rest of this paper is organized as follows. Section 2 introduces the related work in concern with finding a subset of users from social networks, where we also present the limitations of the existing work for solving RUS problem. In Sect. 3, we formulate the problem as RUS problem. Section 4 describes our method KS to mine representative users. In Sect. 5, experiments are conducted to validate the performance of the proposed approach on two real-world datasets. Section 6 concludes the paper.

2   Related Work

There are a small amount of theoretical studies on the problem of extracting representative users from social networks. The related work about extracting users from social networks is divided into the following aspects.

Sampling Representative Users. Recent years, studies on sampling representative users from social networks have emerged. In [14], Maiya et al. proposed a novel method to sample representative subgraphs on the basis of communities structure in the original network. Compared with traditional network measures based on nodes' degrees, [21] extended a statistical learning approach called *skeleton learning* (SKE) to identify representative users from the perspective of information. [18] efficiently used a sampling-based algorithm to explore a user's social network respecting its structure and quickly approximate quantities of interest. Recently, in [23], Tang et al. proposed a novel approach to sample representative users automatically for different topics by greedy algorithm and developed two sampling models: Statistical Stratified Sample (S^3) and Strategic Sampling for Diversity (SSD).

Analyzing Users' Influence. Initial efforts on analyzing users' influence in social networks were primarily based on PageRank [17]. According to the relationships between users as the form of in-degree and out-degree of a vertex, PageRank ranked users' influence. Subsequently, [5] analyzed social influence and selection based on similarities between users. [22] proposed Topical Affinity Propagation (TAP) to perform topic-level influence propagation in topic modeling and network structures. With the purpose of finding a subset of users in social networks who may maximize the spread of influence, [3] proposed an improved greedy algorithm and new degree discount heuristics.

Detecting Communities. Social communities consisted of users with similar interests and topics play an important role in social applications. [10,12,16] tried their best to map users and the relationships between them into graphs such that communities which connect users closely will be found. By this way, users who like similar objects or have similar interests would be gathered into the same community.

Finding Opinion Leaders. In social networks, discovering opinion leaders has been studied. [20] considered opinion leaders in social networks as the representative users who capture the most representative opinions. According to users' influence, they proposed InfluenceRank algorithm to identify opinion leaders. In [11], Amit Goyal et al. proposed a novel frequent pattern mining approach based on users' influence and actions to discover opinion leaders.

In conclusion, despite of the great contributions made by existing studies, none of them focus on preserving both users' characteristics and the network structure while extracting representative users. In this paper, we propose a solution with the consideration of both users' characteristics and topological features to guarantee that representative users have the most information of the original dataset.

3   Problem Definition

In this section, we present symbols in Table 1 and formulate the problem.

Table 1. Definitions of used symbols

Symbol	Definition
D	The original set of users
S	The representative set of users and $S \subseteq D$
k	The desired size of S
$u_i \in D$	$u_i = <u_i^1, u_i^2, ..., u_i^n>$ (u_i^j means i-th user's j-th attribute)
$G = (D, E)$	A network or graph, where D is set of nodes and $E \subseteq D \times D$ is set of edges
	In the graph, we regard users as nodes and relationships between users as edges

Definition 1 *Distance between users' characteristics. Given two user vectors u_i and u_j, the distance between their characteristics is formulated as follows:*

$$dist(u_i, u_j) = \sqrt{\sum_{k=1}^{m}(u_i^k - u_j^k)^2 + \sum_{k=m+1}^{n} Jaccard\_distance(u_i^k, u_j^k)^2} \quad (1)$$

where m and $n - m$ denotes the number of numerical attributes and non-comparable attributes, respectively.

In users' attributes, there are numerical attributes e.g., the number of followers and non-comparable attributes e.g., user's address. We employ Minkowski Distance (we set $p = 2$.) to combine these two types of attributes. Here, we use the Jaccard distance ($Jaccard\_distance(A, B) = 1 - \frac{|A \cap B|}{|A \cup B|}$) to compute the distance between two non-comparable attributes.

Definition 2 *Representative degree of characteristics between two users. The representative degree of characteristics between u_i and u_j is formulated as:*

$$r(u_i, u_j) = 2 \times Sigmoid(1/dist(u_i, u_j)) - 1 \tag{2}$$

We define the representative degree between two users by using $r(u_i, u_j)$, and introduce a sigmoid function $Sigmoid(x) = 1/(1 + e^{-x})$ ($x \in [0, 1]$) to map $r(u_i, u_j)$ into the interval $[0, 1]$. If $dist(u_i, u_j) = 0$, we define $1/dist(u_i, u_j) = 1$. Based on Eq. (2), we can obtain the matrix in Fig. 1.

Definition 3 *Representative degree of characteristics between set and user.*

$$Repre(S, u_j) = max_{u_i \in S, u_j \in D} r(u_i, u_j) \tag{3}$$

For each user u_j in D, we select a user u_i from S that has the highest representative degree $r(u_i, u_j)$ to represent u_j. The representative degree of characteristics of a subset to a user is defined as $Repre(S, u_j)$. $r(u_i, u_j)$ is in the interval $[0, 1]$, so $Repre(S, u_j)$ is also in the interval $[0, 1]$.

Definition 4 *Representative degree of characteristics between two sets. Given a user set D and a subset S, the representative degree of characteristics between S and D is defined as:*

$$R_c(S, D) = \frac{\sum_{u_j \in D} Repre(S, u_j)}{|D|} \tag{4}$$

where $R_c(S, D)$ is also in the interval $[0, 1]$.

Definition 5 *Neighbourhood of set* [14]. *The neighbourhood of S is defined as $N(S) = \{w \in D - S : \exists v \in S, s.t.(v, w) \in E\}$, where E is the set of edges that has been defined in Table 1.*

Definition 6 *Coverage indicator of topological features. Given a user set D and a subset S, the coverage indicator of topological features between S and D is defined as:*

$$R_t(S, D) = \frac{|N(S)|}{|D - S|} \tag{5}$$

where $R_t(S, D)$ is also in the interval $[0, 1]$.

RUS Problem. Given k, which is the number of users to be extracted, our goal is to extract a subset S ($S \subseteq D, |S| = k$) that has high representative degree of both users' characteristics and topological features.

4   Algorithms

In this section, we begin with the proof of the NP-Hardness of RUS problem. Then, we introduce our method KS (K-Selected) which is consisted of a clustering algorithm and a sampling model.

4.1   NP-Hardness of RUS Problem

The NP-Hardness of RUS problem is proved in the way that the decision version of its sub-problem can be reduced from the well-known k-median problem, shown to be NP-Hard in $\mathbb{R}^2$ under Euclidean distance [15].

k-**median problem in 2D**: For points in $\mathbb{R}^2$ as an original set D and given a value $\triangle$, the problem is to find a subset S such that S subjects to $\sum_{v \in D} min_{u \in S} \parallel u - v \parallel \leq \triangle$, where $\parallel u - v \parallel$ means the Euclidean distance between u and v [25].

Lemma 1. *RUS problem is NP-Hard.*

Proof. Recall that RUS problem is to find S whose representative degree of users' characteristics and topological features should be as high as possible at the same time. Next, we will prove that the sub-problem of finding S with the highest representative degree of users' characteristics is NP-Hard which results in the NP-Hardness of RUS problem. Then, we prove the NP-Hardness of the decision version of this sub-problem. This is because if the decision version of this sub-problem is NP-Hard, the sub-problem is also NP-Hard.

The sub-problem is to find a subset S which satisfies that $S = argmax\, R_c(S, D)$ $= argmax \sum_{v \in D} Repre(S, v) = argmax \sum_{v \in D} max_{u \in S} Sigmoid(1/dist(u, v))$. As the function $Sigmoid$ increases monotonously, we change the form of $1/dist(u, v)$ and the sub-problem is equivalent to finding $S =$ $argmin \sum_{v \in D} min_{u \in S} dist(u, v)$. If we replace $dist(u, v)$ with $\|u - v\|$ where $\|u - v\|$ means the Euclidean distance between u and v, it does not affect the proof. Subsequently, k-median problem is equivalent to finding k-subset S of D such that $\sum_{v \in D} min_{u \in S} \|u - v\| \leq \triangle$. Since k-median problem is NP-Hard and the sub-problem can be reduced from k-median problem, the sub-problem is also NP-Hard. In conclusion, RUS problem is NP-Hard.

4.2   KS (K-Selected) Algorithm

Our goal is to select a representative subset of users who have high representative degree both in users' characteristics and topological features, i.e. the representative users should have great coverage both in users' characteristics and topological features. Therefore, to make the representative subset have more characteristics, we firstly cluster all users into different groups such that users in different groups have different characteristics. And then, we sample from each group to guarantee that users in S will have each group's characteristics. Thus, it is reasonable and convincing to guarantee that the representative subset will have more coverage in characteristics by selecting users from each group. Furthermore, compared with sampling from the whole dataset, clustering will improve the efficiency of selecting representative users because we only need to update the result of some specific cluster in each iteration which will be shown in Algorithm 2. We take advantages of K-Medoids [13], a variant of K-Means [7], to cluster users into m groups. Because compared with K-Means, K-Medoids can preserve the real center in each cluster after clustering. As similar as K-Means, the setting of m in K-Medoids is hard to be fixed. Thus, we show the effect of the variation of m to the experimental results in Sect. 5. Though another cluster algorithm DBSCAN [9] does not need to set the initial number of clusters, it still depends on the threshold Eps. Therefore, we choose K-Medoids that works as Algorithm 1.

Algorithm 1. *Clustering Algorithm*

Input: Original DataSet: D; The number of clusters: m; The maximal number of iterations: $max\_iterations$; The function of representative degree of characteristics: r;

Output: All clusters *clusters* consisted of $cluster_i$

1 Randomly select m seeds from D as $medoids = \{d_1, d_2, ..., d_m\}$ (We use $cluster_i$ to represent the cluster to which d_i belongs.);

2 $iter = 0$;

3 **while** *medoids has changed and iter < max_iterations* **do**

4   **for** *each $e \in D - medoids$* **do**

5   put e into $cluster_i$ if d_i has the greatest representative degree to e i.e. $d_i = argmax\ r(d_i, e)$;

6   Update $medoids$. For each $cluster_i$, we choose d_{new} as a new medoid to replace d_i. And d_{new} satisfies that $d_{new} = argmax \sum_{v \in cluster_i} r(d_{new}, v)$;

7      $iter += 1$;

8 **return** $clusters$;

In Algorithm 1, clustering elements costs $\mathcal{O}(m \cdot |D|)$ and updating $medoids$ costs $\mathcal{O}(|cluster_i|^2)$ in each iteration. Thus, if the number of iterations is $iter$, clustering all users costs $\mathcal{O}(iter \cdot (m \cdot |D| + |cluster_i|^2))$.

Next, we propose a sampling model to select users from each cluster to guarantee the subset S would have great coverage both in users' characteristics and topological features. The sampling model is defined as follows:

$$F = \sum_{i=1}^{m} \frac{|cluster_i|}{|D|} \cdot (\gamma \cdot R_c(S, cluster_i) + (1 - \gamma) \cdot \frac{|N(S) \cap cluster_i|}{|cluster_i|}) \qquad (6)$$

where function R_c and $N(S)$ have been defined in Sect. 3 and $\gamma \in [0, 1]$. In this model, we add all results among m clusters to evaluate the general representative degree of S to D. And $\frac{|cluster_i|}{|D|}$ that is related with the size of the cluster is to indicate the "importance" of each cluster. If the size of $cluster_i$ is larger, there will be more representative users from $cluster_i$ in S. In other words, the proportion of representative users of each cluster in S tends to be similar as the proportion of clusters in D.

Because we need to take the representative degree of users' characteristics and the topological features into account simultaneously. The indicator of both users' characteristics and topological features have been integrated into the formula in the first layer of parentheses in F. $R_c(S, cluster_i)$ in the first half of F is the function of representative degree of characteristics and R_c defined in Sect. 3 which is used to evaluate how representative S is to $cluster_i$ in the perspective of characteristics. In the latter part of F, in order to make S have higher representative degree of topological features, $|N(S) \cap cluster_i|$ in F is to guarantee that S would have more neighbours. Based on the discussion in Sect. 3, both $R_c(S, cluster_i)$ and $\frac{|N(S) \cap cluster_i|}{|cluster_i|}$ belong to the interval $[0, 1]$, thus, we take advantages of a ranking aggregation algorithm WBF [2] to aggregate these two metrics and γ is the weight. When γ is higher, the importance of users' characteristics is higher. Otherwise, the importance of topological features is higher. By adjusting γ, we can balance the importance of users' characteristics and topological features. To maximize the objective function F, we give a greedy heuristic algorithm shown in Algorithm 2.

Algorithm 2. *Greedy Algorithm*

Input: Original DataSet: D; The desired size of S: k; The objective function:
  F; The function of representative degree of characteristics: $R_c(S, D)$;
  The function of neighbours of S: $N(S)$;
Output: The representative subset S

1 **while** $|S| < k$ **do**
2 | **for** *each* $v \in D - S$ **do**
3 | | $max\_increment = -1$;
4 | | Compute the increment $increment_v$ of F in each $cluster_i$ when $S \cup \{v\}$;
5 | | **if** $increment_v > max\_increment$ **then**
6 | | | $v^* = v$;
7 | | | $max\_increment = increment_v$;
8 | $S = S \cup \{v^*\}$;
9 | // Only need to update F to $cluster_i$ to which v^* belongs;
10 | Update $R_c(S, cluster_i)$ and $\frac{|N(S) \cap cluster_i|}{|cluster_i|}$ in F;
11 **return** S;

In each iteration, we traverse all users in $D - S$ and add the one that mostly increases the objective function F. For reducing the time complexity, in the end of each iteration, we will update R_c and $N(S)$ in F to each cluster and store the current values of F to each cluster such that when computing each v in $D - S$, we only need to calculate the increment of $R_c(S, cluster_i)$ and $\frac{|N(S) \cap cluster_i|}{|cluster_i|}$ ($v \in cluster_i$). Because when adding each v, it only affects the increment of the cluster which v belongs to. Based on the definition of R_c, we need to search in the set $S \cap cluster_i$ for a representative element to each element in $cluster_i$. Thus, searching for the representative in each cluster will cost $\mathcal{O}(|cluster_i| \cdot |S \cap cluster_i|)$. Because there are m clusters, the total time complexity of Algorithm 2 is $\mathcal{O}(k \cdot m \cdot |D - S| \cdot |cluster_i| \cdot |S \cap cluster_i|)$.

In addition, the quality of Algorithm 1 can be affected by the initial choice of seeds. Thus, in the aim of reducing the effect of selecting seeds randomly when clustering, we will run Algorithm 1 multiple times, each time with different, randomly selected seeds. Then, we run Algorithm 2 on these different results and choose the best.

5   Experiments

In this section, we compare the performance of four methods (XSN [14], PageRank [17], S^3 [23] and KS). We present four evaluation metrics: representative degree of characteristics, representative degree of topological features, representative degree of both these two aspects, and a case study: regarding representative users as a classifier and using the accuracy to evaluate the representative degree of S extracted by each method. From experiments, the subset of representative users extracted by KS preserves the most users' characteristics and topological features of the original dataset.

5.1   Experimental Setup and DataSets

All algorithms are implemented by python 2.7 and benchmarked on a PC with 2TB memory and Intel Xeon 2.20 GHz 64 core processor running CentOS 7. We install MySQL and Neo-4j to store the users' structured information and their relationships. We prepared two real-world datasets.

Twitter DataSet: The first real-world dataset contains 60,071 users crawled from Twitter. Besides users' basic structured information such as the number of followers or friends and location information, we also crawl their tweets. To extract more features of users, we calculate the score of the influence and activity by adding a few users' numerical attributes like the number of followers, the times of tweets being "liked" or "retweeted", the number of friends and so on with a certain weight. The words and phrases with high frequency are extracted from tweets. Finally, we construct each user composed of attributes $<d_1, d_2, d_3, d_4, d_5, d_6, d_7, d_8, d_9>$[1]. And we train a Multinomial Naive Bayes classifier[2] combined with manual annotation to classify users into several specific domains based on their tweets (here we classify them into nine domains). Therefore, there are nine domains in Twitter Dataset and we use XSN, S^3, PageRank and KS respectively to extract representative users from each domain with a sampling rate. Users' structured information is stored into a relational database MySQL and 17,465,370 relationships are stored into a graph database Neo-4j.

---

[1] d_1: the number of user's followers, d_2: the number of user's friends, d_3: the score of user's influence, d_4: the score of user's activity, d_5: the number of tweets, d_6: times of tweets being "liked", d_7: times of tweets being "retweeted", d_8: address, d_9: words and phrases.

[2] http://scikit-learn.org/stable/modules/naive_bayes.html#multinomial-naive-bayes.

SinaWeibo DataSet: We crawled 31,029 users from five domains like actors, directors, singers, players, and ordinary people to validate our method. Similarly, we used the same method to construct users' attributes as Twitter DataSet and crawled their relationships. In the same way, we stored their structured information into MySQL and 181,644 relationships into Neo-4j. Also, we run XSN, S^3, PageRank and KS respectively to extract representative users from each domain with a sampling rate.

5.2   Evaluation Metrics

Representative Degree of Characteristics: We use $R_c(S, D)$ to evaluate the representative degree of users' characteristics of the representative subset S. The larger $R_c(S, D)$ is, the more users' characteristics S can preserve.

Representative Degree of Topological Features: In [14], Maiya et al. proposed the criteria $Composite = \frac{2 \times FRAC \times PART}{FRAC + PART}$ to evaluate the representative degree of community structure in the large network. They employed the partition distance proposed by [6] and the number of communities represented in the sample. $PART$ is the partition distance between the community structure of a sample and the community structure in the larger network. $FRAC$ is the fraction of communities in the sample. We show results of $Composite$ to evaluate the representative degree of topological features of S. The larger $Composite$ is, the more representative the topology of S is.

Representative Degree of Both Characteristics and Topological Features: For the overall evaluation of representative degree of characteristics and topological features, as discussed in the definition of RUS problem in Sect. 3, we use $R = \frac{2 \times R_c(S,D) \times R_t(S,D)}{R_c(S,D) + R_t(S,D)}$ to evaluate the comprehensive representative degree of S.

Accuracy of Classifier: In [23], Tang et al. extracted authors from three major conferences (SIGMOD, ICDM, CIKM) and regarded program committee members of the conferences during 2007–2009 as ground truth. However, because these users are experts in some domains who are different from representative users we desire, this dataset is hard to be used. For our experiments, it is difficult to find a standard social dataset with ground truth. Thus, we design a case study to evaluate whether the subset S is representative both in users' characteristics and topological features. For original dataset D, we use XSN, PageRank, S^3, and KS respectively to extract the representative subset from each domain with a sampling rate α and combine all users in subsets together as S. For any u_j in D, we use S to classify the domain of u_j and the accuracy of classifying is regarded as the metric. When using S to classify users' domain in original dataset, we have considered both users' characteristics and the topology of networks. Specifically, we use the community detection algorithm to find all communities in D. We select CNM algorithm [4] from GN [10], NLE [16] and CNM algorithm as our community detection algorithm because in these three algorithms, only CNM algorithm can be executable on larger networks. Then, we find a subset $I \subseteq S$ in which elements belong to the same community with u_j in the original dataset. Finally, we use u_i in I which is "nearest" to u_j (i.e. $u_i = argmax\ r(u_i, u_j)$) to label u_j, which means that the domain of u_j is equivalent to the domain of u_i. Otherwise, if we cannot find such subset I, we search for the one u_i that has the largest value of $r(u_i, u_j)$ in the whole S and use u_i to label u_j. We compare the accuracy of classifiers composed of the subset of users extracted by all methods. The computation of accuracy proceeds as follows:

$$Accuracy = \frac{|D_{correct}|}{|D - S|} \in [0, 1] \tag{7}$$

where $|D_{correct}|$ denotes the number of users who are classified correctly in $D-S$. The higher accuracy indicates the subset is more representative or has more representative degree to the original dataset. Hence, methods gaining higher accuracy perform better.

5.3 Comparison Methods

To further validate the performance of our methods, we compare the performance of KS with that of following methods: PageRank [17], S^3 [23], and XSN [14].

- PageRank estimates the influence of each node according to their in-degree and out-degree. We use this method to extract top-k users according to the influence score as representative users.
- S^3 is a sampling model that is proposed to sample representative users from several topics with the consideration of users' attributes. This model samples representative users according to a quality function Q through a greedy heuristic algorithm. We mainly compare our method with this algorithm from the perspective of representative degree of users' characteristics.
- XSN is a sampling algorithm that is capable of representing and inferring community structure in the original network. We use this method to extract k nodes by treating each user as a node and relationships as edges. We mainly compare our method with this algorithm from the perspective of representative degree of topological features.

5.4 Experiment Results

Note that we report the best performance of methods XSN, S^3, PageRank, and KS on both datasets Twitter and SinaWeibo in the sequel.

We evaluate the performance of KS in sampling representative users on two real-world datasets. We run all methods to sample representative users from each domain with sampling rate α on both datasets. And we show the average results of R_c, Composite, R and Accuracy among all domains of methods in Fig. 2.

As shown in Fig. 2, KS always achieves the best performance in R_c, R, and Accuracy while Compostie of KS is a little lower than XSN. This is because in order to achieve the best general performance, we adjust parameter γ in F to

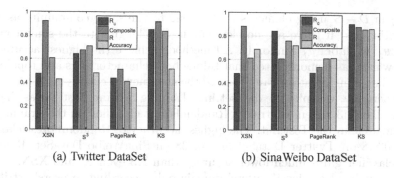

(a) Twitter DataSet   (b) SinaWeibo DataSet

Fig. 2. Results of all methods on metrics

balance the importance of topological features and users' characteristics. If we set γ closer to 0, R_t and *Composite* of KS are closer to XSN. Additionally, KS achieves higher R_c than S^3. This is because KS clusters users based on users' characteristics such that representative users extracted by KS from each group will have more coverage in users' characteristics. Because XSN and PageRank pay more attention to users' relationships or network topology, R_c of them is much lower than KS and S^3. Because the sampling model F in KS has combined the indicator of characteristics with the indicator of topological features, KS achieves the highest R and *Accuracy* that are the general metrics of representative degree. Thus, KS preserves both users' characteristics and topological features.

Table 2. Accuracy under different cases

	Twitter DataSet			SinaWeibo DataSet		
	$Case^1$	$Case^2$	$Case^3$	$Case^1$	$Case^2$	$Case^3$
XSN	0.354	0.316	**0.426**	0.479	0.438	**0.688**
S^3	0.417	0.296	**0.479**	0.663	0.355	**0.72**
PageRank	0.269	0.274	**0.356**	0.42	0.342	**0.612**
KS	0.423	0.333	**0.513**	0.671	0.508	**0.862**

As shown in Fig. 2, the convincing result has been obtained when regarding S as a prototype to classify other users' domain in the original dataset. On both two datasets, KS achieves the highest accuracy when classifying. In Table 2, besides the situation $Case^3$ that we have explained in Sect. 5.2, we also design another two cases to demonstrate that it is beneficial and important to take both users' characteristics and topological features into account at the same time when extracting representative users. $Case^1$ means that we classify users only according to users' characteristics, i.e., label u_j in $D - S$ using $u_i = argmax\ r(u_i, u_j)$. $Case^2$ means that we classify users only according to topological features, i.e., label u_j in $D - S$ using elements that belong to the same community as u_j in S by voting and if there are no elements in S belonging to the same community as u_j, we label u_j randomly. All methods achieve the highest accuracy in $Case^3$, which has demonstrated that taking both characteristics and topological features into consideration can promote the accuracy of classifying.

Furthermore, we split the dataset into Training Set and Test Set as the proportion 8:2 or 7:3. Then, we train a RandomForest classifier[3] in the Training Set and validate it in the Test Set. The highest accuracy of RandomForest classifier is only 0.328 on Twitter DataSet and 0.738 on SinaWeibo DataSet. Random-Forest classifier gains much lower accuracy than KS even S^3 and XSN. This is because it is hard to classify users' domain only according to users' attributes in social networks. Owing to combining users' characteristics with topological features in F, KS achieves higher accuracy.

To obtain the best performance of KS, tuning parameters, such as γ, the initial number of clusters m and sampling rate α, is of critical importance. We therefore study the impact of different parameters as following.

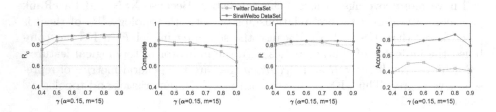

Fig. 3. KS method results varying γ

varying γ: In Fig. 3, the increment of γ makes R_c higher but *Composite* lower. If we set γ larger, which means that users' characteristics play a more important role when sampling representative users, thus, R_c would be larger. KS achieves the highest general performance when γ equals to 0.6 and 0.8 respectively on Twitter DataSet and SinaWeibo DataSet from results of R and *Accurayc*. *Composite* of KS drops faster on Twitter DataSet than that on SinaWeibo DataSet when γ increases. This may be result from that there are more communities on Twitter DataSet such that it is hard to cover more communities when γ increases on Twitter DataSet.

varying m: We show results when $\gamma = 0.6$ on Twitter DataSet and $\gamma = 0.8$ on SinaWeibo DataSet because KS can achieve the best performance. In Fig. 4, when the number of clusters m becomes larger which means that users would be classified into more groups, users' characteristics would be separated in more details and R_c of KS grows. This is because the representative subset would have more coverage in users' characteristics if there are more clusters when

---

[3] http://scikit-learn.org/stable/modules/ensemble.html#random-forests.

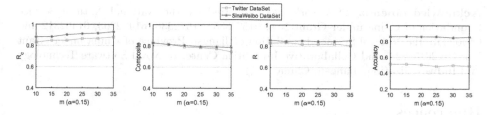

Fig. 4. KS results varying m ($\gamma = 0.6$ on Twitter DataSet, $\gamma = 0.8$ on SinaWeibo DataSet)

selecting users from each group. While m grows higher, KS would achieve lower *Composite*. Because the higher number of clusters would make the decrement of $|N(S) \cap cluster_i|$ when sampling according to F since neighbours of the set $S \cap cluster_i$ may appear in other clusters. As from results of R and *Accuracy*, KS can achieve the best general performance when $m = 15$ or $m = 20$.

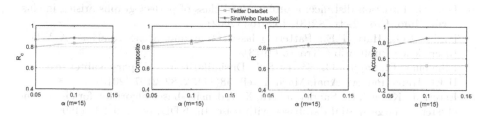

Fig. 5. KS results varying α ($\gamma = 0.6$ on Twitter DataSet, $\gamma = 0.8$ on SinaWeibo DataSet)

varying α: When sampling rate α grows, all methods would achieve better performance because the size of S is larger. In Fig. 5, the performance of KS becomes better when α increases. Due to the limitation of memory and efficiency, α cannot be large. When preserving more users' characteristics and topological features, the smaller α is, the better results are.

6   Conclusion

In this paper, we focus on extracting representative users of social networks towards users' characteristics and topological features. We formulate this problem as RUS problem. We discuss the NP-Hardness of RUS problem and provide the proof. We propose the solution KS combining a clustering algorithm and a sampling model. For the sampling model, a greedy heuristic algorithm is proposed. To validate the performance of the proposed approach, extensive experiments are conducted on two real-world datasets. Results demonstrate that our method outperforms state-of-the-art approaches.

Acknowledgements. This work was supported by the National Natural Science Foundation of China under Grant Nos. 61572335, 61572336, 61472263, 61402312 and 61402313, the Natural Science Foundation of Jiangsu Province of China under Grant No. BK20151223, and Collaborative Innovation Center of Novel Software Technology and Industrialization, Jiangsu, China.

References

1. Anagnostopoulos, A., Kumar, R., Mahdian, M.: Influence and correlation in social networks. In: KDD, pp. 7–15 (2008)
2. Aslam, J.A., Montague, M.: Models for metasearch. In: SIGIR, pp. 276–284 (2001)
3. Chen, W., Wang, Y., Yang, S.: Efficient influence maximization in social networks. In: KDD, pp. 199–208 (2009)
4. Clauset, A., Newman, M.E., Moore, C.: Finding community structure in very large networks. Phys. Rev. E **70**(2), 066111 (2004)
5. Crandall, D.J., Cosley, D., Huttenlocher, D.P., Kleinberg, J.M., Suri, S.: Feedback effects between similarity and social influence in online communities. In: KDD, pp. 160–168 (2008)
6. Dan, G.: Partition-distance: a problem and class of perfect graphs arising in clustering. Info. Proc. Lett. **82**(3), 159–164 (2002)
7. Duda, R.O., Hart, P.E.: Pattern Classification and Scene Analysis. A Wiley-Interscience Publication, Tronto (1973)
8. Elhamifar, E., Sapiro, G., Sastry, S.S.: Dissimilarity-based sparse subset selection. IEEE Trans. Pattern Anal. Mach. Intell. **38**(11), 2182–2197 (2016)
9. Ester, M., Kriegel, H., Sander, J., Xu, X.: A density-based algorithm for discovering clusters in large spatial databases with noise. In: KDD, pp. 226–231 (1996)
10. Girvan, M., Newman, M.E.: Community structure in social and biological networks. Proc. Natl. Acad. Sci. USA **99**(12), 7821 (2002)
11. Goyal, A., Bonchi, F., Lakshmanan, L.V.S.: Discovering leaders from community actions. In: CIKM, pp. 499–508 (2008)
12. Han, Y., Tang, J.: Probabilistic community and role model for social networks. In: KDD, pp. 407–416 (2015)
13. Kaufmann, L., Rousseeuw, P.J.: Clustering by means of medoids. In: Statistical Data Analysis Based on the L1-norm & Related Methods, pp. 405–416 (1987)
14. Maiya, A.S., Berger-Wolf, T.Y.: Sampling community structure. In: WWW, pp. 701–710 (2010)
15. Megiddo, N., Supowit, K.J.: On the complexity of some common geometric location problems. SIAM **13**(1), 182–196 (1984)
16. Newman, M.E.J.: Finding community structure in networks using the eigenvectors of matrices. Phys. Rev. E **74**(3), 036104 (2006)
17. Page, L.: The pagerank citation ranking: bringing order to the web. Stanf. Digit. Libr. Work. Pap. **9**(1), 1–14 (1998)
18. Papagelis, M., Das, G., Koudas, N.: Sampling online social networks. IEEE TKDE **25**(3), 662–676 (2013)
19. Scripps, J., Tan, P., Esfahanian, A.: Measuring the effects of preprocessing decisions and network forces in dynamic network analysis. In: KDD, pp. 747–756 (2009)
20. Song, X., Chi, Y., Hino, K., Tseng, B.L.: Identifying opinion leaders in the blogosphere. In: CIKM, pp. 971–974 (2007)

21. Sun, K., Morrison, D., Bruno, E., Marchand-Maillet, S.: Learning representative nodes in social networks. In: Pei, J., Tseng, V.S., Cao, L., Motoda, H., Xu, G. (eds.) PAKDD 2013. LNCS (LNAI), vol. 7819, pp. 25–36. Springer, Heidelberg (2013). https://doi.org/10.1007/978-3-642-37456-2_3
22. Tang, J., Sun, J., Wang, C., Yang, Z.: Social influence analysis in large-scale networks. In: KDD, pp. 807–816 (2009)
23. Tang, J., Zhang, C., Cai, K., Zhang, L., Su, Z.: Sampling representative users from large social networks. In: AAAI, pp. 304–310 (2015)
24. Ugander, J., Karrer, B., Backstrom, L., Kleinberg, J.M.: Graph cluster randomization: network exposure to multiple universes. In: KDD, pp. 329–337 (2013)
25. Vazirani, V.V.: Approximation Algorithms. Springer, Heidelberg (2003). https://doi.org/10.1007/978-3-662-04565-7
26. Yin, H., Chen, H., Sun, X., Wang, H., Wang, Y., Nguyen, Q.V.H.: SPTF: a scalable probabilistic tensor factorization model for semantic-aware behavior prediction. In: ICDM, pp. 585–594 (2017)
27. Yin, H., Cui, B., Huang, Y.: Finding a wise group of experts in social networks. In: Tang, J., King, I., Chen, L., Wang, J. (eds.) ADMA 2011. LNCS (LNAI), vol. 7120, pp. 381–394. Springer, Heidelberg (2011). https://doi.org/10.1007/978-3-642-25853-4_29
28. Yin, H., et al.: Discovering interpretable geo-social communities for user behavior prediction. In: ICDE, pp. 942–953 (2016)
29. Yin, H., Zhou, X., Cui, B., Wang, H., Zheng, K., Hung, N.Q.V.: Adapting to user interest drift for POI recommendation. TKDE **28**(10), 2566–2581 (2016)

Group Identity Matching Across Heterogeneous Social Networks

Hongchao Qin[1], Ye Yuan[1(✉)], Feida Zhu[2], and Guoren Wang[3]

[1] School of Computer Science and Engineering,
Northeastern University, Shenyang, China
yuanye@mail.neu.edu.cn
[2] School of Information Systems, Singapore Management University,
Singapore, Singapore
[3] School of Computer Science and Technology,
Beijing Institute of Technology, Beijing, China

Abstract. User identity linkage aims to identify and link users across different heterogeneous social networks. In real applications, one person's attributes and behaviors in different platforms are not always same so it's hard to link users using the existing algorithms. In this paper, we discuss a novel problem, namely *Group Identity Matching*, which identifies and links users by an unit of group. We propose an efficient approach to this problem and it can take both users' behaviors and relationships into consideration. The algorithm incorporates three components. The first part is behavior learning, which models the group's behavior distribution. The second part is behavior transfer and it optimizes the behavior distance between groups across the social networks. The third part is relationship transfer and it enhances the similarity of the groups' social network structure. We find an efficient way to optimize the objective function and it convergences fast. Extensive experiments on real datasets manifest that our proposed approach outperforms the comparable algorithms.

Keywords: Group matching · Identity matching
Community matching · Behavior model · Transfer learning

1 Introduction

Today, billions of users are now engaged in multiple online social networks, such as Facebook, Twitter, Foursquare, Weibo, WeChat and so on. In many cases, a given user is simultaneously a member of several different networks. Each social network platform contains different kinds of information and they exhibit different behaviors in each individual network. Unfortunately, the accounts for

Ye Yuan is supported by the NSFC (Grant No. 61572119 and 61622202) and the Fundamental Research Funds for the Central Universities (Grant No. N150402005). Guoren Wang is supported by the NSFC (Grant No. U1401256, 61732003 and 61729201).

© Springer Nature Switzerland AG 2018
H. Hacid et al. (Eds.): WISE 2018, LNCS 11233, pp. 230–246, 2018.
https://doi.org/10.1007/978-3-030-02922-7_16

each networks are independent. So the problem of *User Identity Linkage*, which aims to identify and link users across different social networks, has recently been attracting an increasing amount of attention [2,8,13].

There is an assumption for the problem of *User Identity Linkage* that one user is active at both social networks, and he shares same features in different networks. But we might think that some people will have different features in different services, even they do not have time to participate in many social networks. However, according to the experiments in [5,11,13] and the analysis of our crawled data, the features of linked people in different networks are not very similar so the precision value of the existing algorithms are not high.

Surprisingly, we find that most people tend to fill in similar personal information in different social services and build relationships from the friends' recommendation. Although one user has rare contents in a social network which he uses not so frequently, we can link him through his friends. As we can link more users with considering both the content features and the relationships of users in a group, there are many challenges using the existing techniques.

- The existing techniques only focus on modeling one person's behavior. It is hard to model behaviors of users in a group.
- After modeling behaviors of users in a group, it is hard to define which pair of groups are similar and how to find the most similar groups.
- As behaviors can be modeled by a function and it is easy to compute the distance between behaviors of two given groups, the similarity of relationships is hard to define for the groups.

We study the problem of *Group Identity Matching*, which tries to match groups' identities to get more possible matches. In this way, we can find more latent linkage between social networks even the topic of many persons' contents can not match. We propose a framework of matching group identities across heterogeneous social networks. We not only consider the information and features of every person, but also take the social network structures into account. Figure 1 illustrates a block diagram of the proposed framework. At first, we model the information and features of one group as a behavior distribution and compute the maximum entropy as the learning target to get the best likelihood function and mark it as *Behavior Learning*. Then we consider how to build an objective function to represent the behavior distance across different social networks and call it *Behavior Transfer*. After that we take the social network structures into account and join the most likely related users into the group to get minimum distance and it is the process of *Relationship Transfer*. Finally we get the matching groups of the different social networks.

The contributions of the article are highlighted as follows:

- We propose a behavior learning model based on entropy maximization to learn the activities of a given group of users. Then we can transfer the user behavior of a given group into a new social network platform.
- A novel framework is designed to match the identity of groups across heterogeneous social networks, including the transfer model and some technologies

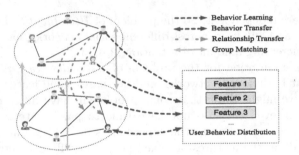

Fig. 1. Group identity matching across different social networks

based on random walk to transfer relationship. The individuals in the group are not only similar in behavior topics but also have same topological relationship, so they are more likely to be the same persons.

- We introduce a novel technique based on improved iterative scaling to optimize the model. And the convergence speed is very fast.
- We conduct experiments on real-world data sets. The experiments demonstrate that our algorithm can find more latent and missing identity linkage and has a better performance on the running time.

The rest of the paper is organized as follows. Section 2 presents the problem model. The algorithm to match the group identities is proposed in Sect. 3. Experimental studies are presented in Sect. 4. We discuss related works in Sect. 5 and conclude the paper in Sect. 6.

2   Problem Definition

In this section, we'll give the formal definitions of many important concepts used in this paper and the formulation of the group identity matching problem.

Definition 1 (Source/Target domain). *The source/target domain $\mathcal{G}^s/\mathcal{G}^t$ can be modeled as a heterogeneous social network. It is a graph $\mathcal{G} = (\mathcal{X}, \mathcal{Y}, \mathcal{E})$, where $\mathcal{X} = \{x_1, x_2...x_N\}$ represents all vertexes, $\mathcal{Y} = \{y_1, y_2...y_N\}$ and every y_i is the content of each vertex corresponding to x_i, $\mathcal{E}$ is the set of all the edges in the graph. The vertexes represent the users and the edges represent the relationships between them.*

Definition 2 (User Behavior Distribution). *Given a group of users' attributes $X = \{x_1, x_2...x_n\}$, the users' contents $Y = \{y_1, y_2...y_n\}$ and a set of all features $\mathcal{F} = \{f_1, f_2..f_F\}$, the user behavior of X in graph $\mathcal{G} = (\mathcal{X}, \mathcal{Y}, \mathcal{E})$ can be defined as a distribution.*

$$p_\Lambda(Y|X) = \frac{1}{Z(\mathcal{X})} exp(\sum_{j=1}^{n}\sum_{i=1}^{F} f_i(x_j, y_j)\lambda_i) \tag{1}$$

Where $Z(\mathcal{X}) = exp(\sum_{j=1}^{N}\sum_{i=1}^{F} f_i(x_j, y_j)\lambda_i)$ *is simply the normalizing factor to ensure a proper probability and* $\Lambda = \{\lambda_1, \lambda_2, ...\lambda_F\}$ *is a set of weights corresponding to each feature.*

Example 1. We can assume that X is $\{Mary, John, Lily, Tom\}$ (marked as $\{x_1, x_2, x_3, x_4\}$), and the users' contents are a set of vectors $\{y_1, y_2, y_3, y_4\}$. The vector space is a set of all meaningful phrases. Then y_1 is a form of $[a_{00}, a_{01}, a_{02}...]^T$, so does $y_2 = [a_{10}, a_{11}, a_{12}...]^T$, y_3... Every item in the vector space is the frequency of the word. If the vector space is $[apple, London, Messi, football, rain...]$, a_{00} is the frequency of word $apple$ in $Mary$'s contents.

$\mathcal{F}$ is a function to represent the features. It means that the labels must contain set of words or phrases. For example, the feature $sport$ (marked as f_1) contains $[football, Messi, Ronaldo, manchester ...]$, then $Mary$'s behavior correspond to f_1 is $(a_{02} + a_{03} + ...)\lambda_1$. And λ_1 is a real-valued weight associated with f_i. So the behavior of X corresponding to f_1 is $(a_{02} + a_{03} + a_{12} + a_{13} + ...)\lambda_1$.

2.1   Behavior Learning

In this section, we can get the behavior distribution of the group X in the social network $\mathcal{G}$. As the user behavior distribution can be calculated in Eq. (1), how to get value for each λ_i is an important task to obtain users' behavior. In order to avoid over-fitting the training data, every λ_i in Λ is often constrained to be near 0 by the use of a regularization term which tries to minimize $\|\Lambda\|_2^2$. Entropy maximization (MaxEnt) is a way of modeling the conditional distribution of labels [1]. Thus the entire expression being optimized is:

$$\max_{\Lambda} \left(\sum_{i=1}^{n} p(x_i, y_i) \, log \, p_\Lambda(y_i|x_i) - \beta\|\Lambda\|_2^2\right) \tag{2}$$

where $\beta > 0$ and it is a parameter controlling the amount of regularization, $p(x_i, y_i) = \frac{sum(y_i)}{\sum_i sum(y_i)}$ is the probability that x_i and y_i count in all contents.

2.2   Behavior Transfer

Once the users' behavior of a given group X^s in $\mathcal{G}^s$ has been learned and it can be presented by Λ^s, the task of behavior transfer is to find a group X^t whose behavior, Λ^t, is similar as Λ^s.

Our methods are motivated by transfer learning [14,19], one recently proposed method for transfer learning in MaxEnt models involves modifying Λ's regularization term. Firstly a model of the source domain, Λ^s, is learned by training on $\{X^s, Y^s\}$. Then a model of the target domain is trained over a limited set of labeled target data X^t, Y^t, but instead of regularizing this target to be near zero by minimizing $\|\Lambda^t\|_2^2$, Λ^t is instead regularized towards the previously learned source values Λ^s by minimizing $\|\Lambda^t - \Lambda^s\|_2^2$. Thus the modified optimization problem is:

$$\max_{\Lambda^t, \Lambda^s} \left(\sum_{i=1}^{n^t} p^t(x_i, y_i) \, log \, p_{\Lambda^t}(y_i|x_i) - \beta \|\Lambda^t - \Lambda^s\|_2^2 \right) \tag{3}$$

where n^t is the number of labeled training examples in the target domain. It should be noted that this model requires Y^t in order to learn Λ^t.

2.3 Relationship Transfer

Since we have got the objective function in Eq. (3), it must be confirmed that the users in group X^s and the users in group X^t are related, which means that they must have more possible same friendships.

Measuring the relatedness of two nodes in the graph can be achieved using the Random Walks with Restarts (RWR) theory. Starting from node v_q, a RWR is performed by following a link to another node at each step. Also, there is a probability to return the node v_q in every step.

$$r^{(n+1)} = \tau S r^{(n)} + (1 - \tau)q \tag{4}$$

where $r^{(n)}$ is a vector with each element $r_i^{(n)}$ denoting the probability that the random walk at step n arrives at node v_i; q is a vector of zero with the element corresponding to the starting node v_q set to 1, i.e., $q[v_q] = 1$; S defines the transition matrix of the random walk, with element S_{ij} denoting the random walking probability from node v_i to node v_j, and it's often set as $1/degree\ (v_i)$.

If all the nodes perform RWR in every step, we can get $r_i^{(n)}$ for every node. After several steps, we can get a stable value of $r_i^{(n)}$ [12]. It is the relatedness of every two nodes, and it can be marked as $\tilde{r}(x_i, x_j)$.

2.4 Problem Formulation

Group Identity Matching. Given a source social network $\mathcal{G}^s$, a group user set $X^s = \{x_1^s, x_2^s..x_n^s\}$ in $\mathcal{G}^s$ and a target social network $\mathcal{G}^t$, the problem of group identity matching (GIM) is to find a group user set X^t in $\mathcal{G}^t$ and the users in X^t are probably corresponding to the same natural persons of X^s. The objective of our transfer learning task is to learn a classifier to find a user set X^t that:

$$max \, (\sum_{i=1}^{n^t} p^t(x_i, y_i) \, log \, p_{\Lambda^t}(y_i^t|x_i^t) - \beta \|\Lambda^t - \Lambda^s\|_2^2) \tag{5}$$

$$s.t. \quad \tilde{r}(x_i, x_j) > \tau \qquad \exists \tau > 0; \{\forall (x_i, x_j) \| x_i, x_j \in X^t\}$$

The objective target is to ensure that they have same behaviors, modeling by Eq. (3). And the subjection means that the users get closely relationships.

3   Proposed Method

3.1   Overview

Our proposed approach consists of four components as following.
Step 1: Setup. At first, the initial user group X^s in $\mathcal{G}^s$ is needed. We can start from one person, and find community containing him in the social network. Many existing methods can be performed, and in this paper we choose the algorithm of online community searching in [3].
Step 2: Learn from the source domain. We need to know the behavior distribution of X^s in the above step, and try to get an optimal Λ^s in Eq. (2).
Step 3: Transfer to the target domain. After getting the Λ^s, we need to choose an initial user group X^t to solve Eq. (3) and get Λ^t.
Step 4: Travel the networks and optimize the objective function. We need to choose the users in user group X^t to let the behavior distance of X^s and X^t be closer. The social network relationships in Eq. (4) are used to rank nodes which are most related to X^t in the target domain. After joining one candidate node into X^t, we repeat step 3 to get the terminate value. If the value increases, the candidate node can be joined into X^t, otherwise we seek for another node. The whole algorithm terminate when all the candidate nodes are visited.

Finally, we will get two groups' matching across the social networks. They have the most same behavior and share the most similarity friends.

3.2   Behavior Learning Method

In this section, we discuss how to learn the user behavior distribution in source domain, and briefly outline an algorithm for calculating the parameters of a maximum entropy classifier. The traditional MaxEnt learns a model consisting of a set of weights corresponding to each class $\Lambda = \{\lambda_1, \lambda_2...\lambda_F\}$ over the features to maximize the conditional likelihood of the training data.

So we have a model of the transfer task in Eq. (2) with some arbitrary set of parameters $\Lambda = \{\lambda_1, \lambda_2...\lambda_F\}$, we can mark it as $L(\Lambda, p)$. We'd like to find a new set of parameters $\Lambda + \Delta = \{\lambda_1 + \delta_1, \lambda_2 + \delta_2...\lambda_F + \delta_F\}$ to get a model of higher log-likelihood. With respect to the given distribution $p_\Lambda(y|x)$, the change in log-likelihood from the given Λ to $\Lambda + \Delta$ is

$$L(\Lambda + \Delta, p) - L(\Lambda, p) = \sum_{x,y} p(x,y) \sum_{i=1}^{F} \delta_i f_i(x,y) - \sum_x p(x) log \frac{Z_{\Lambda'(x)}}{Z_{\Lambda(x)}} - \beta \sum_i^F (\delta_i^2 + 2\lambda_i \delta_i)$$

$$(6)$$

We can use the inequality $- log\ \alpha \geq 1 - \alpha$ (true for all $\alpha > 0$) and Jensen's inequality, $\sum p(x) exp\ q(x) \geq exp \sum p(x) q(x)$, to transfer the equation above

$$L(\Lambda + \Delta, p) - L(\Lambda, p) \geq \sum_{x,y} p(x,y) \sum_{i=1}^{F} \delta_i f_i(x,y) + 1 - \sum_x p(x) \frac{Z_{\Lambda'(x)}}{Z_{\Lambda(x)}} - \beta \sum_i^{F} (\delta_i^2 + 2\lambda_i \delta_i)$$

$$\geq \sum_{x,y} p(x,y) \sum_{i=1}^{F} \delta_i f_i(x,y) + 1 - \sum_x p(x) \sum p_\Lambda(y|x) exp(\sum_i \delta_i f_i(x,y)) - \beta \sum_i^{F} (\delta_i^2 + 2\lambda_i \delta_i)$$

$$\geq \sum_{x,y} p(x,y) \sum_{i=1}^{F} \delta_i f_i(x,y) + 1 - \sum_x p(x) \sum p_\Lambda(y|x) \sum (\frac{f_i(x,y)}{\sum f_i(x,y)}) exp(\delta_i \sum_i f_i(x,y))$$

$$- \beta \sum_i^{F} (\delta_i^2 + 2\lambda_i \delta_i) \tag{7}$$

Differentiating the right part (marked as $\Delta(L)$) with respect to δ_i, we can get

$$\frac{\partial(\Delta(L))}{\partial \delta_i} = \sum_{x,y} p(x,y) \ f_i(x,y) - \sum_x p(x) \sum p_\Lambda(y|x) f_i(x,y) exp(\delta_i \sum_i f_i(x,y)) - 2\beta \sum_i^{F} (\lambda_i + \delta_i) \tag{8}$$

δ_i in the equation above appears alone. So we can solve it for each of the n free parameters $\{\delta_1, \delta_2...\delta_F\}$ individually by differentiating $\Delta(L)$, which suggests an iterative algorithm for find the optimal values of $\{\lambda_1, \lambda_2...\lambda_F\}$. And the process is described in Algorithm 1.

Algorithm 1. BehaviorLearning($\mathcal{G}, X$)

Input: Graph Domain $\mathcal{G}$, a set of users $X = \{x_1, x_2...x_n\}$
Output: The optimal user behavior $\{\lambda_1, \lambda_2...\lambda_F\}$
1 Initialize $Y \leftarrow \{y_1, y_2...y_n\}$ from $\mathcal{G}$ by searching X;
2 Initialize $\mathcal{F} \leftarrow \{f_1, f_2..f_F\}$ by searching the domain $\mathcal{G}$;
3 Each $\lambda_i \leftarrow$ (arbitrary value) in Λ; Each $\delta_i \leftarrow$ (arbitrary value) in Δ;
4 Set $\epsilon > 0$ to control the termination of iterations;
5 **while** *each* $abs(\delta_i) > \epsilon$ **do**
6 $\quad$ Solve $\frac{\partial(\Delta(L))}{\partial \delta_i} = 0$ in Equation. (8) for each δ_i;
7 $\quad$ **for** i *from 1 to F* **do**
8 $\quad\quad$ $\lambda_i = \lambda_i + \delta_i$;
9 **return** $\Lambda = \{\lambda_1, \lambda_2...\lambda_F\}$;

3.3 Behavior Transfer Method

In this section we propose the methods for transfer knowledge of the information in social network, including behavior transfer and relationship transfer.

Behavior Transfer. There is a problem that the joint distribution of the features with labels differs between the source and target domains. In other words, $E^s[f_j]$ (the expectation of features f in $\mathcal{G}^s$) does not necessarily equal $E^t[f_j]$. If the expectations in the train and test datasets are similar, then the Λ learned on the training data will generalize well to the test data.

In Sect. 3.2, we can learn the domain by the model. Consider Eq. (3), the Λ^s can be learned by the existing technique and it can be treated as a constant to get the Λ^t. The Λ^t consider both the information inside the domain $\mathcal{G}^t$ and some knowledge transferred from domain $\mathcal{G}^s$. Same as the process of solving the Λ, the differentiating process contains 3 parameters

$$\frac{\partial(L(\Lambda^t, \Lambda^s + \Delta^s, p^t) - L(\Lambda^t, \Lambda^s, p^t)}{\partial \delta_i^t}$$

$$\geq \sum_{x,y} p^t(x,y) \, f_i^t(x,y) - \sum_x p^t(x) \sum p_{\Lambda^t}^t(y|x) f_i^t(x,y) exp(\delta_i^t \sum_i f_i^t(x,y)) - 2\beta \sum_i^F (\lambda_i^t + \delta_i^t - \lambda_i^s)$$

$$(9)$$

The solution is similar as Algorithm 1 and the only difference is that it differentiates the right part of Eq. (9) and solves it in line 6.

Relationship Transfer. As mentioned in Sect. 3.1, we need to change the users in user group X^t to let the behavior distance of X^s and X^t be closer. In Sect. 2.3, we discuss a PageRank-based model to calculate the relatedness between users. Since the idea of all nodes' random walks for calculating PageRank values is very clear, we can iterate the all nodes' random walks to get a matrix R which records the similarity value for each two nodes in Eq. (4). When the iteration can't change R, we can get relatedness score of nodes x_i and x_j, which can be marked as $\tilde{r}(x_i, x_j)$.

In order to control the density of the constructed network and avoid overfitting, we choose the candidate node x_c for group X by searching neighbors of node in X, for given $0 < \tau, \sigma < 1$, satisfying that the percentage of some similar nodes is larger than τ, and those similar node x in X hold that $\tilde{r}(x, x_c) > \sigma$. Those structure similar nodes can be marked as follows,

$$SN(X) = \{v | \frac{\|\tilde{r}(v,w) > \sigma\|}{\|X\|} > \tau \ for \ all \ w \in X; v \in neighbor(w)\} \quad (10)$$

3.4 Improvement Strategies

Speed Up. In every iteration for behavior learning in Sect. 3.2, we must solve Eq. (8) = 0 in Algorithm 1, which can be marked as $f(\delta_i) = 0$. It is same as the form of $ae^{bx} + cx + d = 0$, so it is not easy to be solved quickly.

We can apply Newton-Raphson method in numerical analysis to solve it. The initial value $\delta_i^{(0)}$ can be set as $log(\frac{\sum_{x,y} p(x,y) \, f_i(x,y)}{\sum_x p(x) \sum p_{\Lambda}(y|x) f_i(x,y)}) / \sum_i f_i(x,y)$. It is calculated by solving the first half of $f(\delta_i)$.

Then the method starts to iterate with a function f, the function's derivative f', and the initial guess $\delta_i^{(0)}$ for a root of the function f. Then a better approximation $\delta_i^{(1)} = \delta_i^{(0)} - \frac{f(\delta_i^{(0)})}{f'(\delta_i^{(0)})}$.

The process is repeated as $\delta_i^{(n+1)} = \delta_i^{(n)} - \frac{f(\delta_i^{(n)})}{f'(\delta_i^{(n)})}$ until a sufficiently accurate value is reached. We can stop the iteration until $f(\delta_i^{(n+1)})$ less than a small number σ which is given to control the accuracy.

The method above can speed up to solve Eq. (8) = 0, and the number of the iterations is often less than 10. Similarly, it can also solve Eq. (9).

Prune Strategy. In this section we discuss some prune strategy when choosing the candidate nodes for X^t in relationship transfer part of Sect. 3.3.

Considering Eq. (3), when a new node x_c joins in X_t, the first half of (3) will increase $p^t(x_c, y_c) \ log \ p_{\Lambda^t}(y_c|x_c)$. The last part of (3) will change by $(-\beta\|\Lambda^t + \Delta^* - \Lambda^s\|_2^2 + \beta\|\Lambda^t - \Lambda^s\|_2^2) = -\beta\sum_i^F(2\lambda_i^t\delta_i^* + \delta_i^2 - 2\lambda_i^s\delta_i^*)$. If node x_c will join in X_t, the value in Eq. (3) increases, that is

$$p^t(x_c, y_c) \ log \ p_{\Lambda^t}(y_c|x_c) - \beta\sum_i^F(2\lambda_i^t\delta_i^* + \delta_i^2 - 2\lambda_i^s\delta_i^*) > 0$$
$$\Rightarrow p^t(x_c, y_c) \ log \ p_{\Lambda^t}(y_c|x_c) > 2\beta\sum_i^F(\lambda_i^t - \lambda_i^s)\delta_i^*$$

So if the cumulative $\Delta^* = \{\delta_1^*, \delta_2^*...\delta_F^*\}$ in Algorithm 1 satisfy that

$$\sum_i^F(\lambda_i^t - \lambda_i^s)\delta_i^* > \frac{p^t(x_c, y_c) \ log \ p_{\Lambda^t}(y_c|x_c)}{2\beta} \tag{11}$$

The iterations will terminate and the node will not join in X_t.

3.5   Group Identity Matching Algorithm

The whole progress of group identity matching is described in Algorithm 2.

At first we can initialize the considering features from $\mathcal{G}^s, \mathcal{G}^t$, learn the behavior Λ^s of the group users in the source domain, using Algorithm 1 and technology in Sect. 3.4 (lines 1–2). Next we find a seed user to join X^t whose behavior is most similar to Λ^s; set arbitrary value for each λ_i^t in Λ^t and δ_i^t in Δ^t; compute the converged relatedness scores (lines 3–5).

After that we begin to travel the target network to get the new group. Our target is to maximum the value in Eq. (3), so we initialize some value to start the iteration (lines 6–7). With one node joins in, after solving Eq. (9)=0, we need to get the optimal Λ^t and a higher value of Eq. (3) (line 12). For each iteration, we travel all the candidate nodes and get the highest likelihood value of Eq. (3) (line 16), join the corresponding node to X^t (line 18). And the prune strategy in Sect. 3.4 is added in finding the node (lines 13–15).

Finally all the candidate nodes are visited and no node's joining in can raise the value in Eq. (3), we get the optimal knowledge distributions of the source/target domain and the most possible group identity matching.

3.6   Model Analysis

We analyse the time complexity of Algorithm 2. At the stage of learning the behavior of X^s in the source domain (line 2), the time is $O(t_1)$ and t_1 is the number of the iteration by Algorithm 1 and technology in Sect. 3.4. Then from line 3–5, the time of finding the node whose behavior is most similar to Λ^s is $O(|V|)$ (line 3) and computing all pairs' PageRank value is $|V|^2\Omega(1/\sigma)$ (line 5) and σ is a given constant [12]. Luckily, the PageRank value can be stored and used for other query X^s. Next from line 8 to 19, the time of finding and

Algorithm 2. GroupMatching($\mathcal{G}^s, X^s, \mathcal{G}^t$)

Input: A source domain $\mathcal{G}^s$, a group user set X^s in $\mathcal{G}^s$ and a target domain $\mathcal{G}^t$
Output: A user set in $\mathcal{G}^t$ which is most likely the same identity as X^s in $\mathcal{G}^s$

1 Initialize $\mathcal{F} \leftarrow \{f_1, f_2..f_F\}$ by searching the domain $\mathcal{G}^s$ and $\mathcal{G}^t$;
2 $\Lambda^s \leftarrow$ BehaviorLearning($\mathcal{G}^s, X^s$);
3 Initialize X^t in $\mathcal{G}^t$ to be one node which is most similar to Λ^s;
4 Each $\lambda_i^t \leftarrow$ (arbitrary value); $\delta_i^t \leftarrow$ (arbitrary value) with i from 1 to F;
5 $\tilde{r}(x_i, x_j) \leftarrow$ relatedness of x_i and x_j, computing by eq.(4);
6 Set $\epsilon > 0$ to control the termination of the iteration;
7 $L^{(0)} \leftarrow 0; n \leftarrow 1; L^{(n)} \leftarrow$ value of eq.(3);
8 **while** *($L^{(n)} - L^{(n-1)} > \epsilon$)* **do**
9 $\quad$ $S \leftarrow \emptyset; SN(X^t) \leftarrow$ nodes satisfying eq(10) by using $\tilde{r}(x_i, x_j)$;
10 $\quad$ **for** v_i in $X_c(X^t)$ **do**
11 $\quad\quad$ Join v_i in X^t;
12 $\quad\quad$ Solve (9) = 0 for the new set $\{v_i \cup X^t\}$, get Δ_i;
13 $\quad\quad$ **if** *eq.(11) not holds* **then**
14 $\quad\quad\quad$ $L_i \leftarrow$ the value of eq.(3) with $\Lambda^t = \Lambda^t + \Delta_i$;
15 $\quad\quad\quad$ S.append($[v_i, \Delta_i, L_i]$);
16 $\quad$ $[v^*, \Delta^*, L^*] \leftarrow$ set in S which L is largest;
17 $\quad$ Update Λ^t with Δ^* ;
18 $\quad$ Update set X^t with the selected v^*;
19 $\quad$ $L^{(n)} \leftarrow L^*; n = n + 1$;
20 **return** X^t;

optimization is $O(t_2|SN(X^t)|)$. t_2 is the number of the iteration by Eq. (9) (the technology is same as algorithm 1, so $t_2 = t_1$). At worst case, $|SN(X^t)|$ is same as $|V|$. But the group size of X^t is always given and a far-away user can not meet the constraints in line 8, so $|SN(X^t)|$ is far less than $|V|$.

The time complexity of Algorithm 2 is $|V|^2 \Omega(1/\sigma)|) + O(t|SN(X^t)|)$, and t is the number of iterations as a substitute for t_1 and t_2. The theoretical convergence rate of the framework in Algorithm 1 is quick, same as the IS demonstrated in Ref. [6]. And in every loop of Algorithm 1, the theoretical convergence rate of technology in Sect. 3.4 is quadratic, so t is small in reality.

4 Experimental Evaluation

We conduct extensive experiments on real datasets to evaluate the performance of the model through case studies. All algorithms are implemented on a server with Intel Xeon (R) Processor E7-4870 v2 @2.30 GHz, 128 GB main memory and all the programs are executed in the Python environment with JIT compiler.

4.1   Experimental Setup

Data Sets. Different social network platforms are detailed below:

- Twitter: We gather and select a set of 160,338 Twitter users in Singapore, associated with 2,405,628 social relationships. The data set consists of user profiles, follow relationships, tweets and so on from 2014.11 to 2016.01.
- Foursquare: We extract check-ins performed by Foursquare users who are also Twitter users in our dataset. We gather and select friends of these users and crawled the profiles of 76,503 users in Singapore associated with 1,531,357 social relationships, as well as their check-ins and contents from 2014.11 to 2016.01.

We manually annotate the linkage ground-truth for Twitter and Foursquare and selected 15,281 ground-truth matched user pairs as declared by the users in these two networks. Thanks to the technical support from fullcontact.com, the ground-truth of the linkage of each user across all the platforms are provided by selecting each user's accounts, e-mails or telephone numbers, national ID number, IP address and home address, all of which collectively serve as the most reliable data to uniquely identify a natural person and link all the different accounts.

Data Analysis. We analyse the data by availability and consistency.

In Twitter, the user generated content contains: screen name; display name; bio (summary); location; birthday; tweet content; place; activity time. In Foursquare, the user generated content contains: full name; bio (summary); home city; venue likes; venue check-in; venue category; activity time.

The availability of attributes depends on the social network, for example Twitter does not ask users about their full name while Foursquare does. We count up the profile which is not null and the number of contents, activities, friendships which is not zero. All the data are crawled by breadth first searching and the availability of those data can be shown at Table 1.

Table 1. The availability of the data

Dataset	name	friends	bio	location	birthday	tweets	place(GPS)
Twitter	100%	65.2%	80.4%	45.4%	10.4%	63.2%	33.4%

Dataset	full name	friends	bio	location	venue likes	check-in
Foursquare	100%	64.3%	26.1%	91.5%	27.3%	13.2%

The consistency of attributes depends on the similarity metrics for profile attributes and word attributes. We borrow a set of standard metrics to compute similarity between the values of attributes [5]: the Jaro distance to measure the similarity between names; the geodesic distance to measure the similarity between locations; and the percentage of common friends between two person; the text similarity between the bio text and the word-to-vector similarity with all the fixed word features [9]. We randomly select 1000 pairs of matching users from the ground-truth and compute the similarity marks for every metric. We set critical value for each mark and compute the percentage of pair whose mark is higher than the critical value. Such percentages for non-matching users are also computed. The results can be shown at Table 2.

Table 2. The consistency of the data

Pairs	Name	Locations	Bio	Friends	Fixed features
Matching users	76.4%	55.2%	60.5%	46.3%	64.3%
Non-matching users	35.4%	14.5%	32.2%	13.4%	45.3%

Competing Algorithms. We compare our algorithm with the following state-of-the-art user identity matching approaches.

- *MOBIUS:* a behavior-modeling approach to link users across social media platforms [17].
- *Alias-disamb:* an unsupervised approach to decide whether cross-platform user identities with same username belongs to same natural person [10].
- *Ulink:* a profile-based approach to match users' identification focused on profile attributes including name, gender, birthday, city and so on [13].
- *HYDRA:* a semi-supervised multi-objective framework jointly modeling heterogeneous behaviors and structure consistency [11].

Experiment Settings. For our proposed algorithm Group Identity Matching (GIM), the parameters we need to control are (a) the set of features $\mathcal{F} = \{f_1, f_2..f_F\}$ given in Eq. (1); (b) β in Eq. (3); (c) the community scale of the group $\|X^s\|$ in $\mathcal{G}^s$ and $\|X^t\|$ in $\mathcal{G}^t$ in Algorithm 2.

For (a), we choose several category like user-profiles, news, finance, technology, sports, entertainment, car, video, property, fashion, education, travel and game. In every category, we choose some subcategories, and they make up the feature space. For (b), based on our observations of the content, we set $\beta = 0.6$ in our experiments. For (c), we suppose that the community scale $n = \|X^s\| = \|X^t\|$ and it is a variable factor using the method in [3].

Based on the ground truth, we evaluate the algorithm using precision and recall. We need to repeat the GIM algorithm for k times, choosing different X^s in $\mathcal{G}^s$. And in every single instance, we compute the precision P_i and recall R_i for the given X_i^s and X_i^t. Precision is the fraction of the user pairs that are correctly matched in partial ground truth. Recall is the fraction of partial ground truth matched users that appear among the all users in X_i^s and X_i^t.

So a complete process of our algorithm can be defined as $GIM(n, k)$, where n is the scale of the matched groups, k is the repeat times for different groups of the source domain. And we evaluate the average precision and recall of the matching numbers.

4.2   Self Evaluation

In this section, we evaluate *GIM(n, k)* using different n and k. In real social networks, the number of users in one community is usually very small [3], so we range n from 5 to 40. In this experiment, we set $k = 1000$ so the output value are the average precision and recall for 1000 times. In Fig. 2a, we can find that when the number of the matched group grows from 5 to 25, the precision and recall grow fast. Then n increases from 25 to 40, the precision decrease a lot. It's the reason that the group is too large and they may not have the same behavior.

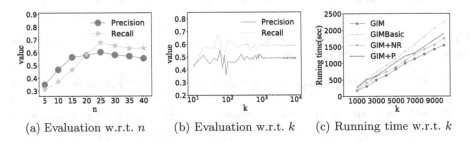

(a) Evaluation w.r.t. n (b) Evaluation w.r.t. k (c) Running time w.r.t. k

Fig. 2. Evaluation and running time of *GIM(n, k)* for different n and k

Since the reasonable value for n maybe 25, we need to range k to get a reasonable repeating time. In Fig. 2b, the X-axis zooms to $log_{10}(x)$, k ranges from 10^1 ro 10^4, and the precision and recall value become stable when k near 10^3. So we can set $k = 1000$ in the following experiments.

Figure 2c illustrate that the running time increases linearly with the repeated times. We record the computation time with k ranging from 1000 to 10000. *GIM-Basic* is the algorithm without the improvement strategies in Sect. 3.4. *GIM+NR* is the algorithm of adding the Newton-Raphson method into *GIMBasic*. *GIM+P* is the algorithm of adding the prune strategy into *GIMBasic*. *GIM* is the final algorithm with all the strategies. As can be seen, *GIM* outperforms better than the other algorithms and we do the experiments below using *GIM*.

4.3   Effectiveness Evaluation

Since our problem focus on matching the groups, we use the competing algorithms to give linkages of the whole graph and then compare with *GIM(25, 1000)* at every repeating time. That means that once we change the X^s in $\mathcal{G}^s$, we compute the precision and recall of the source group X^s and the target group X^t in both *GIM* and other algorithms. Then we record the average precision and recall of the matching numbers.

We transfer the datasets and design three tasks for matching identities in the datasets of Twitter and Foursquare. The first task is to link the real users in Twitter and Foursquare and it can be marked as $T - F$. The users in Twitter and

Foursquare have different relationships and different behaviors. The second task is to link the users in Twitter who have same links but different behaviors. We randomly collect part of the users' features and set the extraction probability of every word as 0.8. Then we record two samples of the dataset Twitter and mark them as $T - T$. The third task is same as the second task but the sampled dataset is Foursquare. And this task can be marked as $F - F$. In $T - T$ and $F - F$, the ground-truth is easy to get since the users are same but they have different contents.

In the experiments on $T - T$ and $F - F$, we compare *GIM* with *MOBIUS*, *Ulink*, *Alias-disamb* and *HYDRA*. As showed in Fig. 3a and d, *GIM* performs well in the competing algorithms. The precision and recall of the matching results is higher than the others. *MOBIUS* and *Alias-disamb* are much less than the other algorithms because they use name or some contents to do the linking and they didn't model the user behavior. In Fig. 3b and e, our algorithm *GIM* also performs the best among the algorithms. Figure 3c and f present the real matching results among Twitter and Foursquare, due to the contents of the real datasets are different, the precision and recall is much lower than the construction datasets but our algorithm has a greater advantage while linking the real matching. Since *Alias-disamb* only uses names as the most important feature, it has a poor performance. *Ulink* and *HYDRA* build a whole model for the users' behavior space, so they perform not much less than *GIM*. Our algorithm considers the social relationships of the users, so *GIM* performs the best. All the results demonstrate that our algorithm can get exactly and more missing connections across different social networks.

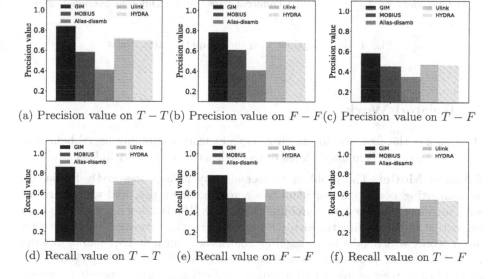

(a) Precision value on $T - T$ (b) Precision value on $F - F$ (c) Precision value on $T - F$

(d) Recall value on $T - T$ (e) Recall value on $F - F$ (f) Recall value on $T - F$

Fig. 3. Evaluation with the competing algorithms

4.4 Efficiency Evaluation

GIM can match one given user or a given group across different networks just considering users of the given group, but the other competing algorithms are performed to get the whole graph matching candidates after a whole iteration. Figure 4a presents the running time of matching just one given group on $T - F$. Since we store the PageRank value and every nodes' behavior distributions at the offline process, *GIM* is much quick than the other algorithms.

We generate some graphs with different scale and average degree to observe scalability of *GIM* with the setting of *GIM(25, 1000)*. Figure 4b shows that the size of the graph make no difference to the running time (except the offline process), since *GIM* is a local-based search algorithm.

The average degree will influence the $|SN(X^t)|$ in Sect. 3.6, and the changing scale of $|SN(X^t)|$ results will make a difference to the running time of *GIM*. Figure 4c demonstrate that when the average degree of nodes is increasing, the running time of *GIM* is growing linearly. And it shows that *GIM* has good scalability.

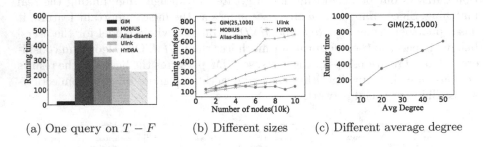

(a) One query on $T - F$ (b) Different sizes   (c) Different average degree

Fig. 4. Running time with the competing algorithms

5 Related Work

Our work is the first one to match group's identity combining users' behaviors and relationships. It mainly relates to researches of user identity linkage problem.

Content Model. The problem of user identity linkage was firstly formalized as connecting corresponding identities across communities, and was addressed with an approach based on searching user-names [16]. It takes names as the most important feature to link the identities across the social networks. There are some other works which take more tagging information, such as user-name, description, hometown, profession, educational background and so on [7,18]. So we can call them user-profile-based approaches. Next some user-generated-content-based methods are proposed, they collect both users' profiles and personal identifiable information from user personal reading records or user-generated content [10,15].

User Behavior Model. Recent approaches for user identity linkage are user-behavior-model-based, which analyze behavior patterns and build feature models from user names, language and writing styles. Liu et al. [11] propose a large-scale social identity linkage framework via heterogeneous behavior modeling which learns the mapping function by multi-objective optimization. Mu et al. [13] explore a new concept of latent user space to model the relationship between the underlying real users and their observed projections onto the varied social platforms. Zhong et al. [19] introduce different implementations, corresponding to different ways to transfer knowledge from composite social relations. Gao et al. [4] propose an unsupervised method to link users across heterogeneous social networks. Kong et al. [8] design an unsupervised approach to match entities, which utilizes the locality sensitive hashing schema to reduce the candidate tuples and speed up the matching process.

6 Conclusion

We study the problem of group identity matching, which identify and link users by an unit of group. Different from traditional studies that link users with every single user's profile and content, we proposed a novel framework to gather the users in a group and link the identities by the group's behavior model. The proposed algorithm contains three parts including behavior learning, behavior transfer and relationship transfer. So it can take both the groups' behavior and relationship into consideration to get more matched links. We perform extensive experiments to evaluate the proposed algorithm, and the results demonstrate the performance of our algorithm.

References

1. Arnold, A., Nallapati, R., Cohen, W.W.: A comparative study of methods for transductive transfer learning. In: ICDMW 2007, pp. 77–82. IEEE (2007)
2. Chen, W., Zhu, F., Zhao, L., Zhou, X.: When peculiarity makes a difference: object characterisation in heterogeneous information networks. In: Navathe, S.B., Wu, W., Shekhar, S., Du, X., Wang, X.S., Xiong, H. (eds.) DASFAA 2016. LNCS, vol. 9643, pp. 3–17. Springer, Cham (2016). https://doi.org/10.1007/978-3-319-32049-6_1
3. Cui, W., Xiao, Y., Wang, H., Lu, Y., Wang, W.: Online search of overlapping communities. In: SIGMOD, pp. 277–288. ACM (2013)
4. Gao, M., Lim, E.P., Lo, D., Zhu, F., Prasetyo, P.K., Zhou, A.: CNL: collective network linkage across heterogeneous social platforms. In: ICDM, pp. 757–762. IEEE (2015)
5. Goga, O., Loiseau, P., Sommer, R., Teixeira, R., Gummadi, K.P.: On the reliability of profile matching across large online social networks. In: SIGKDD, KDD 2015, pp. 1799–1808. ACM, New York (2015)
6. Huang, F.L., Hsieh, C.J., Chan, K.W., Lin, C.J.: Iterative scaling and coordinate descent methods for maximum entropy models. J. Mach. Learn. Res. **11**, 815–848 (2010)
7. Iofciu, T., Fankhauser, P., Abel, F., Bischoff, K.: Identifying users across social tagging systems. In: ICWSM (2011)

8. Kong, C., Gao, M., Xu, C., Qian, W., Zhou, A.: Entity matching across multiple heterogeneous data sources. In: Navathe, S.B., Wu, W., Shekhar, S., Du, X., Wang, X.S., Xiong, H. (eds.) DASFAA 2016. LNCS, vol. 9642, pp. 133–146. Springer, Cham (2016). https://doi.org/10.1007/978-3-319-32025-0_9
9. Kusner, M.J., Sun, Y., Kolkin, N.I., Weinberger, K.Q.: From word embeddings to document distances. In: ICML, pp. 957–966 (2015)
10. Liu, J., Zhang, F., Song, X., Song, Y.I., Lin, C.Y., Hon, H.W.: What's in a name?: an unsupervised approach to link users across communities. In: WSDM, pp. 495–504. ACM (2013)
11. Liu, S., Wang, S., Zhu, F., Zhang, J., Krishnan, R.: HYDRA: large-scale social identity linkage via heterogeneous behavior modeling. In: SIGMOD, SIGMOD 2014, pp. 51–62. ACM, New York (2014)
12. Lofgren, P.A., Banerjee, S., Goel, A., Seshadhri, C.: FAST-PPR: scaling personalized PageRank estimation for large graphs. In: SIGKDD, pp. 1436–1445. ACM (2014)
13. Mu, X., Zhu, F., Lim, E.P., Xiao, J., Wang, J., Zhou, Z.H.: User identity linkage by latent user space modelling. In: SIGKDD, KDD 2016, pp. 1775–1784. ACM, New York (2016)
14. Pan, S.J., Yang, Q.: A survey on transfer learning. IEEE TKDE **22**(10), 1345–1359 (2010)
15. Yang, Y., Sun, Y., Tang, J., Ma, B., Li, J.: Entity matching across heterogeneous sources. In: SIGKDD, pp. 1395–1404. ACM (2015)
16. Zafarani, R., Liu, H.: Connecting corresponding identities across communities. ICWSM **9**, 354–357 (2009)
17. Zafarani, R., Liu, H.: Connecting users across social media sites: a behavioral-modeling approach. In: SIGKDD, KDD 2013, pp. 41–49. ACM, New York (2013)
18. Zhang, J., Kong, X., Yu, P.S.: Transferring heterogeneous links across location-based social networks. In: WSDM, pp. 303–312. ACM (2014)
19. Zhong, E., Fan, W., Yang, Q.: User behavior learning and transfer in composite social networks. ACM Trans. Knowl. Discov. Data **8**(1), 6:1–6:32 (2014)

NANE: Attributed Network Embedding with Local and Global Information

Jingjie Mo[1,2,3], Neng Gao[2,3(✉)], Yujing Zhou[1,2,3], Yang Pei[1,2,3], and Jiong Wang[1,2,3]

[1] School of Cyber Security, University of Chinese Academy of Sciences,
Beijing, China
{mojingjie,zhouyujing,peiyang2,wangjiong}@iie.ac.cn
[2] State Key Laboratory of Information Security, Chinese Academy of Sciences,
Beijing, China
gaoneng@iie.ac.cn
[3] Institute of Information Engineering, Chinese Academy of Sciences, Beijing, China

Abstract. Attributed network embedding, which aims to map structural and attribute information into a latent vector space jointly, has attracted a surge of research attention in recent years. However, existing methods mostly concentrate on either the local proximity (i.e., the pairwise similarity of connected nodes) or the global proximity (e.g., the similarity of nodes' correlation in a global perspective). How to learn the global and local information in structure and attribute into a same latent space simultaneously is an open yet challenging problem. To this end, we propose a Neural-based Attributed Network Embedding (**NANE**) approach. Firstly, an affinity matrix and an adjacency matrix are introduced to encode the attribute and structural information in terms of the overall picture separately. Then, we impose a neural-based framework with a pairwise constraint to learn the vector representation for each node. Specifically, an explicit loss function is designed to preserve the local and global similarity jointly. Empirically, we evaluate the performance of **NANE** through node classification and clustering tasks on three real-world datasets. Our method achieves significant performance compared with state-of-the-art baselines.

Keywords: Attributed social networks · Deep learning
Local and global information · Pairwise constraint

1 Introduction

Social networks are ubiquitous in our daily lives, ranging from online social-networking sites such as Facebook and Weibo, biological gene-disease networks [6,18], to citation networks [14,16]. Network embedding, which maps the information of each node into a latent space, has grown up to be an effective method in social network data mining. As a result, various data mining methods, e.g.,

© Springer Nature Switzerland AG 2018
H. Hacid et al. (Eds.): WISE 2018, LNCS 11233, pp. 247–261, 2018.
https://doi.org/10.1007/978-3-030-02922-7_17

group clustering [27], link prediction [21], anomaly detection [3], node classification [2,21], can be directly conducted in the latent space.

Early works(e.g., DeepWalk [24], LINE [28], node2vec [9]) explore the effect of local and global structure similarity [8,12,17,31] on network embedding. Local structure similarity indicates that a pair of nodes with edges are more prone to be similar. On the other hand, global structure similarity reveals node status in the network. For instance, a pair of nodes with semblable structural neighbors tend to be closer. However, these methods need to be improved for lack of abundant nodes' attribute information. Recently, several algorithms pay attention to attributed network embedding leveraging structural and attribute information both. LANE [10] only takes the local attribute similarity into consideration which means that nodes in the network with similar attribute are more likely to be in a same community. UPP-SNE [35] is more prone to capturing global structural and attribute similarity under the framework of DeepWalk. How to extract the local and global information in structure and attribute jointly is still an arduous problem.

In this paper, we propose a **N**eural-based **A**ttributed **N**etwork **E**mbedding framework named **NANE** to address the aforementioned problem. In our approach, various information is comprehensively considered, including the local and global information in structure and attribute. We impose an autoencoder model to encode the global attribute and structural information into a latent space with a pairwise constraint to preserve the local information. Specifically, we regard the attribute similarity of two nodes as an uncertain link and the similarity indicates the possibility of connecting two nodes. On top of that, an affinity matrix is introduced to represent the attribute global proximity.

In conclusion, the main contributions can be summarized as follows:

- We propose a generic neural-based framework **NANE** to represent attributed social network capturing both structural and attribute non-linear similarity. To our best of knowledge, our model is the first attempt to capture the local and global similarity in structure and attribute jointly. Specifically, we introduce an affinity matrix to measure the global attribute similarity among nodes.
- We empirically conduct experiments on three real-world datasets with multi-class classification of vertices and node clustering tasks. Comparing with cutting-edge network embedding algorithms, we evaluate the effectiveness of **NANE**.

The rest of this paper is organized as follows. Firstly, we discuss the related work in Sect. 2, and then we show some definitions in our work in Sect. 3. On top of that, we propose the **NANE** algorithm in Sect. 4, followed by experiments in Sect. 5. Finally, Sect. 6 concludes the paper and visions the future work.

2   Related Work

In this section, we briefly summarize the development of network embedding methods which we can simply divide into two parts. The first part is named structure-based network embedding which focuses on structural information only. The latter is named content-aware network embedding that combines additional information with network structure for a better performance on graph representation.

2.1 Structure-Based Network Embedding

Some earlier works utilize manifold learning to capture structure proximity [1, 25, 29]. Recently, inspired by Skip-Gram [19, 20], DeepWalk [24] imposes random walks to generate sequences of nodes, and feeds them into Skip-Gram to capture the similarity of nodes. However, DeepWalk is prone to preserving the global structural proximity since the limit of random walks. On top of that, node2vec [9] introduces biased random walks to balance the global and local structural information. LINE [28] designs an explicit objective function to capture both global and local structural information while it lacks further fusion.

On the other hand, some researchers try to introduce neural networks to extract node high non-linearities. DNGR [5] adopts a random surfing model to capture network structure and reduces it into a limit dimension by a SDAE model. It happens that there is a similar case. SDNE [33] also exploits a deep neural network with a supervised component to extract the structural information.

All the methods mentioned above are based on network structure only without considering attribute information which is extremely beneficial to graph representation.

2.2 Content-Aware Network Embedding

Some recent efforts have been devoted to leveraging additional information for a better performance.

TADW [34], a text-associated DeepWalk model, creatively explores the contribution of nodes contents in network embedding. It imposes matrix factorization to encode text features into network representation learning. TriDNR [22] exploits DeepWalk and Doc2Vec [13] to extract structural information and capture content information respectively. Relying on a late fusion which is a series of weighted sums only, it ignores the correlation between structure and contents since it lacks further convergence. Further, these two models which concern about node contents only cannot handle noisy attribute information.

LANE [10] is the first attempt to utilize label information and node attributes. It utilizes three matrices to measure networks: the network adjacent matrix, node content-level similarity matrix and node label-level similarity matrix, and then maps them into the same vector space by dimension reduction. LANE projects node embedding with matrix factorization which cannot reflect the non-linear correlation between nodes. UPP-SNE [35] is the first attempt to take user profiles into consideration. The basic idea is to extract the similarity of nodes by random walks and embed user profile information via a non-linear mapping into a low-dimensional latent space. However, UPP-SNE is more prone to extracting the global proximity while it ignores the local information.

3   Problem Definition

We first summarize some notations used in this paper. We denote scalars as lowercase alphabets(e.g., n) and represent vectors as boldface lowercase alphabets(e.g., $\mathbf{s}$). Moreover, matrices are represented by boldface uppercase alphabets(e.g., $\mathbf{S}$). $\mathbf{s}_i$ is the ith row of a matrix $\mathbf{S}$, and the $(i,j)^{th}$ element of it is denoted by s_{ij}. We list the main notations in Table 1.

We regard a social network as a homogeneous attributed network which indicates that there is only one relationship between nodes, each edge is undirected and some nodes have their attributes. Under these assumptions, we denote a social graph as $\mathcal{G} = (\mathcal{V}, \mathcal{E}, \mathbf{U})$, where $\mathcal{V}$ is a set of n nodes, $\mathcal{E}$ is a set of edges and $\mathbf{U}$ is the node attribute matrix. $\mathbf{A}$ is the adjacent matrix, and a_{ij} is defined as 1 if there is a edge between node i and node j.

The aim of this paper is to project the nodes into a low-dimensional vector space while preserving structural and attribute information jointly. It is vital to map the matrices $\mathbf{U},\mathbf{A}$ into the same latent space jointly. In this end, we propose a neural-based attributed network embedding model to learn a comprehensive representation.

Table 1. Main notations and descriptions

Notations	Descriptions
n	Total number of nodes in the network
m	Total number of attribute categories of all nodes
d	The dimension of embedding space
t	The dimension of the output of self-feedforward layer
K	The number of deep autoencoder layers
$\mathbf{A} \in \mathbb{R}^{n \times n}$	The adjacent matrix
$\mathbf{U} \in \mathbb{R}^{n \times m}$	The user attribute information matrix
$\mathbf{S} \in \mathbb{R}^{n \times n}$	The node attribute similarity matrix
$\mathbf{X} \in \mathbb{R}^{n \times d}$	The final embedding matrix
$\hat{\mathbf{R}} \in \mathbb{R}^{n \times 2n}$	The output of reconstruction layer
$\mathbf{h}_i^{(j)}$	j^{th} Hidden layer of node i
$\mathbf{W}^{(k)},\mathbf{b}^{(k)}$	The kth hidden layer weight matrix and biases
$\mathbf{W}_{struc},\mathbf{W}_{attr}$	The weight matrix of structural information and attribute information

4   Methodology

In this section, we first elaborate the framework of our final model **NANE** preserving structural and attribute information jointly. And then we give the explicit loss function and its optimization in detail.

4.1 Overall Architecture

In this paper, we propose a neural-based attributed network embedding model to capture structural and attribute information jointly. As shown in Fig. 1, **NANE** consists of three major steps. Specifically, we first convert structural and attribute information to a set of binary features, and then leverage a self-feedforward layer to measure the weight of each feature. We then concatenate the output of self-feedforward layer and feed it into a neural network structure. The early fusion operation makes it possible to optimize all parameters simultaneously. To capture the highly non-linear node correlation, we utilize a deep autoencoder [26] model to reconstruct overall information and design a specific loss function to preserve both local and global similarity in the end. The details of each step are elaborated as follows.

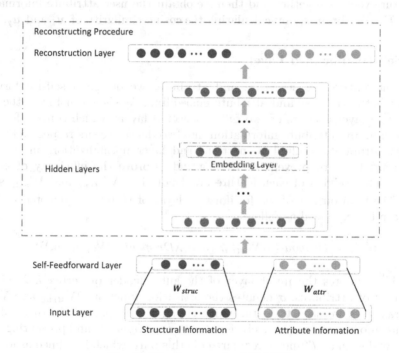

Fig. 1. The framework of **NANE**

4.2 Global Information Encoding

To economize the global attribute and structural information, we introduce an affinity matrix **S** and an adjacent matrix **A** respectively. In this paper, we take a row of the adjacency matrix to represent for structural information. For example, a_i is the structural information of vertex i. Furthermore, we utilize a cosine similarity to represent the global attribute information. We simply take a row of the similarity matrix $\mathbf{S}$ to represent the attribute information, and s_i is the attribute similarity information of vertex i.

To generate the attribute affinity matrix $\mathbf{S}$, we need to define the user attribute information matrix $\mathbf{U}$ in the first place. As is well-known, attribute information is rich and diverse in social networks. However, it is inevitably incomplete and noisy due to its heterogeneity and feature of manual filling. To tackle this problem, we first convert all discrete attributes, e.g., user demographics [4], user interest [23], to a set of binary features by one-hot encoding. For instance, the marital status attribute has four values {married, single, divorced, widowed}, we can encode a married user as the vector $\mathbf{v} = \{1, 0, 0, 0\}$, where the first binary feature of value 1 represents married. For continuous attributes, e.g. age, we normalize it to reduce the impact of value by Max-Min Normalization. For missing attribute values, we set the feature vector to all zeros. Thus, we aggregate all the feature vectors together and then we obtain the user attribute information matrix $\mathbf{U}$. We obtain s_{ij} by calculating the cosine similarity of u_i and u_j.

4.3   Self-Feedforward Layer

To weight features in structure and attribute, we design a self-feedforward layer for both structure and attribute embedding. As shown in Fig. 1, the self-feedforward layer consists of two fully connected layers, which convert features in structural and attribute information into two dense vectors respectively. And we apply dropout, residual connections and layer normalization on the self-feedforward layer to prevent overfitting and improve the efficiency of neural network. The weights of each feature are denoted as $\mathbf{W}_{struc}$ and $\mathbf{W}_{attr}$ separately. The final output of self-feedforward layer of vertex i is denoted as $\mathbf{h}_i^{(0)}$, which can be expressed as follows:

$$\mathbf{h}_i^{(0)} = \left[Dropout(\sigma(\mathbf{W}_{struc} \cdot \mathbf{a}_i)), \lambda Dropout(\sigma(\mathbf{W}_{attr} \cdot \mathbf{s}_i)) \right] \tag{1}$$

where $\mathbf{h}_i^{(0)}$ denotes the input layer of the autoencoder of vertex i, λ adjusts the impact of attributes, σ denotes the activation function, $\mathbf{W}_{struc}$ and $\mathbf{W}_{attr}$ are parameters we need to learn, which measure the weights of structural and attribute information respectively. For a better convergence and preventing overfitting, we design an $\mathcal{L}2$-norm regularizer in this part, which is defined as follows:

$$\mathcal{L}_{F-reg} = \|\mathbf{W}_{struc}\|_F^2 + \|\mathbf{W}_{attr}\|_F^2 \tag{2}$$

where $\|\cdot\|_F^2$ denotes the square of F-norm.

4.4   Reconstructing Procedure

To capture the local and global proximity in structure and attribute jointly, we extend a deep autoencoder model to measure the interplay between structure and attribute and encode them into a latent space. A deep autoencoder consists of two parts, i.e. the encoder and decoder. At the encode step, multilayer non-linear function $f(\cdot)$ is applied to map the input data into a low-dimensional vector space, which is also called embedding space. On the contrary, there are also multilayer mirrored non-linear function $g(\cdot)$ is extended to map the embedding space to the reconstruction space. In this paper, the input data of deep autoencoder is the output of self-feedforward layer $\mathbf{h}_i^{(0)}$, and the representation of hidden layers can be denoted as follows:

$$\mathbf{h}_i^{(k+1)} = \sigma(\mathbf{W}^{(k)} \cdot \mathbf{h}_i^{(k)} + \mathbf{b}^{(k)}) \tag{3}$$

where $W^{(k)}$, $b^{(k)}$ is the kth hidden layer weight matrix and biases.

The output of reconstruction layer is denoted as $\widehat{\mathbf{R}}$. Our goal of this step is to minimize the reconstruction error of output and input. The input data $\mathbf{R}$ that need to be reconstructed is shown as follows:

$$\mathbf{R} = [\mathbf{A}, \lambda\mathbf{S}] \tag{4}$$

And the loss function is denoted as follows:

$$\mathcal{L} = \left\| \widehat{\mathbf{R}} - \mathbf{R} \right\|_F^2 \tag{5}$$

However, due to the sparsity of the input data, the number of zero elements in $\mathbf{R}$ is much larger than that of non-zero ones, which means that it is more prone to reconstruct the zero elements rather than non-zero ones. However, it is contrary to our intention. To address this problem, we impose an offset coefficient matrix to reset the weights of different elements, and the redesigned loss function is denoted as follows:

$$\mathcal{L}_{global} = \left\| (\widehat{\mathbf{R}} - \mathbf{R}) \odot \mathbf{B} \right\|_F^2 \tag{6}$$

where $\mathbf{B}$ is the offset coefficient vector, $b_{ij}=1$ when $r_{ij}=0$ while $b_{ij} = \beta > 1$ when $r_{ij}=1$, and $\odot$ means the Hadamard product. With the help of the deep autoencoder, the vectors of the vertexes which have semblable features would have similar representation. However, it is not only necessary to capture the global similarity, but also vital to preserve the local similarity. A pair of nodes with edges should also have semblable embedding i.e. the local structure similarity. Therefore, we aggrandize a loss function for local structure similarity, and the expression is shown as follows:

$$\mathcal{L}_{struc-local} = \sum_{i,j=1}^{n} a_{ij} \left\| \mathbf{x}_i - \mathbf{x}_j \right\|_2^2 \tag{7}$$

where a_{ij} is an element of adjacency matrix, $a_{ij} = 1$ when there is a link between node i and node j. x_i denotes the final embedding of node i. The local structure loss function aims to make the embedding of two connected nodes closer.

Similar with the local structure similarity, we also design a loss function to constraint the local attribute similarity. A pair of nodes with semblable attributes should also have semblable embedding. The loss function for local attribute similarity is denoted as follows:

$$\mathcal{L}_{attr-local} = \sum_{i,j=1}^{n} s_{ij} \|\mathbf{x}_i - \mathbf{x}_j\|_2^2 \tag{8}$$

where s_{ij} is an element of attribute similarity matrix. The greater value of s_{ij}, the attributes of node i and node j are more similar. Akin to the local structure loss function, the local attribute loss function aims to make the embedding of two nodes which have similar attributes closer.

4.5   Loss Functions and Optimization

To preserve the global and local similarity simultaneously, we combine the afore-mentioned loss functions in a integrated framework and propose a unified structure and attribute preserving model **NANE**. The joint loss function is denoted as follows:

$$\mathcal{L} = \alpha \mathcal{L}_{global} + \gamma \mathcal{L}_{struc-local} + \theta \mathcal{L}_{attr-local} + \eta \mathcal{L}_{reg}$$

$$= \alpha \left\| (\hat{\mathbf{R}} - \mathbf{R}) \odot \mathbf{B} \right\|_F^2 + \gamma \sum_{i,j=1}^{n} a_{ij} \|\mathbf{x}_i - \mathbf{x}_j\|_2^2 \tag{9}$$

$$+ \theta \sum_{i,j=1}^{n} s_{ij} \|\mathbf{x}_i - \mathbf{x}_j\|_2^2 + \eta \mathcal{L}_{reg}$$

where $\alpha, \gamma, \theta, \eta$ are four hyper-parameters to adjust the weights of each part. More-over, $\mathcal{L}_{reg}$ is the total regularization which consists of two parts: the regulariza-tion of self-feedforward layer $\mathcal{L}_{F-reg}$ and the regularization of deep autoencoder $\mathcal{L}_{ae-reg}$. We define $\mathcal{L}_{reg}$ as

$$\mathcal{L}_{reg} = \mathcal{L}_{F-reg} + \mathcal{L}_{ae-reg}$$

$$= \|\mathbf{W}_{struc}\|_F^2 + \|\mathbf{W}_{attr}\|_F^2 + \sum_{k=1}^{K} \left(\left\| \mathbf{W}^{(k)} \right\|_F^2 + \left\| \mathbf{b}^{(k)} \right\|_F^2 \right) \tag{10}$$

To optimize the aforementioned framework, we apply RMSProp to minimize the objective in Eq.(9), which is able to adjust the learning rate for each param-eter. We feed mini-batch into our model each time to accelerate the speed of training. Besides, in order to prevent falling into a local optimum solution, it is essential to find a good set of initialized parameters. Therefore, we adapt Restricted Boltzmann Machine to pre-train the parameters at first, which is a classic method of parameter initialization in neural network [7]. The integrated algorithm is presented in Algorithm 1.

Algorithm 1. NANE model

Input: The network $\mathcal{G} = (\mathcal{V}, \mathcal{E}, \mathbf{U})$ with the adjacency matrix $\mathbf{A}$ and hyper-parameters
Output: Network embedding $\mathbf{X}$
1: Extract the feature vectors and calculate the attribute similarity matrix $\mathbf{S}$;
2: Feed matrix $\mathbf{A,S}$ into self-feedforward layer, and merge them together;
3: Pre-train the parameters of deep autoencoder through RBM to get initial parameter values;
4: Feed the output of self-feedforward layer into deep autoencoder;
5: **repeat**
6:    Based on the input data and the weights of each layer, apply Eq.(1) to generate the reconstruct matrix $\widehat{\mathbf{R}}$;
7:    minimize Eq. (9) by RMSProp, and update parameters at each epoch;
8: **until** converge
9: Acquire the network embedding $\mathbf{X} = \mathbf{H}^{(K/2)}$

---

5 Experiments

In this section, we evaluate our method by performing experiments on three real-world network datasets and compare it with several state-of-the-art baseline algorithms.

5.1 Datasets

We conduct experiments on three real-world networks: Facebook, Hamilton and Rochester. The statistics of the three datasets is summarized in Table 2.

Ego-Facebook is an ego-network which was collected from survey participants using the Facebook app. The dataset contains 1403-dimensional node features, 4039 nodes and 88234 edges. Besides, people education type is used as group label [15].

Hamilton and **Rochester** are two datasets collected by Adam D'Angelo of Facebook, consists of nodes from the Facebook networks at each of 100 American institutions [30]. Each node contains 7 attributes: student/faculty status flag, gender, major, second major/minor, dorm/house, year, and high school, which is described by a 144-dimensional and 235-dimensional feature vector respectively, and student/faculty status flag is selected as group class. Two datasets include 2314 nodes, 192788 edges and 4563 nodes, 322808 edges separately.

Table 2. The statistics of the three datasets

Dataset	Node	Edge	Attribute
Facebook	4039	88234	1403
Hamilton	2314	192788	144
Rochester	4563	322808	235

5.2   Baseline Methods

We compare **NANE** with several state-of-the-art network embedding methods, which are divided into two categories.

Structure-Based NRL Methods

- DeepWalk [24] generates node sequences by random walks, and feed them into Skip-Gram model to learn network embedding.
- node2vec [9] introduces biased random walks to DeepWalk, which aims to capture the local and global structure jointly.
- LINE [28] imposes two separate objective functions to extract the first-order and the second-order proximity both.
- SDNE [33] leverages a deep neural network which is a non-linear mapping operation to exploit structural information.

Attribute-Aware NRL Methods

- LANE [10] fuses structural, attribute and label information together to preserve node similarity. Here, we only use the version without utilizing label information.
- UPP-SNE [35] generates random walks to capture node pairwise similarity and embeds user profile information into a low-dimensional vector space.

5.3   Parameter Settings

For a fair comparison, we set the embedding dimension to 256 for all methods. In DeepWalk, node2vec and UPP-SNE, we set the window size t to 10, the walk length l to 80. In node2vec, we empirically set the return hyperparameter p to 2.0, and the in-out hyperparameter q to 0.5. In LINE, we set the first-order vector dimension to 128, and the second-order vector dimension to 128 in the same way. In SDNE, we set the weight γ, α, β to 1, 500, 100 respectively, and the learning rate is 0.01. In LANE, we tune the hyper-parameters of β_1, β_2 by grid search. In UPP-SNE, we set the number of iterations to 20. For our method, we use a grid search to find the best parameters. We set α to 1000, β to 100, and γ, θ and η to 1. The rest parameters are given in Table 3.

Table 3. Parameter settings of the three datasets

Dataset	Batch size	λ	t
Facebook	600	1.5	256
Hamilton	400	1.5	1000
Rochester	600	1	256

5.4   Node Classification

We first evaluate the effectiveness of **NANE** by multi-class node classification. To be fair, we range the training size from 15% to 75% by taking 10% as a step, and apply a rbf-kernel svm classifier with $\gamma = 100$ to all of the generated node representations. For each training size, we split train and test set in a random way for 10-times, and then conduct 5-fold cross-validation and output the average micro F1-score of node classification on three different datasets.

Table 4. Node classification F1-score(%) on Facebook

Train size	15%	25%	35%	45%	55%	65%	75%
LINE	68.93	68.93	68.56	68.87	69.73	68.78	68.56
DeepWalk	68.31	68.98	69.76	69.98	69.14	68.88	69.01
node2vec	68.84	69.34	69.5	69.89	69.36	69.34	69.01
SDNE	63.51	63.73	64.09	64.32	63.32	64.01	63.13
LANE	70.03	70.2	69.9	70.2	69.86	69.66	70.1
UPP-SNE	85.53	85.87	86.41	86.81	87.18	87.13	87.82
NANE	**88.62**	**88.92**	**89.16**	**89.18**	**89.03**	**89.79**	**90.97**

Table 5. Node classification F1-score(%) on Hamilton

Train size	15%	25%	35%	45%	55%	65%	75%
LINE	90.04	90.45	91.18	91	91.41	91.56	92.36
DeepWalk	92.32	92.86	92.62	92.77	93.19	92.72	92.75
node2vec	92.43	92.86	92.56	92.46	92.9	93.09	93.09
SDNE	91.2	91.65	92.36	92.46	92.71	92.59	92.4
LANE	79.36	79.26	79.27	82.88	87.04	89.88	92.26
UPP-SNE	93.86	93.55	93.52	93.17	93.75	93.45	94.25
NANE	**94.76**	**94.47**	**94.35**	**94.27**	**94.32**	**94.44**	**95.16**

Tables 4, 5 and 6 show the average classification accuracy of all the methods on Facebook, Hamilton and Rochester, where the best results are bold-faced. It is not hard to see that our method consistently yields the best classification results among all the baselines. Extraordinary accuracy compared with structure-based NRL algorithms demonstrates that our method works well on the fusion of attribute information. Especially on Facebook, **NANE** achieves more than 30% improvement over all the structure-based NRL baselines, pointing to the significant performance on node classification. Moreover, **NANE** also outperforms all the attribute-aware NRL baselines, demonstrating its effectiveness on capturing the local and global similarity in structure and attribute. Our method imposes an early fusion of structural and attribute information, which

Table 6. Node classification F1-score(%) on Rochester

Train size	15%	25%	35%	45%	55%	65%	75%
LINE	86.22	86.58	86.73	87.28	87.79	87.65	87.56
DeepWalk	88.1	88.29	88.57	88.97	88.9	89.74	88.69
node2vec	88.32	88.14	88.71	89.4	89.29	89.36	88.52
SDNE	86.08	86.21	86.92	86.57	87.2	86.73	85.89
LANE	80.97	81.27	81.56	81.35	80.87	81.1	80.89
UPP-SNE	89.46	89.89	90.26	90.08	89.87	89.80	88.96
NANE	**89.53**	**90.61**	**90.86**	**90.58**	**90.8**	**90.61**	**90.62**

enables our method to exploit the interplay between structure and attribute. On top of that, our method economizes the global information with a pairwise constraint, resulting in remarkable consequences.

5.5 Parameter Sensitivity

In this section, we explore the parameter sensitivity of our method. We select three crucial parameter batch size, λ and t to conduct our experiments and investigate how these parameters affect the results. The train ratio is selected as 55%. We report the accuracy of node classification on different datasets with disparate parameters. In turns, we study the effect of one parameter with fixing the rest. And the results of our experiments are shown in Fig. 2.

We first show how the batch size affects the performance in Fig. 2(a). As we can see, the performance of our method on Hamilton and Rochester is stable with different values. However, performance on facebook is much fluctuant. The best value of batch size on facebook is 600. We then show how the weight of attribute information λ influence the result in Fig. 2(b). It not hard to see that when the weight of attribute information approach to 1, which means structural information is as important as attribute information, the performance is much better. However, the performance of our method falls badly when *lambda* is larger than 10. It indicates that it may leads to the very opposite effect when the influence of attributes is too large. At last, how the dimension of the output of self-feedforward layer t affects the result is shown in Fig. 2(c). We can see that when t is close to the embedding dimension, the performance is much better in most case.

5.6 Node Clustering

For node clustering task, we apply K-Means++ to network embedding on three datasets and leverage ARI [11] and NMI [32] to evaluate the consequence of clustering. We set the number of clusters same as the group number, calculate the indexes 10 times to reduce occasionality. the average results are shown in Fig. 3.

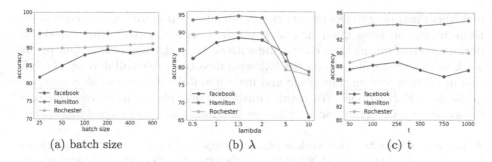

Fig. 2. The impact of parameters batch size, λ, and t

As we can see, **NANE** achieves the best performance among almost all the cutting-edge baselines. Similar clusterings have a positive ARI, whereas negative values indicate poor performance. Furthermore, values of exactly 0 on NMI present purely independent label assignments. On Facebook, the values of ARI are lower than zero for LINE and SDNE, verifying that the representations of LINE and SDNE have independent labelings. Specifically, **NANE** achieves 0.125 improvement compared with the second best results. On Hamilton, **NANE** also yields the best clustering results, outperforming 0.127 and 0.178 lift on ARI and NMI respectively. It firmly demonstrates the better performance of **NANE** on unsupervised learning task.

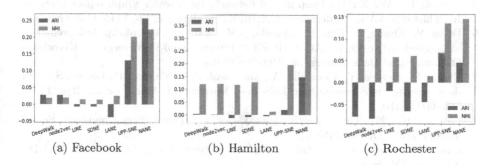

Fig. 3. The results of node clustering on three datasets

6 Conclusion and Future Work

Attributed network, due to its inherent heterogeneity and sparsity, presents new challenges for many tasks. In this paper, we propose a Neural-based Attributed Network Embedding method to capture structural and attribute information jointly. We deem that it is crucial to explore the global and local similarity simultaneously for a better representation. Firstly, we impose a fully-connected layer with some dropout to extract the relevance in attribute and structure respectively. And then we leverage a deep autoencoder model which is a non-linear mapping to reconstruct the global information with a local constraint. Experiments on three different real-world datasets demonstrate the remarkable performance of our method comparing with cutting-edge algorithms. Our future work in this area will focus on the following directions. We will develop **NANE** to a multi-task learning algorithm and make it suitable for large-scale industrial-grade data. Furthermore, We are also interested in exploring how to capture user behaviors in time series and user images and map them into embedding space.

Acknowledgements. This work is partially supported by National Natural Science Foundation of China (No.U163620068) and National Key Research and Development Program of China.

References

1. Belkin, M., Niyogi, P.: Laplacian Eigenmaps and spectral techniques for embedding and clustering. In: International Conference on Neural Information Processing Systems: Natural and Synthetic, pp. 585–591 (2002)
2. Bhagat, S., Cormode, G., Muthukrishnan, S.: Node Classification in Social Networks. Springer, New York (2011)
3. Bhuyan, M.H., Bhattacharyya, D.K., Kalita, J.K.: Network anomaly detection: methods, systems and tools. IEEE Commun. Surv. Tutor. **16**(1), 303–336 (2014)
4. Burger, J.D., Henderson, J., Kim, G., Zarrella, G.: Discriminating gender on twitter. In: Conference on Empirical Methods in Natural Language Processing, pp. 1301–1309 (2011)
5. Cao, S., Lu, W., Xu, Q.: Deep neural networks for learning graph representations. In: Thirtieth AAAI Conference on Artificial Intelligence, pp. 1145–1152 (2016)
6. Dong, Y., Zhang, J., Tang, J., Chawla, N.V., Wang, B.: Coupledlp: link prediction in coupled networks. In: ACM SIGKDD International Conference on Knowledge Discovery and Data Mining, pp. 199–208 (2015)
7. Erhan, D., Bengio, Y., Courville, A., Manzagol, P.A., Vincent, P., Bengio, S.: Why does unsupervised pre-training help deep learning? J. Mach. Learn. Res. **11**(3), 625–660 (2010)
8. Granovetter, M.: The strength of weak ties: a network theory revisited. Sociol. Theory **1**(6), 201–233 (1983)
9. Grover, A., Leskovec, J.: node2vec: Scalable feature learning for networks. In: KDD 2016, pp. 855–864 (2016)
10. Huang, X., Li, J., Hu, X.: Label informed attributed network embedding. In: Tenth ACM International Conference on Web Search and Data Mining, pp. 731–739 (2017)
11. Hubert, L., Arabie, P.: Comparing partitions. J. Classification **2**(1), 193–218 (1985)
12. Kossinets, G., Watts, D.J.: Origins of homophily in an evolving social network1. Am. J. Sociol. **115**(2), 405–450 (2009)
13. Le, Q.V., Mikolov, T.: Distributed representations of sentences and documents, vol. 4, p. II-1188 (2014)
14. Leskovec, J.: Graphs over time: densification laws, shrinking diameters, explanations and realistic generators. In: KDD, pp. 177–187 (2005)
15. Mcauley, J., Leskovec, J.: Learning to discover social circles in ego networks. In: International Conference on Neural Information Processing Systems, pp. 539–547 (2012)

16. Mccallum, A.K., Nigam, K., Rennie, J., Seymore, K.: Automating the construction of internet portals with machine learning. Inf. Retr. **3**(2), 127–163 (2000)
17. Mcpherson, J.M., Smith-Lovin, L.: Homophily in voluntary organizations: status distance and the composition of face-to-face groups. Am. Sociol. Rev. **52**(3), 370–379 (1987)
18. Menche, J., et al.: Disease networks. Uncovering disease-disease relationships through the incomplete interactome. Science **347**(6224), 1257601 (2015)
19. Mikolov, T., Chen, K., Corrado, G., Dean, J.: Efficient estimation of word representations in vector space. In: Computer Science (2013)
20. Mikolov, T., Sutskever, I., Chen, K., Corrado, G., Dean, J.: Distributed representations of words and phrases and their compositionality, vol. 26, pp. 3111–3119 (2013)
21. Narayanan, A., Chandramohan, M., Chen, L., Liu, Y., Saminathan, S.: subgraph2vec: learning distributed representations of rooted sub-graphs from large graphs (2016)
22. Pan, S., Wu, J., Zhu, X., Zhang, C., Wang, Y.: Tri-party deep network representation. In: International Joint Conference on Artificial Intelligence, pp. 1895–1901 (2016)
23. Pennacchiotti, M., Popescu, A.M.: Democrats, republicans and starbucks afficionados: user classification in twitter. In: ACM SIGKDD International Conference on Knowledge Discovery and Data Mining, pp. 430–438 (2011)
24. Perozzi, B., Al-Rfou, R., Skiena, S.: Deepwalk: online learning of social representations. In: ACM SIGKDD International Conference on Knowledge Discovery and Data Mining, pp. 701–710 (2014)
25. Roweis, S.T., Saul, L.K.: Nonlinear dimensionality reduction by locally linear embedding. Science **290**(5500), 2323–6 (2000)
26. Salakhutdinov, R., Hinton, G.: Semantic Hashing. Elsevier Science Inc., Amsterdam (2009)
27. Tang, J., Liu, J., Zhang, M., Mei, Q.: Visualizing large-scale and high-dimensional data, vol. 37(7), pp. 287–297 (2016)
28. Tang, J., Qu, M., Wang, M., Zhang, M., Yan, J., Mei, Q.: Line:large-scale information network embedding, vol. 2, pp. 1067–1077 (2015)
29. Tenenbaum, J.B., Silva, V.D., Langford, J.C.: A global geometric framework for nonlinear dimensionality reduction. Science **290**(5500), 2319 (2000)
30. Traud, A.L., Mucha, P.J., Porter, M.A.: Social structure of facebook networks. Physica A Stat. Mech. Appl. **391**(16), 4165–4180 (2012)
31. Tsur, O., Rappoport, A.: What's in a hashtag? Content based prediction of the spread of ideas in microblogging communities. In: ACM International Conference on Web Search and Data Mining, pp. 643–652 (2012)
32. Vinh, N.X., Epps, J., Bailey, J.: Information Theoretic Measures for Clusterings Comparison: Variants, Properties, Normalization and Correction for Chance (2010). JMLR.org
33. Wang, D., Cui, P., Zhu, W.: Structural deep network embedding. In: ACM SIGKDD International Conference on Knowledge Discovery and Data Mining, pp. 1225–1234 (2016)
34. Yang, C., Liu, Z., Zhao, D., Sun, M., Chang, E.Y.: Network representation learning with rich text information. In: International Conference on Artificial Intelligence, pp. 2111–2117 (2015)
35. Zhang, D., Yin, J., Zhu, X., Zhang, C.: User profile preserving social network embedding. In: Twenty-Sixth International Joint Conference on Artificial Intelligence, pp. 3378–3384 (2017)

Topical Authority-Sensitive Influence Maximization

Xiaoqing Xiong, Ruixuan Li, Yuhua Li[✉], Xiwu Gu, and Tianan Liang

School of Computer Science and Technology,
Huazhong University of Science and Technology, Wuhan 430074, China
idcliyuhua@hust.edu.cn

Abstract. Influence maximization has been widely studied in social network analysis. However, most existing works focus on user's global influence while ignoring the fact that user's influence varies with different topics. Even though a few works take topics into consideration, they all neglect the authority of users on a specific topic which is also a very important indicator when selecting seed nodes. In this paper, we propose a new Topical Authority-sensitive Independent Cascade model (TAIC) by introducing user's authority on a given topic, and based on which, a Topical Authority-sensitive Greedy algorithm (TAG) is presented to try to find the most influential nodes on the given topic. We also propose a new metric, Influence Spread of seed set on a Given Topic ($ISGT$) to evaluate the performance of our proposed model and algorithm. Experiments on two real-world datasets show that our proposed TAG algorithm performs continuously better in finding nodes with higher influence on a given topic and has better time efficiency than other baseline algorithms.

Keywords: Social network · Influence maximization
Topical authority-sensitive model

1 Introduction

The influence maximization problem [1] originates from social network analysis, that aims to find a K-nodes seed set through which the number of finally affected users will be maximized. In recent years, the rapid development of online social networks, e.g. Twitter and Facebook, has greatly promoted the research of this problem as its commercial value in wide areas, such as rumor control and viral marketing.

However, most existing works [2–5] do not take user's topic information into consideration, which means they regard a node's influence as a constant, even for totally different topics. However, user's influence will obviously vary from different topics, for example, a person may be quite influential when discussing "sports", but has no influence on "music". Thus, we should select different influential users for different topics. The choice of influential users depends on the propagation model, which determines the probability of information propagation

© Springer Nature Switzerland AG 2018
H. Hacid et al. (Eds.): WISE 2018, LNCS 11233, pp. 262–277, 2018.
https://doi.org/10.1007/978-3-030-02922-7_18

between adjacent pairs of nodes in the network. Previous works always view the propagation probability as a constant. Hence, no matter how the topic changes, the nodes found at the end are the same, which is obviously not in line with the actual needs. Therefore, in order to achieve the "match" between topics and influential users, it is necessary to incorporate topic information into the propagation model so that when the topic changes, the propagation probability between adjacent nodes in the network may also change.

Even though a few works [6–8] take topics into consideration, they all neglect an important fact that users' authority on a given topic will also affect the propagation probability between adjacent nodes. For example, if three users a,b and c like watching movies, but user a and b are movie appreciation experts and have a high authority in the movie appreciation field, while user c does not understand movie appreciation. Then if user a and c recommend a movie to user b at the same time, whose recommendation do you think is more likely to be accepted by user b? In fact, since user a is also a movie appreciation expert in the eyes of user b, compared to user c, user b will be more likely to accept the movie recommended by user a. Therefore, we also need to consider the topical authority of users in the propagation model.

In this paper, we study the influence maximization problem with the consideration of user's topical authority. We use historical propagation data of users to try to identify influential nodes on a given topic in social networks. The main contributions of this paper are as follows:

- We calculate user's authority for different topics. Inspired by Topical HITS [9], we first use Latent Dirichlet Allocation (LDA) [10] to extract user-topic matrix, based on which Topical HITS is utilized to calculate the topic-authority vector of each user, and hence model user's authority for different topics.
- We extend the classic Independent Cascade (IC) model to be topical authority sensitive, and then use a greedy algorithm to give an approximate solution of the influence maximization problem for a given topic.
- We propose a new metric, Influence Spread of seed set S on a Given Topic T ($ISGT$), to test the effectiveness of our model.
- Our experiments on real-world social networks show that the topical authority-sensitive influence propagation model outperforms the traditional "topical authority -blind" IC model in finding more influential nodes on a given topic, and has better time efficiency.

The rest of the paper is organized as follows. Section 2 reviews the related work; Sect. 3 gives some preliminaries; Sect. 4 introduces the topical authority-sensitive influence maximization; Experimental studies are presented in Sect. 5, and we conclude this paper in Sect. 6.

2   Related Work

We categorize the related works into two types, namely traditional influence maximization and topic-sensitive influence maximization.

Traditional Influence Maximization. Kempe et al. [1] formally format the influence maximization problem and propose the traditional greedy algorithm (GA) that can approximate the optimal solution within a factor of $(1 - 1/e - \epsilon)$, where e is the base of natural logarithm, and ϵ depends on the accuracy of the Monte-Carlo estimate of the influence spread. After that, lots of works have been done to tackle this problem as GA suffers from poor time efficiency. [2] makes further use of the sub-modularity property to improve the efficiency of the traditional GA. Ohsaka et al. [4] achieve this by pruning the Monte Carlo simulation. [5,11] take different ways respectively to generate reverse reachable set in order to speed up. Cohen et al. [12] propose a greedy Sketch-based Influence Maximization(SKIM) algorithm scales to graphs with billions of edges, with one to two orders of magnitude speedup over the best greedy methods. In addition to improve the GA directly, many heuristics based algorithms are also studied to solve the efficiency issue, such as degree [13], propagation path of influence [3]. In recent years, some researchers start to study this problem with consideration of geographical distance [14], discounts provided to users [15], opinion of users [16] or the dynamic nature of influence [17].

The algorithms described above are all neglecting the topical features of information content which is very important in information diffusion.

Topic-Sensitive Influence Maximization. Tang et al. [6] first propose a Topical Affiliation Propagation(TAP) model to extract the topic-sensitive influence strength between pairs of nodes. Barbieri et al. [7] propose a topic-aware influence propagation model whose parameters, influence probabilities and topic distribution of items, are estimated by real propagation logs. Inspired by the work in [7], Chen et al. [8] develop a faster topic-sample-based method to solve the topic-aware influence maximization problem. Since these works are based on user's action log, so we can't compare our model with them as our experimental data don't contain logs.

Although topic-sensitive influence maximization has been studied before, to the best of our knowledge, we are the first to take user's topical authority into account to mine influential users for a given topic, and that's the origination and reason of our work.

3   Preliminary

In this section, we first discuss the classic Independent Cascade model and then formally define the topical authority-sensitive influence maximization problem.

3.1   Classic Diffusion Model

Diffusion model is used to propagate influence in social networks. Most of the works in influence maximization focus on the Independent Cascade (IC) model. In the IC model, each node is in one of the two states: active or inactive. Before the beginning of the diffusion process, all the seed nodes will be first activated. Once a node u becomes active in step $t - 1$, then at step t, for node u, it is given a single chance to try to activate each of its inactive neighbors v, and succeeds with probability $p_{u,v}$. And if node v is successfully activated, it becomes active at step t and will also try to activate its neighbors in the next step. This diffusion process will stop until no nodes can be activated.

3.2 Problem Definition

Definition (Topical Authority-Sensitive Influence Maximization). A social network is a bidirectional graph $G = (V, E)$ (as we consider the collaboration relationship between nodes). V represents the set of nodes in the network while E is the set of edges between nodes. Given a social network G, a number K, a topic T and a topical authority-sensitive influence diffusion model, topical authority-sensitive influence maximization aims to seek a K-nodes seed set S that can maximize the influence spread $\sigma_T(S)$ under topic T in the network.

$$S^* = argmax\{\sigma_T(S), |S| = K, S \subseteq V\} \tag{1}$$

Kempe et al. [1] proves that under IC model, this problem is NP hard. If a greedy algorithm is needed to guarantee the $(1 - 1/e - \epsilon)$ approximation performance, the objective function must hold non-negative, monotone and submodular properties. We say a function $f(\cdot)$ is submodular if it satisfies a natural "diminishing returns" property: i.e., $\forall v \in V, S \subseteq M \subseteq V$, $f(S \cup \{v\}) - f(S) \geq f(M \cup \{v\}) - f(M)$.

4 Topical Authority-Sensitive Influence Maximization

In this section, we first introduce how to extract the topic information of users, and explain how to use Topical HITS algorithm to generate the user-authority vector for a given topic, based on which we propose our topical authority-sensitive IC model. Then we formally introduce the topical authority-sensitive greedy algorithm, and explain how it works to mine seed nodes. Finally, the approximation guarantee of our proposed algorithm will be presented at the end of this section. Table 1 lists some notations used in this paper.

4.1 Topic Distillation

For each user, we integrate all the content information published by this user into a document, and each of these documents will be regarded as a mixture of various topics, topic distillation aims to automatically identify the topic mixture of all these documents. In this paper, we use LDA [10] to extract user's topics. LDA takes the document-word frequency matrix as input (we also need to specify the number of extracted topics N), and the output of LDA are two matrices, γ and β. γ is an $N \times W$ topic-word matrix, W is the number of words in the corpus and β is a $|V| \times N$ (document) user-topic matrix. Through matrix β we can obtain an N-dimensional vector for each user. Namely, each node $v \in V$ is associated with an N-dimensional vector $\theta_v = (\theta_v^1, \theta_v^2, ..., \theta_v^N)$, where $\sum_i \theta_v^i = 1$, θ_v^i is the value of the i-th dimension. The matrix β will be the input of next Topical HITS algorithm.

Table 1. Notations

Terms	Descriptions
$G = (V, E)$	A social graph G with vertex set V and edge set E
K	Number of seed nodes
R	Number of rounds of Monte Carlo simulations
S	The K-nodes seed set
T	A given topic
N	Number of topics
p	The uniform propagation probability
$p_{u,v}$	The propagation probability that node u activates node v
$UA(T)$	The user-authority vector under topic T
A_u^T	The authority value of node u on topic T
$IS(S)$	Influence spread of seed set S
$RS(S)$	Resulting set of influence cascade originated from seed set S
$ISGT(S, T)$	Influence spread of seed set S on a given topic T

4.2 Topical Authority Calculation

Here, we use Topical HITS [9] algorithm to obtain the topical authority for each node in the network. In order to obtain the user-authority vector, first we need to get the user-topic matrix β by using LDA, and it is also necessary to get the node link relationship through G. Then for each node i in G, we use the following Eqs. 2 and 3 to calculate the authority and hubness vectors over the N topics:

$$H_{i,T} = \sum_{j \in O_i} \frac{\alpha A_{j,T} + (1 - \alpha)\beta_{j,T} \sum_{T \in TS} A_{j,T}}{|I_j|} \tag{2}$$

$$A_{i,T} = \sum_{j \in I_i} \frac{\alpha H_{j,T} + (1 - \alpha)\beta_{j,T} \sum_{T \in TS} H_{j,T}}{|O_j|} \tag{3}$$

In the above equations, TS means the topic set extracted by LDA, α is a propagating factor, $H_{i,T}$ and $A_{i,T}$ are the hubness and authority values of node i on topic T respectively, I_i and O_i are the incoming and outgoing neighbors of node i respectively, the authority and hubness values are calculated alternately (see [9] in detail). When the calculation is terminated, each node will have an N-dimensional topic-authority vector, and if we specify a topic T for all users, we can finally get the user-authority vector under topic T, namely $UA(T)$. $UA(T)$ will be used as a parameter of next TAIC model.

4.3 Topical Authority-Sensitive IC Model (TAIC)

The traditional IC model assigns a propagation probability for all pairs of nodes. The probability is the same to all edges which can't actually capture the authority of users on different topics. However, our TAIC model considers the propagation probability is not only in relation to influence frequency but also related to user's authority on a specific topic. In other words, the higher authority on topic T by node u and v, the more likely the influence is going to happen. According to the description above, we now give the definition of $p_{u,v}$ as follows:

$$p_{u,v} = p \cdot \delta(A_u^T, A_v^T) \tag{4}$$

This new cascade model is called Topical Authority-sensitive Independent Cascade model (TAIC). As we can see in Eq. 4, the original propagation probability p is weighted with a user's topical authority function δ, the authority values of A_u^T and A_v^T are coming from the $UA(T)$ which is computed by Topical HITS. As for the instantiation of δ function, we choose the arithmetic mean of A_u^T and A_v^T, namely,

$$\delta(A_u^T, A_v^T) = \frac{A_u^T + A_v^T}{2} \tag{5}$$

TAIC model will be used in the next TAG algorithm to find the most influential nodes for topic T. If we need to mine the seed nodes for other different topics, it can be easily done by adjusting the given topic T to get the $UA(T)$ of other topic, thus we are able to calculate the propagation probability under other topic.

4.4   Topical Authority-Sensitive Greedy Algorithm (TAG)

Based on the TAIC model described above, now we propose our Topical Authority-sensitive Greedy algorithm (TAG). Before starting to search seed nodes, we first use TAIC model together with the user authority vector $UA(T)$ to calculate the propagation probability on the given topic T for all pairs of nodes in the network. Then TAG takes the social graph G, the seed set size K as input to find a seed set S of K-nodes with a view to maximize the final influence spread generated by seed set S under the topic T. Algorithm 1 describes the details of TAG.

The basic idea of TAG is to run for K rounds, and search the node with the largest marginal gain as the seed node for each round. In Algorithm 1, we first calculate the propagation probability of each edge in G by utilizing the user-authority vector $UA(T)$, based on the TAIC model(lines 3–5); Then the marginal gain as well as the flag value will be calculated for each single vertex in V, flag value here is used to identify whether the current marginal gain is valid. Each vertex will be inserted into a priority queue, which will sort all nodes by it's marginal gain value in descending order(lines 7–15); Then we start to find the K-nodes seed set(we use the search strategy of CELF [2]). At the beginning of each round, the head node of the queue will be taken out to check whether the flag value of this node is valid, if valid, it means we have found the seed node for this round (lines 19–22). Otherwise, we will update marginal gain of this node, reset its flag value, then insert this node back into the queue (lines 23–27). This updating process will continue until the next valid head node is found. Finally, a K-nodes seed set S will be returned.

Algorithm 1. Topical authority-sensitive greedy algorithm

Input: The social graph $G = (V, E)$, number of seed nodes K, a topic T,
          user-authority vector $UA(T)$
Output: S(a K-size seed set)
1 **Initialization:** An empty set $S = \emptyset$, an empty priority queue $Q = \emptyset$,
    $R = 10000$
2 //Calculate the propagation probability of each edge in G
3 **for** *each edge* $(u, v) \in E$ **do**
4 $\quad\mid\quad p_{u,v} = p \cdot \delta(A_u^T, A_v^T)$
5 **end**
6 //Calculate the marginal gain of each single vertex $v \in V$
7 **for** *each vertex* $v \in V$ **do**
8 $\quad\mid\quad mg = 0$
9 $\quad\mid\quad flag = 0$
10 $\quad\mid\quad$ **for** $i = 1$ *to* R **do**
11 $\quad\mid\quad\mid\quad IS(\{v\})+ = |RS(\{v\})|$
12 $\quad\mid\quad$ **end**
13 $\quad\mid\quad mg = IS(\{v\})/R$
14 $\quad\mid\quad$ insert $node(v, mg, flag)$ into Q
15 **end**
16 //Get the K-nodes seed set
17 **while** $|S| < K$ **do**
18 $\quad\mid\quad node = Q.poll()$
19 $\quad\mid\quad$ **if** $node.flag == |S|$ **then**
20 $\quad\mid\quad\mid\quad S = S \cup \{node.v\}$
21 $\quad\mid\quad\mid\quad IS(S)+ = node.mg$
22 $\quad\mid\quad$ **end**
23 $\quad\mid\quad$ **else**
24 $\quad\mid\quad\mid\quad node.mg = |RS(S \cup \{node.v\})| - IS(S)$
25 $\quad\mid\quad\mid\quad node.flag = |S|$
26 $\quad\mid\quad\mid\quad$ insert $node(node.v, node.mg, node.flag)$ into Q
27 $\quad\mid\quad$ **end**
28 **end**

4.5 Approximation Guarantee

Our proposed TAG algorithm is based on TAIC model, an edge-weighted IC model, which Kempe et al. [1] prove the resulting objective function is submodular, and evidently, the objective function in Eq. 1 is also non-negative and monotone. We can use our TAG algorithm to get a $(1 - 1/e - \epsilon)$ approximation.

5 Experiments

In this section, we conduct extensive experiments to verify the effectiveness and efficiency of our proposed model and algorithm. We first introduce the experimental setup, and then present the results.

5.1 Experimental Setup

Datasets. We use two real-world datasets Aminer[1] and NetHEPT[2]. Aminer is an open social network dataset, which contains data extracted from DBLP, ACM, and other sources. NetHEPT is an academic collaboration network that extracted from the "High Energy Physics - Theory" part of arXiv. For every author in each dataset, we assign a unique ID. We think two cooperative authors interact with each other. If there is a cooperative relationship between two authors, there will be a two-way edge between them. The basic statistics of the two datasets are shown in Table 2.

Table 2. The basic statistics of datasets

Datasets	#Vertices	#Edges	#Avg. degree	#Max. degree
Aminer	20 K	90.5 K	4.52	73
NetHEPT	15.1 K	51.7 K	3.42	53

Since the TAIC model proposed in this paper needs to use the topic information of users, before we use LDA to extract user's topic distribution, for each author in the above two datasets, we integrate all the abstracts of their published papers into a document, and preprocess all these documents by removing disturbing information like punctuations, url strings or stop words. After getting the user-topic matrix β, we take the β and the link relationship between users as input of Topical HITS to obtain the topic-authority vector for each user.

Algorithms. We compare our TAG algorithm with the following five baseline algorithms:

- **CELF:** The original greedy algorithm on the IC model with the lazy-forward optimization [2].
- **PageRank:** A very popular algorithm used to rank web pages [18]. Here we set the restart probability to be 0.15.
- **Topic-Sensitive PageRank (TSPR):** TSPR is able to bias the calculation to increase the rank value for some pages that contain a particular topic [19]. Here the restart probability is still 0.15 and we set the bias to be 0.1.
- **SingleDiscount:** A degree based heuristics [13], the key idea of this algorithm is that if a node is selected as a seed node, then the degree of its neighbors will be reduced by 1.
- **Degree:** The traditional degree based algorithm which selects the top$-K$ seed nodes simply according to their degrees. Here we choose the K nodes with the highest out-degrees as seed nodes.

---

[1] http://resource.aminer.org.

[2] http://www.arXiv.org.

Evaluation Metrics. Since the traditional metric $IS(S)$ does not consider topic information, we also propose a new metric $ISGT(S, T)$ to evaluate the performance of our proposed algorithm in the context of topical authority.

- **$IS(S)$** : Influence Spread of seed set S. When given a seed set S, $IS(S)$ is defined as the number of finally activated nodes when the diffusion process ends.
- **Running time:** The running time is defined as the time it takes to select the $K-$nodes seed set.
- **$ISGT(S, T)$** : Influence Spread of seed set S on a Given Topic T. It is defined in Eq. 6

$$ISGT(S, T) = \sum_{u \in RS(S)} (A_u^T)^2 \tag{6}$$

We set $R = 10000$ for all models to get more accurate simulation results, $N = 10$ for extracted topics and vary K from 5 to 50 as different seed set size.

5.2   Experimental Results and Analysis

Before demonstrating the results of various algorithms, we first give the top ten words with the highest probability of occurrence on the topic we choose to study on the two datasets. For Aminer, the top ten words are *performance, memory, system, applications, parallel, paper, cache, data, design, storage*. For NetHEPT, the top ten words are *model, conformal, boundary, field, equation, algebra, function, matrix, limit, theory*. From the words listed above, we can roughly infer what these two topics are about. The topic of Aminer is most likely about *improvement of computational performance or computer architecture*, while the topic of NetHEPT may be about *mathematical physics*. We now compare TAG with other baseline algorithms on the above two topics respectively.

IS vs $ISGT$ of All Algorithms on the Given Topic. Figure 1 presents the IS and $ISGT$ of all algorithms on Aminer and NetHEPT respectively. We can see from Fig. 1(a) and (c) that on the two datasets the IS of TAG is the smallest, while CELF holds the largest one. Since CELF does not consider the topic information, it tends to find the most influential nodes across all topics, while TAG takes the topical authority into account to find the most influential nodes on a give topic T. However, a node with large influence on a single topic does not necessarily have large influence on all other topics, so TAG will have a poor performance under the IS metric that does not consider topic.

However, even though the IS of TAG is the smallest, TAG can obtain the largest $ISGT$ over all other algorithms as we can see from Fig. 1(b) and (d). Obviously, the performance of TAG is much better than other algorithms on the two datasets. This is because TAG can find the most influential nodes on a given topic while other algorithms can not. And when K = 50, the performance of TAG on Aminer is about 256% higher than CELF, and about 265% higher than CELF on NetHEPT. It can also be seen that on the Aminer dataset, since TSPR takes topics into consideration, its performance is better than other heuristic

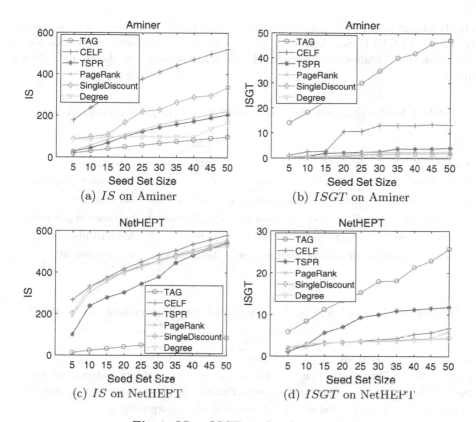

Fig. 1. *IS* vs *ISGT* on the given topic

algorithms, and TSPR even outperforms CELF on the NetHEPT dataset, when $K = 50$, the performance of TSPR is about 74% higher than CELF. In addition, SingleDiscount performs well for the *IS* metric on both datasets. On Aminer, when $K > 15$, the performance of SingleDiscount is much better than Degree. On NetHEPT, it is obvious that the performances of PageRank, SingleDiscount and Degree are quite close. This is mainly due to the difference in graph structure. The distribution of nodes in the Aminer is relatively uniform, while in NetHEPT dataset, there will be a large number of nodes connected to each other, making the distribution of nodes very non-uniform, so the strategy based on degree discount used by SingleDiscount will be "failed" in this case, which makes its performance very similar to the Degree algorithm. This kind of uneven distribution of nodes in the NetHEPT dataset will also have a great impact on the running time of algorithms, which we will verify later.

Top−5 Seed Nodes Found by TAG and CELF for the Given Topic. Now, we try to observe the performance differences between our proposed TAG algorithm and the CELF algorithm in a more intuitive way. We will compare the top−5 seed nodes found by these two algorithms on the two datasets and check whether these nodes fit in with the topics we specified before. For the nodes found by TAG and CELF, we demonstrate the real name and some keywords when searching these names on internet, the detailed results are shown in Tables 3 and 4.

Table 3. Seed nodes found by TAG and CELF for the given topic on Aminer

TAG		CELF	
Name	Keywords	Name	Keywords
K. Sankaralingam	Computer architecture	T. A. DeFanti	Computer graphics
Sadasivan Shankar	Large-scale computation	James Diebel	3D mapping
Doug Burger	Computer architecture	W. Emmerich	Software engineering
Ron K. Cytron	Program optimization	Z. Shi	Modern physics
Christos Kozyrakis	Computer architecture	John Geddes	Neuroscientist

Table 4. Seed nodes found by TAG and CELF for the given topic on NetHEPT

TAG		CELF	
Name	Keywords	Name	Keywords
P. Mathieu	Mathematical physics	Edward Witten	String theory and quantum gravity
Paul A. Pearce	Mathematical physics	Sergio Ferrara	Supersymmetry and supergravity
Tetsuji Miwa	Mathematical physics	Renata Kallosh	Supergravity
G. Mussardo	Statistical physics	C. N. Pope	Strings and gravity
M. Jimbo	Mathematical physics	S. P. Sorella	Quantum field theory

It is quite obvious that the seed nodes found by TAG do conform to the topic we specified. In Table 3, three of the nodes found by TAG are in the area of computer architecture while other two are about computational optimization and in Table 4, the five nodes are all in the area of mathematical physics. This indicates that TAG can indeed find seed nodes that are more influential under a given topic. However, it can also be seen that for the CELF algorithm, since it is based on the IC model that does not consider topic information, none of the nodes found by CELF conforms to our given topic on both datasets, which also explains why the CELF algorithm with the best performance under the IS metric performs poorly under the $ISGT$ metric.

IS vs $ISGT$ of TAG on Different Topics. To further study the effectiveness of $ISGT$ and IS when considering user's topical authority, we conduct topic-level experiments on both datasets. We first choose two different topics for the two datasets, then use TAG to find the seed nodes for each topic. Taking Aminer as an example, the two topics we select are PO(Performance Optimization) and CN(Computer Network), and the seed sets found by TAG for these two topics are po and cn respectively. We compare the IS and $ISGT$ of the two seed sets, the results are shown in Fig. 2(a) and (b). We do the same operation on NetHEPT, the two topics we choose are ST(Space Theory) and MP(Mathematical Physics), the seed sets found by TAG for them are st and mp, the IS and $ISGT$ of the two seed sets are shown in Fig. 2(c) and (d).

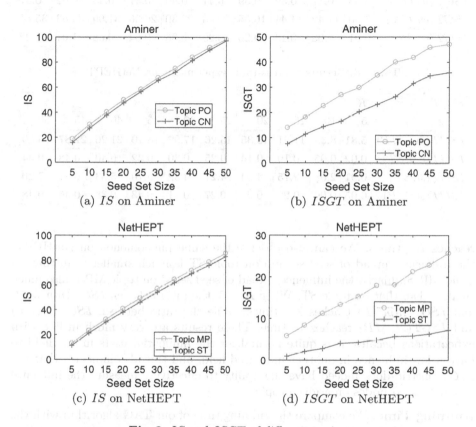

(a) IS on Aminer   (b) $ISGT$ on Aminer

(c) IS on NetHEPT   (d) $ISGT$ on NetHEPT

Fig. 2. IS and $ISGT$ of different topics

It is quite obvious that, for the two datasets, even if we have selected two completely different topics, the seed nodes found under these two topics are very close to the traditional IS metric, while the $ISGT$ metric on different topics are significantly different. We further compare the influence of the two seed sets on different topics, Tables 5 and 6 present the results on two datasets respectively.

From Tables 5 and 6, it can be clearly seen that the $ISGT$ metric is very sensitive to the topic change. Taking Aminer as an example, the influence spread of seed set po on topic CN is much smaller than that on topic PO. Similarly, the influence spread of seed set cn on topic PO is also much smaller than that on topic CN. When $K = 5$, the ratio between $ISGT(po, PO)$ and $ISGT(po, CN)$ reaches 262 times, while the ratio between $ISGT(cn, CN)$ and $ISGT(cn, PO)$

Table 5. Results of cross-topic experiments on Aminer

$ISGT$	K									
	5	10	15	20	25	30	35	40	45	50
$ISGT(po, PO)$	13.12	18.17	22.71	27.05	29.92	34.85	40.08	41.86	45.85	47.04
$ISGT(po, CN)$	0.05	0.17	0.39	0.39	0.41	0.60	0.47	0.43	0.62	0.67
$ISGT(cn, CN)$	7.60	11.42	14.44	16.55	20.43	23.30	26.36	31.59	34.63	35.81
$ISGT(cn, PO)$	0.01	0.03	0.13	0.20	0.29	0.36	0.64	0.68	1.01	1.22

Table 6. Results of cross-topic experiments on NetHEPT

$ISGT$	K									
	5	10	15	20	25	30	35	40	45	50
$ISGT(mp, MP)$	5.81	8.35	11.21	13.35	15.26	17.92	18.10	21.26	22.87	25.69
$ISGT(mp, ST)$	0.02	0.05	0.10	0.14	0.25	0.30	0.32	0.40	0.48	0.54
$ISGT(st, ST)$	0.91	2.08	2.75	4.04	4.27	4.82	5.70	5.83	6.27	7.20
$ISGT(st, MP)$	0.05	0.09	0.22	0.20	0.27	0.31	0.43	0.38	0.46	0.48

reaches 760 times. We can also observe the same phenomenon on NetHEPT, the influence spread of seed set mp on topic ST is much smaller than that on topic MP. Similarly, the influence spread of seed set st on topic MP is also much smaller than that on topic ST. When $K = 5$, the ratio between $ISGT(mp, MP)$ and $ISGT(mp, ST)$ reaches 290 times, while the ratio between $ISGT(st, ST)$ and $ISGT(st, MP)$ reaches 18 time. These results are very much in line with expectations because it is quite normal for the influential users in one field to have a low influence in another unrelated field, and this also indicates that the $ISGT$ metric does indeed have the ability to effectively measure the influence spread of seed nodes on a given topic.

Running Time. We compare the running time of our TAG algorithm with the CELF algorithm, as both of the two algorithms consider propagation probability when selecting seed nodes while other four baseline algorithms are heuristic thus not comparable with TAG. Figure 3 presents the result.

Apparently, it takes much less time of TAG to find the seed nodes, while CELF spends quite a long time to finish. This is because our TAG takes topic into account, which greatly reduces the range of nodes that can be searched, thus the time it takes to run the TAG will be decreased accordingly. However, CELF needs to find the influential nodes across all topics, so the range of nodes that need to be searched will be greatly increased. It is worth noting that when running CELF on NetHEPT, the time spent in it is surprisingly long, actually the reason behind this we've described before. Since lots of high-degree nodes are linked to each other, once one of them is selected as the seed node, it will lead to the marginal gain of many nodes need to be recalculated, which directly results to the increase of running time.

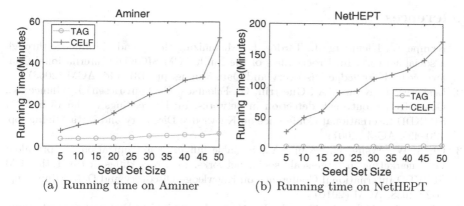

(a) Running time on Aminer   (b) Running time on NetHEPT

Fig. 3. Running time comparison of TAG and CELF

6   Conclusion and Future Work

In this paper, we first propose a topical authority-sensitive IC model, based on which the TAG algorithm is put forward to try to find the most influential nodes on a given topic. To evaluate the effectiveness and efficiency of our proposed model, we conduct extensive experiments on the performance of influence spread, influence spread on a given topic as well as the running time of our TAG and other baseline algorithms. The experimental results show that our proposed TAG algorithm performs consistently better than other baseline algorithms in finding seed nodes with higher influence on a given topic and has better time efficiency at the same time. For our future work, we will try to extend our model to be multi-topical authority sensitive and study the influence maximization problem under the multi-topics scenario. Meanwhile, we are also going to expand the size of datasets to test the performance of our proposed model and algorithm under larger datasets.

Acknowledgment. This work is supported by the National Key Research and Development Program of China under grants 2016QY01W0202 and 2016YFB0800402, National Natural Science Foundation of China under grants 61572221, 61672254, U1401258, 61433006 and 61502185, Major Projects of the National Social Science Foundation under grant 16ZDA092, and Guangxi High level innovation Team in Higher Education Institutions-Innovation Team of ASEAN Digital Cloud Big Data Security and Mining Technology.

References

1. Kempe, D., Kleinberg, J., Tardos, É.: Maximizing the spread of influence through a social network. In: Proceedings of the Ninth ACM SIGKDD International Conference on Knowledge Discovery and Data Mining, pp. 137–146. ACM (2003)
2. Leskovec, J., Krause, A., Guestrin, C., Faloutsos, C., VanBriesen, J., Glance, N.: Cost-effective outbreak detection in networks. In: Proceedings of the 13th ACM SIGKDD International Conference on Knowledge Discovery and Data Mining, pp. 420–429. ACM (2007)
3. Chen, W., Wang, C., Wang, Y.: Scalable influence maximization for prevalent viral marketing in large-scale social networks. In: Proceedings of the 16th ACM SIGKDD International Conference on Knowledge Discovery and Data Mining, pp. 1029–1038. ACM (2010)
4. Ohsaka, N., Akiba, T., Yoshida, Y., Kawarabayashi, K.I.: Fast and accurate influence maximization on large networks with pruned Monte-Carlo simulations. In: AAAI, pp. 138–144 (2014)
5. Tang, Y., Shi, Y., Xiao, X.: Influence maximization in near-linear time: a martingale approach. In: Proceedings of the 2015 ACM SIGMOD International Conference on Management of Data, pp. 1539–1554. ACM (2015)
6. Tang, J., Sun, J., Wang, C., Yang, Z.: Social influence analysis in large-scale networks. In: Proceedings of the 15th ACM SIGKDD International Conference on Knowledge Discovery and Data Mining, pp. 807–816. ACM (2009)
7. Barbieri, N., Bonchi, F., Manco, G.: Topic-aware social influence propagation models. In: 2012 IEEE 12th International Conference on Data Mining (ICDM), pp. 81–90. IEEE (2012)
8. Chen, S., Fan, J., Li, G., Feng, J., Tan, K.I., Tang, J.: Online topic-aware influence maximization. Proc. VLDB Endowment 8(6), 666–677 (2015)
9. Nie, L., Davison, B.D., Qi, X.: Topical link analysis for web search. In: Proceedings of the 29th Annual International ACM SIGIR Conference on Research and Development in Information Retrieval, pp. 91–98. ACM (2006)
10. Blei, D.M., Ng, A.Y., Jordan, M.I.: Latent dirichlet allocation. J. Mach. Learn. Res. 3, 993–1022 (2003)
11. Tang, Y., Xiao, X., Shi, Y.: Influence maximization: near-optimal time complexity meets practical efficiency. In: Proceedings of the 2014 ACM SIGMOD International Conference on Management of Data, pp. 75–86. ACM (2014)
12. Cohen, E., Delling, D., Pajor, T., Werneck, R.F.: Sketch-based influence maximization and computation: Scaling up with guarantees. In: Proceedings of the 23rd ACM International Conference on Conference on Information and Knowledge Management, pp. 629–638. ACM (2014)
13. Chen, W., Wang, Y., Yang, S.: Efficient influence maximization in social networks. In: Proceedings of the 15th ACM SIGKDD International Conference on Knowledge Discovery and Data Mining, pp. 199–208. ACM (2009)
14. Wang, X., Zhang, Y., Zhang, W., Lin, X.: Distance-aware influence maximization in geo-social network. In: ICDE, pp. 1–12 (2016)
15. Yang, Y., Mao, X., Pei, J., He, X.: Continuous influence maximization: what discounts should we offer to social network users? In: Proceedings of the 2016 International Conference on Management of Data, pp. 727–741. ACM (2016)
16. Galhotra, S., Arora, A., Roy, S.: Holistic influence maximization: combining scalability and efficiency with opinion-aware models. In: Proceedings of the 2016 International Conference on Management of Data, pp. 743–758. ACM (2016)

17. Wang, Y., Fan, Q., Li, Y., Tan, K.L.: Real-time influence maximization on dynamic social streams. Proc. VLDB Endowment **10**(7), 805–816 (2017)
18. Page, L., Brin, S., Motwani, R., Winograd, T.: The pagerank citation ranking: bringing order to the web. Technical report, Stanford InfoLab (1999)
19. Haveliwala, T.H.: Topic-sensitive PageRank. In: Proceedings of the 11th International Conference on World Wide Web, pp. 517–526. ACM (2002)

Microblog Data Analysis

SensorTree: Bursty Propagation Trees as Sensors for Protest Event Detection

Jeffery Ansah[✉][iD], Wei Kang[iD], Lin Liu[iD], Jixue Liu[iD], and Jiuyong Li[iD]

School of Information Technology and Mathematical Sciences,
University of South Australia, Adelaide, SA 5095, Australia
{jeffery.ansah,wei.kang,lin.liu,jixue.liu,jiuyong.li}@unisa.edu.au

Abstract. Protest event detection is an important task with numerous benefits to many organisations, emergency services, and other stakeholders. Existing research has presented myriad approaches relying on tweet corpus to solve the event detection problem, with notable improvements over time. Despite the plethora of research on event detection, the use of the implicit social links among users in online communities for event detection is rarely observed. In this work, we propose SensorTree, a novel event detection framework that utilizes the network structural connections among users in a community for protest event detection. SensorTree tracks information propagating among communities of Twitter users as propagation trees to detect bursts based on the sudden changes in size of these communities. Once a burst is identified, SensorTree uses a latent event topic model to extract topics from the corpus over the burst period to describe the event that triggered the burst. Extensive experiments performed on real-world Twitter datasets using qualitative and quantitative evaluations show the superiority of SensorTree over existing state-of-the-art methods. We present case studies to further show that SensorTree detects events with fine granularity descriptions.

Keywords: Burst · Event detection · Propagation trees
Social media · Twitter

1 Introduction

With the advancement of Web 2.0 technologies, social media sites such as Twitter, Facebook and Weibo have become a viable source for monitoring and analyzing the rich continuous flow of information for protest event detection. To detect events, one of the predominant approaches is to model events in text streams as bursts with keywords rising sharply in frequency as an event emerges. The basic assumption is that some related words will exhibit a sudden increase in their usage when an event is happening. An event is therefore conventionally represented by a number of keywords showing burst in appearance counts.

While these event detection approaches have gained successes, some challenges still prevail. First choosing the right set of predefined keywords to track is

© Springer Nature Switzerland AG 2018
H. Hacid et al. (Eds.): WISE 2018, LNCS 11233, pp. 281–296, 2018.
https://doi.org/10.1007/978-3-030-02922-7_19

a herculean task and usually requires knowledge from domain experts. Secondly, once the right keywords are chosen, another hurdle is understanding the correct context of these bursty keywords for event detection. For example, if we have two tweets of the form: *"Reduce tax and import duties or get ready for protest"* and *"Too much protest from Arsenal fans in this game"*. We observe that the keyword "protest" is used in both tweets. However, the context clearly shows that the ongoing conversation is about two separate events. Keyword based models that use the counts of keywords for event detection may not be able to distinguish different events using similar predefined keywords.

A promising solution is to leverage the social network structure and follower relationship among users for protest event detection. The intuition here is that tweets which are sent between a tightly knit group of users in a community may be more indicative of a particular event of interest than a set of tweets being propagated by a random set of users who do not have any form of social network connections. This suggests that a discussion is more likely to become active during or even before an event among Twitter users who follow each other. Such implicit relationship among users in a community can be captured using propagation trees. Propagation trees have been proven [4,8] to be effective in depicting how communities of online users connected via social links (follower relationship) discuss topics of interest. These trees are able to capture the structure and temporal growth of communities as information propagates. For example, Fig. 1 shows a plot of the size distribution of propagation trees built on the hashtag "freeport" to capture series of events on coal mining in Indonesia. These trees show bursts (represented as spikes) that correspond to increasing online user activity prior to real world protest events of interest. The spikes 1, 3 and 5 correspond to days of protest events. Spikes 2 and 4 were days of significant events leading to protests. These dynamics clearly show that propagation trees are useful in capturing the discussion of new topics, the occurrence of new events, etc. in an online community, and can be used for our event detection task.

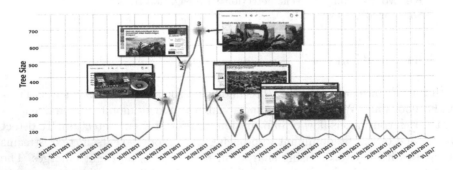

Fig. 1. Propagation tree size on the Freeport event (Blue dots (1, 3, 5) indicate main protest event days and green dots (2, 4) indicate sub-event days (sub-events are phenomena that are precursors to future protest events)). (Color figure online)

However, two main challenges exist: (1) Detecting bursts in online communities. This task requires accurately capturing the tree growth in continuous time to detect the period of sudden increase (burst) in online user communities. This problem is not trivial since there is currently no measure of quantifying bursts in propagation trees for event detection. (2) Event inference from trees. Once a burst is detected, the second challenge is to infer the protest event that has triggered the burst. Current propagation tree formulations make it impossible to infer the semantic context of the information propagation (i.e. what topic is being discussed at which time) for protest event detection. This is mainly because these information propagation studies [4,8] are only interested in capturing the numeric structural and temporal features such as the size, depth, growth rate, etc., from information propagating through an online community.

To address the above challenges, we propose a novel event detection framework called SensorTree. SensorTree framework is developed upon the semantic propagation tree proposed in this paper. The semantic propagation tree does not only capture the numeric structural and temporal features but also has the capability of capturing the corpus (textual component) of the ongoing discussion in an online community of Twitter users. SensorTree builds semantic propagation trees and efficiently computes the changes in acceleration of the tree size. The change in acceleration is then used as a means of quantifying the period of a sudden change in growth of online communities as the burst for event detection. Once the period of a burst is detected, a tensorized latent event topic model is triggered to infer and provide extra textual information on the associated events causing the burst. SensorTree is efficient and effective in detecting protest events as well as discovering topics of discussion within an online community on a protest event.

To the best our knowledge, this is the first work to perform event detection using the change in acceleration in the growth of online communities. Our main contributions can be summarized as:

- We propose a novel event detection method which leverages burst in information propagation from online communities to detect protest events using the change in acceleration of the size of a semantic propagation tree.
- We develop an event inference approach for describing detected protest events in bursty online communities using semantic propagation trees.
- We develop SensorTree, a novel framework based on the above proposed methods. Experiments on real world datasets show that SensorTree is language independent, does not require a predefined set of keywords and capable of detecting and describing events with fine granularity.

2   Related Work

Event Detection is a vibrant research area with evolving interest in news, blogs, and social media. This line of research has been extensively studied by the text mining community in the context of Topic Detection and Tracking [7,14]. For the purpose of our study, we review related works on event detection using social media, which can be categorized into two active lines of research:

1. Document clustering and semantic similarity: This line of work clusters data points on the basis of some textual similarity measures for event detection. The underlying assumption is that documents are somehow related to a number of undiscovered events. Events are identified by cluster associated word frequency or topic distribution using traditional LDA [7] topic models and its variants.

The authors of [5] explored multi-feature similarity techniques and proposed a novel framework that leverages normalized mutual information for online event clustering. To achieve a similar goal of event detection, the authors of [6] distinguished events from non-events using aggregated statistics of topically similar clusters obtained using Term Frequency-Inverse Document Frequency (tf-idf) weighting measures. EDCoW [17] combines wavelet analysis with clustering techniques to build signals for event detection. The signals are then filtered using time-series autocorrelation measures and a modularity-based graph partition technique is used to detect events. While the work in [5,6] was focused on general event detection, the authors of [9,16] studied specific events such as protest events, earthquakes, and other disasters. In [16], a tweet-based classifier with semantic analysis was used to detect earthquakes in real-time by modeling tweets as sensor information. The classifier utilized keywords in tweets as features and inferred event location using Bayesian Kalman filters. A non-parametric heterogeneous graph scan statistics was proposed by [9] to detect protest events. The authors modelled tweets, retweets, and hashtags as a graph that senses anomalous neighbourhood clusters to detect events. A notable similarity of most of the techniques in this domain is the use of textual features for event detection.

2. Event detection using bursty terms: In this line of research, a burst is defined as a sudden rise in the frequency, size, volume etc. of some keywords or data points. The intuition here is that a sudden rise in the frequency of keywords or data points can be attributed to an important event taking place.
Initial efforts by the authors of [13] was an infinite-state automaton approach to model data stream in which bursts appear as state transitions. Once a burst is detected, a nested representation of the burst is evaluated using the hierarchical structure of the overall stream. In [12], the authors presented an alternative perspective of bursts in real time as a time varying Markov modulated Poisson process. Another real-time online event topic detection framework is TopicSketch [18]. The authors proposed a bursty topic detection framework using the velocity and acceleration of words by drawing inspiration from earlier work in [10], along with hashing dimension reduction techniques to achieve scalability.

All the aforementioned works leverage textual information together with some clustering or semantic similarity measures to detect events. Table 1 shows the uniqueness of our work in this paper in comparison to closely related work. We differentiate SensorTree from these works in the sense that, SensorTree uses the growth of online communities (users connected by follower relationships) for event detection rather than predefined keywords.

Table 1. A comparison of SensorTree to existing works in social media event detection

Event Detection using Social Media		
	Keywords/Text (Tweets)	Social Context
Clustering/Document Similarity	[5, 6, 9, 16, 17]	[2, 9]
Burst Detection	[1, 10, 13, 15, 18]	SensorTree (Our approach)

3   Problem Definition

Twitter provides a functionality for users to follow other users. This enables users to receive information from those they follow on their timeline. Generally, information propagates on Twitter in the following manner: a Twitter user Alice posts a tweet on a protest event on a given day, in the form of retweet, @mentions, normal tweet etc.; Bob, a follower of Alice also posts a tweet on the same protest event after seeing the original post by Alice. As this process continues, information propagates through an online community of Twitter users. The main objective of this paper is to utilize the network structure and social links of information propagation among users in online communities for protest event detection. The intuition is that a sudden increase in activity (burst) in a community is a result of the emergence of an event of interest, hence we capture burst to detect events. Our main problem can thus be formulated as:

Problem Definition: *Given some information propagating in the form of tweets through a community of online Twitter users, our goal is to detect the periods of sudden changes (bursts) as well as the event that triggers the burst.*

To address this problem, we break it into two sub-problems as discussed below.

Sub-Problem Formulation

Propagation trees [4], have been used to effectively capture information propagating among online user communities; a feature that we need for solving the above defined research problem. Figure 2 shows a tree under construction at time t_1 using tweets and the Twitter follower network [4]. The nodes **A**, **B**, and **C** represent users in a community who have already posted tweets on some protest event of interest. A new user **D**, who is a follower of **B** posts a message on the same protest at time t_2, we add node **D** and a directed edge from **B** to **D** to the tree. The tree grows as this process continues.

To be able to detect bursts, we need to capture the growth of these trees in continuous time. For example, Fig. 3 shows the timeline of two propagation trees. The green tree shows a sudden growth (burst) between time t_3 and t_4 while the red tree shows a gradual growth in its timeline. Studying these growth dynamics will help us to effectively capture burst for event detection. Burstiness in propagation trees may be a result of breaking news information or the emergence of compelling events which attract a lot of attention and rouse people to

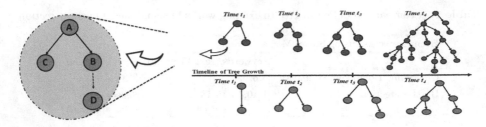

Fig. 2. Propagation tree

Fig. 3. The growth of two propagation trees over time

tweet about them. Therefore by making use of the propagation tree intuition, we can break our research problem into the following two sub-problems:

Sub-Problem 1: Burst Detection in Online Communities
Given tweets and the Twitter follower network, our goal is to build propagation trees and capture the period of burst W_b in continuous time as the tree grows.
This task requires modeling the increase in growth of online communities, storing of the tree size tree in continuous time and developing a method to effectively capture the period of burst.

Once we have been able to detect a burst, our second research challenge is to infer the protest event that triggers the sudden changes in the growth of the community. This leads to our next sub-problem.

Sub-Problem 2: Event Inference from Tree Bursts
Given a W_b in a tree, the goal is to infer the event which triggered the burst.
This can be formulated as a latent topic inference problem based on the tweet corpus in W_b. Existing formulations of propagation trees [4,8] can only capture structural and temporal features of information propagation which is only useful for sub-problem 1. To address sub-problem 2, our major challenge is to redefine propagation trees to preserve both the structural-temporal links as well as the semantic context of information propagation for event detection. This formulation will help us not only detect bursts but also describe the associated latent topics in a burst period.

4   SensorTree Framework

To solve the research problem in this work, we develop SensorTree, which tracks information propagating in an online Twitter community in continuous time to capture the burst for event detection. Figure 4 gives an overview of the SensorTree framework, which contains: (1) The tree construction phase to build the propagation trees, (2) The Tree Data Gridding (TDG) phase to store the growth dynamics of a tree in sliding time windows, (3) The Burst Sensoring phase to track changes in the acceleration in the TDG to identify bursty windows for event detection, (4) The Latent Event Topic Modeling phase which infers latent

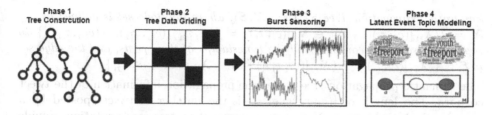

Fig. 4. Workflow of SensorTree framework

topics in a bursty time window and reports the detected event. The four phases are described in detail below.

Phase 1. Tree Construction: The first step of SensorTree is to capture information propagation among users in an online community. Propagation trees have been proven to be useful [4,8] to effectively capture how communities of online users connected via social links (follower relationship) discuss topics of interest. These trees are constructed using tweets and the Twitter follower network as described below [4].

Let the directed graph $\mathbf{G} = \langle \mathbf{V}, \mathbf{E} \rangle$ represent the Twitter follower network, where $\mathbf{V} = \{X_1, X_2, ..., X_N\}$ is the set of N Twitter users, and $\mathbf{E} = \{X_i \to X_j | X_i, X_j \in \mathbf{V}, i \neq j\}$ is the set of directed edges representing that user X_j is a follower of user X_i on Twitter. Information therefore propagates from X_i to X_j if an edge $X_i \to X_j$ exists in $\mathbf{G}$. Let $\mathcal{C}$ be the tweet corpus posted on a given day, and $p = (X, c, \tau)$ the tweet posted by user $X \in \mathbf{V}$ at time τ with content $c \in \mathcal{C}$. A time indexed tweet series is denoted by $\mathcal{P} = \langle p_1, p_2, ..., p_K \rangle$ s.t. $p_i.\tau \leq p_j.\tau$ if $i \leq j$, where $i, j \in \{1, 2, ..., K\}$ and K is the number of tweets posted on the current day.

Definition 1 (Propagation Tree (PT) [4]). *Given a Twitter follower network* $\mathbf{G}$ *and time indexed tweet series* $\mathcal{P}$ *for the current day, let* τ_i *be the time of the first post of* X_i *in* $\mathcal{P}$ *and* τ_j *be the time of the first post of* X_j *in* $\mathcal{P}$, *a Propagation Tree* $\mathbf{PT} = \langle \mathbf{V}', \mathbf{E}' \rangle$ *where* $\mathbf{V}' = \{(X_i, \tau_i) \mid X_i \in \mathbf{V}\}$ *and* $\mathbf{E}' = \{(X_i, \tau_i) \to (X_j, \tau_j) \mid X_i \to X_j \in \mathbf{E}, \tau_i \leq \tau_j\}$ *is a set of directed edges.*

The nodes of a *PT* represent users and the timestamps of their posts in the community and the edges represent the follower links between these users.

Recall from Sub-problem 2, that we are also interested in inferring the event that triggers the burst, which means that we need the corpus from a burst period to describe the event. However, the above existing propagation tree representation does not retain the corpora from ongoing discussion in online communities. Also, the *PT* framework captures just the first post of a user for the tree construction. Thus we introduce Semantic Propagation Tree, which is able to capture the growth features of a tree, as well as all the posts generated by the users in an online community.

Definition 2 (Semantic Propagation Tree (SPT)). *Given a Twitter follower network* $\mathbf{G}$ *and time indexed tweet series* $\mathcal{P}$ *for the current day, a*

Semantic Propagation Tree $SPT = \langle \mathcal{X}, \mathcal{E} \rangle$, *where the node-set is defined as* $\mathcal{X} = \{(X_i, [(c_i, \tau_i)]) \mid X_i \in \mathbf{V}, [(c_i, \tau_i)] = [(c_{i_1}, \tau_{i_1}), (c_{i_2}, \tau_{i_2}) \cdots (c_{i_Q}, \tau_{i_Q})]\}$ *to represent users, the contents and the timestamps of their posts, and the edge set is* $\mathcal{E} = \{(X_i, [(c_i, \tau_i)]) \rightarrow (X_j, [(c_j, \tau_j)]) \mid X_i \rightarrow X_j \in \mathbf{E}, \tau_{i_1} \leq \tau_{j_1}\}$.

The semantic propagation tree stores the propagation information of the tweet corpora. The length of $[(c_i, \tau_i)]$ i.e. i_Q is the number of tweets posted by a user X_i. Such additions to the existing propagation tree representation provide the capability for inferring latent events that trigger bursts in the growth of communities. We use the same criteria as [4] to construct semantic propagation trees. To build a tree, the first Twitter user who posts a tweet on a protest event on a given day is selected as the source node.

Following **Criterion 1** (tree growth) from [4], assuming we $(X_m, [(c_m, \tau_m)])$ represents a new node at $\tau_m \; \forall (X_i, [(c_i, \tau_i)]) \in SPT, \tau_m > \tau_i$. If $(X_i \rightarrow X_m) \in \mathbf{G}$, the follower network, we grow the tree by adding the node $(X_m, [(c_m, \tau_m)])$ and a directed edge from $(X_i, [(c_i, \tau_i)]) \rightarrow (X_m, [(c_m, \tau_m)])$. As an extension to **Criterion 1**, if a user who is already in the current tree under construction posts a message, we add his or her new post to the tree by extending the sequence $[(c_i, \tau_i)]$. By this, we are able to capture all the user's contribution to an ongoing discussion in an online community.

As information propagates in a community of online users in the form of a tree, there is a likelihood that users who belong to different communities will also share some information. This brings us to the use of **Criterion 2** (Emergence of new Propagation Trees) [4]. Given a semantic propagation tree SPT under construction with node set $\mathcal{X}$, let the follower list of X be $\mathcal{F}(X)$. If a user X has posted a tweet and X is not a follower of any of the users included in any of the existing $SPTs$, a new SPT is created with node $(X_i, [(c_i, \tau_i)])$ as the root. An SPT is terminated using (**Criterion 3 (Tree Termination)**)[4]. Assuming $(X_i, [(c_i, \tau_i)])$ is the last node added to the current SPT in a given day, the SPT is terminated if none of the followers of X_i posts a tweet after X_i's post.

Phase 2. Tree Data Gridding (TDG): Recall from Sub-problem 1 that being able to capture the changes in community growth in continuous time is non-trivial in order to detect bursts. We design a Tree Data Grid (TDG) to keep track of the growth of trees. With the TDG, we segment the timeline of tree growth into discrete time windows $W_{T_1}, W_{T_2}, ..., W_{Tk}$. We use $\mathcal{T}$ to represent the window size. $\mathcal{T}$ is varied using fixed time windows (e.g. $\{\mathcal{T} = 15, 30, 45\}$) to measure different granularities for capturing tree growth. The TDG is synonymous to an **m x n** matrix $Z_{[\mathbf{m,n}]}$, where each row is a tree and each column represents a time window. For example, a matrix element $Z_{[A, W_{T_i}]}$ in the TDG will return the value measured from tree A in time window W_{T_i}.

Phase 3. Burst Sensoring: The term burst can be viewed in different ways. We adopt a definition of burst involving kinetics and Newtonian motion [10, 18]. From a physics perspective, we define a burst as a notable change in the velocity of the tree growth. This rate of change has a natural interpretation as a kind of 'acceleration' or 'force', leading us to an intuitive physical model of bursts here as dramatic accelerations of the tree growth.

Using a basic construct in physics, a tree has an associated quantity $K(t)$ at time t. An example could be the number of tweets or the number of users in a tree. In this work, we use the size of a tree $|\mathcal{X}|$ as a basic quantity to model burst in online communities because of its ability to produce strong signals for event detection as shown earlier in Fig. 1. Generally, the velocity is obtained by finding the first derivative $\frac{dK(t)}{dt} = \frac{d|\mathcal{X}|(t)}{dt}$. We define $|\mathcal{X}|(t)$ at discrete points, thus we estimate $v(t) = \frac{\Delta|\mathcal{X}|(t)}{\Delta t}$. After obtaining the velocity, we can calculate the acceleration by finding the first derivative of $v(t)$ or the second derivative of $|\mathcal{X}|$ with respect to t. This is mathematically expressed as:

$$a(t) = \frac{d^2|\mathcal{X}|(t)}{dt^2} = \frac{v_{(t)} - v_{(t-\mathcal{T})}}{\mathcal{T}} \tag{1}$$

Once $a(t)$ is obtained, we can measure bursts in terms of positive accelerations beyond a parametric threshold θ. The threshold θ operates with the z-score $z_{a(t)} = \frac{a(t)-\mu}{\sigma}$, where μ is the average acceleration and σ is the standard deviation. For every time window W_{T_i} whose acceleration $a(t)$ has a z-score higher than the parametric threshold θ, we consider that window to be a bursty window.

Phase 4. Latent Event Topic Modeling (LETM): Recall from sub-problem 2 that once a burst is detected in a tree, we are interested in inferring the event associated with the burst. This requires that we utilize the conversations in the community during the burst period in order to solve our event inference problem. We extract the corpus over the bursty window and propose to employ latent event topic model using tensor decompositions [3] for event inference. Given the tweet corpus in W_b, a latent event variable $\phi = (\phi_1, \phi_2, ..., \phi_k)$ represents the proportion of k event topics where each ϕ_i is a distribution over an exchangeable bag of words. We denote $|Voc|$ as the vocabulary size in W_b from which words on ϕ_i are sampled from a generalized Dirichlet distribution [7] with concentration parameter $\alpha = [\alpha_1, \alpha_2, ..., \alpha_k]$. We propose to use a single topic model over the corpus in W_b which is a special case with $\alpha_0 = 0$. Learning latent topics can be carried out efficiently via tensor-based techniques with low sample and computational complexities which achieve better performance [3,11]. By using tensor decomposition [3], we can derive the first, second and third order moments, and reduce the model to moment forms to recover ϕ. An event is represented in the **LETM** as topics and its associated word descriptions.

A summary of SensorTree's framework is presented in Algorithm 1. The algorithm accepts tweets and the Twitter follower network relationships as input, and then carries out the four phases sequentially. **TreeConstruct** constructs semantic propagation trees as described in Phase one. The **TDG** stores the size of each tree based on the time window size $\mathcal{T}$. For each tree, the **Burst-Sensor** computes $a(t)$ using Eq. (1) to detect burst. Once a burst is detected in a tree, the **LETM** extracts tweets over the burst period. The **LETM** then builds a tensorized topic model to describe the event associated with the burst period. While we focus on protest events in this work, it is worth mentioning that SensorTree can be modified and extended in the context of general event detection. SensorTree outputs detected events as topics and word descriptions.

Algorithm 1. SensorTree

1: **procedure** : **Input:** *time indexed tweet series* $\mathcal{P}$, *follower relationship from* **G**
2: **Output:** *treeID, Event-topics, associated word descriptions,*
3: **TreeConstruct:** *Select* $p_1.X \in \mathcal{P}$ *as the source node*
4: for *every other* $p \in \mathcal{P}$
5: → *Use* **Criterion 1 and 2** *to construct trees* \\following tree algorithm in [4]
6: → *Terminate tree using* **Criterion 3**
7: **TDG:** *Store* $|\mathcal{X}|$ *for each* W_T \\vary $\mathcal{T}$ to observe optimum threshold
8: **BurstSensor:** Compute $a(t)$ *using Eq. (1)*
9: → Estimate $z_{a(t)} = \frac{a(t)-\mu}{\sigma} > \theta$ to detect W_b
10: **LETM:** *Extract Corpus in* W_b *for topic modelling using tensor decomposition [3]*
11: → *Construct and estimate empirical 2nd and 3rd order moments*
12: → *Whiten the data via SVD and extract the eigenvectors*
13: → *Use stochastic gradient descent to estimate the spectrum of whitened tensor*
14: → *Post-processing: Use power iteration in pairs to calculate topic-word matrix*

---

5   Experiments and Evaluation

We present a detailed description of how the experiments were conducted and the results obtained in comparison with existing state-of-the-art models.

5.1   Datasets and Settings

We use two different Twitter datasets for our experiments as shown in Table 2.
(I). Freeport Dataset: The Freeport dataset contains tweets published from Jan 2017–April 2017 on the coal protests in Indonesia. During this period, there were series of protest actions towards Indonesian mining giants with protesters calling for the mines to be shut down.
(II). NewCastle Habour Blockade Dataset (NHB): The NHB-dataset contains tweets published in Australia from Feb.-May 2016 on conversations regarding the NewCastle habour blockade and its related protest activities.

For both datasets, we collect the follower lists of all users who posted tweets during the periods of observation using the Twitter API. We use different time window settings ($\{\mathcal{T} = 5, 10, 15, ..., 120\}$) to empirically observe different time granularities of detecting bursts using Equation (1). We implemented and conducted all experiments using Python 3.2 and 3.6 on a Windows machine with 8 GB Memory and a 64-bit Linux virtual machine with 6 GB memory all running on an Intel(R) Core(TM) i5-4310m CPU 2.70 Ghz processor.

Table 2. Datasets description

Data	Tweets	Size of network community (# of Users)
Freeport Dataset	200,800	12,453,643
Newcastle Habour Blockade	4,900,305	52,424,200

5.2   Comparison Methods

We compare SensorTree to **LDA** [7] **tensorLDA, EvenTweet** [1], and **KW-Freq. EvenTweet** is a state-of-the-art event detection framework that extracts bursty keywords from tweets for event detection. We follow strictly the authors' implementation based on the published work [1]. For **LDA** and **tensorLDA**, we build the model over the entire corpus on daily basis and extract the event topics. We vary the number of topics (from $k = 1$ to $k = 5$) to achieve the most meaningful results for both models. We present the results of **tensorLDA** as it achieved better performance as compared to LDA in our experiments. As a baseline model, we also build **KW-Freq** as a naive event detection model that relies on the daily counts of keyword frequency in the entire datasets.

5.3   Evaluation Metrics

For the Freeport dataset we have ground truth data which we refer to as Gold Standard Record (GSR). The GSR contains coded protest event information extracted from major news sources, blogs and articles on real-world protest events compiled by news analysts. The event information includes the protest event date, headline description and news source with sample entries shown in Table 3. The headline description of the GSR contains a summary of protest events both in English and Indonesian.

Table 3. GSR protest event entries

Date	Event Id	News source	Event headline description
2017-02-05	9856112330	Kompas	*Freeport to reduce Indonesian Mining Activities. Tuntut Izin Ekspor, Kayawan Freeport Gelar Aksi di Kantor Bupati*
2017-02-24	996533162	Republika	*GMNI encourages government to take over freeport GMNI Dukung Pemerintah Ambil Alih*

We remove stopwords and manually select keywords token from the GSR event headline description. The goal is to use the keywords token to measure the performance of the various methods. Thus, for every GSR description, we expect the models to detect some keywords in its top words that match the GSR tokens. The following metrics are used in evaluating the results obtained from the Freeport dataset.

1. **Topic Intrusion Score** (I_T)**:** This score measures the quality of event description generated by a model. Denoting $\hat{c}_a$ and $\hat{c}_b$ as the word vectors for the GSR word tokens and tokens generated an event detection method respectively. The topic intrusion score is given by: $I_T(\hat{c}_a, \hat{c}_b) = 1 - \frac{\hat{c}_a \cdot \hat{c}_b}{|\hat{c}_a|^2 + |\hat{c}_b|^2 - \hat{c}_a \cdot \hat{c}_b}$. As events are represented by words belonging to a topic, a higher I_T score implies that a model's output has more intruding words which differ from the GSR tokens and unrelated to the event detected.

2. Topic Coherence (C_T)**:** Cosine similarity is one of the most popular similarity measures applied to text documents. We use the cosine similarity to measure the coherence of the word tokens to the GSR word tokens. A high C_T signifies that a model is able to concisely describe events detected at a fine granularity. The topic coherence C_T is given by: $C_T(\hat{c}_a, \hat{c}_b) = \frac{\hat{c}_a \cdot \hat{c}_b}{|\hat{c}_a| \times |\hat{c}_b|}$.

3. Precision: We measure precision as: $Precision = \frac{\#EventMatches}{Total\#GSREvents}$. A model's output is classified as a match if more than 60% of its top 20 (highest topic probabilities) word descriptions matches the word tokens of a GSR event. In our experiments, we select the 20 most representative keywords from each topic extracted daily and compare them with the GSR event descriptions.

5.4   Results and Discussion

5.4.1   Burst in Trees for Event Detection

After we build the propagation trees, we study the effect of using different time window settings in the detection of burst in propagation trees. We observe that bursts are generally very sparse and the number of detected bursts increases with increasing $\mathcal{T}$. We filter out trees with $|\mathcal{X}| \leq 10$ since their average accelerations over the minimum $\mathcal{T}$ is < 1. Using different time granularities, we show the timeline of the 10 trees with the largest size in Fig. 5 on the Freeport dataset. It can be observed that burstiness is very sparse across the timeline. We achieved similar results with varying $\mathcal{T}$ on both datasets as shown in Fig. 6. We also

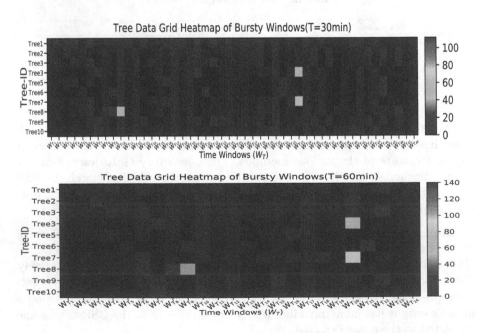

Fig. 5. Tree timeline showing growth dynamics for different time window size

observed in Fig. 6 that a reasonable window size helps in effectively capturing all the burst.

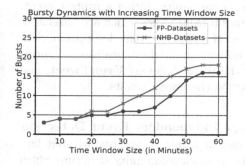 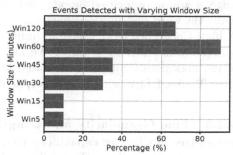

Fig. 6. Bursts dynamics on different datasets.

Fig. 7. Percentage of event detected using different window sizes.

We are also interested in evaluating the prowess of SensorTree in detecting real world protest events. We evaluate the percentage of real protest events detected by SensorTree using different time granularities in Fig. 7. The results show that SensorTree is able to be detect 96% of ground truth protest events with $\mathcal{T} = 60$ minutes. Hence, we set $\mathcal{T} = 60$ in the rest of our experiments.

5.4.2   Event Topic Evaluation

To compare the results of SensorTree with competing models, we show the performance of the various models on the Freeport dataset using the evaluation metrics (I_T, C_T and Precision) in Fig. 8. SensorTree outperforms comparison models in all the evaluation metrics.

On topic Intrusion, recall that a higher I_T means that the model has more intruding words and is unable to concisely describe events. It is not surprising that the **KW-Freq** recorded the highest value for I_T. This observation can be attributed to the fact that many events are discussed daily on social media and

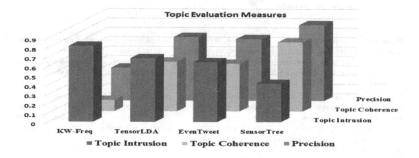

Fig. 8. 3D Bar Plots of Model Performance on Freeport dataset

the highest 20 keywords in a given day may be a distribution of words from different unrelated events. **TensorLDA** and **EvenTweet** achieved similar results on topic intrusion. SensorTree records the least I_T value showing that conversations are more focused among users in a community during a protest event. Thus the word distributions have less intrusion from topics which are unrelated. A high Coherence score C_T is an indication of a concise event description. SensorTree achieves the C_T value. We also achieve the highest precision in terms of matches with GSR token, followed by the state-of-the-art **EvenTweet**, and then **TensorLDA**, with **KW-Freq** achieving the lowest results.

5.4.3 Fine Granularity and Language Independent Event Detection

SensorTree has further capabilities to detect at a fine granularity without language restrictions. SensorTree relies on the network community structure which is blind to the use of a specific language. Thus we are able to capture ongoing conversations between users in non-English speaking online communities. Results on the Freeport dataset in Table 4 shows that SensorTree is able to detect events irrespective of the language at a finer granularity than comparison models. We highlight the words that match ground truth event descriptions in blue.

5.4.4 Case Studies

We performed a number of case studies on both datasets. Due to space limitations, we discuss some notable events that were detected by SensorTree which

Table 4. Comparing protest event descriptions from various models with news reports

Date	Event headline description	SensorTree	EvenTweet	TLDA
16/02/17	Demanding export mine permit freeport employees stage demonstration at Regent's office. Tuntut Izin Ekspor Karyawan Gelar Aski di Kantor Bupati	tuntut, freeport, halt, export, gelar, kantor, mine, union, Izin, destruction	freeport, googlebox, iphone, scandal, halt, mine, music lovers	freeport, Indonesia, topcharts, ranking, now playing
24/02/17	Demo di Freeport dan ESDM GMNI Minta Demonstration at Freeport and the Ministry of Energy. National Students Movement urges government to be firm	freeport, students energy, demonstration, government, destruction, Indonesia	movement, energy freeport, royalband concert, showdown trending, tickets	student, concert, freeport, energy, crowd, music, tickets, fighters
07/03/17	Berpotensi Bentrok dengan Karyawan Demo anti airport bubar. Clashes with Freeport employees and Anti Freeport Protests	employees, freeport, Karyawan, berpotensi, mine, bubardengan, clash	employees, freeport movie, trailer, mine, reviews, clash, sold-out	employees, mine, freeport, trailer, sold-out, album, clash
13/03/17	PRD Unjuk Rasa di Depan Kementerian Dozens of members of the Peoples Democratic Party (PRD) held a freeport demonstration in front of the Ministry of Energy	PRD,demonstration Ministry, freeport, front,energy, destruction	demonstration, export, union, NBA, playlist, stocks, hit-maker	freeport, energy export, union, Kong, playlist, NBA, hipop

thc othcr comparison models were not able to detect. The news[1] coverages of these events are shown in Fig. 9.

(i) NHB Dataset: We detect a series of events using this dataset. A notable one was the Santos Pilliga coal protest. There were hundreds of protesters on Feb. 21st, 2016 at Santos coal seam gas waste water plant in the Pilliga Forest. The protest event followed earlier protest attempts where 29 people were charged with various trespassing offences. SensorTree detected this protest with event description words such as *Pilligia, coal, csg, protest, Narrabi, arrest, police, etc.*

(ii) Freeport Dataset: SensorTree was able to detect events which were not actual protest events, but sub-events that led to other protest events. On the 19th of Feb. 2017, the Youth of Muhammadiyah placed a wake up call on the Indonesian government not to lose on ongoing Freeport lawsuits. This was not an actual protest event, but was a precursor to a series of protest events some days later. SensorTree accurately detected this sub-event. This added capability of SensorTree in detecting precursors is useful for forecasting future protest events.

Fig. 9. Events detected from Case Studies

6   Conclusion

In this paper, we have presented SensorTree, an event detection framework for protest event detection. SensorTree models information propagation within a community of Twitter users as a sensor to detect a period of burst as events. Once burst is detected, SensorTree infers the details (topics) of the event that triggered the burst using an event topic model. We have performed a set of experiments with real world datasets and compared SensorTree to competing event detection models using various evaluation metrics. The results show that SensorTree outperforms the comparison models. The case studies presented show the capabilities of SensorTree to accurately detect protest events with fine granularity and no language restrictions.

Acknowledgements. We acknowledge Data to Decisions CRC (D2DCRC), Cooperative Research Centres Programme, and the University of South Australia for funding this research. The work has also been partially supported by ARC Discovery project DP170101306.

---

[1] https://goo.gl/LRdwd6 (left image) and https://goo.gl/c5LCZ2 (right image).

References

1. Abdelhaq, H., Sengstock, C., Gertz, M.: EvenTweet: online localized event detection from twitter. PVLDB **6**(12), 1326–1329 (2013)
2. Aggarwal, C.C., Subbian, K.: Event detection in social streams. In: Proceedings of the 2012 SDM, pp. 624–635. SIAM (2012)
3. Anandkumar, A., Ge, R., Hsu, D., Kakade, S.M., Telgarsky, M.: Tensor decompositions for learning latent variable models. J. Mach. Learn. Res. **15**(1), 2773–2832 (2014)
4. Ansah, J., Kang, W., Liu, L., Liu, J., Li, J.: Information propagation trees for protest event prediction. In: Phung, D., Tseng, V.S., Webb, G.I., Ho, B., Ganji, M., Rashidi, L. (eds.) PAKDD 2018. LNCS (LNAI), vol. 10939, pp. 777–789. Springer, Cham (2018). https://doi.org/10.1007/978-3-319-93040-4_61
5. Becker, H., Naaman, M., Gravano, L.: Learning similarity metrics for event identification in social media. In: 3rd ACM WSDM, pp. 291–300. ACM (2010)
6. Becker, H., Naaman, M., Gravano, L.: Beyond trending topics: real-world event identification on twitter. ICWSM **11**(2011), 438–441 (2011)
7. Blei, D.M., Ng, A.Y., Jordan, M.I.: Latent Dirichlet allocation. J. Mach. Learn. Res. **3**(Jan), 993–1022 (2003)
8. Cadena, J., Korkmaz, G., Kuhlman, C.J., Marathe, A., Ramakrishnan, N., Vullikanti, A.: Forecasting social unrest using activity cascades. PloS one **10**(6), e0128879 (2015)
9. Chen, F., Neill, D.B.: Non-parametric scan statistics for event detection and forecasting in heterogeneous social media graphs. In: Proceedings of the 20th ACM SIGKDD, pp. 1166–1175. ACM (2014)
10. He, D., Parker, D.S.: Topic dynamics: an alternative model of bursts in streams of topics. In: Proceedings of 16th ACM SIGKDD, pp. 443–452. ACM (2010)
11. Huang, F., Niranjan, U., Hakeem, M.U., Anandkumar, A.: Online tensor methods for learning latent variable models. J. Mach. Learn. Res. **16**(1), 2797–2835 (2015)
12. Ihler, A., Hutchins, J., Smyth, P.: Adaptive event detection with time-varying poisson processes. In: Proceedings of 12th ACM SIGKDD, pp. 207–216 (2006)
13. Kleinberg, J.: Bursty and hierarchical structure in streams. Data Min. Knowl. Discov. **7**(4), 373–397 (2003)
14. Kontostathis, A., Galitsky, L.M., Pottenger, W.M., Roy, S., Phelps, D.J.: A survey of emerging trend detection in textual data mining. In: Berry, M.W. (ed.) Survey of Text Mining, pp. 185–224. Springer, New York (2004). https://doi.org/10.1007/978-1-4757-4305-0_9
15. Li, J., Wen, J., Tai, Z., Zhang, R., Yu, W.: Bursty event detection from microblog: a distributed and incremental approach. Concurr. Comput. Pract. Exp. **28**(11), 3115–3130 (2016)
16. Sakaki, T., Okazaki, M., Matsuo, Y.: Earthquake shakes twitter users: real-time event detection by social sensors. In: 19th WWW, pp. 851–860. ACM (2010)
17. Weng, J., Lee, B.S.: Event detection in twitter. ICWSM **11**, 401–408 (2011)
18. Xie, W., Zhu, F., Jiang, J., Lim, E.P., Wang, K.: TopicSketch: real-time bursty topic detection from twitter. IEEE TKDE **28**(8), 2216–2229 (2016)

Claim Retrieval in Twitter

Wenjia Ma[1], Wenhan Chao[1], Zhunchen Luo[2(✉)], and Xin Jiang[1]

[1] School of Computer Science and Engineering, Beihang University, Beijing, China
{mawenjia,chaowenhan,xinjiang}@buaa.edu.cn
[2] Information Research Center of Military Science,
PLA Academy of Military Science, Beijing, China
zhunchenluo@gmail.com

Abstract. Controversial topics, especially the new emerging ones are widely discussed and searched in social medias like Twitter. When people are interested in topics and search on Twitter, high quality tweets are expected to appear at the top. Since it is only argumentation that truly reasons things out, we believe that high quality tweets are those with argumentation that consists of claim and evidence. Moreover, claim is the heart of argumentation, we concentrate on claim retrieval in Twitter. Based on a learning-to-rank framework, we integrate Twitter structural information and topic-independent claim-related lexicon to re-rank the relevant tweet list pre-retrieved by BM25 scores. We also automatically construct topic-dependent claim-oriented lexicons to further elevate the retrieval performance. Additionally, our model can be easily adapted to new topics without any manual process or external information, which guarantees the practicability of our model.

Keywords: Claim retrieval · Twitter structural information
Claim-oriented lexicon · Topic adaptable

1 Introduction

Since controversial topics, especially the new ones, are widely discussed in Twitter, the search tool of Twitter is frequently used by people. However, the retrieved tweets which only reflect tweeters' opinions or just general support or oppose these controversial topic are not meaningful enough. Argumentation is known as the most convincing structure, which is often used in law, persuasive essay, and debate domain and has been researched for decades. Among diverse argumentation definitions [3,4,10,16,19], a widespread one is claim and evidence [12]. Due to the short texts, Christian and Iryna [16] point out that argumentation structure is rare, or likely to be incomplete in social media. It means that some tweets may contain only claims, while others may contain only evidences or both claims and evidences. Specifically, the heart of every argumentation lies in a single claim, which is a assertion the argumentation aims to prove [5]. Moreover, only when the claim is confirmed, can the evidences make sense. To help users swiftly obtain many pre-eminent claims about the query topic, there is a pressing need for tools that can automatically retrieve claim-oriented tweets.

© Springer Nature Switzerland AG 2018
H. Hacid et al. (Eds.): WISE 2018, LNCS 11233, pp. 297–307, 2018.
https://doi.org/10.1007/978-3-030-02922-7_20

Table 1. Examples for tweets separately relevant to two topics, "abortion" and "animal testing". "Y" means it contains a claim and "N" means it does not.

Topic: Abortion (should abortion be allowed)	
T1	RT @nelsonhardiman: Supreme Court strikes down Texas abortion restrictions: https://t.co/xsRz8IHiIK#SCOTUS#abortion#SupremeCourt#Texas	N
T2	@patrickmadrid she support abortion I say abortion is murder. Before they were even born	Y
T3	@okeyjames i.e. therapeutic abortion is allowed in Nigeria	N
T4	Like omfg how does someone else getting an abortion affect you in any way. If you're pregnant; want an abortion, get an abortion +	N
Topic: Animal testing (should animal testing be allowed)	
T5	I've just watched a disgusting video about animal testing and tomorrow I'm throwing all my none cruelty free makeup out	Y

Hence, given a topic, our task aims to retrieve a list of claim-oriented tweets. We assume a claim-oriented tweet should meet three criteria: (1) the tweet should be topic-related; (2) the tweet clearly supports or opposes the topic; (3) the tweet provides an arguable reason[1] for its stance. For examples, as shown in Table 1: **T1** is a piece of news which contains no stance; **T2** is clearly against the topic, and contains an explicit disputable reason, *"abortion is murder"*; **T3** is a objective truth which is not in dispute (seems like an evidence); **T4** just has an opposing stance without showing a reason; **T5** contains an implicit claim, *"animal testing used by cosmetics is cruel"*. Consequently, **T2** and **T5** are claim-oriented tweets that we need to retrieve.

Previous studies of predicting whether a document contains claims use supervised learning approaches [5,15], parse tree measures [6], and more recent works concentrating on neural networks [2]. There are two major challenges rendering these approaches not suitable for our task.

Chaotic Twitter. Tweets are short and often contain specific conventions. For instance, in the first sample in Table 1, tweet contains hashtags, URLs, and retweet (RT@), while the textual content are really short. Cleaning these Twitter specific conventions using NLP techniques will cause incomplete sematic of the tweet. Therefore, these chaotic elements in Twitter represent an open challenge for standard claim detection approaches.

Vague Claim. In fact, the majority of online users do not really need to present a well-formed argumentation or their proposition. As a consequence, claims made by the users will often be unclear, ambiguous, vague, or simply poorly worded [17]. For example, people need background knowledge *"cosmetics often use*

---

[1] This is to distinguish from "evidence" or "data" which is essential prerequisite for world knowledge [19].

animal testing" to recognize that **T5** in Table 1 contains an implicit claim "*animal testing used by cosmetics is cruel*", which is clearly challenging.

In this paper, we explore both Twitter structural information and claim-oriented information to address the above issues. Twitter structural information refers to hashtags, URLs, re-tweet (RT@), etc. And the claim-oriented information denotes indicative words whose appearances represents that the tweet is likely to contain claim. First, We utilize a learning-to-rank framework to learn a ranking function that uses both Twitter structural information and topic-independent claim-related information[2] in addition to traditional topic-related information and stance information. And then we elevate the performance by automatically generate topic-dependent claim-oriented lexicons and use them in a lexicon-based approach. Additionally, since the topic-dependent claim-oriented lexicon can be constructed using unlabeled topic-relevant tweets, our model can be easily adapted to new topics which guarantees the practicability of our model.

The contributions of this work can be summarized as follows:

(1) We define a novel claim-oriented tweet retrieval task. We construct a real-world dataset for this task.
(2) Our method integrates both topic-independent and topic-dependent claim-oriented information and achieves portability to all controversial topics.
(3) Experimental results show that best performance of our ranking model is significantly better than baselines.

2   Related Work

The task of automatic claim-oriented document detection was first introduced by Levy et al. [5] who used a supervised learning approach to detect context dependent claims in Wikipedia articles. Lippi and Torroni [6] focused on the rhetoric structure of claims and relied on the ability of Partial Tree Kernels to generate the feature set. More recently, Roitman et al. [15] proposed a two-step retrieval approach to do claim-oriented document retrieval task, and they concentrated on retrieving as many relevant claims as possible from wikipedia corpus. Our experimental results show that claim-oriented document retrieval features do not perform well in Twitter.

Our task shares relationship with argument mining in Twitter or online forum [1,11,18,20]. Theodosis et al. [18] did not distinguish between domain entities and claims, since they thought the claims are not expressed literally. However, in our opinion, both explicit and implicit claims are contained in tweets, and only when the claim is confirmed, can the evidences make sense. Other examples often considered argument as evidence. Addawood and Bashir [1] used a supervised classifier trained with different kinds of features to capture the evidence types in social media. To conclude, none of the work mentioned above concentrated on claim mining in Twitter.

---

[2] Claim-related information refers to words whose appearance can make information gain for detecting whether a tweet contains claim.

Since we define the claim-oriented tweet should contain a clear stance, stance detection in Twitter is also important for our task. Saif et al. [9] proposed a state-of-art stance detection system using a SVM classifier along with distant supervision techniques. We use their features to measure whether there are stances in tweets.

3   Methodology

To generate a good function which ranks the tweets according to our principle for finding claim-oriented tweets, we investigate the features concerning topic relevance, stance existence and arguable reason inclusion of a tweet. In general, we use a learning-to-rank framework to integrate topic-related feature, stance detection features, Twitter structural features and topic-independent claim-related features. To further elevate the retrieval performance, we use a topic-dependent claim-oriented lexicon to score whether each tweet contains arguable reasons.

3.1   Learning to Rank Method

Learning-to-rank is a data driven approach that effectively incorporates a bag of features into the retrieval process. To generate a general model for all kinds of controversial topics, we develop topic-independent features into a learning-to-rank scenario. In the remainder of this section, we will focus on these topic-independent features.

Relevance Feature. We use the Okapi BM25 [14] to measure the relevance between topics and tweets.

Stance Features. Since the claim-oriented tweets need to express a clear stance toward the given controversial topic, we use a feature set *TwitStan* integrated in a state-of-art classifier which is proposed by Saif et al. [9] to address the SemEval-2016 task on stance detection in Twitter. The features used for our method include n-grams, sentiment, target, POS, encodings, and word embeddings trained on large collections of tweets in November 2015 using Glove [13].

Twitter Structural Features. Compared to traditional media data, Twitter has many specific structural information, such as URLs, hashtags, etc. Some of them have been proved to have significant influence on Twitter retrieval [7,8]. However, most argument mining works in Twitter treat tweets as plain texts by removing them [1]. This may lead to the information loss of tweets. To explore the relationship between Twitter structural information and claim-oriented tweets, we use them as binary features.

"*RT @*" indicates copying and rebroadcasting of the original tweet, we assume that persuasive tweets containing clear propositions are more likely to be broadcasted. *URL* indicates the links to out side content. Observationally, advertisements and news that are unlikely to contain a claim in Twitter often contain a URL. Inspired by the assumption that high quality claims arise in debates or quarrels, we use *"reply"* which describes whether this tweet is a comment or a reply.

Topic-Independent Claim-Related Features. Some claim-oriented tweets expressed arguable reasons explicitly, and they often express in general patterns, for instance,

> *(1) @mmfa Abortion is not a choice, abortion is the killing of an innocent life*
> *(2) RT @hailey stiegel: MAKING ABORTION ILLEGAL IS NOT GET-TING RID OF ABORTION, IT IS GETTING RID OF SAFE ABOR-TION*

"`A is not B, it is C`" pattern appears in these explicit claim-oriented tweets. In order to capture these claim-oriented patterns, which involve be verbs, modal verb, we utilize an information gain based method to calculate the claim score of each word.

Table 2. Table for information gain. $C_{1*} = C_{11} + C_{12}$; $C_{2*} = C_{21} + C_{22}$; $C_{*1} = C_{11} + C_{21}$; $C_{*2} = C_{12} + C_{22}$; $C = C_{11} + C_{12} + C_{21} + C_{22}$.

	t	$\neg t$	Row total
Claim_Oriented. set	C_{11}	C_{12}	C_{1*}
Non_Claim. set	C_{21}	C_{22}	C_{2*}
Col. total	C_{*1}	C_{*2}	C

C_{ij} in Table 2 indicates the number of tweets having/not-having term t in the claim-oriented/non-claim set respectively. For example, C_{11} is the number of claim-oriented tweets which contain term t. Then, we give definitions of some concepts: $H(X)$ is the entropy of X. For each topic, the total claim entropy is called $H(C) = -\sum_{i=1}^{2} p_{i*} \log_2 p_{i*}$, where $p_{i*} = \frac{C_{i*}}{C}$ is the probability of the C_{i*}. For each term t, we compute the entropy of claim on the term t $H(C|t)$ as follows:

$$H(C|t) = -p_t \sum_{i=1}^{2} p(C_i|t) \log_2 p(C_i|t) - p_{(\neg t)} \sum_{i=1}^{2} p(C_i|\neg t) \log_2 p(C_i|\neg t) \quad (1)$$

$IG(C, t) = H(C) - H(C|t)$ calculates the information gain about claim of term t. The number of claim-oriented tweets varies from topics. For example, there are 40 tweets containing claims in topic **"abortion"**, but only 2 tweets contain claims on topic **"Trump"**. Therefore, tweets about topic **"abortion"** are more likely to contain claims. If term scores are calculated without considering the topic, insignificant topic words will score higher and be seen as claim-oriented words. For instance, "abortion", "woman" (high frequency words on topic **"abortion"**) etc. To avoid this situation, term scores are calculated separately according to topics. For each term t, we use $H(t|K) = \sum_{i=1}^{n} p_{k_i} H(t|K = k_i)$ to represent t's distribution under the topic set K.

If term t is a topic-independent claim indicator, it should be evenly distributed under various topics. And this situation will cause $H(t|K)$ to increase. Therefore, t's score $Claim_{TI}(t)$ which used to indicate claim relatedness is calculated as follows:

$$Claim_{TI}(t) = \sum_{k \in K} \frac{IG_k(C,t) \cdot H(t|K)}{TN_k} \tag{2}$$

where TN_k is the number of tweets about topic k. The highest score terms are selected to form the **Topic-Independent Claim-Related Lexicon** *TICRLex* and will be used as topic-independent claim-related features.

3.2   Lexicon Method

Some arguable reasons in claim-oriented tweets are expressed implicitly. For instance, there are 2 tweets of topic **"death penalty"**:

(1) @mmellmmar because death penalty treats you better if you are rich and guilty than if you are poor and innocent..
(2) Death penalty should not exist, esp because it is against those who are poor.#deathpenalty

They expressed the claim that *"the death penalty for the poor and the rich is different"*, which requires background knowledge to identify. We find that these implicit claim-oriented tweets often contain some topic-dependent words, like "poor", "rich" with topic **"death penalty"**. To capture these words, we develop a approach to automatically generate topic-dependent claim-oriented lexicons using unlabeled topic-related tweets.. Additionally, since it is impossible to train a supervised model for every topic, we use topic-dependent claim-oriented lexicons in a lexicon-based method. We estimate the claim-oriented score of each tweet by calculating the average claim-oriented score over certain terms.

Topic-Dependent Claim-Oriented Lexicon. We suppose that if term t often appear with topic-independent claim-oriented words simultaneously, then term t is likely to be a claim-oriented word. In the above two examples, we suppose that term "because" is a topic-independent claim-oriented word. The term "poor" appear with "because" twice in these two tweets. Since we suppose that topic-dependent claim-oriented and topic-independent claim-oriented words are often united, term "poor" can be seen as a claim-oriented word of topic **"death penalty"**.

First, suppose we have already got the topic-independent claim-related lexicon *TICRLex*. To distinguish claim-oriented terms in the claim-related lexicon, we introduce a signal function $Sgn(t)$ for each term t:

$$Sgn(t) = \begin{cases} -1 & \frac{C_{11}}{C_{*1}} \leq \frac{C_{1*}}{C} \\ 1 & \frac{C_{11}}{C_{*1}} > \frac{C_{1*}}{C} \end{cases} \tag{3}$$

$Claim_{TI}(t)$ is the term t's claim score in **TICRLex**. Then we compute the new score $Claim_{TI}(t)^+ = Claim_{TI}(t) \cdot Sgn(t)$ of each term t in **TICRLex**. If $Claim_{TI}(t)^+ > 0$, means term t is **positively** related to claim, we add t to a new **Lexicon** called **posLex**.

$CoT(w_i, t)$ represents the co-occurrence frequency of term t in topic-related tweet set TS with the term w_i in **posLex**. TN_t is the number of tweets containing term t. t's topic-dependent claim-oriented score $Claim_{TD}(t)$ is then defined as the weighted sum of $CoT(w_i, t)$:

$$Claim_{TD}(t) = \sum_{w_i \in posLex} \frac{Claim_{TI}(w_i)^+ \cdot CoT(w_i, t)}{TN_t} \tag{4}$$

The highest score terms are selected to form the **Topic-Dependent Claim-Oriented Lexicon** **TDCOLex**.

4 Experiments

4.1 Datasets

We construct a real-world dataset for our claim-oriented tweet retrieval task[3]. We crawled and indexed about 90 million tweets using the Twitter API in 2016 and reserve the English tweets. Using these tweets we implemented a search engine based on ElasticSearch[4]. We collected 30 debate topics from debate website[5] as the queries. Given a query the search engine would present a list of relevant tweets ranked based on the Okapi BM25 [14] score. A native English speaker and two experienced annotators with NLP background were hired to identify whether the tweet contains a claim following the criteria we proposed (in Sect. 1) by assigning binary labels to every tweet. The inter-annotator agreement was 90.1% for topic-relevance, 78.2% for clear stance and 75.2% for arguable reason[6]. The high consistency of the annotation proves our claim-oriented criteria are easy to convey to human labelers. We marked an instance with a claim only if at least 2 annotators labeled them as containing claim. Totally, 2520 tweets were selected for study and 586 tweets were identified as containing claims.

4.2 Experimental Settings

For learning to rank, SVM light[7] which implements the ranking algorithm is used. To avoid overfitting, we perform 10 fold cross-validation in our dataset.

---

[3] https://sourceforge.net/projects/claimretrieval/files/corpus/download.

[4] https://www.elastic.co/products/elasticsearch.

[5] www.procon.org.

[6] The overall inter-annotator agreement was calculated by averaging the agreements on all tweets in the dataset. For each tweet, the inter-annotator agreement was calculated as the number of annotators who agree over the majority label divided by the total number of annotators for that tweet.

[7] http://www.cs.cornell.edu/people/tj/svm_light/svm_rank.html.

We use Mean Average Precision (MAP), Precision@5, and Precision@10 as evaluation metrics.

4.3   Baselines

We investigate the features used by previous similar tasks, and separately develop these bags of features into a learning-to-rank scenario as our baselines.

BM25 Similarity. We use BM25 similarity as a basic measure. The Okapi BM25 scoring shows the relevance between query topic and the tweet.

TwitStan. TwitStan is a feature set used in a state-of-art stance classifier for tweets [9]. We combine the BM25 as the relevance feature.

WikiClaim. WikiClaim is a claim-discovery feature list from Roitman et al. [15]. Considering tweets do not have title or headers, we only use the content features. We combine the BM25 as the relevance feature.

TwitArgument. Since claim and evidence are all argumentative components, we also use TwitArgument which is a feature set used by argument identification tasks in Twitter [18]. We combine the BM25 as the relevance feature.

4.4   Results

Experiment I: Baselines. Table 3 gives the performance of the baselines. Due to the particularity of corpus, $LTR_{WikiClaim}$ which is effective on Wikipedia corpus do not perform well. The results also show that $LTR_{TwitArgument}$ is much worse than $LTR_{TwitStan}$. Because argument mining in Twitter tends to find different

Table 3. Results for baselines. A significant improvement over the $BM25$ with $^{\triangle}$ and $^{\blacktriangle}$ (for p < 0.05 and p < 0.01).

id	Baselines	MAP	P@5	P@10
1	$BM25$	0.299	0.253	0.260
2	$LTR_{TwitStan}$	**0.500**$^{\blacktriangle}$	**0.513**$^{\blacktriangle}$	**0.436**$^{\blacktriangle}$
3	$LTR_{WikiClaim}$	0.291	0.280	0.283
4	$LTR_{TwitArgument}$	0.328$^{\triangle}$	0.313	0.336$^{\triangle}$

types of evidence, which is usually described objectively and it is difficult to see the stance of tweeter. However, the claim needs the tweeter to clearly express his stance. So our following experiment is on the basis of $LTR_{TwitStan}$.

Experiment II: Topic-Independent Features. The first column of Table 4 presents the effect of using Twitter structural features and topic-independent claim-related features. Each feature is combined with the $LTR_{TwitStan}$ and evaluated separately. Among these Twitter features, **re-tweet** ("RT @"), **reply**, **structure** (re-tweet+URLs+reply) intuitively perform better than others, which serve as useful proofs to conceive that some Twitter specific features really have correlation with claims. The improvement of ranking result using **re-tweet** feature is very possible because of the high forward frequency of valuable claim. As for the **reply**, it is probably because the argumentation always occurs during the discuss or quarrel. Besides, some features' combination may greatly improve the performance. For example, News in Twitter presents a specific structure as it contains both **re-tweet** and **URLs**, and it rarely contains a claim. For comparison, we use a controversy lexicon (CL) that has been proved useful for document claim-oriented retrieval [15]. However, the 7th case in Table 4 shows that CL is not very effective in Twitter. This may be because the text of tweets is different from documents.

Experiment III: Topic-Dependent Lexicon. Table 5 gives claim-related terms in the *TICRLex* and the claim-oriented terms in *TDCOLex* of topic "abortion". Apparently, the terms in *TICRLex* are some modal verbs, linking verbs, conjunction, negative words and punctuation which often do not have an exact meaning but are used to form a sentence pattern. However, words in *TDCOLex* tend to be content words. For example, when it comes to **Abortion**, "rights", "murder", "control" are included. Part of the reason can be that abortion supporters often think that abortion is part of women rights, while *"abortion is murder"*, *"abortion is not birth control"* are claims widely accepted by opponents. The 8th case in Table 4 shows that topic-dependent lexicons provide further boost to a model on the basis of topic-independent features. It shows that our lexicon does capture important topic-dependent claim-oriented information.

Finally, both effective topic-independent and topic-dependent elements including BM25, features in *TwitStan*, Re-tweet, Reply, Urls, *TICRLex*(best), *TDCOLex*(best) have been added to build our best model $LTR_{TI}+[TD]$ which improved the MAP by 95.7% compared with solely BM25, and 17% compared with $LTR_{TwitStan}$.

Table 4. Experiment results (structure:re-tweet+URLs+reply, TI: structure + *TICRLex*, TD: *TDCOLex*). A significant improvement over the $LTR_{TwitStan}$ with $\triangle$ and ▲ (for p < 0.05 and p < 0.01).

id	Twitter features	MAP	P@5	P@10	id	Claim-oriented lexicons	MAP	P@5	P@10
1	$LTR_{TwitStan}$	0.500	0.513	0.436	1	$LTR_{TwitStan}$	0.500	0.513	0.436
2	+re−tweet	0.557▲	0.513	0.436	9	+ *[TDCOLex]*	0.542△	0.520▲	0.443△
3	+URLs	0.530△	0.526△	0.446△	10	$LTR_{TI}+[TD]$	**0.585▲**	**0.533▲**	**0.480▲**
4	+reply	0.536▲	0.531▲	0.473▲					
5	+structure	0.550▲	**0.540▲**	**0.480▲**					
6	+TICRLex	0.533▲	0.533▲	0.450▲					
7	+CL	0.514	0.513	0.436					
8	LTR_{TI}	**0.558▲**	0.532▲	0.450▲					

Table 5. Comparison of the claim terms in *TICRLex* and *TDCOLex* of topic "abortion".

TICRLex	:, is, will, a, ,, ..., if, were, more, and, in, are, who, even, be, have, ?, they, ;, would, you, this, but, all, on, we, no, want, than, that, !, because, those, thus, was
TDCOLex	murder, cheerleader, supported, failed, dangerous, excuses, LGBTQ, stop, healthyLiving, rights, control, catholic, proabortion

5   Conclusion and Future Work

We define a novel claim-oriented tweet retrieval task which will be certainly helpful in the development of public opinion research. We utilize the Twitter structural information to deal with the chaotic Twitter problem, and leverage claim-oriented lexicons to solve the vague claim problem. The topic-dependent claim-oriented lexicon can be generated using a large number of unlabeled topic-related tweets. Hence, our model can be easily adapted to new emerging topics in Twitter. We construct a real-world dataset. The best performance of our model improves the MAP by 95.7% compared with $BM25$ baseline, and 17% compared with $LTR_{TwitStan}$ baseline.

The main future work is threefold: first, we plan to use our automatic method to get an extended corpus and leverage deep learning techniques to learn more claim-oriented features. Second, we will diversify the searched claims and detect the relevant evidence of the known claim to generate a complete argumentation structure in Twitter. Third, we will study how to assess the quality of a claim.

Acknowledgments. We appreciate the comments from anonymous reviewers. This work is supported by National Key Research and Development Program of China (Grant No. 2017YFB1402400) and National Natural Science Foundation of China (No. 61602490).

References

1. Addawood, A., Bashir, M.: "What is your evidence?" A study of controversial topics on social media. In: The Workshop on Argument Mining, pp. 1–11 (2016)
2. Eger, S., Daxenberger, J., Gurevych, I.: Neural end-to-end learning for computational argumentation mining. In: ACL (2017)
3. Freeley, A., Steinberg, D.: Argumentation and debate (2008)
4. Habernal, I., Gurevych, I.: Argumentation Mining in User-Generated Web Discourse. MIT Press, Cambridge (2016)
5. Levy, R., Bilu, Y., Hershcovich, D., Aharoni, E., Slonim, N.: Context dependent claim detection. In: COLING, pp. 1489–1500 (2014)
6. Lippi, M., Torroni, P.: Context-independent claim detection for argument mining. In: International Conference on Artificial Intelligence, pp. 185–191 (2015)

7. Luo, Z., Osborne, M., Petrovi, S., Wang, T.: Improving Twitter retrieval by exploiting structural information. In: Twenty-Sixth AAAI Conference on Artificial Intelligence (2012)
8. Luo, Z., Osborne, M., Wang, T.: An effective approach to tweets opinion retrieval. World Wide Web **18**(3), 545–566 (2015). https://doi.org/10.1007/s11280-013-0268-7
9. Saif, M., Parinaz, S., Svetlana, K.: Stance and sentiment in tweets. ACM Trans. Internet Techn. **17**(3), 26:1–26:23 (2017). https://doi.org/10.1145/3003433
10. Ma, W., Chao, W., Luo, Z., Jiang, X.: CRST: a claim retrieval system in Twitter. In: COLING (2018)
11. Mihai, D., Elena, C., Serena, V.: Argument mining on Twitter: arguments, facts and sources. In: EMNLP, pp. 2307–2312 (2017)
12. Palau, R., Moens, M.: Argumentation mining: the detection, classification and structure of arguments in text. In: International Conference on Artificial Intelligence and Law, pp. 98–107 (2009)
13. Pennington, J., Socher, R., Manning, C.: Glove: global vectors for word representation. In: Proceedings of the 2014 Conference on Empirical Methods in Natural Language Processing (EMNLP), pp. 1532–1543 (2014)
14. Robertson, S., Walker, S., Hancock-Beaulieu, M., Gull, A., Lau, M.: Okapi at TREC. In: Text Retrieval Conference, pp. 21–30 (1992)
15. Roitman, H., Hummel, S., Rabinovich, E., Sznajder, B., Slonim, N., Aharoni, E.: On the retrieval of Wikipedia articles containing claims on controversial topics. In: International Conference Companion on World Wide Web, pp. 991–996 (2016)
16. Christian, S., Iryna, G.: Parsing argumentation structures in persuasive essays. CoRR, abs/1604.07370 (2016)
17. Jan, S.: Social media argumentation mining: the quest for deliberateness in raucousness. CoRR, abs/1701.00168 (2017)
18. Theodosis, G., Christos, L., Georgios, P., Vangelis, K.: Argument extraction from news, blogs, and social media. Int. J. Artif. Intell. Tools 287–299 (2015)
19. Toulmin, S.: The uses of argument. Ethics **10**(1), 251–252 (1958)
20. Wei, Z., Liu, Y., Li, Y.: Is this post persuasive? Ranking argumentative comments in online forum. In: Meeting of the Association for Computational Linguistics, pp. 195–200 (2016)

PUB: Product Recommendation with Users' Buying Intents on Microblogs

Xiaoxuan Ren[1,2(✉)], Tianshu Lyu[1,2], and Yan Zhang[1,2]

[1] Department of Machine Intelligence, Peking University, Beijing 100871, China
{renxiaoxuan,lyutianshu,zhy.cis}@pku.edu.cn
[2] Key Laboratory of Machine Perception (MOE), Beijing 100871, China

Abstract. Recommendation systems mostly rely on users' purchase records. However, they may suffer problems like "cold-start" because of the lack of users' profiles and products' demographic information. In this paper, we develop a method called PUB, which detects users' buying intents from their own tweets, considers their needs, and extracts their demographic information from their public profiles. We then recommend products for users by constructing a heterogeneous information network including users, products, and attributes of both. In particular, we consider users' shopping psychology, and recommend products that better meet their needs. We conduct extensive experiments on both direct intent recommendation and additional product recommendation. We also figure out users' potential preference which can help to recommend a great varied types of products.

Keywords: Product recommendation · Buying intent detection
Heterogeneous information network embedding

1 Introduction

Most existing product recommendation systems use users' historical transaction records or interactions like rating or clicking, and mostly rely on techniques like collaborative filtering [3,9,15]. However, the mere collaborative filtering technique may suffer from the "cold-start" problem for the lack of information of a new user. Furthermore, collaborative filtering cannot precisely discover users' real needs either, especially the ad hoc ones. Actually a lot of information on OSN platforms (e.g. the tweets) contain users' buying intents [11], which can help to fill up the entrenchments.

In our work, we make a bridge between the e-commerce websites and social networks and develop a method, PUB, Product recommendation with Users' Buying intent. Figure 1 is the framework of PUB. PUB identifies users' buying intents, needs for the products and product adopters by a bootstrap based method from Sina Weibo[1], the Chinese largest OSN platform. Then PUB selects

---

[1] http://weibo.com.

© Springer Nature Switzerland AG 2018
H. Hacid et al. (Eds.): WISE 2018, LNCS 11233, pp. 308–318, 2018.
https://doi.org/10.1007/978-3-030-02922-7_21

the most appropriate product for the specific user from JD.com[2], one of the largest B2C e-commerce websites in China. We also develop a heterogeneous information network with nodes of users, products and attributes. Finally, PUB recommends products for users which fit their needs and also some other kinds of products that the users may like.

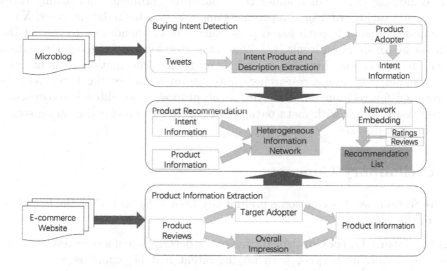

Fig. 1. Framework of PUB.

The main contributions of our method are as follows: (1) We accurately detect buying intents from users' real-time tweets, and extract users' needs and demographic information. (2) We develop a heterogeneous information network to link together e-commerce websites and social networks, and also consider users' shopping psychology to meet the users' demand. (3) We conduct extensive experiments on both direct product recommendation and additional product recommendation, and the results show that our method outperforms other baselines on extracting users' intents as well as recommending products.

2   Related Work

Social Network Extraction. Kröll *et al.* introduce the idea of intent analysis, and evaluate intent profiles from speeches [5]. Hollerit *et al.* detect commercial intent in tweets [4], and their work is considered a first step to bring together buyers and sellers. The most related work is from Wang *et al.* [11], in which they identify users' intents and classify them into six categories. However, they fail to extract users' needs and don't recommend products for the users at last. Meanwhile, there exist some works on extracting users' demographic characteristics.

---

[2] http://jd.com.

Bachrach *et al.* study how users' public profiles on Facebook relate to their personalities [1]. There are also studies on extracting users' public profiles in Weibo and using these information for product recommendation [17], but they fail to extract the potential preference of users.

Product Recommendation. Most recommendation systems rely on collaborative filtering [3,9], which suffer the "cold-start" problem, and mainly takes into account ratings of users on products and users with similar interests. Xu *et al.* propose a semantic path based personalized recommendation to predict the rating scores of users on items [13]. Zhao *et al.* develop a novel recommendation system based on matching the users' demographic information with product demographics [17]. Some researchers have begun to notice the heterogeneous information for recommendation. Shi *et al.* propose a weighted heterogeneous information network with meta path called SemRec to predict the rating scores of users on items [10].

3   Preliminary Concepts

In this Section, we describe some preliminary knowledge as well as the notations used in this paper.

Buying Intent Detection. According to the description of a commercial intent tweet [4], we define a tweet with buying intent if it (1) contains at least one verb and (2) explicitly describes the intention to buy a certain product (3) in a recognizable way. Here is an example:

Please recommend a bright light lipstick, my girlfriend wants to buy one.

The tweet above contains buying intent which satisfies all the three conditions in the definition. However, we can figure out that the **product adopter** is not the user but the user's girlfriend. Inspired by the definition on intent-indicator and intent-keyword [11], we define intent-indicator, intent-product and intent-description as below. **Intent-indicator** is a verb phrase or infinitive phrase that express the users' intent to buy something. "Lipstick" in the example is an **intent-product**, which is a noun in most cases, and is the product that the user wants to buy. **Intent-description** is an adjective or noun between the intent-indicator and intent-product, or appearing alone.

User and Product Demographics. User demographics describe the intent buyers' characteristics. In our work, we only consider the attributes of age and gender, for users are not willing to provide other attributes in most cases. The gender is classified into male and female, while the age is grouped into eight clusters: 0–3, 4–6, 7–17, 18–29, 30–40, 41–55, 56–64, 65+. The product demographics are extracted from product reviews, which describe the characteristics of the buyers of the product.

Heterogeneous Information Network. Heterogeneous information network have been proposed as a general data representation for many different types of data [16]. A heterogeneous information network has different types of objects and different links representing different relations, which can better fit in most scenarios, especially in product recommendation. Similar to [14], we define heterogeneous information network as follows:

Let $G = (V, E)$ denotes a graph with an object type mapping function ϕ : $V \to A$ and a link type mapping function $\psi : E \to R (|A| > 1$ or $|R| > 1)$. Each object $v \in V$ belongs to one object type $\phi(v) \in A$, and each link $e \in E$ belongs to a relation type $\psi(e) \in R$.

4 Methodology

4.1 Buying Intent Detection and Users' Demographic Information Extraction

Let u and $U = \{u_t\}_{t=1}^{M}$ denote a user and the whole user set. Similarly, let p and $P = \{p_t\}_{t=1}^{N}$ denote a product and the whole product set. To detect $u's$ buying intents on Weibo, we first filter out irrelevant tweets with a list of key words, then we use a bootstrap based method proposed in Algorithm 1.

First of all, we input a microblog tweet sentence corpus T, and a seed set of intent-indicators, such as "want to buy". The output is an extension of the intent-indicators set $\mathcal{I}$ and intent-products set $\mathcal{P}$, as well as the set of intent-descriptions $\mathcal{D}$. The function **ExtractIntentProduct**(i, t) and **ExtractIntentDescription**(i, t) aim to extract users' intent-products and intent-descriptions according to the intent-indicators. We consider the positions of the intent-products may be the preceding n_1 tokens of the intent-indicators, the middle n_2 tokens or the following n_3 tokens. In order to better obtain users' preference, we also extract the intent-descriptions, which often appear before the intent-products or alone.

With the intent-products that frequently co-occur with the intent-indicators, we in turn use these intent-products to find more intent-indicators in function **ExtractIntentIndicator**(i, t). We repeat the previous steps, until no more intent-indicators or intent-products can be found.

The product adopters is extracted similar with buying intents extraction. Then we extract users' demographic characteristics from their public profiles.

The product adopter, needs for the product and $u's$ demographic characteristics are all regarded as the $u's$ attributes, denote as a_u and $A_u = \{a_t^u\}_{t=1}^{L}$.

4.2 Products Demographic Information Extraction

We use the online products reviews from JD.com to extract products demographic information. First of all, we find the target adopter of product p through the bootstrap based method mentioned in Sect. 4.1, and infer the age and gender. After that, we learn the buyers' overall impression of the products. To achieve it, all the reviews of product p are merged into a single document. We segment Chinese streams into words and extract the keywords from the reviews of one product with TextRank [6].

Algorithm 1. Bootstrap based algorithm for buying intent detection.

Input: microblog tweet sentence corpus T; seed intent-indicator patterns
Output: an extension set of intent-indicator patterns $\mathcal{I}$; a set of intent-products $\mathcal{P}$; a
    set of intent-descriptions $\mathcal{D}$

1: $\mathcal{I} \leftarrow$ seed intent-indicator patterns
2: $\mathcal{I}' \leftarrow$ seed intent-indicator patterns
3: $\mathcal{P} \leftarrow \varnothing$
4: $\mathcal{D} \leftarrow \varnothing$
5: **repeat**
6:     $\mathcal{P}' \leftarrow \varnothing$
7:     $\mathcal{D}' \leftarrow \varnothing$
8:     **for** each pattern $i \in \mathcal{I}'$ **do**
9:         **for** each sentence $t \in T$ **do**
10:           **if** i exists in t **then**
11:             $\mathcal{P}' \leftarrow \mathcal{P}' \cup$ ExtractIntentProduct(i, t);
12:             $\mathcal{D}' \leftarrow \mathcal{D}' \cup$ ExtractIntentDescription(i, t);
13:           **end if**
14:         **end for**
15:     **end for**
16:     $\mathcal{I}' \leftarrow \varnothing$
17:     **for** each product $p \in \mathcal{P}'$ **do**
18:         **for** each sentence $t \in T$ **do**
19:           $\mathcal{I}' \leftarrow \mathcal{I}' \cup$ ExtractIntentIndicator(i, t);
20:         **end for**
21:     **end for**
22:     $\mathcal{I}' \leftarrow$ ExtractTopFrenquentIndicator$(\mathcal{I}')$;
23:     $\mathcal{I} \leftarrow \mathcal{I} \cup \mathcal{I}'$
24:     $\mathcal{P} \leftarrow \mathcal{P} \cup \mathcal{P}'$
25:     $\mathcal{D} \leftarrow \mathcal{D} \cup \mathcal{D}'$
26: **until** no new pattern is identified;
27: **return** an extension set of intent-indicator patterns $\mathcal{I}$;a set of intent-products $\mathcal{P}$;a
    set of intent-descriptions $\mathcal{D}$

Let $G = (V, E)$ be a graph with the set of vertices V and set of edges E. Let $In(V_i)$ be the set of vertices that point to vertex V_i, and $Out(V_i)$ be the set of vertices that V_i points to. Taking into account the edge weights, the score associated with a vertex in the graph can be defined as:

$$WS(V_i) = (1 - d) + d \sum_{V_j \in In(V_i)} \frac{w_{ji}}{\sum_{V_k \in Out(V_j)} w_{jk}} WS(V_j) \qquad (1)$$

Where d is a damping factor that can be set between 0 and 1, w_{ij} is the weight between two edges.

We filter the words with Part-of-Speech (POS) including noun, adjective and descriptive. We also add some stop words. Let the words be the vertices of the graph, and build the edges with $co - occurrence$ relation: if two vertices' corresponding lexical units co-occur within a window of maximum N words, where N can be set anywhere from 2 to 10 words, then the two vertices are connected. We calculate the score for each word, and collapse the sequences of adjacent keywords into a multi-word keyword. At last, we export the top-10 words as the impression keywords.

The impression keywords, target adopter with age and gender are seen as the product $p's$ attributes, denote as a_p and $A_p = \{a_t^p\}_{t=1}^R$.

4.3   Graph Embedding for Product Recommendation

We create a heterogeneous information network consists of users, products, and their attributes. An illustration of our product recommendation in heterogeneous information network is shown in Fig. 2. What needs to be pointed out is that the attributes of users and products often overlap, then we let $A = A_u \cup A_p$ to denote the whole attribute set.

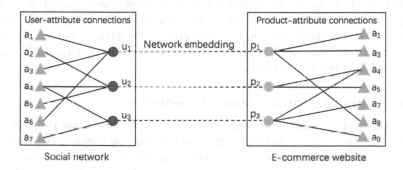

Fig. 2. A heterogeneous information network.

In our work, we use Deepwalk [7], which learns a latent representation of adjacency matrices using deep learning techniques developed for language learning. In order to better satisfy the users' needs, we also merge the words with similar meaning, like "dark" and "black". At last, we predict which products should be recommended to the user by calculating the cosine similarity between the given user node and the products. Then the input and output of the network are defined as:

Input: A social network domain with $\{U, A_u\}$; an e-commerce domain with $\{P, A_p\}$; and $A = A_u \cup A_p$.

Output: A personalized recommendation list for each user u with products by ranking the similarity.

We also consider the ratings and the numbers of reviews, because according to a new study from Psychological Science [8], people prefer more-reviewed items (even with poor ratings). We calculate the users' preference for a given product and denote it as Pr. The equation for calculating the preferred value between the user and the product is as follows:

$$Pr = Sim \cdot (\alpha \log_m(\#CM_p + 1) + \beta R_p) \tag{2}$$

Where $\#CM_p$ is the number of the reviews of the product p, R_p is the rating for the product (range from 0 to 5), α and β are weights of every term. The log base m can be calculated as below:

$$m = \mathbb{E}(\#CM)^{\frac{1}{\mathbb{E}(R)}} \tag{3}$$

where $\mathbb{E}(\#CM)$ is the expectation of the number of the reviews of the products, and $\mathbb{E}(R)$ is the expectation of ratings.

With a heterogeneous information network, we can not only recommend direct products that the users want (with the category label that the users' intents belong to), but also recommend additional products with all categories.

5   Experiment

5.1   Experimental Setup

We crawl tweets from Weibo which contains at least one keyword expressing buying intent, like "want to buy" or "please recommend", and get 6,095 tweets ultimately. We also crawl the users' demographic information of age and gender. We use the dataset from [12] which has 246,444 products and 12,127,267 users with 138,905,740 records of reviews.

5.2   Evaluation on Buying Intent Detection

Because there is no existing ground truth, we invite two annotators who are familiar with the language habits on microblogs, and ask them to label whether the tweets contain buying intent with the description in Sect. 3. If the two annotators have different ideas on one tweet, then we abandon this tweet. At last, we obtain 1,090 tweets with buying intents, and 4,867 tweets without buying intents. The Cohen's Kappa agreement coefficient between annotators is 93.5%, which is really satisfied.

We consider the following comparison methods in our experiments for the buying intent detection:

Wang's: Proposed in [11], which only consider the intent-products just following the intent-indicators with a bootstrap based method, and fail to consider the intent-descriptions.

SVM: We train the Support Vector Machine (SVM) with linear kernel, and use the bag-of-words of the tweets as the input characteristics for classifier.

Ours: Our proposed method for buying intent detection, which considers the intent-indicators, different possible positions for intent-products, and can extract users' needs for the products at the same time.

The results are shown in Table 1. From the results we can tell that our method outperforms the two baseline methods in terms of precision, recall and F_1 Measure. In particular, the improvement of recall is significant compared with Wang's method by 19%. We may conclude that considering the positions of intent-products and the existence of intent-indicators, we can extract more tweets with buying intents.

Table 1. Performance comparison on buying intent detection (%).

Methods	Prec.	Rec.	F_1
Wang's	74.07	27.03	39.60
SVM	70.45	41.89	52.54
Ours	82.93	45.95	59.14

5.3   Evaluation on Product Recommendation

We choose three frequently mentioned intent product types in the Weibo dataset, camera, lipstick, and foundation (makeup). The statistics of the three products are summarized below in Table 2.

Table 2. The statistics of the three given products.

Types	#brands	#products	#comments	#intents
Carema	41	2,341	623,081	30
Lipstick	89	476	432,326	74
Foundation	126	606	258,096	40

Then we find the target adopter, and learn the overall impression every product. Here is an example of the overall impression we learn from the reviews of a Bobbi Brown lipstick shown in Table 3.

Table 3. The overall impression of a lipstick.

Hydrated, wife, orange, light, warm colour, good quality, reasonable price, anthentic, pleasing colour, matte

However, the overall impression from JD.com of the same product is only one word "not bad", even with 62 reviews. We may infer that our learned overall impression are more suitable for the recommendation system, as it better conclude the image of the given product.

We invite five annotators to evaluate which ways recommended products better fit the intent needs. We give each annotator the needs for the product, the user's age and gender, the product's demographic information, the product's overall impression and the rating as well as the number of reviews. We also break the order of the three methods for the three products.

We consider three methods for product recommendation in our experiments:

MART [2]: A pointwise learning-to-rank approach. In this approach we only consider the profiles of age and gender extracted based on a given user and a candidate product.

PUB$_{Sim}$: This is a graph embedding method but only consider the similarity between the given user node and the products.

PUB$_{All}$: This is the proposed method which not only consider the similarity, but the ratings and the numbers of reviews as well.

Table 4. Performance comparison for chosen the most suitable product for the given intent from five annotators.

Annotators	Camera			Lipstick			Foundation		
	MART	PUB$_{Sim}$	PUB$_{All}$	MART	PUB$_{Sim}$	PUB$_{All}$	MART	PUB$_{Sim}$	PUB$_{All}$
A	6	9	15	21	22	31	9	10	21
B	10	7	13	16	22	36	11	10	19
C	10	8	12	19	27	28	13	12	15
D	6	8	16	25	18	31	6	11	23
E	12	10	8	18	24	32	13	10	17
Avg	8.8	8.4	12.8	19.8	22.6	31.6	10.4	10.6	19

From the results in Table 4 we can see that the products recommended from our final method outperforms others. We may conclude that the using of graph embedding can extract the users' latent preference. However, the learning to rank method only consider the relation between the query (buying intent tweet) and the relevant document (an adopted product).

5.4 A Case Study on Direct Product Recommendation and Additional Product Recommendation

Direct Product Recommendation. Firstly, we give an example on direct product recommendation, which means recommend the kind of products that the users want. We use the buying intent described in Sect. 5.2, that a girl wants to buy an orange lipstick, with the age between 18–29 and the gender as female. The recommended products' demographic information of the three methods are shown in Table 5.

Table 5. The products' demographic information recommended by three methods.

Information	MART	PUB$_{Sim}$	PUB$_{All}$
Age	18–29	18–29	18–29
Gender	Female	Female	Female
#reviews	39	15	784
Rating$_{avg.}$	4.74	4.87	4.65
Impression	Hydrated, girlfriend, cheap, light	Hydrated, orange, light, good quality, matte	Hydrated, orange, cheap, light, pleasing color, authentic

The product recommended by MART only satisfies the age and gender, while the products recommended by two proposed methods seems to be similar, but the $Ours_{All}$ method's product have more reviews and lowest rating.

Additional Product Recommendation. After the experiments mentioned above, we then build a heterogeneous information network with the whole 1,095 buying intents and 246,444 products, and try to figure out what will recommend to the user with a settled intent.

Table 6. The other product recommended for the girl demand for orange lipstick.

Information	Bag	Ear phone	Luggage
Age	18–29	18–29	18–29
Gender	Female	Female	Female
#reviews	934	592	845
Rating$_{avg.}$	4.28	4.57	4.32
Impression	Cheap, orange, able to hold things, portable	Fashion, good-looking, orange, good sounds	Durable, size suitable, light, orange

In Table 6, we can see that the method recommends some other products for the girl which are suitable for young ladies, and satisfy her own preference. This is quite useful in the real world, for it extracts the latent rations between different types of products, recommends the products that others in the same conditions may like, and people may want to buy a bag which meet her needs even though what she really wants at that time is a lipstick.

6   Conclusion and Future Work

In this paper, we develop a method called PUB, which can detect users' buying intents and profiles from microblogs, and recommend products from e-commerce websites. We use a graph embedding method, and construct a heterogeneous information network. We conduct experiments on both buying intent detection and product recommendation. We do the recommendation job on both direct product and additional product recommendation, to figure out the potential preference. As a future work, we want to incorporate more features considered into our method, especially those implicit features. We also plan to add sentiment analysis for the product reviews or users' tweets. The graph embedding method can be improved and optimized as well.

Acknowledgment. This work is supported by 973 Program under Grant No. 2014CB340405, NSFC under Grant No. 61532001, and MOE-ChinaMobile under Grant No. MCM20170503.

References

1. Bachrach, Y., Kosinski, M., Graepel, T., Kohli, P., Stillwell, D.: Personality and patterns of facebook usage. In: Proceedings of ACM Web Science (2012)
2. Friedman, J.H.: Greedy function approximation: a gradient boosting machine. Ann. Stat. **29**, 1189–1232 (2001)
3. He, X., Liao, L., Zhang, H., Nie, L., Hu, X., Chua, T.S.: Neural collaborative filtering. In: Proceedings of WWW (2017)
4. Hollerit, B., Kröll, M., Strohmaier, M.: Towards linking buyers and sellers: detecting commercial intent on Twitter. In: Proceedings of WWW (2013)
5. Kröll, M., Strohmaier, M.: Analyzing human intentions in natural language text. In: Proceedings of K-CAP 2009 (2009)
6. Mihalcea, R., Tarau, P.: TextRank: bringing order into text. In: EMNLP (2004)
7. Perozzi, B., Al-Rfou, R., Skiena, S.: DeepWalk: online learning of social representations. In: Proceedings of ACM SIGKDD (2014)
8. Powell, D., Yu, J., DeWolf, M., Holyoak, K.J.: The love of large numbers: a popularity bias in consumer choice. Psychol. Sci. **28**, 1432–1442 (2017)
9. Sarwar, B., Karypis, G., Konstan, J., Riedl, J.: Item-based collaborative filtering recommendation algorithms. In: Proceedings of WWW (2001)
10. Shi, C., Zhang, Z., Luo, P., Yu, P.S., Yue, Y., Wu, B.: Semantic path based personalized recommendation on weighted heterogeneous information networks. In: Proceedings of ACM CIKM (2015)
11. Wang, J., Cong, G., Zhao, W.X., Li, X.: Mining user intents in Twitter: a semi-supervised approach to inferring intent categories for tweets. In: AAAI (2015)
12. Wang, J., Zhao, W.X., He, Y., Li, X.: Leveraging product adopter information from online reviews for product recommendation. In: ICWSM (2015)
13. Xu, G., Gu, Y., Dolog, P., Zhang, Y., Kitsuregawa, M.: SemRec: a semantic enhancement framework for tag based recommendation. In: AAAI (2011)
14. Yu, X., et al.: Personalized entity recommendation: A heterogeneous information network approach. In: Proceedings of WSDM 2014 (2014)
15. Zhang, H., Shen, F., Liu, W., He, X., Luan, H., Chua, T.S.: Discrete collaborative filtering. In: Proceedings of ACM SIGIR (2016)
16. Zhao, H., Yao, Q., Li, J., Song, Y., Lee, D.L.: Meta-graph based recommendation fusion over heterogeneous information networks. In: Proceedings of ACM SIGKDD (2017)
17. Zhao, X.W., Guo, Y., He, Y., Jiang, H., Wu, Y., Li, X.: We know what you want to buy: a demographic-based system for product recommendation on microblogs. In: Proceedings of ACM SIGKDD (2014)

Learning Concept Hierarchy from Short Texts Using Context Coherence

Abdulqader Almars[1,2]([✉]), Xue Li[1]([✉]), Ibrahim A. Ibrahim[1]([✉]),
and Xin Zhao[1]([✉])

[1] The University of Queensland, Brisbane, QLD, Australia
{a.almars,xueli,i.ibrahim,x.zhao}@uq.edu.au
[2] Taibah University, Madinah, Saudi Arabia

Abstract. Uncovering a concept hierarchy from short texts, such as tweets and instant messages, is critical for helping users quickly understand the main concepts and sub-concepts in large volumes of such texts. However, due to the sparsity of short texts, existing hierarchical models fail to learn the structural relations among concepts and discover the data more deeply. To solve this problem, we introduce a new notion called *context coherence*. *Context coherence* reflects the coverage of a word in a collection of short texts. This coverage is measured by analyzing the relations of words in whole texts. The major advantage of context coherence is that it aligns with the requirements of a concept hierarchy and can lead to a meaningful structure. Moreover, we propose a novel non-parametric context coherence-based model (CCM) that can discover the concept hierarchy from short texts without a pre-defended hierarchy depth and width. We evaluate our model on two real-world datasets. The quantitative evaluations confirm the high quality of the concept hierarchy discovered by our model compared with those of state-of-the-art methods.

Keywords: Hierarchical structure · Ontology learning
Context coherence · Short texts

1 Introduction

In recent years, short texts such as Twitter and Weiboa have become a popular form of information on the web. Short texts contain different latent concepts of a product or topic that can be hierarchically discovered. For example, in smartphone-related tweets, users discuss the main concepts of a smartphone, such as the overall design, battery capacity, screen size, and camera. Constructing a concept hierarchy from short texts can help users understand the contents implied at different granularity levels and can facilitate many applications, such as recommendation [19], summarization [1,5], and sentiment analysis [8] applications.

Hierarchical topic models have been previously proposed to effectively extract the hidden structures from traditional texts [3,6,7]. However, applying these

© Springer Nature Switzerland AG 2018
H. Hacid et al. (Eds.): WISE 2018, LNCS 11233, pp. 319–329, 2018.
https://doi.org/10.1007/978-3-030-02922-7_22

models on short texts may result in less effective performance due to the sparsity of text. In short texts, a few studies have addressed the problem of discovering a concept hierarchy from short texts [1,15,16], but these approaches do not fully exhibit the following three characteristics of an optimal tree. First, a concept on a high level, close to the root node, should cover a wider range of sub-concepts than those on a lower level. Second, a concept in a tree should be organized as parent and children concepts, where the parent concept is semantically related to its children rather than to its non-children . Third, the depth and width of the tree should be automatically inferred from the data.

To fill the gap in the current research, we propose context coherence-based model (CCM), a top-down recursive model, to learn concept hierarchies from short texts by analyzing the relations between words. To achieve this, we introduce a novel measurement called *"context coherence"* to estimate the coverage of words in the whole document. *Context coherence is measured by the number of words that are related to a given a word.* A greater number of related words in a document implies that a word covers a wider range of aspects and that the size of the sub-hierarchy rooted in this word is relatively large. Unlike in the existing models, the parameters of CCM (e.g., depth and width) can be automatically learned from the data. The hierarchy shape is inferred by the average context coherence of words in each level. Indeed, we define a minimum threshold to limit the number of children concepts and control the depth of the tree.

Most current hierarchical models apply subjective methods, such as surveys, to evaluate the hierarchies they generate [1,15]. Consequently, the results are dependent on the participants' experience, and the preciseness and fairness of subjective evaluations are not convincing. In this paper, we suggest objective methods to evaluate the quality of a concept hierarchy. The main contributions of this paper are as follows:

- We introduce a new measurement, namely, context coherence, to measure the containment relationship of words for concept hierarchy construction.
- We propose a new algorithm, the CCM, to learn the concept hierarchy from short texts without a pre-defended hierarchy shape.
- We use objective criteria to evaluate the quality of the concept hierarchy. Comprehensive experiments demonstrate the effectiveness of our proposed method in comparison with other methods.

The rest of the paper is organized as follows: in Sect. 2, we give a brief review of related works. Section 3 introduces our hierarchical model for short text, and discuss its implementation in Sect. 4. Experimental results are presented in Sect. 5. Finally, conclusions are made in the last section.

2   Related Works

Hierarchical topic models have been proposed to discover hidden structures in documents. Several approaches have been proposed to address the problems of hierarchical extraction. For example, the hierarchical Pachinko allocation model (hPAM) [9] was developed to capture the correlations between topics. hPAM produces multiple levels of super- and sub-topics. Each topic in the tree is a mixture of distributions over words. However, the hierarchical structure of hPAM is predetermined. Blei et al. [2] proposed a generative probabilistic model known as the nested Chinese restaurant process (nCRP) to hierarchically learn latent structures from data. Kim et al. [7] applied an HASM, a more advanced model that automatically discovers the structure of aspects with corresponding sentiment polarity. However, the main limitation of these existing models [3,7,17] is that they have only been developed to deal with normal texts. Directly applying these models to short texts might produce an incorrect and incomplete result. A few studies have been conducted to effectively reveal the latent structures from short texts [1,15,19]. Zhao and Li [19] developed an algorithm based on a formal concept analysis (FCA) to extract *hot features* and organize them hierarchically in a tree. Wang et al. [13,15] proposed a novel phrase mining approach to recursively construct topical from a content-representative document (the title). Moreover, in our previous works [1], we developed an LDA-based method called Structured Sentiment Analysis (SSA) approach which recursively learns the hierarchical tree of the topics from the short text. The drawbacks of current methods is that the depth and shape of the tree are manually specified. Hence, our aim in this paper is to propose an effective approach to model the topic hierarchy over short texts.

3 Problem Formulation

Below we first introduce some closely related concept, and then define the CCM problem.

Definition 1. (Coherent Words). The basic unit of a concept hierarchy is a word. Coherent word cw is referred to a word with high coherence (coverage) score to other words in vocabulary V.

Definition 2. (Concept). A concept, t, in a tree, T, is represented by either a single coherent word, cw_i, or as groups of coherent words, $t = \{cw_1, cw_2, cw_3...\}$, where every words $cw_1 \in t$ are refer to the same thing. A coherent word can appear in multiple concepts, though it will have a different order based on the coherence score in each concept. The number of words is decided by the merge operation (see Sect. 4.2).

Definition 3. (Concept Hierarchy). A concept hierarchy is defined as T where each node in the tree is a concept. Every non-leaf concept, t_i, has a number of children, defined as $chn^{t_i} = \{chn_1^{t_i}, chn_2^{t_i}, ...\}$. All children concepts should be semantically related to their parent concept.

Problem 1. Given a collection of short texts about a specific topic, $D = \{d_1, d_2, ..., d_l\}$, where $|D|$ is the length of D, our task is to extract a coherent concept hierarchy T with unbounded depth and width.

4   Proposed Approach

Existing hierarchical models [12,14,18] learn concepts by observing document-level word co-occurrence, whose performance will be significantly influenced in the case of short texts. To address this problem, we propose a novel CCM that learns concept hierarchies from short texts.

4.1   Concept Extraction

Concept extraction is the main task of the CCM, where concepts are defined as either single words or groups of words. In this paper, we introduce the notion of *context coherence* to extract concepts. The idea behind context coherence is to measure the coverage of a given word by analyzing the associations between the words in the entire text. The coverage of a given word is calculated by identifying the number of words that are related to it. The relatedness reflects the similarity between the given word to the other words in the texts. More related words implies that this word covers a wider range of sub-words. The difference between CCM and frequency-based models is that in frequency-based models, a word's frequency implies the importance of this word in the whole text. While our model assumes that a word is important if it covers a high number of words.

Given whole collections of short texts, D, we first measure the similarity between w_i and w_j. Specifically, pointwise mutual information (PMI) [4] is employed to calculate the similarity of pairs as shown in Eq. 1, where $P(w_i, w_j)$ indicates the probability that two words, w_i and w_j, occur together in texts, while $P(w_i)$ and $P(w_j)$ indicate the occurrence probability of w_i and w_j in the texts, respectively.

$$W_{i,j} = log \frac{p(w_i, w_j)}{p(w_i)p(w_j)} \tag{1}$$

To compute the context coherence, we represent the text as a term-term matrix M in which each row and column stands for all unique terms in V. Each cell describes the word-pair similarity score in short texts. The context coherence of a given word w_i is calculated by taking the average similarity score with other words using the equation below:

$$CC(w_i) = \frac{1}{n} \sum_j W_{i,j} \tag{2}$$

where n is the number of words in D. Our model uses the average PMI for term-term analysis because in the vocabulary of short texts, most pairs of words do not appear together frequently. That is, the PMI between most pairs of words is negative. The average PMI of a word is decided based on how many words are not related to it. This aligns with the definition of a word's context coherence. Hence, average PMI is a good approximation of context coherence.

4.2   Hierarchical Extraction

Our hierarchical extraction function consists of two main components: a splitting process and a merging process. Splitting is performed by a recursive algorithm that is responsible for generating a hierarchical tree, while the merging process is responsible for grouping similar concepts under a new concept.

Splitting Operation. The first goal of the CCM is to create a flexible tree in which each parent concept has better relatedness to its children concepts than non-children concepts. The CCM's second aim is to build a hierarchy in which concepts are general near the root and specific near the leaves. To achieve this aim, the splitting operation takes the extracted candidate concepts as inputs to recursively build a tree. It recursively partitions a current concept into a number of sub-concepts. For example, if concept t_1 talks about a *camera* and concept t_2 talks about a *headphone*, then the whole document will be partitioned into two sub-documents. In this way, all concepts in the same path should be semantically related and refer to the same thing. Another advantage of our model is that the shape of the tree (e.g., its depth and width) is automatically determined from the data. The number of concepts in each level is inferred by removing the children concepts whose context coherence is less than a predefined threshold. For the depth, the CCM stops the splitting process when the average coherent score of the concepts is less than the threshold. Notice that not all candidate concepts are considered in the tree. Only the concepts that exceed the specified threshold are kept.

Algorithm 1. Splitting Operation

   Data: $D, minS$
   Result: Build a concept hierarchy T.

1	**Function** *Recursive(D)*
2	**foreach** $w_i \in V$ **do**
3	**foreach** $w_j \in V$ **do**
4	**if** $w_1 \neq w_2$ **then**
5	$CC(w_i) = \frac{1}{n} \sum_j W_{i,j}$;
6	**if** $CC(w_i) > minS$ **then**
7	$overlap(t1, t2)$;
8	Add w_i to T ;
9	$Recursive(split(D, w_i))$;
10	**end**
11	**end**
12	**end**
13	**end**

Given a document D and stopping criteria $minS$, the specific recursive process of our approach can be described as follows (i.e., Algorithm 1). For each word, $w_i \in D$, calculate the context coherence $CC(w_i)$ to the other words in the document using Eqs. (1) and (2) (Line 5). If the coherence score exceed the a predefined threshold $minS$, we do one of the following: (1) Add a word as a concept, t to T and then split the whole document into a number of sub-concepts (sub-documents), (2) Create a new concept by merging similar concepts (Line 7). Section 13 explains the merge operation in more detail. We recursively apply the same process again for every generated sub-document to extract a concept hierarchy.

Merging Operation. All concepts created by the splitting operation contain a single word. The task of the merging operation is responsible to find similar words and group them under a new concept. For the merging operation, there are three situations where concepts need to be merged. First, people often use different words to refer to the same concept (i.e., synonyms), such as *photo*, *pic* and *picture*. Those words, rarely appear next to each other in the same text. The merging operation aims to find these kinds of words and combine them under the same concept. Second, the CCM also tries to group words that share the same context, such as *"front"* and *selfie*. Third, in some situations, we have some concepts that may appear twice in two branches, such as *screen* → *case* and *case* → *screen*. The CCM handles such duplications by removing one of them from a concept in the tree. For cases 1 and 2, our algorithm finds the common sub-concepts for concepts t_i and t_j and then merges them into a new concept t, either if $chn^{t_i} \subset chn^{t_j}$ or if the overlap score of two concepts exceeds the predefined threshold $minM$. In our experiment, we set the overlap threshold to 0.60. The overlap score of two concepts is measured using Jaccard similarity measure.

$$overlap(t_i, t_j) = \frac{|chn^{t_i} \cap chn^{t_j}|}{|chn^{t_i} \cup chn^{t_j}|} \tag{3}$$

where chn^{t_i} and chn^{t_j} are the children of concepts t_i and t_j. For case 3, our algorithm checks if concepts $t_i \in chn^{t_j}$ and vice versa. Then, the common concept will be deleted from one of them.

5   Experiments

In this section, we first introduce the dataset and the methods used for evaluation and then demonstrate the experimental results.

5.1   Datasets

Our method is tested on two real-world short-text corpora. In the following section, we give brief descriptions of them.

- **Smartphone.** A collection of more than 68,000 distinct tweets crawled from Twitter API. This dataset has been used in a previous study on concept hierarchies [1]. The raw data of datasets are very noisy. Hence, we performed the following preprocessing on this dataset: (1) converting letters to lowercase; (2) removing all non-alphabetic characters, stop words, and URLs ; and (3) removing words with fewer than 2 characters.

– **DBLP**. A collection of 33,313 titles was retrieved from a set of recently published papers in computer science in six research areas: data mining, computer vision, databases, information retrieval, natural language processing, and machine learning. This dataset was prepared and has been previously used in [15,20].

5.2 Methods for Comparison

We mainly compare our approach with three typical models of hierarchical construction.

– **rCRP** [6]. A non-parametric hierarchical model that recursively infers the hierarchical structure of topics from discrete data.
– **hPAM** [10]. A parametric hierarchical model that takes a document as input and generates a specific number of super-topics and sub-topics.
– **SSA** [1]. This is a recursive state-of-the-art hierarchical model that extracts a tree with a specified depth and width.
– **HASM** [6]. A non-parametric hierarchical aspect sentiment model that discovers a aspects with the corresponding sentiment polarity.
– **CCM**. For evaluation of our model , we set the CCM's parameters approximately to generate the same tree. For all methods above, we tune its hyperparameters to discovers the same number of concepts as other methods.

5.3 Evaluation Measures

In this paper, we introduce three measures to quantitatively evaluate the quality, concept coherence, coverage, and parent-child relatedness. We then use these metrics to compare the characteristics of a concept hierarchy constructed by our model with baseline methods.

Table 1. Average coherence score

	SmartPhone			DBLP		
	Level 1	Level 2	Level 3	Level 1	Level 2	Level 3
CCM	−1.58	−1.99	−2.42	−2.91	−3.01	−3.10
rCRP	−3.38	−3.14	−2.54	−3.18	−3.23	−3.14
hPAM	−1.84	−2.85	-	−2.99	−3.05	-
HASM	−2.60	−2.03	-	−3.17	-	-
SSA	-	-	-	-	-	-

Quality of Concepts. In this paper, we use the measure coherence concept proposed by Mimno et al. [11] to evaluate the quality of concepts. The coherence concept is based on the idea that all words in this concept should be semantically related. Suppose a concept t is characterized using a list $t = \{w_1^t, w_2^t, ..., w_n^t\}$ of n words. The coherence score of t is given by:

$$Coherence(t) = \sum_{i=2}^{n} \sum_{j=1}^{i-1} log \frac{D(w_i^t, w_j^t) + 1}{D(w_j^t)}. \tag{4}$$

where $D(w_i, w_j)$ is the number of documents containing both w_i and w_j. $D(w_j)$ is the number of documents containing a word, w_j. In our experiments, we set the number of words in each concept to five. Since our model can produce concepts with less than five words, we only evaluate concepts that contain five words. To evaluate the overall quality of a concept set, we calculate the average coherence score for each method. Here, we only show the score related to three levels of a concept hierarchy. The results are illustrated in Table 1. A higher coherence score indicates a better quality concept. For both datasets, the results show that the CCM achieved significant improvements compared with the rCRP and hPAM. However, due to data sparsity and the shortness of the texts, the HASM failed to discover a comprehensive concept hierarchy. In the SSA mode , we did not evaluate the concept coherence because the concept in the tree was represented by a single word.

Table 2. Average coverage score

	SmartPhone			DBLP		
	Level 1	Level 2	Level 3	Level 1	Level 2	Level 3
CCM	−0.80	−0.92	−0.98	−0.66	−0.84	−0.90
rCRP	−0.65	−0.66	−0.72	−0.32	−0.50	−0.43
hPAM	−0.84	−0.72	-	−0.64	−0.51	-
HASM	−0.78	−0.83	-	-	-	-
SSA	−0.88	−0.94	−0.96	−0.72	−0.83	−0.87

Coverage. The second important characteristics of the concept hierarchy is the coverage of the concepts which reflects that the concepts near the root node should have a higher coverage than those close to the leaf nodes. Given the N top words of a concept, $t_z = \{w_1, w_2, w_3, ...w_N\}$, we replace the top five words in the whole document with the first word. We assume that all words under the same concept talk about the same thing. The coverage score is calculated as follows:

$$Coverage(L) = \frac{1}{z} \sum_z PMI(t_z). \tag{5}$$

$$PMI(t_z) = \frac{1}{n} \sum_j log \frac{p(w_1^{t_z}, w_j)}{p(w_1^{t_z})p(w_j)}. \tag{6}$$

where z is the number of concepts in level L. The default value of N is set to 5 in our experiments. A higher coverage score indicates a better quality concept. The results are illustrated in Table 2. For all datasets, the CCM and SSA clearly show a decrease in the coverage score when the depth of the tree increases, which means the concepts near the root nodes are general concepts, while those near the leaf-nodes are specific concepts. Unlike our model, the patterns in rCRP and the hPAM are different.

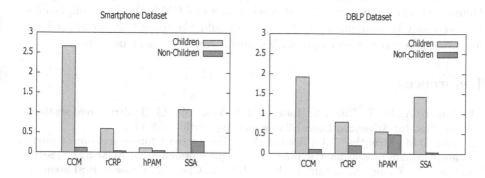

Fig. 1. Parent-child relatedness.

Parent-Child Relatedness. The third evaluation is meant to assess parent-child relatedness. In other words, we assume that parent concept t should be more similar to its direct children than to the children that descend from other concepts. For simplicity, we only compute the relatedness score for a parent concept at the second level, with the children concepts at the third level. Given a concept t, we measure the concept's relatedness score to its children and compare it to that of its non-children using Eqs. (5) and (6):

$$Children(t) = \frac{1}{k} \sum_k \frac{D(t, chn_k^t)}{D(t), D(chn_k^t)} \tag{7}$$

$$Non-Children(t) = \frac{1}{k} \sum_z \sum_k \frac{D(t, chn_k^z)}{D(t), D(chn_k^z)}, z \neq i \tag{8}$$

where $D(t, chn_k^t)$ is the number of times parent concept t appears with its child concept, chn_k^t and $D(t, chn_k^z)$ is the number of times parent concept t appears with its non-child concept chn_k^{tz}. The overall parent-child relatedness is measured by taking the average score of relatedness to children and non-children concepts for all parent concepts at the second level. Figure 1 illustrates the parent-child relatedness of four models. The higher relatedness scores for children indicate that a parent is more similar to its direct children, compared to non-children nodes at the same level. Both the CCM and SSA show significant differences between children and non-children, while the nCRP and hPAM do not. The relatedness of the HASM is not calculated for the smartphone dataset since it produced duplicate children.

6   Conclusion

Discovering concept hierarchies from short texts is an significantly critical because of the prevalence of short texts on the internet. In this paper, we propose a non-parametric CCM for short texts. The CCM can automatically discover a concept hierarchy by observing and analyzing the relations between words in whole texts. This can be done by the proposed new measurement, *context coherence*. The results demonstrate that our approach can discover higher quality trees than previous methods can. Another advantage of CCM is that it is simple, effective and easy to implement. In future, we would like to extend our approach to automatically extract concepts with cosponsoring sentiment polarity.

References

1. Almars, A., Li, X., Zhao, X., Ibrahim, I.A., Yuan, W., Li, B.: Structured sentiment analysis. In: Advanced Data Mining and Applications (2017)
2. Blei, D.M., Griffiths, T.L., Jordan, M.I.: The nested Chinese restaurant process and bayesian nonparametric inference of topic hierarchies. J. ACM **57**, 7 (2010)
3. Chen, P., Zhang, N.L., Liu, T., Poon, L.K.M., Chen, Z., Khawar, F.: Latent tree models for hierarchical topic detection. Artif. Intell. **250**, 105–124 (2017). https://doi.org/10.1016/j.artint.2017.06.004
4. Church, K.W., Hanks, P.: Word association norms, mutual information, and lexicography. Comput. Linguist **16**, 22–29 (1990)
5. Gerani, S., Carenini, G., Ng, R.T.: Modeling content and structure for abstractive review summarization. Comput. Speech Lang. **2016**, 7 (2016)
6. Kim, J.H., Kim, D., Kim, S., Oh, A.: Modeling topic hierarchies with the recursive Chinese restaurant process. In: Proceedings of the 21st ACM international conference on Information and knowledge management (2012)
7. Kim, S., Zhang, J., Chen, Z., Oh, A.H., Liu, S.: A hierarchical aspect-sentiment model for online reviews. In: AAAI (2013)
8. Kontopoulos, E., Berberidis, C., Dergiades, T., Bassiliades, N.: Ontology-based sentiment analysis of twitter posts. Expert. Syst. Appl. **40**, 4065–4074 (2013)
9. Li, W., McCallum, A.: Pachinko allocation: dag-structured mixture models of topic correlations. In: ICML 2006 (2006)
10. Mimno, D., Li, W., McCallum, A.: Mixtures of hierarchical topics with pachinko allocation. In: ICML 2007 (2007)
11. Mimno, D., Wallach, H.M., Talley, E., Leenders, M., McCallum, A.: Optimizing semantic coherence in topic models. In: EMNLP 2011 (2011)
12. Teh, Y.W., Jordan, M.I., Beal, M.J., Blei, D.M.: Sharing clusters among related groups: hierarchical dirichlet processes. In: Advances in Neural Information Processing Systems, vol. 17 (2005)
13. Wang, C., Danilevsky, M., Liu, J., Desai, N., Ji, H., Han, J.: Constructing topical hierarchies in heterogeneous information networks. In: ICDM 2013 (2013)
14. Wang, C., Liu, X., Song, Y., Han, J.: Scalable and robust construction of topical hierarchies. ArXiv e-prints (2014)
15. Wang, C., et al.: A phrase mining framework for recursive construction of a topical hierarchy. In: KDD 2013 (2013)
16. Wang, C., Liu, X., Song, Y., Han, J.: Towards interactive construction of topical hierarchy: a recursive tensor decomposition approach. In: KDD 2015 (2015)

17. Xu, Y., Yin, J., Huang, J., Yin, Y.: Hierarchical topic modeling with automatic knowledge mining. Expert. Syst. Appl. **103**, 106-117 (2018)
18. Yao, L., Mimno, D., McCallum, A.: Efficient methods for topic model inference on streaming document collections. In: KDD 2009 (2009)
19. Zhao, P., Li, X., Wang, K.: Feature extraction from micro-blogs for comparison of products and services. In: WISE (2013)
20. Zuo, Y., et al.: Topic modeling of short texts: a pseudo-document view. In: KDD 2016 (2016)

Graph Data

Eliminating Temporal Conflicts in Uncertain Temporal Knowledge Graphs

Lingjiao Lu[1], Junhua Fang[1], Pengpeng Zhao[1], Jiajie Xu[1], Hongzhi Yin[2],
and Lei Zhao[1(✉)]

[1] School of Computer Science and Technology, Soochow University, Su Zhou, China
ljlu@stu.suda.edu.cn, {jhfang,ppzhao,xujj,zhaol}@suda.edu.cn
[2] School of Information Technology and Electrical Engineering Brisbane,
The University of Queensland, Brisbane, Queensland, Australia
db.hongzhi@gmail.com

Abstract. In the real world, a majority of facts are not static or immutable but highly ephemeral. Each fact is valid for only a limited amount of time, or it stands in temporal dependencies. In addition, facts with time information are usually accompanied by a real-valued weight which witnesses the possibility of a fact. However, most of existing Knowledge Graphs (KGs) focus on static data thus impeding the comprehensive solution for the management of uncertain and temporal facts in KGs. To fill this gap, we emphasize the characteristics of time and propose a coherent management framework ETC (Eliminate Temporal Conflicts) for temporal consistency. ETC is based on maximum weight clique to detect temporal conflicts in uncertain temporal knowledge graphs and eliminate them to achieve the most probable knowledge graph according to related constraints. Constraint graphs with detailed description have first been proposed to identify temporal constraints for the conflict detection. Also, implicit constraints and weight conversion have been propose for conflict resolution. Experiments over two different temporal knowledge graphs demonstrate the high recall rate and precision rate of our framework.

Keywords: Uncertain temporal knowledge graphs
Temporal conflicts · Constraint graphs

1 Introduction

Large-scale Knowledge Graphs (KGs) have been widely used in recent years for knowledge-based question answering (KB-QA) [17,21], financial field, social sphere and so on. Facts in those knowledge graphs are usually expressed in the form of triple (*subject, relation, object*). KGs as DBpedia [3], Wikidata, YAGO [20], Google's Knowledge Graph, NELL (Never Ending Language Learning) [13] are automatically constructed by Open Information Extraction (OIE) which will

© Springer Nature Switzerland AG 2018
H. Hacid et al. (Eds.): WISE 2018, LNCS 11233, pp. 333–347, 2018.
https://doi.org/10.1007/978-3-030-02922-7_23

extract a significant amount of incorrect, incomplete or even inconsistent factual knowledge. So the knowledge is often summarized under the term of uncertain facts. Nowadays, OIE is used to extract dynamic facts [15] from datasets which also result in temporal inconsistency. For example, an extractor might erroneously extract two temporal facts that Neymar played for Brazil club from 2013 to 2017 and he played for Santos club from 2009 to 2014. But a player cannot play for two different clubs at the same time. It affects the credibility of knowledge of Neymar's career. We can express that only one of the two above facts may be true in the real world. Without an explicit constraint, which puts these two facts into conflict with each other, there is no formal inconsistency in a knowledge graph containing the above two facts.

So cleaning knowledge graphs from noisy temporal facts ought to be the first concern. A limitation of existing methods [7,16,19] are only focused on static facts that could be identified by being encoded to binary relations ignoring the possibility of temporal inconsistency changes in time. In addition, little has been done in terms of techniques to debug uncertain temporal KGs. The previous method [8] is to provide consistency rules based on the real-world module where knowledge not satisfied with rules is conflict. Most of the existing methods use first-order logic Horn formulae with temporal predicates to express temporal constraints [4,18]. Besides, most of the existing methods(e.g.,[6]) focus on finding conflicts by MAP reference.

However, [4,18] are limited to a small set of temporal patterns and have not taken all types of temporal conflicts into consideration. [6] is more inclined to numerical comparisons and lacks the uniqueness of the knowledge represented by certain temporal predicates such as birth information. Moreover, to get a consistent knowledge graph, constraints used in these methods are mainly handmade which have not taken the incompleteness of the given constraints into consideration. In general, obtaining a conflict-free temporal knowledge graph is faced with two main challenges. The first one is expressing temporal conflicts and detecting conflict facts from the given knowledge graph. Conflicts between facts are many-to-many because knowledge facts might depend on multiple constraints. This can be drawn that the truth of the fact depends on a number of other facts. So the second challenge is to decide which facts need to be removed from the given uncertain temporal knowledge graph to keep it consistent.

To tackle these barriers, we define constraint graphs to represent conflicts. ETC framework has been proposed to eliminate temporal conflicts in uncertain temporal knowledge graph. Our framework considers five kinds of temporal constraints for conflicts detection: *Precedence, Overlap, Inclusion, disjointedness, Mutex*. These kinds of temporal constraints are the expansion of Allen [1] which considers thirteen categories temporal relations. In ETC framework, we are dedicated to finding implicit constraints to detect more conflicts. We map the conflict facts to an undirected graph and solve the maximum weight clique of it to get the most possibility subgraph of the uncertain temporal knowledge graph.

The main contributions of this paper can be highlighted as below: (*i*) the most possible and conflict-free uncertain temporal knowledge graph has been declared in details, (ii) expanded temporal constraints have been taken into consideration, (iii) constraint graphs have been proposed to represent temporal constraints and implicit constraints have been dug to identify conflict facts more accurately, (iv) a series of experiments based on two real-world knowledge graphs have been carried out to evaluate the the performance of ETC framework.

Organization: The remainder of this paper is organized as follows. In Sect. 2, related work of temporal knowledge graphs is introduced. In Sect. 3, we provide a formal of our data model, constraint types, constraint graphs and the definition of our problem. In Sect. 4, we resolve temporal conflicts in uncertain temporal knowledge graphs. Our experimental results are shown in Sect. 5 and we conclude our work in Sect. 6.

2   Related Work

Extending knowledge base using open domain information extraction always leads to uncertainty facts. The previous work [14] is committed to narrow down, identify and explain likely errors from a knowledge graph which text along with optional source documents, provenance information, and confidence scores by linguistic analytic and entity linking. In addition, some works aim to debug erroneous facts by using a set of function constraints [7,12]. They consider using a set of functional constraints to debug conflicts which still is limited to deal with static and literal facts.

Formal semantics for temporal knowledge graphs have first been offered in [11]. In [2,5], OWL is extended in order to enable temporal reasoning for supporting temporal queries. The authors define SWRL rules that are compatible with a reasoner that supports Dl-safe rules in order to detect inconsistencies. However, their system can only determine whether the knowledge base is consistent and cannot resolve the existing conflicts.

Besides, in an earlier version [8] proposes a model for reasoning temporal conflicts in RDF knowledge bases for consistency constraints and queries which used first-order logic Horn formulas with time to express constraints. It defines the optimization problem as a scheduling task and introduces an approximation of a scheduling algorithm. While it is only able to deal with a large-scale knowledge base and can only solve a little temporal conflicts. Also, the approach does not explicitly incorporate terminological knowledge when resolving the conflicts. In [6], it proposes an approach to resolving conflicts in temporal knowledge graphs. The idea is to highlight the use of Markov Logic Networks to debug temporal conflicts with handcrafted temporal constraints and knowledge graphs are mapped to first-order logic. However, this study does not consider all kinds of temporal conflicts in knowledge graphs and the authors have not taken the incompleteness of handcraft constraints into consideration.

3 Preliminaries

3.1 Data and Representation Model

Knowledge Graphs. Facts in these knowledge graphs are expressed in the form of triple (*subject, relation, object*) abbreviated as (s, r, o) where s, $o \in \mathcal{E}$, $r \in \mathcal{R}$, $\mathcal{E}$ is a set of entities and $\mathcal{R}$ is a set of relations.

Uncertain Temporal Knowledge Graph. Uncertain temporal knowledge graphs (short for UTKGs) are extension of knowledge graphs with weight and temporal element. It has been shown that a knowledge graph can be extended with a temporal information by labeling each triple in the graph with a temporal element and weight information [10]. The temporal element represents the time period in which the triple may be valid. The weight information can be interpreted as how likely is for the fact to hold.

We assign time domain T as a linearly ordered finite sequence of time points. A time interval is an ordered pair $[t_b, t_e]$, with $t_b <= t_e$ and $t_b, t_e \in T$. The time pair denotes the closed interval from t_b to t_e and we will work with the interval-cased temporal domain for defining our data model. For presentation purposes, we will denote interval as if they range over years, like the interval [2009, 2014] which begins in 2009 and finishes in 2014. Note that point-based temporal domain t can be converted into interval-based domain by translating t into $[t, t]$.

Definition 1 (*Uncertain Temporal knowledge graph*). *An uncertain temporal knowledge graph $\mathcal{G}$ is a set of temporal knowledge facts $\mathcal{F}$ that each fact (s, r, o) in the knowledge graph is extended with a valid time interval $[t_b, t_e]$, $t_b \leq t_e$ and a positive constant w which values from 0 to 1. We refer to f as an uncertain temporal fact, $f \in \mathcal{F}$.*

$$f = (s, r, o, [t_b, t_e], w) \tag{1}$$

where w represents how likely f is to be true and the greater w is, the more likely f is true.

For example, Fig. 1 represents an uncertain temporal knowledge graph that facts are related to *Neymar*.

f_1 : (Neymar, birthIn, Brazil,[1992,1992], 0.9)
f_2 : (Neymar, birthIn, America,[1992,1992], 0.3)
f_3 : (Neymar, playFor, Brazil Club, [2003,2013], 0.6)
f_4 : (Neymar, playFor, Brazil Club, [2014,2019], 0.7)
f_5 : (Neymar, playFor, Santos Club, [2009,2014], 0.6)
f_6 : (Neymar, winPrize, Brazil Cup champion, [2010,2010], 0.7)
f_7 : (Neymar, studyAt, Santos Clara university, [1990,1992], 0.2)

Fig. 1. The content of temporal facts of Neymar da Silva Santos Jnior.

3.2   Constraint Types

More clear constraints are, more conflict facts can be identified. Depending on the choice of constraints, the combinatorial complexity of resolving conflicts is varying, making it crucial to decide which constraints we allow to be formulated. In the following, we introduce five constraint types that would enable identifying temporal conflicts: (i) *Precedence* (ii) *Overlap* (iii) *Inclusion* (iv) *Disjointedness* (v) *Mutex*

We denote C_t as a set of all constraint types, $C_t = \{Precedence, Overlap, Inclusion, Disjointedness, Mutex\}$. For ease of expressing, two temporal facts in a UTKG we considered are $f_i < s_i, r_i, o_i, [t_i, t_i'], w_i >$, $f_j < s_j, r_j, o_j, [t_j, t_j'], w_j >$, $i \neq j$. Figure 2 shows four constraint types without *Mutex*. We will introduce it at the last of this subsection because of the special characteristics of *Mutex*.

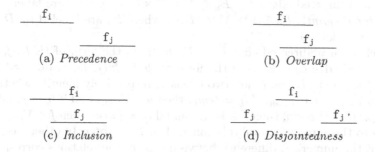

(a) *Precedence*   (b) *Overlap*

(c) *Inclusion*   (d) *Disjointedness*

Fig. 2. Four types of constraints

Precedence defines f_i finishes before f_j begins ($t_i' \leq t_j$) as shown in Fig. 2(a). In particular, if $t_i' = t_j$, we call f_i meets f_j. Otherwise, f_i before f_j. *Overlap* means two facts share a common time interval ($t_i < t_j < t_i' < t_j'$) as shown in Fig. 2(b). *Inclusion*, shown in Fig. 2(c), means the time intervals of f_i, f_j are wholly involved where ($t_i \leq t_j \leq t_j' \leq t_i'$). In particular, if $t_i = t_j$ and $t_i' = t_j'$, we call f_i equal f_j. If only $t_i = t_j$, we call f_i starts f_j or if only $t_i' = t_j'$, we call f_i finishes f_j. In other cases, f_i contains f_j. *Disjointedness* is to express the time intervals of two facts f_i, f_j are non-overlapping when they share the same relation and subject. As shown in Fig. 2(d), f_j and f_j' are two kinds of consistent circumstances when given fact f_i. We know a player can only play for one club at a time, so f_5 is conflict with f_3 and f_4 in Fig. 1.

Mutex, as the last type of constraints we consider, defines a set of facts which are all in conflict with each other, regardless of time. In general, a relation r with a differing argument must not occur.

Example 1. *A natural constraint in the domain of people is that a person only have one birthplace and corresponding specific date. Then f_1 and f_2 from Fig. 1 are in conflict.*

3.3   Constraint Graph

Conflicts between facts need to be detected by the consistency constraints. Consistency constraints in our researching framework is a graph named with constraint graph (denoted as G_c). A constraint graph is a more compact representation of the consistency constraints than Horn formula. Meanwhile, it represents a higher level of abstraction than considering temporal conflicts among actual facts.

A constraint graph $G_c = (V, E)$ is a pair consisting of vertices labeled $V \subseteq \mathcal{R} \times \{begin, finish\}$ and labeled edges $E \subseteq E_u \cup E_d$. V can be partitioned into two sets $V = V_{begin} \cup V_{finish}$, $V_{begin} \cap V_{finish} = \emptyset$. $v \in V$ is to identify relation r, where $r \in \mathcal{R}$. If $v \in V_{begin}$, v represents the time when r happens or if $v \in V_{finish}$, v represents the time when r_i finishes. So each vertex v has a label to indicate it in V_{begin} or V_{finish}. The set of edges E is composed of undirected edges E_u and directed edges E_d. $E_u \subseteq V \times V \times U$ where E_u are labeled by U, $U = \{mutex, disjoint\}$. $E_d \subseteq V \times V \times \{D, \delta\}$ where E_d are labeled by D and δ, $D = \{before, order\}$, $\delta = \{0, 1\}$.

We define an undirected edge $e_u \in E_u$ to be in the form of (b, l_b, f, l_f, U) and a directed edge $e_d \in E_d$ is in the form of $(b, l_b, f, l_f, D, \delta)$ directed from b to f where $b, f \in V$, l_b and l_f are two labels to respectively identify whether b, f belongs to V_{begin} or V_{finish}. If $l_b = begin$, then $b \in V_{begin}$ and if $l_b = finish$, then $b \in V_{finish}$. If $l_f = begin$, then $f \in V_{begin}$ and if $l_f = finish$, then $f \in V_{finish}$. If b, f belongs to the same temporal relation r, then $D = order$. In other cases, $D = before$. δ is the numerical difference between time of the relations corresponding to the connected nodes. If $\delta = 0$, then the time of b ought to as same as that of f; if $\delta = 1$, then the time of b ought to no more than that of f. Note that, if $D = order$, we define $\delta \equiv 1$ because the time interval of facts is $[t_b, t_e]$ that $t_b \leq t_e$.

In Fig. 3(a), $birthIn$ is represented by the directed edge $(birthIn, begin,, birthIn, finish, order, 1)$. The directed edge $(playFor, begin, diedIn, begin, before, 1)$, shown in Fig. 3(b), represents that $diedIn$ ought to happen after the beginning of $playFor$.

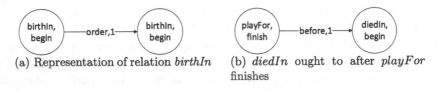

(a) Representation of relation $birthIn$ (b) $diedIn$ ought to after $playFor$ finishes

Fig. 3. Subgraphs of G_c

3.4   Problem Definition

The broad goal of this paper is to eliminate conflict facts in uncertain temporal knowledge graphs. Given an uncertain temporal knowledge graph $\mathcal{G}$ with weighted temporal facts $\mathcal{F}$, and a constraint graph G_c, we aim to find a UTKG

$\mathcal{G}'$ that facts set $\mathcal{F}'$ in it is the subset of $\mathcal{F}$. In addition, facts in $\mathcal{F}'$ are consistent with each other according to G_c.

However, resolving conflict facts in a UTKG $\mathcal{G}$ by selecting a consistent subset of facts $\mathcal{F}$ may get several consistent subsets with the same cardinality. So we extend our study by requiring that the possibility of $\mathcal{G}'$ is maximized, as it is expressed by formula (2).

$$p(\mathcal{G}') = \max_{Satisfy(G_c,\mathcal{F}')\equiv true} \sum_{i=1}^{n} w(f_i) \tag{2}$$

where $Satisfy$ is the logical evaluation of all constraints in G_c by setting all facts in $\mathcal{F}'$ to $true$ and all facts in $\mathcal{F} \setminus \mathcal{F}'$ to $false$, n is the number of facts in $\mathcal{F}'$ and $w(f_i)$ is the weight of f_i where $f_i \subset \mathcal{F}'$.

4   Algorithm

This section is structured in accordance with the general flow of ETC framework as described in Algorithm 1.

Algorithm 1. ETC Framework

Input: A constraint graph G_c and a UTKG $\mathcal{G}$ with the set of facts $\mathcal{F}$
Output: A conflict-free temporal knowledge graph with maximum probability
1: Initialize $Res = \emptyset$, $CS = \emptyset$, $S^f = \emptyset$
2: Get all constraints from G_c and put them to CS
3: **if** G_c has implicit constraints **then**
4:     Put implicit constraints to CS
5: **end if**
6: **for all** $cs \in CS$ **do**
7:     $S^f = S^f \cup FindConflictFacts(cs)$
8: **end for**
9: $Q = \mathcal{F} - S^f$
10: $Res = MaximumProbabilitySet(S^f)$
11: **return** $Res \cup Q$

As a first step, we analysis G_c to get all constraints (line 2). In Line 4, we mine implicit constraints (Sect. 4.1) from the given G_c for detecting conflict more accurately. We denote CS as the set of (implicit) constraints we collected. According to each constraint, we define a set of conflict facts as S_f which contains all conflict facts from $\mathcal{G}$. Then we put facts which are not consistent into set S_f (line 7). Then we calculate the complementary set Q of S^f that all facts in Q are consistent with all facts in the given UTKG.

Turning to obtain a conflict-free uncertain temporal knowledge graph, in line 9, function $MaximumProbabilitySet$ is used to map conflict facts to an undirected graph where vertices and edges correspond to facts and conflicts (Sect. 4.2).

The function based on solving maximum weight clique will return a subset Res of S^f that facts in Res are consistent with each other. Also, Res has the maximum possibility of all subsets of S^f. Finally, the conflict-free and most possible knowledge graph is consisted of facts in Res and Q.

4.1 Detect Conflicts from the Given UTKG

In this section, we aim to detect conflict facts according to constraint graphs. Before introducing the approach to identify conflict facts, we need to extend the use of membership ($\vdash$) as follows.

Given a set of constraint types $C_t = \{Precedence, Overlap, Inclusion, Disjointedness, Mutex\}$, $c_t \in C_t$, r_1, r_2 is two relations. We denote $(r_1, r_2) \vdash (c_t, \alpha)$, if and only if the time intervals of r_1 and r_2 satisfy the c_t constraint and α is a set of labels to identify the numerical difference between time intervals of r_1 and r_2. If $c_t = Disjointedness$ or $Mutex$, $\alpha = \emptyset$; if $c_t = Precedence$, α contains one label; in other cases, α contains two labels.

Identity Constraints. All constraints is given by a constraint graph $G_c = (V, E)$ which shows the relationships between time intervals. By observing G_c, we conclude five constraint graph models corresponding to constraint types to identify all constraints.

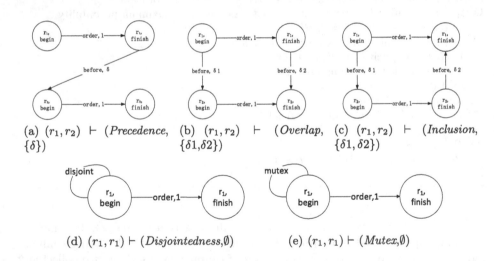

(a) $(r_1, r_2) \vdash (Precedence, \{\delta\})$ (b) $(r_1, r_2) \vdash (Overlap, \{\delta 1, \delta 2\})$ (c) $(r_1, r_2) \vdash (Inclusion, \{\delta 1, \delta 2\})$

(d) $(r_1, r_1) \vdash (Disjointedness, \emptyset)$ (e) $(r_1, r_1) \vdash (Mutex, \emptyset)$

Fig. 4. Graph models

(i) $(r_1, r_2) \vdash (Precedence, \{\delta\})$ as shown in $Fig.4(a)$ has an edge $(r_1, finish, r_2, begin, before, \delta)$. If $\delta = 0$, the time when r_1 finishes ought to be the time when r_2 happens. If $\delta = 1$, r_2 has to happen after r_1 finishes. (ii) $(r_1, r_2) \vdash (Overlap, \{\delta 1, \delta 2\})$ is rare to be required in the real-word while it's still necessary to explicate in detail. Figure 4(b) shows edges $(r_1, begin, r_2, begin, before, \delta 1)$ and $(r_1, finish, r_2, finish, before, \delta 1)$. If $\delta 1 = \delta 2 = 1$, r_1 ought to finish after r_2 begins. (iii) $(r_1, r_2) \vdash (Inclusion, \{\delta 1, \delta 2\})$ as shown in Fig. 4(c) has edges $(r_1, begin, r_2, begin, before, \delta 1)$ and $(r_2, finish, r_1, finish, before, \delta 2)$. If $\delta 1 = \delta 2 = 0$, r_1 and r_2 have to happen and finish at the same time; If $\delta 1 = 0$, $\delta 2 = 1$, r_1 share r_2 with same beginning time and r_2 finishes after r_1; If $\delta 1 = 1$, $\delta 2 = 0$, r_1 share r_2 with same finishing time and r_2 begins after r_1 begins; If $\delta 1 = \delta 2 = 1$, time interval of r_1 contains that of r_2. (iv) $(r_1, r_1) \vdash (Disjointedness, \emptyset)$ as shown in Fig. 4(d) has a edge $(r_1, begin, r_1, begin, disjoint)$. Then facts which have relation r_1 have to satisfied with *Disjointedness* constraint. (v) $(r_1, r_1) \vdash (Mutex, \emptyset)$ shows in Fig. 4(e) that the graph has an edge $(r_1, begin, r_1, begin, mutex)$.

After giving graph models of five constraint types, we can obtain all constraints from G_c. Firstly, by identifying each vertex in G_c which have undirected edge labeled by *mutex* or *disjoint*, we can learn which relations ought to satisfy *Mutex* or *Disjointedness* constraints. Secondly, we compare all edges in G_c corresponding to the relations of facts with the edges in *Precedence, Overlap, Inclusion* constraint graph models to obtain all constraints.

Complete Constraint Graphs. Most handmade constraints are mostly incomplete. Therefore, if conflict detection is performed only based on given constraints, it is a great possibility that part of the conflict facts will be undetected. For example, a knowledge graph contains the following temporal facts.

f_8 (Amy, birthIn, America, [2013, 2013], 0.6)

f_9 (Amy, studyAt, Yale University, [2003,2006], 0.8)

f_{10} (Amy, playFor, Brazil club, [2001,2003], 0.3)

If we use the constraint graph as Fig. 5, we can identify (birthIn, studyAt) $\vdash (Precedence, 1)$, (studyAt, playFor) $\vdash (Precedence, 1)$ that a person ought to study at university after her born and he could play for a club only after she finished his study. So f_9 is conflict with f_8 and f_{10}. To get a most possible consistent knowledge graph, we will remove f_9. However, owning to the continuity of time, the UTKG is still inconsistent that a person cannot play for a club before his born. This constraint is not revealed in Fig. 5 but is also important for conflicts detection. *Precedence* and *Inclusion* constraints have been discovered to be transitivity. We define these constraints as implicit constraints. Implicit constraints are used to dig out all the implied chronological order of the given constraint graph.

Definition 2 *(Implicit Constraints).* *Given a constraint type $c_t \in \{Precedence, Inclusion\}$, a constraint graph G_c. If $(r_1, r_2) \vdash (c_t, \alpha 1)$ and $(r_2, r_3) \vdash (c_t, \alpha 2)$ can be obtained from G_c, then $(r_1, r_3) \vdash (c_t, \alpha 3)$ is an implicit constraint. $\alpha 3$ is valued as follows:*

(i) If $c_t = Precedence$, $\alpha 1 = \{\delta 1\}$, $\alpha 2 = \{\delta 2\}$, then $\alpha 3 = \{\min(\delta 1, \delta 2)\}$.

(ii) If $c_t = Inclusion$. $\alpha 1 = \{\delta 1, \delta 2\}$, $\alpha 2 = \{\delta 3, \delta 4\}$, then $\alpha 3 = \{\min(\delta 1, \delta 3), \min(\delta 2, \delta 4)\}$.

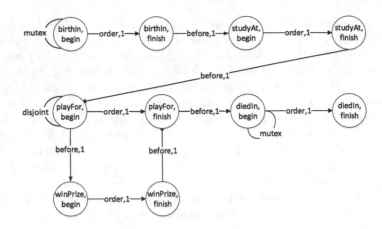

Fig. 5. Constraint graph example

Example 2. *If we identify (birthIn, studyAt) ⊢ (Precedence, 1), (studyAt, playFor) ⊢ (Precedence, 1) for a constraint graph, then an implicit constraint (birthIn, playFor) ⊢ (Precedence, 1) can be deduced.*

4.2   Solution of Temporal Conflicts

Although it has been shown from the previous section that facts are affected by other facts, facts which are not in a temporal conflict with each other have independence that they have no influence on conflicts. This characteristic allows us to only decide on a different assignment between conflict facts. We consider to divide the set of conflict facts S^f into several sets that each set is called a related conflict facts set (denoted as S^c).

Definition 3 *(Related Conflict Facts Set). Given a set of conflict facts S^f and a constraint graph G_C, $f_i, f_j, f_k \in S^f$, a related conflict fact set S^c must meet the following requirements:*

(i) $\forall\, f_i \in S^c$, $\exists\, f_j \in S^c$ that f_j is conflict with f_i according to G_c.

(ii) $\forall\, f_i \in S^c$, if $f_k \notin S^c$, then f_i is consistent with f_k.

Therefore, to solve the maximum probability of conflict-free knowledge graph is to solve the maximum weight consistence facts in each related conflict facts set. We map S^f to an undirected graph G_{Sf} that the facts are thought as vertices and the conflicts are thought as the edges. This is a bijection between facts and vertices. We can conclude that a set of related conflict facts S^c corresponds the set of vertices in a connected subgraph (denoted as G_{S^c}) of G_{Sf}. We solve the complement of G_{S^c} as G_m so that getting the most possible conflict-free subset of S^c is computing the maximum weight clique of G_m.

Definition 4 *(Maximum Weight Clique). For an undirected graph $G_m = (V, E)$ and a set $V' \subseteq V$, $G_m(V')$ denotes the subgraph of G_m induced by V'. If any two vertices in $G_m(V')$ are adjacent to each other, V' is called a clique in G_m. For a vertex $v \in V$, let $w(v)$ be the weight of v. For $V' \subseteq V$, let $w(V') = \sum_{v \in V'} w(v)$. Give a graph G_m and vertex-weight $w(.)$, the maximum weight clique problem is to find a clique V' in G_m of the maximum weight.*

Computing maximum weight clique is finding a clique with the maximum weight in a graph, which is a prominent combinatorial prosecutorial optimization problem. It can be solved in large-scale real-world graphs, and do not be limited by the sum of branches of maximal nodes.

Converting Weight of Facts to Credibility. From Sect. 4.1, we can get a related conflict facts set $S^c = \{f_8, f_9, f_{10}\}$ where f_9 is conflict with f_8 and f_{10} without taking implicit constraints into consideration. To get a maximum probability of temporal facts, we will remove f_9. However, the weight of f_9 is 0.8 which means it is more probably a true fact compared with the weight of f_8 (weight is 0.6) and f_{10} (weight is 0.3).

In order to reserve facts in given UTKG $\mathcal{G}$ with higher weight as much as possible, we convert the weight of facts to confidence scores $c(.)$ as shown in formula (3).

$$c(f_i) = \tan(B \times w(f_i))/e^A \tag{3}$$

where $f_i \in \mathcal{G}$, $w(f_i)$ is the weight of f_i ranging from 0 to 1 and A, B is the normalization constant that A is to make sure that the value of $c(f_i)$ is no more than 1 and B makes $c(f_i)$ positive.

Algorithm 2. *MaximumProbabilitySet*

Input: A set of conflict facts S^f
Output: A set of consistent facts *Res*
1: Initialize Res $= \emptyset$
2: For each $f_i \in S^c$, let $c(f_i) = \tan(B \times w(f_i))/e^A$
3: Replace $w(f_i)$ with $c(f_i)$
4: Map all conflict facts $f_1, f_2 ... f_n$ to vertices $v_{f_1}, v_{f_2}, ... v_{f_n}$ in G_{Sf}
5: **for all** $f_i, f_j \in S^f$ and $f_i < f_j$ **do**
6:     **if** f_i has conflict with f_j **then**
7:         Add an edge (v_{f_i}, v_{f_j}) to G_{Sf}
8:     **end if**
9: **end for**
10: Solve all connected subgraphs G of G_{Sf}
11: **for all** Each connected subgraph $G_{S^c} \in G$ **do**
12:     $G_m = GenComplement(G_{S^c})$
13:     $V' = MWCPsolve(G_m)$
14:     Map V' to a set of facts F
15:     $Res = Res \cup F$
16: **end for**
17: **return** *Res*

The Algorithm 2 shows the pseudo-code of resolving conflict facts which are extracted from the given UTKG and return a set of consistent facts. In the first step, we convert the weight of facts to confidence scores by formula (3) (line 2) and use confidence scores to compute maximum possibility. Then we map the conflict facts to an undirected graph G_{Sf} by considering facts as vertices and add an edge between them if they are in conflict (line 4–9). In the next step, a set of all connected components G_{Sc} from G_{Sf} have been solved by *disjoint-set equivalence relational* (line 10). All vertices in one connected component represent all facts in one related conflict facts Set S^c according to Definition 3. After that, complementary graph G_m is computed by *GenComplement* (line 12). *MWCPsolve* is used to compute the maximum weight clique of G_m and return the vertices set V' (line 13). Because it is a bijection relation between vertices and facts, we map each vertex to the corresponding fact in turn and put these facts into *Res* (line 14–15). Finally, *Res* is the set of all consistent facts.

5   Experiments and Analysis

In this section, we experimentally evaluate the precision rate (P) and recall rate (R) of proposed *ETC* framework. We ran the experiments by Python 3.2 on Linux. For the convenience of operation, we use Neo4j graph database to store temporal knowledge graphs and the corresponding constraint graphs.

Dataset and Constraint Graph. At present, uncertain temporal facts are not wildly used. As result, we use expanded FootballDB [6] and DBpedia as our dataset. FootballDB has been extracted about football players. We use five relations as *playFor, birthIn, winPrize, studyAt, diedIn*. The size of this dataset we used is almost 100k temporal facts. DBpedia contains structured temporal information obtained from various sources using OIE and it only involves two temporal relations *birthIn, diedIn*. So we additional extracted *playFor* relations. Finally we use over 1 million temporal facts.

By default, we assume facts in the knowledge base to be *true*, and (implicitly) all facts not contained in the knowledge base to be *false*, an approach generally known as *closed-world assumption*. Thus we add synthetic facts to create conflicts specifying the *playFor, birthIn, winPrize, studyAt, diedIn* relations for FootballDB and *birthIn, diedIn, playFor* for DBpedia. First of all, we set facts in FootballDB and DBpedia with random assigned weights in range [0.7, 1.0]. Then we inject a fraction of 20%, 40%, 60%, 80% and 100% erroneous facts to FootballDB and DBpedia separately. For instance, injecting 20% additional erroneous facts to FootballDB means that we add 20% additional wrong facts for each of the five relations. In addition, We separately assigned weight in the range of [0, 1] and [0.5, 1] to the newly added facts.

As the constraint graph used in these experiments, we employ the graph as shown in Fig. 5

Parameters and Tool. The only free parameters are A and B in formula (3). After several experiments we fix $A = 2.68$, $B = 1.5$ that the confidence of facts with low weight is much lower than facts with high weight.

We used the state-of-art maximum weight clique solver, namely WLMC [9]. WLMC consists of two main components an efficient preprocessing process to initialize the vertex ordering and reduce the size of the graph and the BnB algorithm enables incremental vertex weighting to reduce the number of branches in the search space. It can solve high dense graph in the real word.

Competitor. We use N-Rocket [6] as a comparative test. N-Rocket can obtain a conflict-free knowledge graph with inference rules which mainly focuses on given numerical constraints.

Performance. In this experiment, we used ETC framework to detect conflict facts on the given dataset with uncertain facts. We give the precision rate and recall rate when adding erroneous temporal facts with the different ranges of weight. We repeat each experiment 10 times for the random of weight and present average rate.

In the first step, we evaluate the performance of implicit constraints and converting weights of facts to credibility by formula (3) when used in ETC framework. Implicit constraints have been used in Method 1, formula (3) has been used in Method 2 and both of these are used in Whole Method.

Table 1(a) and (b) show the performance in DBpedia and FootballDB. We respectively inject 20% and 50% additional error facts with weight in range of [0,1] into these knowledge graphs. As we can see, Method 1 and Method 2 can increase precision and recall rate and Method 1 increases more than Method 2. When executing the whole method, the precision rate and recall rate in both knowledge graphs increase considerably. So we can conclude that implicit constraints and converting weight of facts to credibility can improve recall rate and precision rate.

Table 1. Accuracy evaluation

(a) DBpedia

Process	20%		50%	
	P	R	P	R
None	75.5%	99.5%	53.2%	99.0%
Method 1	85.5%	99.8%	69.3%	99.3%
Method 2	82.9%	99.6%	57.3%	99.3%
Whole Method	86.8%	99.9%	73.1%	99.8%

(b) FootballDB

Process	20%		50%	
	P	R	P	R
None	65.9%	95.0%	50.1%	91.0%
Method 1	70.1%	97.3%	65.6%	95.8%
Method 2	68.9%	97.5%	53.9%	95.4%
Whole Method	80.4%	98.1%	72.7%	97.0%

In the second step, we evaluate the accuracy of N-Rocket and our framework which uses the Whole Method. Shown in Fig. 6, ETC-[0,1] and N-Rocket-[0,1] represent that we randomly assign error facts with weights in the range [0,1] to do experiments. Meanwhile, ETC-[0.5,1] and N-Rocket-[0.5,1] means that we inject error facts with weights in range of [0.5,1]. We give the precision rate and recall rate of our approach and N-Rocket when adding erroneous temporal facts with the different range of weight.

Figure 6 depicts that when injecting error facts with weight in range [0,1], ETC performs better than N-Rocket in both DBpedia and FootballDB. Even in a setting where we add 100% erroneous temporal facts, we are still able to achieve a precision of 63% in DBpedia and 65% in Football. When arranging error facts with weight in the range of [0.5, 1] that reducing the gap of weight between correct and error facts, the recall rate of ETC is little worse than N-Rocket owning to implicit constraints have not been considered in N-Rocket. However, the precision rate is still much higher than that of N-Rocket. We still achieve a precision of 65% for ETC in DBpedia and 52% in FootballDB when adding 100% erroneous facts.

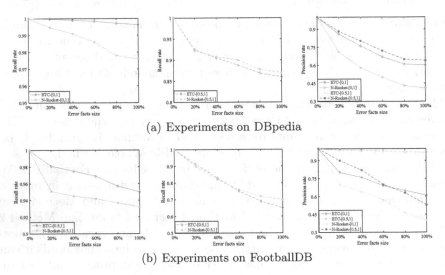

(a) Experiments on DBpedia

(b) Experiments on FootballDB

Fig. 6. Accuracy comparison with N-Rocket

6   Conclusion and Future Work

We have proposed ETC framework to eliminate temporal conflicts in uncertain temporal knowledge graphs. Under the premise of uncertain temporal knowledge graph definition, we propose five temporal constraint types. Then, we used constraint graphs to detect conflict facts in UTKGs. Implicit constraints have first been put forward to detect conflicts as much as possible. We achieve a conflict-free knowledge graph with maximum possibility by solving maximum weight clique. We carried out experiments on a state-of-art tool and real-world knowledge graphs to observe the feasibility of our approach. As for future work, we aim to discuss the temporal facts fusion between temporal knowledge graphs.

Acknowledgements. This work was supported by the National Natural Science Foundation of China under Grant Nos. 61572335, 61572336, 61472263, 61402312 and 61402313, the Natural Science Foundation of Jiangsu Province of China under Grant No. BK20151223, and Collaborative Innovation Center of Novel Software Technology and Industrialization, Jiangsu, China.

References

1. Allen, J.F.: Maintaining knowledge about temporal intervals. Commun. ACM **26**(11), 832–843 (1983)
2. Anagnostopoulos, E., Batsakis, S., Petrakis, E.G.M.: CHRONOS: a reasoning engine for qualitative temporal information in OWL. In: KES, pp. 70–77 (2013)
3. Auer, S., Bizer, C., Kobilarov, G., Lehmann, J., Cyganiak, R., Ives, Z.G.: Dbpedia: a nucleus for a web of open data. In: ISWC, pp. 722–735 (2007)
4. Bao, J., Duan, N., Yan, Z., Zhou, M., Zhao, T.: Constraint-based question answering with knowledge graph. In: COLING, pp. 2503–2514 (2016)
5. Batsakis, S., Stravoskoufos, K., Petrakis, E.G.M.: Temporal reasoning for supporting temporal queries in OWL 2.0. In: KES, pp. 558–567 (2011)
6. Chekol, M.W., Pirrò, G., Schoenfisch, J., Stuckenschmidt, H.: Marrying uncertainty and time in knowledge graphs. In: AAAI, pp. 88–94 (2017)
7. Chen, Y., Wang, D.Z.: Knowledge expansion over probabilistic knowledge bases. In: SIGMOD, pp. 649–660 (2014)
8. Dylla, M., Sozio, M., Theobald, M.: Resolving temporal conflicts in inconsistent RDF knowledge bases. In: BTW, pp. 474–493 (2011)
9. Fang, Z., Li, C., Xu, K.: An exact algorithm based on maxsat reasoning for the maximum weight clique problem. J. Artif. Intell. Res. **55**, 799–833 (2016)
10. Gutiérrez, C., Hurtado, C.A., Vaisman, A.A.: Temporal RDF. In: ESWC, pp. 93–107 (2005)
11. Gutierrez, C., Hurtado, C.A., Vaisman, A.A.: Introducing time into RDF. IEEE Trans. Knowl. Data Eng. **19**(2), 207–218 (2007)
12. McCusker, J.P., Dumontier, M., Yan, R., He, S., Dordick, J.S., McGuinness, D.L.: Finding melanoma drugs through a probabilistic knowledge graph. PeerJ Comput. Sci. **3**, e106 (2017)
13. Mitchell, T.M., Cohen, W.W., Jr., E.R.H.: Never-ending learning. In: AAAI, pp. 2302–2310 (2015)
14. Padia, A.: Cleaning noisy knowledge graphs. In: ISWC (2017)
15. Saeeda, L., Kremen, P.: Temporal knowledge extraction for dataset discovery. In: ISWC (2017)
16. Schlobach, S., Huang, Z., Cornet, R., van Harmelen, F.: Debugging incoherent terminologies. J. Autom. Reason. **39**(3), 317–349 (2007)
17. Singh, S.P., Markovitch, S. (eds.): Conference on Artificial Intelligence. AAAI Press, San Francisco (2017)
18. Singla, P., Domingos, P.M.: Lifted first-order belief propagation. In: AAAI, pp. 1094–1099 (2008)
19. Sirin, E., Parsia, B., Grau, B.C., Kalyanpur, A., Katz, Y.: Pellet: a practical OWL-DL reasoner. J. Web Sem. **5**(2), 51–53 (2007)
20. Suchanek, F.M., Kasneci, G., Weikum, G.: Yago: a core of semantic knowledge. In: WWW, pp. 697–706 (2007)
21. Zou, L., Özsu, M.T., Chen, L., Shen, X., Huang, R., Zhao, D.: gstore: a graph-based SPARQL query engine. VLDB J. **23**(4), 565–590 (2014)

Renovating Watts and Strogatz Random Graph Generation by a Sequential Approach

Sadegh Nobari[1], Qiang Qu[2(✉)], Muhammad Muzammal[2], and Qingshan Jiang[2]

[1] Rakuten, Inc., Tokyo, Japan
sadegh.nobari@rakuten.com
[2] Shenzhen Institutes of Advanced Technology,
Chinese Academy of Sciences, Beijing, China
{qiang,muzammal,qs.jiang}@siat.ac.cn

Abstract. Numerous data intensive applications call for generating gigantic random graphs. The Watts-Strogatz model is well noted as a fundamental, versatile yet simple random graph model. The Watts-Strogatz model simulates the "small world" phenomenon in real-world graphs that includes short average path lengths and high clustering. However, the existing algorithms for the Watts-Strogatz model are not scalable. This study proposes a sequential algorithm termed ZWS that generates *exact* Watts-Strogatz graphs with fewer iterations than the state-of-the-art Watts-Strogatz algorithm and therefore faster. Given the so-called edge *rewiring* probability p ($0 \leq p \leq 1$) of the Watts-Strogatz model and m neighbouring nodes of v nodes, ZWS needs $p \times m \times v$ random decisions while the state-of-the-art Watts-Strogatz algorithm needs $m \times v$, such that for large graphs with small probability, ZWS is able to generate Watts-Strogatz graphs with substantially less iterations. However, the less iterations in ZWS requires complex computation which avoids ZWS to achieve its full practical speedup. Therefore, we further improve our solution as *PreZWS* that enhances ZWS algorithm through precomputation techniques to substantially speedup the overall generation process practically. Extensive experiments show the efficiency and effectiveness of the proposed scheme, e.g., *PreZWS* yields average speedup of 2 times over the state-of-the-art algorithm on a single machine.

Keywords: Random graph generation
Exact Watts-Strogatz graphs · Web graphs
A sequential approach · High performance

1 Introduction

Generating random graphs is gaining attention in many data intensive applications due to its wide uses, e.g., evaluating algorithms [2,42,43] and simulating

© Springer Nature Switzerland AG 2018
H. Hacid et al. (Eds.): WISE 2018, LNCS 11233, pp. 348–363, 2018.
https://doi.org/10.1007/978-3-030-02922-7_24

processes [14]. Furthermore, a random-graph generation process may also serve sampling [24,29]. Thanks to the versatility of graphs as data representation model, random graph models have been widely utilized in plentiful fields. For instance, Crisóstomo et al. [6] utilize random graphs to study the propagation of probabilistic information. The authors in the study [9] investigate the dispersal of viruses in synthetic networks because the epidemic dispersing behavior determines the dispersing behavior of graphs following a power law. In the study [19], uniform random graphs are used to explore data search and replication in peer-to-peer networks. The authors conclude that uniform random graphs perform the best. Other examples and applications of random graph models can also be found in biology [20,23] and social network studies [8,13,25].

Random graph generation process has been employed for growing real graphs as well. Graph growth models not only provide theoretical abstractions for studying natural complex graphs, but also can be used to *generate* synthetic graphs for further analysis and simulation. In the last two decades, various randomized algorithms have been developed to generate synthetic graphs for different growth models. The primary motivation for recent interest in generating synthetic graphs is that, real-world networks are hardly accessible to researchers due to privacy concerns. In addition, synthetic graph generators can be used to create multiple independent graphs with similar statistical properties, which is very useful for statistical tests, e.g. significance tests and cross-validation [18]. Recently, research is being shifted towards modeling very large networks, because small input graphs may not exhibit the same properties that can be found in large-scale graphs [1, 16].

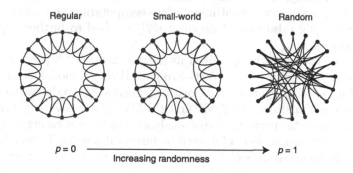

Fig. 1. Watts-Strogatz networks for varying probabilities p, borrowed from [39].

The random graph models can model directed and undirected graphs with and without self loops, as well as bi- and multipartite graphs. Among fundamental, versatile yet simple random graph models Erdős-Rényi and Watts-Strogatz can be noted. Erdős-Rényi is a pure random model with no structural property guarantee while the Watts-Strogatz model addresses the limitation of the Erdős-Rényi model, e.g. the local clustering and triadic closures [39]. The Watts-Strogatz model simulates the "small world" phenomenon in real graphs that includes short average path lengths and high clustering. Generally, the Watts-Strogatz model first constructs a ring in which each vertex is linked to its m nearest neighbors. Then it uniformly, i.e. with the same probability, rewires each edge of the ring to a non existing edge in order to produce a new graph with small-world properties. The structural property guarantees and yet simplicity of the Watts-Strogatz model among the synthetic graphs, allow this specific model to find applications ranging from physics [38] and biology [34] to networks [21] and sociology [35]. We thus focuses on the study of the Watts-Strogatz model.

Despite the omnipresence of random graphs as a data representation tool, previous research has not paid due attention to the question of efficiency in random graph generation. Naïve implementations of the basic Watts-Strogatz graph generation process do not scale well to very large graphs. This research proposes novel sequential algorithms, namely from WS to PreZWS, for random graph generation under the $WS_{v,m,p}$ model in which v is the number of nodes, m is the number of neighbours of each node and p is the rewiring probability. We outline our full algorithm, i.e. PreZWS, via a progression for the Watts-Strogatz model, bringing in additional features. Each algorithm in this progression forms a random graph generation in its own right, and the effect of the additional feature can be evaluated via a comparison to its predecessor. All our algorithms can be tuned to generate specific graphs (e.g., directed, undirected, with or without self-loops, and multipartite) by an orthogonal decoding function.

Our baseline algorithm, *WS*, employs the random process of Gilbert's model to the Watts-Strogatz random generation process. The second algorithm *ZWS* exploits the possibility to skip the expected number of edges in a geometric approach. Finally, we propose *PreZWS* to improve *ZWS* by avoiding the expensive computation of logarithms via pre-computation and using an acceptance/rejection [33] (also named Ziggurat [22]) method to further improve the skipping techniques.

The rest of the study is organized as follows. Section 2 briefly presents the principles and properties of the Watts-Strogatz $WS_{v,m,p}$ model along with two corresponding baseline generation algorithms. Section 3 introduces the proposed sequential algorithms, namely ZWS and PreZWS. Extensive experiments are conducted to study the performance of all algorithms for generating graphs under Watts-Strogatz model in Sect. 4 followed by further discussion in Sect. 5. Related work is briefly reviewed in Sect. 6, and we conclude the paper in Sect. 7.

2   Watts-Strogatz Model

In this section, we describe the Watts-Strogatz random graph generation model in detail. The Watts-Strogatz model generates undirected graphs in two steps with 3 parameters, $WS_{v,m,p}$, in which v is the number of nodes, m is the number of neighbours of each node, and p is the rewiring probability. This model first draws a ring lattice, n vertices each connected to its m nearest neighbors by undirected edges. Then, the model selects an edge, i.e. sampling the edges with probability p, from the so-called ring and rewires the edge. Rewiring intends to manipulate the structure of the regular ring lattice to a random network. The rewiring procedure for every vertex v selects the first Nearest Neighbor (NN) u in the clockwise direction, with probability p, and reconnects v to a vertex selected uniformly out of all the vertices, excluding v and the vertices that v is currently connecting. In order to select a vertex for rewiring, we uniformly select a vertex out of all the vertices. If the selected vertex produces a self-loop or parallel edges, we select another vertex uniformly until we find a valid one. Next, similarly for every vertex, we consider the second NN for rewiring and then the third until the $m/2$ NN. Figure 1 illustrates how varying p from 0 to 1 can result to the regular ring lattice, an intermediate network, and a totally random network.

Given v as the number of vertices, the Watts-Strogatz model generates a graph of v vertices and $\frac{v \times k}{2}$ edges. This model first constructs the so-called ring lattice, then iteratively rewires each edge of this ring with probability p. The process can be straightforwardly implemented by a loop over $\frac{v \times k}{2}$ edges, rewiring an edge (i, j), if a random number between 0 and 1 drawn uniformly is smaller than p, to an edge (i, j') such that j' is selected uniformly at random among $\{j' \in [0, j-1] | j' \neq i \wedge (i, j') \notin E\}$.

Algorithm 1. Watts-Strogatz

Input: v : number of vertices, indexed 0 to $v - 1$; p : inclusion probability; m : average degree

Output: G: a Watts and Strogatz graph

1   $G = \emptyset$;
2   **for** $i = 0$ *to* $v - 1$ **do**
3   **for** $j = i + 1$ *to* $i + \frac{m}{2}$ **do**
4       $G \leftarrow (i, j \bmod v)$;
5   **for** $i = 0$ *to* $v - 1$ **do**
6   **for** $j = i + 1$ *to* $i + \frac{m}{2}$ **do**
7   *Draw a uniform random number* $\theta \in [0, 1)$;
8   **if** $\theta < p$ **then**
9         $G = G - (i, j \bmod v)$;
10   *Draw a uniform random number* $j' \in [0, v - 1]$;
11   **while** $j' = i \vee (i, j') \in G$ **do**
12   *Draw a uniform random number* $j' \in [0, v - 1]$;
13         $G \leftarrow (i, j')$;

The pseudocode outlined by Algorithm 1 is to generate graphs under the Watts-Strogatz model and it is designed for a particular desired type of graph, i.e., a undirected graph of v vertices without self-loops. However, an algorithm for the Watts-Strogatz model can be implemented without prejudice to particular desired graph types. Given the number of possible edges E, in a certain graph, the algorithm suffices to generate indices between 0 and $E - 1$. The indices can be decoded with regard to graph types. For instance, the edge with the index 13 for a directed graph of 5 vertices (labeled 0 to 4) with self-loops is the edge between vertices 2 and 3.

Algorithm 2 produces the decoding. Note that similar coding and decoding functions are available for directed and undirected graphs, with and without self loops, as well as for multipartite graphs.

Algorithm 2. Edge Decoding
___
Input: v : number of vertices; ind : edge index
Output: decoded edge (i, j) from vertex i to vertex j
1 $i = \lfloor \frac{ind}{v} \rfloor$;
2 $j = ind \mod v$;

The decoding is orthogonal to the algorithm performance. Thus, any decoding can be relegated to the stage after graph generation. For sampling, the decoding is rendered irrelevant, as a particular graph to be sampled is given. As a result, without loss of generality, we do not present particular decoding functions, which are *the same* for all the proposed algorithms in this study, yielding the same further speedup. They are inconsequential to the performance study. We now rewrite the WS algorithm using a single loop instead of two nested loops over the vertices. Algorithm 3 shows the decoding for the Watts-Strogatz model when we iterate through edges, serving as our WS algorithm.

3   The Proposed Methods

The definition of the aforementioned model recalls a known geometric distribution corresponding to the Bernoulli process. In this section, we propose how to exploit this distribution feature and pre-computing to substantially scale the generation ($\equiv$ sampling) process. To this end, we introduce two algorithms to explain each of the mentioned ideas, namely ZWS and PreZWS for Watts-Strogatz model.

3.1   Skipping Edges in ZWS

The WS algorithm employs a procedure of successive edge processing, i.e., selecting and rewiring the edges. It is a Bernoulli process with probability p and sequence length E as the total possible edges in a graph. The number of processed edges follows the distribution $B(E, p)$, hence the mean is:

$$\mu = p \times E, \tag{1}$$

and the standard deviation is:

$$\sigma = \sqrt{p \times (1 - p) \times E}. \tag{2}$$

Let k be the number of skipped edges before the next edge processed for rewiring. k is following a geometric distribution with respect to the parameter p. The probability that k edges are skipped at any step of the sequence is:

$$f(k) = (1 - p)^k \times p. \tag{3}$$

Algorithm 3. Watts-Strogatz (single loop)

Input: v : number of vertices, indexed 0 to $v - 1$; p : inclusion probability; m : average degree

Output: G: a Watts and Strogatz graph

1   $G = \emptyset$;

2   $LastE = \frac{v \times p}{2}$;

3   **for** $j = 0$ *to* $LastE$ **do**

4      $u_1 = \frac{j}{\frac{m}{2}}$;

5      $u_2 = j \mod \frac{m}{2}$;

6      $i = u_1 \times v + u_2$;

7      $G \leftarrow e_i$;

8   **for** $j = 0$ *to* $LastE$ **do**

9   *Draw a uniform random number* $\theta \in [0, 1)$;

10   **if** $\theta < p$ **then**

11        $u_1 = \frac{j}{\frac{m}{2}}$;

12       $u_2 = j \mod \frac{m}{2}$;

13       $i = u_1 \times v + u_2$;

14       $G = G - e_i$;

15   *Draw a uniform random number* $j' \in [0, v - 1]$;

16       $i = u_1 \times v + j'$;

17   **while** $j' = i \vee e_i \in G$ **do**

18   *Draw a uniform random number* $j' \in [0, v - 1]$;

19         $i = u_1 \times v + j'$;

20       $G \leftarrow e_i$;

The corresponding cumulative distribution function is:

$$F(k) = \sum_{i=0}^{k} f(i) = 1 - (1 - p)^{k+1} \tag{4}$$

with the mean $\frac{1-p}{p}$ and the standard deviation $\frac{\sqrt{1-p}}{p}$.

As argued in the study [2], the per se computation of each skipped edge during the Bernoulli process can be avoided. Instead, at the beginning of the process, and when an edge has been processed, we can randomly generate a value k (i.e., the number of skipped edges), and thus directly process the next $(k + 1)^{th}$ edge. To generate the value, we follow the below method. Let α be a random number drawn uniformly in $(0, 1]$. Then, the probability that α falls in the interval $(F(k - 1), F(k)]$ is $F(k) - F(k - 1) = f(k)$. In other words, the probability that k is the smallest positive integer such that $\alpha \leq F(k)$ is $f(k)$. Next, to assure that each possible value of k is generated with probability $f(k)$, it suffices to generate α and then calculate k as the *smallest* positive integer such that $F(k) \geq \alpha$, hereby $F(k-1) < \alpha \leq F(k)$ (or zero if no such positive integer exists). Set $\varphi = 1 - \alpha$, which is equivalent to:

$$1 - (1-p)^k < \alpha \leq 1 - (1-p)^{k+1} \Leftrightarrow \tag{5}$$

$$(1-p)^{k+1} \leq \varphi < (1-p)^k, \tag{6}$$

where φ is drawn uniformly at random in $[0, 1)$. According to Eq. 6, we have

$$k = \max(0, \lceil \log_{1-p} \varphi \rceil - 1) \tag{7}$$

Theoretical Speedup. In the Watts-Strogatz graph generation model, we want to rewire E edges with probability p. The straightforward approach operates on every edge, so the algorithm repeats the random process for $|E|$ times, i.e. $\frac{|V| \times m}{2}$, in a $WS_{v,k,p}$ model. However, thanks to the Eq. 7, we can reduce the number of iterations of the random process from E to the number of times we need to calculate skips (k) in order to produce a graph of E edges. Therefore, the theoretical speedup(s) of this technique is $s = \frac{E}{\mathcal{E}(n)}$, where $\mathcal{E}(n)$ is the expectation of n. $\mathcal{E}(n)$ depends on the expected value of k, which is

$$\mathcal{E}(k) = \lim_{l \to +\infty} \sum_{i=0}^{l} i \times (1-p)^i \times p = \frac{1-p}{p}. \tag{8}$$

As k edges are skipped, the expected value of the offset of the next processed edge $k+1$ is $\mathcal{E}(k+1) = \frac{1-p}{p} + 1 = \frac{1}{p}$. If n skips are required to process all E edges, we have $\sum_{i=1}^{n} \mathcal{E}(k+1) = n \times \frac{1}{p} = E$. Thus, the expectation of skips is $\mathcal{E}(n) = p \times E$. Hereby, the theoretical speedup is $s = \frac{E}{\mathcal{E}(n)} = \frac{1}{p}$ times. For instance, when the given probability of a random graph generation is 10%, we expect to have 10× speedup.

ZWS: Skipping Edges in WS. By the preceding discussion, we can eschew the generation of a random number for each rewiring; instead, we compute the offsets of edge-skipping steps. The WS algorithm consists of two phases, drawing a regular ring lattice and rewiring the edges of the ring. In the rewiring phase of the WS algorithm, we can take advantage of the proposed edge-skipping technique. Because the edge-skipping process is similar with the Z Reservoir sampling algorithm [37], the algorithm is named ZWS that differs from WS by leveraging the skip technique in the rewiring phase. Therefore, ZWS, in contrast to WS, instead of drawing a random number for every edge, in order to decide whether rewire the edge, computes a skip value and jumps from the edges that do not require rewiring with the given probability p. Algorithm 4 is the pseudocode of skipping edges in the WS algorithm, i.e. ZWS.

3.2   Pre-computing Techniques in PreZWS

The logarithm computation prevents the edge skipping technique, e.g. ZWS, from achieving its full potential. If we generate a 16-bit random number

---

Algorithm 4. ZWS

Input: v : number of vertices, indexed 0 to $v - 1$; p : inclusion probability; m :
average degree

Output: G: a Watts and Strogatz graph

1 $G = \emptyset$;
2 $LastE = \frac{v \times p}{2}$;
3 **for** $j = 0$ *to* $LastE$ **do**
4      $u_1 = \frac{j}{\frac{m}{2}}$;
5      $u_2 = j \mod \frac{m}{2}$;
6      $i = u_1 \times v + u_2$;
7      $G \leftarrow e_i$;
8 $j = -1$;
9 **while** $j < LastE$ **do**
10   *Draw a uniform random number* $\varphi \in [0, 1)$;
11   *Calculate the skip value* $s = \max(0, \lceil \log_{1-p} \varphi \rceil - 1)$;
12      $j = j + s + 1$;
13      $u_1 = \frac{j}{\frac{m}{2}}$;
14      $u_2 = j \mod \frac{m}{2}$;
15      $i = u_1 \times v + u_2$;
16      $G = G - e_i$;
17   *Draw a uniform random number* $j' \in [0, v - 1]$;
18      $i = u_1 \times v + j'$;
19   **while** $j' = i \vee e_i \in G$ **do**
20   *Draw a uniform random number* $j' \in [0, v - 1]$;
21          $i = u_1 \times v + j'$;
22      $G \leftarrow e_i$;
23 *Discard the last edge*;

$\varphi \in (0, 1]$, φ may have $2^{16} = 65,536$ different values. We might have a small number of calls of the logarithm function (line 11 of ZWS). For example in WS, assume that we produce a graph of $v = 1,000,000$ vertices and $m = 100$, hence we have $E = 10^8$ candidate edges under probability $p = 0.1$. Then, we expect to have $\mathcal{E}(n) = p \times E = 10,000,000$ logarithm calls. As a result, the pre-computation of the logarithm values of all the possible $65,536$ random numbers shall improve the performance. Such pre-computation can be especially useful when we generate multiple random graphs. We can achieve the improvement if we assume 16-bit random numbers. However, this improvement does not hold any more if random numbers are of higher precision. The findings of empirical experiments show that when we use 32-bit, the pre-computation is only competitive for generating very large graphs with high probability values p. Still, we desire our algorithm to be of general utility for reasonably-sized graphs and high random-number precision. Subsequently, we design methods benefiting from the above observation to prevent costly logarithm computation.

By Eq. 3, PreZWS pre-computes the $m + 1$ breakpoints of the intervals in which random number α is most likely to fall. Then, α is uniformly generated at random in the interval $[0, 1)$, and we compare it with the cumulative probability $F(k)$ for $k = 0$ to m.

Algorithm 5. PreZWS

Input: v : number of vertices; p : inclusion probability; m : average degree
Output: G: a Watts and Strogatz graph

1 $G = \emptyset$;
2 **for** $t = 0$ *to* m **do**
3 | *Compute the cumulative probability* $F[t]$;
4 $LastE = \frac{v \times p}{2}$;
5 **for** $j = 0$ *to* $LastE$ **do**
6 | $u_1 = \frac{j}{\frac{m}{2}}$;
7 | $u_2 = j \mod \frac{m}{2}$;
8 | $i = u_1 \times v + u_2$;
9 | $G \leftarrow e_i$;
10 $j = -1$;
11 **while** $j < LastE$ **do**
12 | *Draw a uniform random number* $\alpha \in (0, 1]$;
13 | $r = 0$;
14 | **while** $r \leq t$ **do**
15 | | **if** $F[r] > \alpha$ **then**
16 | | | *Set the skip value* $k = r$; **Break**;
17 | | $r = r + 1$;
18 | **if** $r = t + 1$ **then**
19 | | *Compute the skip value* $s = \lceil \log_{1-p}(1-\alpha) \rceil - 1$;
20 | $j = j + s + 1$;
21 | $u_1 = \frac{j}{\frac{m}{2}}$;
22 | $u_2 = j \mod \frac{m}{2}$;
23 | $i = u_1 \times v + u_2$;
24 | $G = G - e_i$;
25 | *Draw a uniform random number* $j' \in [0, v - 1]$;
26 | $i = u_1 \times v + j'$;
27 | **while** $j' = i \vee e_i \in G$ **do**
28 | | *Draw a uniform random number* $j' \in [0, v - 1]$;
29 | | $i = u_1 \times v + j'$;
30 | $G \leftarrow e_i$;
31 *Discard the last edge*;

The above method is to pre-compute all the needed logarithm values such that we do not need to compute the same value twice. However, it would be more effective if we totally avoid the logarithm computation. Indeed, instead of pre-computing the logarithm values, we pre-compute the breakpoints of the cumulative distribution $F(k)$ as follows.

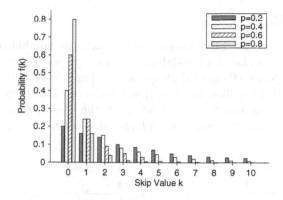

Fig. 2. f(k) for varying probabilities p.

By Eq. 3, the probability $f(k)$ decreases with k. Figure 2 illustrates it for several values of p. Thus, the value of k is likely to be smaller than some sufficiently large and fixed integer t. Next, instead of computing k as a function of $\varphi = 1 - \alpha$ over each iteration, we can pre-compute $t + 1$ breakpoints of the intervals from which k is easily determined. Then, we can set k to tbe the smallest value, thus $F(k) > \alpha$ or otherwise, if $F(t) \leq \alpha$, calculate k by calling a logarithm computation as in the algorithm ZWS. The value of t is decided in terms of the requirement of a particular application.

Algorithm 5 shows the PreZWS algorithm, where we also further improve the skipping techniques by an acceptance/rejection [33] (also named Ziggurat [22]) method given the pre-computation afterwards.

4   Experimental Study

This section shows the comprehensive study on the performance of all the proposed algorithms for generating the Watts-Strogatz graphs under the following configuration.

4.1   Settings

The experiments are conducted on a 2.33 GHz Core 2 Duo CPU machine with 4 GB RAM and a PCI Express ×16 bus. All the algorithms were implemented in Visual C++.

By default, the algorithms are set to generate directed random graphs with self-loops having 10, 000 vertices, thus at most 100, 000, 000 edges. Execution time is measured by user time. All the results are averaged over 10 runs.

The parameter r for the algorithm PreZWS is set to 9, which empirically yields the best performance shown in Sect. 4.3.

4.2   Size Scalability

For hitherto, we have used two graph sizes N (i.e., the number of vertices $|V|$): 1 and 10 million. For clarity of scalability, we put graphs of various sizes beside each other. In Figs. 3 to 6, the (a) charts are graphs with 1 million vertices and the (b) charts with 10 million vertices. The results confirm the previous findings on larger graphs. The results show that given the same parameters (i.e. m, k and p), PreZWS generates graphs with larger size faster.

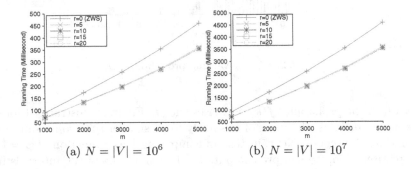

(a) $N = |V| = 10^6$ (b) $N = |V| = 10^7$

Fig. 3. Average runtime of PreZWS for $p = 0.5$ and varying neighbours (m)

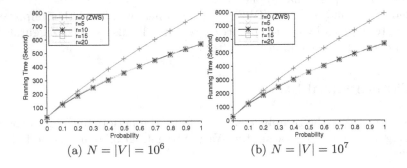

(a) $N = |V| = 10^6$ (b) $N = |V| = 10^7$

Fig. 4. Average runtime of PreZWS for $m = 5000$ and varying probability (p)

4.3   Performance Tuning

An experimental study is also conducted for our tuning decisions. In PreZWS, we need to pre-compute a fixed number of cumulative probabilities, r. In order to find the best r, we vary r values given various m and p, and we measure the whole execution time. Figures 3 and 4 illustrate the running time of the PreZWS algorithm when varying m and p, respectively. The results show that only marginal benefits can be gained for r values larger than 9, due to the trade-off discussed before. Therefore, we use the value $r = 9$ in our experiments.

4.4   Overall Comparison

We compare the performance of all the algorithms in the study, namely WS, ZWS and PreZWS on a single machine. Figures 5 and 6 illustrate the running time of all the algorithms when varying m and p, respectively. The findings show that when the probability is small, like $p = 0.1$, the proposed algorithms are about 2 times faster than the state of the art (WS) on average, of which PreZWS is the outperforming algorithm, thanks to the precomputation together with the skipping edges techniques. Given $p = 0.5$, which is big, PreZWS is still obviously the winner observed from Fig. 5. The runtimes of all the algorithms linearly increase with the number of neighbours (m). Furthermore, we can observe that although ZWS requires less iterations to generate the same graph as WS, the computation of logarithm in practice avoids its superiority. However, the precomputation technique in PreZWS overcomes logarithm copmutation all together. To be noted, tasks are likely to generate graphs with small p in reality since the larger the p is the less property guarantee that we would have.

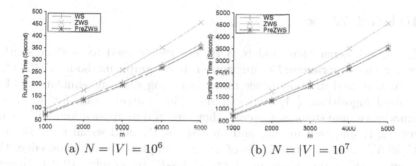

(a) $N = |V| = 10^6$ (b) $N = |V| = 10^7$

Fig. 5. Average runtime of all algorithms when $p = 0.5$ and varying neighbours

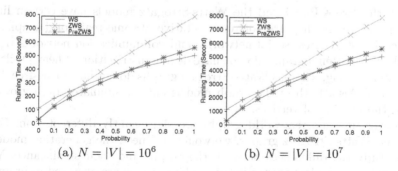

(a) $N = |V| = 10^6$ (b) $N = |V| = 10^7$

Fig. 6. Algorithm average runtime when $m = 5000$ and varying probability

5   Further Discussion

A possible objection to the significance of the study is that the improvement is achieved over an algorithm that is already relatively efficient compared to the time of writing the generated graphs into disk. Indeed, the cost to write individual or buffered results into disk dominates that of the generation in RAM. However, this is not the case when a generated graph remains in main memory. Furthermore, the relevance and applicability of the proposed methods would be enhanced by the increasing availability of flash memory storage and by the proliferation of graph-based techniques, systems, and simulation requirements in modern data management tasks.

The proposed algorithms are also efficient for generating large graphs in parallel or in a streaming fashion. This is even more the case when the generated graphs remain in the memory of the processing units such as GPU for further processing. For the sake of a fair comparison, the experiments have been conducted with all the results written to main memory. In addition, some applications require loops over the generation of graphs [10, 26].

6   Related Work

Random graph generation models [3, 10] are widely used to produce synthetic graphs for the experimental evaluation of many graph methods [17, 32, 43, 44]. They are also used in many other tasks, including scientific simulation [14, 40], randomized algorithms [11], and sampling [36]. Furthermore, for information and knowledge management systems, they are used in applications such as data mining [12, 41, 43], web mining and analysis [4, 15], and social network analysis [25, 30, 31], among a multitude of other examples. To our knowledge, there exist some studies on generating large Erdős-Rényi graphs [10, 27]. However, few exist to scale the Watts-Strogatz model in generating exact Watts-Strogatz graphs.

Both the Erdős-Rényi and the Watts-Strogatz models have certain limitations, and therefore they are not representative of some real-world graphs such as molecular structures, social networks, and communication networks [7]. For example, Erdős-Rényi graphs do not have degree distribution of heavy tails, and have low clustering, whereas Watts-Strogatz graphs do not have scale-free degree distribution. Also, both of the graph models cannot simulate graph growth as they fix the number of vertices.

However, if some real-world graphs are significantly different from Erdős-Rényi and Watts-Strogatz graphs, we would still need the generation models to further study the differences between the graphs and their significance. Moreover, certain topological properties of random graphs are observed in brain networks [5, 28] and thereby being used for brain simulation by neuroscientists.

7   Conclusion

We proposed novel algorithms to generate random graphs by the Watts-Strogatz model $WS_{v,m,p}$. In order to motivate, explain, and evaluate this work, we have outlined a succession of algorithm leading to our flagship contribution. To the best of our knowledge, PreZWS is the fastest known sequential algorithm for producing *exact* Watts-Strogatz graphs. The proposed algorithm yields average speedup of 2 times over the state-of-the-art algorithm on a single machine. The findings show that the proposed algorithms enable significant efficiency gains in modern data management tasks, applications, and simulations, whenever they need the generation model as a component.

Acknowledgements. The work was partially supported by the CAS Pioneer Hundred Talents Program under grant number Y84402.

References

1. Alam, M., Khan, M., Marathe, M.: Distributed-memory parallel algorithms for generating massive scale-free networks using preferential attachment model. In: HPC (2013)
2. Batagelj, V., Brandes, U.: Efficient generation of large random networks. Phys. Rev. E **71**(3), 036113 (2005)
3. Bollobas, B.: Random graphs, 2nd edn. Academic Press (2001)
4. Broder, A.Z., et al.: Graph structure in the web. Comput. Netw. **33**(1–6), 309–320 (2000)
5. Bullmore, E., Sporns, O.: The economy of brain network organization. Nat. Rev. Neurosci. **13**(5), 336–349 (2012)
6. Crisóstomo, S., Schilcher, U., Bettstetter, C., Barros, J.: Analysis of probabilistic flooding: how do we choose the right coin. In: IEEE ICC (2009)
7. Erdös, P., Rényi, A.: On random graphs, i. Publicationes Mathematicae (Debrecen), vol. 6, pp. 290–297 (1959)
8. Fowler, J.H., Dawes, C.T., Christakis, N.A.: Model of genetic variation in human social networks. PNAS **106**(6), 1720–1724 (2009)
9. Ganesh, A., Massoulié, L., Towsley, D.: The effect of network topology on the spread of epidemics. In: INFOCOM, pp. 1455–1466 (2005)
10. Hadian, A., Nobari, S., Minaei-Bidgoli, B., Qu, Q.: ROLL: fast in-memory generation of gigantic scale-free networks. In: Proceedings of the 2016 International Conference on Management of Data, SIGMOD Conference 2016, San Francisco, CA, USA, 26 June–01 July 2016, pp. 1829–1842 (2016)
11. Hanhijärvi, S., Garriga, G., Puolamäki, K.: Randomization techniques for graphs. In: SDM, pp. 780–791 (2009)
12. Inokuchi, A., Washio, T., Motoda, H.: Complete mining of frequent patterns from graphs: mining graph data. Mach. Learn. **50**(3), 321–354 (2003)
13. Ioannides, Y.M.: Random graphs and social networks: an economics perspective. Technical report 0518, Department of Economics, Tufts University (2005)
14. Kaiser, M., Martin, R., Andras, P., Young, M.P.: Simulation of robustness against lesions of cortical networks. Eur. J. Neurosci. **25**(10), 3185–3192 (2007)

15. Kleinberg, J.M., Kumar, R., Raghavan, P., Rajagopalan, S., Tomkins, A.: The web as a graph: measurements, models, and methods. In: COCOON, pp. 1–17 (1999)
16. Leskovec, J.: Dynamics of large networks. Ph.D. thesis, CMU (2008)
17. Liu, S., Qu, Q.: Dynamic collective routing using crowdsourcing data. Transp. Res. Part B Methodol. **93**, 450–469 (2016)
18. Looz, M., Staudt, C., Meyerhenke, H., Prutkin, R.: Fast generation of dynamic complex networks with underlying hyperbolic geometry. CoRR (2015)
19. Lv, Q., Cao, P., Cohen, E., Li, K., Shenker, S.: Search and replication in unstructured peer-to-peer networks. In: SIGMETRICS, pp. 258–259 (2002)
20. Maayan, A., Lipshtat, A., Iyengar, R., Sontag, E.: Proximity of intracellular regulatory networks to monotone systems. Syst. Biol. **2**(3), 103–112 (2008)
21. Majumdar, S.: Application of scale free network on wireless sensor network. Ph.D. thesis, Jadavpur University (2014)
22. Marsaglia, G., Tsang, W.W.: The ziggurat method for generating random variables. J. Stat. Softw. **5**(8), 1–7 (2000)
23. McDonald, D., Waterbury, L., Knight, R., Betterton, M.: Activating and inhibiting connections in biological network dynamics. Biol. Direct **3**(1), 49 (2008)
24. Motwani, R., Raghavan, P.: Randomized Algorithms. Cambridge University Press (1995)
25. Newman, M.E.J., Watts, D.J., Strogatz, S.H.: Random graph models of social networks. PNAS **99**(Suppl 1), 2566–2572 (2002)
26. Nobari, S., Karras, P., Pang, H., Bressan, S.: L-opacity: linkage-aware graph anonymization. In: EDBT, pp. 583–594 (2014)
27. Nobari, S., Lu, X., Karras, P., Bressan, S.: Fast random graph generation. In: EDBT, pp. 331–342 (2011)
28. Oh, S.W., et al.: A mesoscale connectome of the mouse brain. Nature **508**(7495), 207–214 (2014)
29. Pettie, S., Ramachandran, V.: Randomized minimum spanning tree algorithms using exponentially fewer random bits. ACM Trans. Algorithms **4**(1), 5:1–5:27 (2008)
30. Qu, Q., Chen, C., Jensen, C.S., Skovsgaard, A.: Space-time aware behavioral topic modeling for microblog posts. IEEE Data Eng. Bull. **38**(2), 58–67 (2015)
31. Qu, Q., Liu, S., Yang, B., Jensen, C.S.: Integrating non-spatial preferences into spatial location queries. In: SSDBM, pp. 8:1–8:12 (2014)
32. Qu, Q., Liu, S., Zhu, F., Jensen, C.S.: Efficient online summarization of large-scale dynamic networks. IEEE Trans. Knowl. Data Eng. **28**(12), 3231–3245 (2016)
33. Robert, C.P., Casella, G.: Monte Carlo Statistical Methods. Springer Texts in Statistics. Springer, New York (2005). https://doi.org/10.1007/978-1-4757-4145-2
34. Rudolph-Lilith, M., Muller, L.E.: Neural graphs: small-worlds, after all? BMC Neuroscience **15**(Suppl 1), O13 (2014)
35. Song, H.F., Wang, X.J.: Simple, distance-dependent formulation of the watts-strogatz model for directed and undirected small-world networks. Phys. Rev. E **90**(6), 062801 (2014)
36. Tsourakakis, C.E., Kang, U., Miller, G.L., Faloutsos, C.: DOULION: counting triangles in massive graphs with a coin. In: KDD, pp. 837–846 (2009)
37. Vitter, J.S.: Random sampling with a reservoir. ACM Trans. Math. Softw. **11**(1), 37–57 (1985)
38. Wang, Y., Xu, X.: Quantum transport with long-range steps on Watts-strogatz networks. Int. J. Mod. Phys. C **27**(2), 1650015 (2015)
39. Watts, D.J., Strogatz, S.H.: Collective dynamics of 'small-world' networks. Nature **393**, 440–442 (1998)

40. Xu, Y., Liu, P., Li, X.: Discovering the influences of complex network effects on recovering large scale multiagent systems. Sci. World J. (2014)
41. Yang, J., Leskovec, J.: Defining and evaluating network communities based on ground-truth. Knowl. Inf. Syst. **42**(1), 181–213 (2015)
42. Zhou, F., Qu, Q., Toivonen, H.: Summarisation of weighted networks. J. Exp. Theor. Artif. Intell. **29**(5), 1023–1052 (2017)
43. Zhu, F., Zhang, Z., Qu, Q.: A direct mining approach to efficient constrained graph pattern discovery. In: SIGMOD, pp. 821–832 (2013)
44. Zhu, F., Qu, Q., Lo, D., Yan, X., Han, J., Yu, P.S.: Mining top-k large structural patterns in a massive network. PVLDB **4**(11), 807–818 (2011)

Which Type of Classifier to Use for Networked Data, Connectivity Based or Feature Based?

Zan Zhang[1,2(✉)], Jiuyong Li[2], Hao Wang[1], Lin Liu[2], and Jixue Liu[2]

[1] School of Computer Science and Information Engineering,
Hefei University of Technology, Hefei, Anhui, China
zhangzan99@163.com
[2] School of Information Technology and Mathematical Sciences,
University of South Australia, Adelaide, Australia

Abstract. Multi-label classification of social network data has become an important problem. Two types of information have been used to classify nodes in a social network: characteristics of nodes, and the connectivity between nodes. Existing classification methods can be categorized to two types too, feature based methods, and connectivity based methods. We observe that there are no one size fits all classification methods, since the performance is data dependent, but in general node's class labels are determined by two factors, personal preference and peer influence. However, some data sets are personal preference dominated and are suitable for feature based methods, whereas some data sets are peer influence dominated and are suitable for connectivity based methods. The challenge then is how to judge if a data set is personal preference dominated or peer influence dominated, so a suitable classification method can be selected for its classification. In this paper, we develop a causality based criterion to determine the characteristics of a data set. Experiments on real-world data sets demonstrate the criterion can predict the suitability of a classification method for a data set.

Keywords: Networked data · Multi-label classification · Causal analysis · Propensity score

1 Introduction

Nowadays, with the development of social networks, increasing attention has been paid to multi-label classification of networked data. For example, in a social media network, people may be related to multiple groups simultaneously. In a document network, one document may be related to multiple topics. In a research collaboration network, a researcher usually has multiple research interests. In all these cases, each node in the network is associated with multiple class labels representing the node's social groups, document topics, or research interests. Multi-label classification of networked data is more complex than the traditional single-label classification where a node is only associated with a single label.

Multi-label classification in general is a well studied topic in machine learning and data mining [22], but networked data has posed new challenges to the study as instances

H. Hacid et al. (Eds.): WISE 2018, LNCS 11233, pp. 364–380, 2018.
https://doi.org/10.1007/978-3-030-02922-7_25

(nodes) in networked data are interconnected. This means that traditional classification approaches, which usually assume that the individuals or instances are independently identically distributed, may not perform well [5]. Hence, some new models and algorithms have been developed for classifying networked data based on node connectivity.

Traditional multi-label classification methods, such as SOCDIM [1] or MLKNN [2], use nodes' features to classifying networked data. These feature based methods usually assume that nodes with similar features tend to have similar class labels. On the other hand, connectivity based methods, such as the relational learning method and the random walk based methods [3, 4] assume that connected nodes are more likely to have similar class labels.

The connectivity based methods have better performance than the feature based methods with some data sets [5], but the results are reversed with other data sets. None of the two types of algorithms is good for all data sets, indicating the results of the algorithms are data dependent.

The question then arises as to how to tell which algorithm is suitable for which type of data or network. We consider a node's class labels are determined by two factors, *personal preference* and *peer influence*. Personal preference in this paper refers to the personality or characteristics of an individual (instance or node) in a social network, hence we see that feature based methods are based on personal preference for the classification. Peer influence is defined as how one's behaviors change with the change of his/her friends' behaviors [6]. Hence, connectivity based methods classify a node based on the labels of this node's neighbors.

Therefore, in practice, for data sets with stronger peer influence among connected nodes, comparing to the impact of personal preference, connectivity based methods should have better performance than feature based methods, while feature based methods are expected to outperform connectivity based methods on the data sets where nodes show strong personal preference. Therefore, estimating the impacts of these two factors and choosing appropriate classification methods accordingly are very important for multi-label classification of networked data.

However, give a social network, it is difficult to estimate the contributions of peer influence and personal preference on a node's labels. For example, take the adoption of a certain product as a label of a node in a social media network, peer influence is associated with the presence of adopters of the same product among the node's friends (called adopter friends hereafter). The influence of personal preference is associated with other adopters (not friend) with similar personal characteristics. The impacts of peer influence and personal preference are intertwined, it is difficult to estimate how much one's behavior is due to the influence of her friends and how much is a result of personal preference (influence of like-mined non-friends) only.

In this paper, we propose a causation-based criterion to distinguish Peer influence from Personal preference (CPP) in a social network data, and link the two factors to different types of classification models. We propose a criterion to differentiate a data set for its suitability for different types of classification methods. We also show that an integration of different types of methods is promising for data sets neither peer influence nor personal preference dominated.

2 Related Work

Multi-label learning has been successfully applied to various applications in the past decades, such as text classification [7, 25], image classification [8–11], bioinformatics [12] and music [26, 27]. In recent years, many researchers have paid attention to multi-label classification on social network data, and two types of methods have been used in multi-label classification of networked data.

The first type is the traditional feature based methods, such as the Social-Dimension-based (SocDim) method [1], MLKNN [2] and MLNB [11]. These methods use nodes' features to classify networked data. SocDim captures different affiliations among nodes in a network by extracting latent social dimensions via modularity maximization. Nodes' social dimensions are used as features to construct a discriminative SVM model to predict the node labels. MLKNN is an adaptation of the k-nearest neighbor (KNN) algorithm for multi-label data. MLKNN uses the KNN algorithm separately for each label. It finds the k nearest examples to the test instance in the feature space. MLKNN can also be used to classify networked data based on node features. MLNB adapts the naïve Bayes classifiers for multi-label classification. To improve performance, feature selection strategies based on principal component analysis and genetic algorithms are incorporated into the method.

The second type consists of connectivity based methods. In a social network, nodes are linked by edges for some social connection. Linked nodes may have higher probability to have the same labels, compared to the unlinked nodes. Some methods have been proposed to classify social network data using the connectivity.

Neville and Jensen [13] presented an iterative classification procedure to exploit the relationships between nodes. This approach uses a simple Bayesian classifier in an iterative fashion. Heatherly et al. [14] modified the naïve Bayes classifier to make use of the link type information for classification. The weighted-vote relational neighbor (wvRN) algorithm [3] classifies a node based on its neighboring nodes' labels. It calculates the class membership probability of a node as a weighted average of the estimated class membership probabilities of the neighboring nodes.

Lin and Cohen proposed the Multi Rank Walk [15] approach, which is based on the principle of random graph walks. The structural neighborhood based classifier (SNBC) [4] builds model to classify social network data by random walk based on the links between nodes. When classifying a node, it takes a random walk from the node and makes a decision based on how nodes in the respective kth-level neighborhood are labeled.

3 Problem Definition

Let $G = (V, E, C, F)$ represent a social network, where $V = \{v_1, v_2, \ldots, v_n\}$, is the set of nodes denoting the objects or variables in the network; E is the set of edges between the nodes; $C = \{c_1, c_2, \ldots, c_m\}$, is the set of node labels; $F = \{F_i = (f_1, f_2, \ldots, f_q), 1 \leq i \leq n\}$, is a set of vectors each denoting the features of a node. N_i is the set of neighbors of v_i, i.e. nodes directly linked to v_i.

The labels of a node $v_i \in V$ can be described by a binary vector, $L_i = \left(l_i^{c_1}, l_i^{c_2}, \ldots, l_i^{c_m} \right) \in \{0, 1\}^m$, where $l_i^{c_k} = 1$ if c_k is a label of v_i; otherwise $l_i^{c_k} = 0$. For instance, suppose that $C = \{c_1, c_2, c_3\}$, i.e. in a social network, three labels are considered, e.g. buying a novel, a magazine, or a biography. $L_i = (0, 1, 0)$ indicates that v_i has bought a magazine.

Classification on the networked data can be stated as follows. Given G and a training set $V' \subset V$, where the label vector L_i for every node in V' is known. Network data classification is to learn a model in V' to predict unknown L_j for every node $v_j \in (V - V')$. That is to predict nodes' labels based on the observed behaviors of other nodes in the same network.

Peer influence and personal preference are the two major factors impacting the labels of a node in a network. Peer influence (or peer pressure) [6] is the direct influence on people by peers, or the effect on an individual who gets encouraged to follow their peers by changing their attitudes, values or behaviors to conform to those of the influencing group or individual.

An example of a sports club network is shown in Fig. 1(a). In this network, members are represented by nodes, and two members are linked together if they are friends. We use basketball and football as labels. A member has a label if he/she likes a sport and each member can have multiple labels.

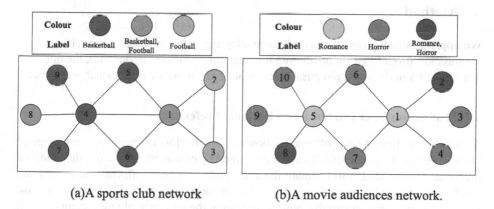

(a)A sports club network   (b)A movie audiences network.

Fig. 1. Exemplar peer influence and personal preference dominated networks

The network in Fig. 1(a) is peer influence dominated. In this network, most people have the same label with their friends. People play basketball and football with their friends, and they join a club just because of a friend's recommendation.

Connectivity information is very important for classifying nodes' labels in a peer influence dominated network, since connected nodes are more likely to have similar labels. Therefore, connectivity based methods, which makes use of node connectivity in classification, are suitable for the classification for this type of networks.

Personal preference is the influence on people by their own characteristics that affect their cognitions, emotions, motivations, and behaviors.

An example of a movie audience network is shown in Fig. 1(b). In this network, audiences are represented by nodes, and two audiences are linked together if they are friends. We use two movie types, romance and horror as labels. An audience has a label if he/she likes a movie type and an audience can have two labels.

The network in Fig. 1(b) is personal preference dominated. In this network, many people have different labels from their friends, as people's preference for movie types is usually determined by their personalities or experience, not by peer influence.

Nodes' features are crucial for classifying nodes' labels in a personal preference dominated network. Nodes with similar features describing their preferences tend to have same labels. Therefore, feature based methods, which makes use of node features in classification, are suitable for the classification for this type of networks.

For a network in which neither of the two factors dominates, it would be desirable to combine the strength of both connectivity based and feature based methods for classifying a node.

The question is how we know a networked data set is peer influence dominated, personal preference dominated or neither, so we can choose a right method for its classification. Our aim is to develop a criterion to judge if a networked data set is peer influence dominated, personal preference dominated or neither.

4   Method

We apply a causality based approach to develop the criterion since causal relationships are the most proper tool for attributing the outcome to factors. In our case, the outcome is a label of a node, and the (causal) factors are peer influence or personal preference.

4.1   Estimating Peer Influence and Personal Preference

A causal definition of peer influence is how one's behaviors change with the change of his/her friends' behaviors [6]. It can be quantified by casual effect. We use the potential outcome model [16–18] to estimate the causal effect of peer influence on one's class labels. Comparing to correlation analysis, causal approaches assess the effect of a cause variable on the outcome variable while eliminating the influence of other factors, which enables us to have a "purer" estimation of peer influence.

For each label $c \in C$ and each node $v_i \in V$, we consider having neighbors with label c (peer influence) as the treatment to v_i, and v_i's label as the outcome. That is, we define T^c, the treatment variable as:

$$T^c = \begin{cases} 1, & v_i \text{ has one or more neighbors with label } c \\ 0, & \text{otherwise} \end{cases}$$

Following the potential outcome model, v_i has two potential outcomes, Y_i^1, the potential outcome of v_i receiving the treatment ($T^c = 1$), and Y_i^0, the potential outcome of v_i without receiving the treatment ($T^c = 0$), then the causal effect of the treatment is:

$$CE_i(T^c) = Y_i^1 - Y_i^0 \qquad (1)$$

When we aggregate the causal effects over all nodes, we have the average causal effect (ACE) of T^c as:

$$ACE(T^c) = E(Y_i^1) - E(Y_i^0) \qquad (2)$$

However, in reality, only one potential outcome is observed. We cannot observe the labels of the same node with and without a neighbor at the same time. We can only observe either Y_i^1 or Y_i^0 for each node.

To estimate causal effects, matching methods [19] are often used to obtain a treated group (individuals with $T^c = 1$) and a control group (individuals with $T^c = 0$) such that the distributions of the covariates in the two groups are similar. In this way, nodes in the two groups have similar features (described by F) except their status of getting treatment T^c, hence the effects of the covariates on the outcomes can be eliminated, and hence we can estimate the ACE of T^c (ACE^c for short) as the difference in the average (observed) outcomes of the two groups:

$$ACE^c = E(Y^{obs}|T^c = 1) - E(Y^{obs}|T^c = 0) \qquad (3)$$

where Y^{obs} is the observed outcome.

We use the simple example in Fig. 2 to show the advantage of the matching method. In this example, we want to estimate the peer influence on buying a novel, i.e. the effect of one's friends buying a specific novel on the person's behavior of buying the same novel. Figure 2(a) is the original data set, showing that 7 people (ID: 1, 2, 3, 4, 9, 10, 11) have adopter friends (who adopted a specific novel), so they are considered treated ($T = 1$).

The other 7 people (ID: 5, 6, 7, 8, 12, 13, 14) do not have adopter friends, so they are untreated ($T = 0$). Y_i is the outcome variable, representing a person buying a specific novel ($Y_i = 1$) or not ($Y_i = 0$). f_1, f_2, f_3 and f_4 are these persons' characteristics.

When we use this data set to estimate the average causal effect of friends' adoption, denoted as ACE_o, we have:

$$ACE_o = \frac{1}{7} \times (1 + 1 + 0 + 1 + 1 + 0 + 1) - \frac{1}{7} \times (1 + 0 + 1 + 1 + 0 + 1 + 1) = 0$$

That is, a person's friends buying a specific novel has no effect on people's adoption of the same novel.

Now let us use matching method to estimate the causal effect. For each treated sample, we select the matched control sample based on the similarity of their characteristics (f_1, f_2, f_3 and f_4). From Fig. 2(a), we find that Samples 6, 7 and 12 are the best matches for Samples 1, 3, and 4 respectively.

ID	T	Y_i	f_1	f_2	f_3	f_4
1	1	1	1	0	1	1
2	1	1	0	1	1	0
3	1	0	1	0	0	1
4	1	1	1	1	0	1
5	0	1	0	0	0	1
6	0	0	1	0	1	1
7	0	1	1	0	0	1
8	0	1	0	0	1	0
9	1	1	0	1	0	1
10	1	0	1	1	0	0
11	1	1	1	1	1	0
12	0	0	1	1	0	1
13	0	1	0	1	1	1
14	0	1	0	0	1	1

ID	T	Y_i	f_1	f_2	f_3	f_4
1	1	1	1	0	1	1
3	1	0	1	0	0	1
4	1	1	1	1	0	1
6	0	0	1	0	1	1
7	0	1	1	0	0	1
12	0	0	1	1	0	1

(a)Original data set   (b)Treatment and Control groups

Fig. 2. An example of matching process

We obtain the treatment group which contains Sample 1, 3 and 4, and the control group which contains Samples 5, 7 and 8 (see Fig. 2(b)). Using the matched samples to estimate the average causal effect, denoted as the ACE_m, we have:

$$ACE_m = \frac{1}{3} \times (1 + 0 + 1) - \frac{1}{3} \times (0 + 1 + 0) \approx 0.33$$

Therefore, a person's friends buying a specific novel has a positive effect on the person's adoption of the same novel.

In real world social networked data, the dimension of the features (i.e. F) is normally high. A matching method suffers from the curse of dimensionality. Instead of matching all feature values, propensity score summarizes the feature value of a node into one single value (a scalar). Therefore, we use propensity score matching [20, 24], which has shown to be an effective method for causal effect estimation for high-dimensional data.

Propensity score is defined as follows. Suppose that T^c is a binary treatment, and F is the set of covariates. The propensity score of an individual v_i (for whom $F = F_i$) is defined as the conditional probability of the individual receiving the treatment given the covariate values:

$$PS_i^c = P(T^c = 1 | F = F_i) \tag{4}$$

Propensity scores are commonly estimated by using logistic regression.

When applying propensity score matching, for each label c and a treated node v_i, i.e. a node has one or more neighbors with label c, we choose an untreated individual v_j, which has no neighbor with label c, such that among all the untreated nodes, its propensity score is the closest to v_i's propensity score. Then we add v_i to the treated group $(T^c = 1)$ and v_j to the control group $(T^c = 0)$.

Because we are interested in the contributions of peer influence and personal preference instead of the absolute causal effect value, we estimate the weights of peer influence, W^c_{PI} using the ratio of ACE^c (causal effect purely due to peer influence) to the observed average outcome in the treated group, $E(Y^{obs}|T^c = 1)$:

$$w^c_{PI} = \frac{ACE^c}{E(Y^c|T^c = 1)} = \frac{E(Y^c|T^c = 1) - E(Y^c|T^c = 0)}{E(Y^c|T^c = 1)} \tag{5}$$

After the peer influence is estimated, we consider the remaining contribution to one's behavior is from personal preference. Therefore, we estimate the weight of personal preference as the remaining portion of the overall contribution:

$$w^c_{PP} = 1 - w^c_{PI} = 1 - \frac{ACE^c}{E(Y^{obs}|T^c = 1)} = \frac{E\left(Y^{obs}|T^c = 0\right)}{E(Y^{obs}|T^c = 1)} \tag{6}$$

4.2   Criterion for Selecting a Classifier

For a data set with m labels, we calculate the average contribution of peer influence (w^c_{PI}) and personal preference (w^c_{PP}) of all labels as:

$$w_{PI} = \frac{1}{m} \sum_{c=1}^{m} w^c_{PI}$$

$$w_{PP} = \frac{1}{m} \sum_{c=1}^{m} w^c_{PP} \tag{7}$$

Given a threshold γ indicating the significance of the difference,

1. If $(w_{PI} - w_{PP}) \geq \gamma$, we say that the data set is peer influence dominated and is suitable for a connectivity based method.

2. If $(w_{PP} - w_{PI}) \geq \gamma$, we say that the data set is personal preference dominated and is suitable for a feature based method.

3. If $|w_{PI} - w_{PP}| < \gamma$, we say there is no significance difference in the estimated contributions of peer influence and personal preference in the data, it is recommended to apply both types of classification method to the data set, then integrate the result by using weighted probabilities.

Let $P^{cb}\left(l^c_i\right)$ stand for the probability of v_i having label c estimated by a connectivity based method, and $P^{fb}\left(l^c_i\right)$ stand for the probability of v_i having label c estimated by a connectivity based method. We estimate the $P(l^c_i)$, the probability of v_i having label c as the following:

$$P\left(l^c_i\right) = w^c_{PI} P^{cb}\left(l^c_i\right) + w^c_{PP} P^{fb}\left(l^c_i\right) \tag{8}$$

4.3   Algorithm

Based on the method described in the previous section, in this section, we present the CPP algorithm, which employs the causality based method to estimate the impact of peer influence and personal preference, to guide the selection of classification methods for a social network.

The input of CPP includes the networked data G (the set of nodes V, the set of edges E, the set of labels C and the set of features F for all nodes), the threshold γ. The output is the type of classifiers to be used, which can be feature based, connectivity based, or integrated according to Eq. 8. The steps of the CPP are presented in Algorithm 1.

Lines 1–12 of Algorithm 1 are to estimate w_{PI}^c and w_{PP}^c using the nodes in V', i.e. nodes whose labels are given. For each label c, we define two empty set $V'_{T^c=1}$ and $V'_{T^c=0}$ for treated group and control group, respectively (Lines 1–2). Then we compute the propensity score for each node in V' according to Eq. (4) (Lines 3).

We divide V' into two disjoint parts V'_1 and V'_2 based on the value of the treatment variable by putting nodes having 1 or more neighbors with label c into V'_1, and nodes having no neighbors with label c into V'_2 (Line 4).

For $v_i \in V'_1$, we choose $v_j \in V'_2$ whose propensity score is the closest to the propensity score of v_i, then we put v_i into $V'_{T^c=1}$ and v_j into $V'_{T^c=0}$ (Lines 5–8). Then we can estimate w_{PI}^c and w_{PP}^c according to Eqs. (5) and (6) (Line 9–11). w_{PI} and w_{PP} can be calculated according to Eq. (7) (Line 12).

If $\left(w_{PI} - w_{PP} \right) \geq \gamma$, we choose connectivity based methods to classify data (Line 13–14). If $\left(w_{PP} - w_{PI} \right) \geq \gamma$, we choose feature based methods to classify data (Line 15–16). When $\left| w_{PI} - w_{PP} \right| < \gamma$, we choose combined methods to classify data according to Eq. (8) (Line 17–18).

Algorithm 1. *CPP algorithm*

Input: $G = \{V, E, C, F\}, \gamma$, *where* V *is the set of nodes,* E *is the set of edges,* C *is the set of labels,* F *is the set of features and* γ *is the threshold*

Output: *type of classifiers to be used*

Steps:

1. *for each label* $c \in C$
2. $V'_{T^c=1} = \emptyset;\ V'_{T^c=0} = \emptyset$
3.   *compute propensity score* PS_i^c *for each* $v_i \in V'$ *according to Eq.(4)*
4.   *divide* V' *into two disjoint parts:* V'_1 *and* V'_2 *based on the value of the treatment variable*
5.   *for each* $v_i \in V'_1$
6.   *choose* $v_j \in V'_2$ $(j = argmin_j(|PS_i^c - PS_j^c|))$
7. $V'_{T^c=1} \leftarrow V'_{T^c=1} \cup \{v_i\};\ V'_{T^c=0} \leftarrow V'_{T^c=0} \cup \{v_j\}$
8.   *endfor*
9.   *compute* w_{PI}^c *according to Eq.(5)*
10.   *compute* w_{PP}^c *according to Eq.(6)*
11. *endfor*
12. *compute* w_{PI} *and* w_{PP} *according to Eq.(7)*
13. *if* $(w_{PI} - w_{PP}) \geq \gamma$
14.   *choose connectivity based methods to classify data*
15.   *else if* $(w_{PP} - w_{PI}) \geq \gamma$
16.   *choose feature based methods to classify data*
17.   *else if* $|w_{PI} - w_{PP}| < \gamma$
18.   *choose combined methods to classify data according to Eq.(8)*
19.   *endif*
20. *endif*

---

5   Experiments

5.1   Data Sets for Experiments

We use three real-world multi-label relational data sets, DBLP, IMDb and YouTube to evaluate the classification performance of our proposed framework.

The DBLP data set[1] is collected by Wang and Sukthankar [5]. This data set contains a network consisting of 8865 authors who have co-authorship. In this network, an author is represented by a node, and two authors are linked together if they have co-authored at least two papers. The DBLP data set uses 15 representative conferences as labels. An

---

[1] http://ial.eecs.ucf.edu/Data/SCRN-Data.zip.

author has a label if he/she has published a paper in a conference and each author can have multiple labels. Our task is to classify authors.

The IMDb data set is also collected by Wang and Sukthankar [5]. This data set contains a network consisting of 11746 movies and TV shows released between 2000 and 2010. In this network, a movie is represented by a node, and two movies are linked together if they are directed by the same director. The IMDb data set use 27 movie genres as labels. A movie can be assigned to a subset of 27 different labels in the database. Our task is to classify movies.

The YouTube data set[2] is collected by Tang [28]. This data set contains a network consisting of 11942 users. In this network, a user can subscribe to different interest groups and add other users as his/her contacts. The YouTube data set use 47 interest groups as labels. A user can be assigned to a subset of 47 different labels in the database. Our task is to classify users.

A summary of the three data sets can be found in Table 1.

Table 1. Data set summary

Data	DBLP	IMDb	YouTube
Number of nodes	8,865	11,746	11,942
Number of links	12,989	323,892	69,319
Maximum degree	86	290	851
Average degree	2.9	55.2	11.6
Number of labels	15	27	47
Average number of labels per node	2.3	1.5	1.9

5.2 Evaluation Measures

In this section, we explain the evaluation criteria: Macro-F1 score and Micro-F1 score [21]. Given the data set G, let $y_i, \hat{y}_i \in \{0, 1\}^m$ be the true and predicted label sets, respectively, for a node $v_i (1 \le i \le n)$. For a label $c_k \in C$, if P^k and R^k denote the precision and the recall, respectively, F1 score is defined as the harmonic mean of precision and recall:

$$F_1^k = \frac{2P^k R^k}{P^k + R^k} = \frac{2\sum_{i=1}^n y_i^k \hat{y}_i^k}{\sum_{i=1}^n y_i^k + \sum_{i=1}^n \hat{y}_i^k} \tag{9}$$

Macro-F1 score is the averaged F1 score over categories:

$$\text{Macro-F1} = \frac{1}{m} \sum_{k=1}^m F_1^k \tag{10}$$

Macro-F1 score is sensitive to the performance of rare labels, since all labels are weighted evenly.

---

[2] http://leitang.net/social_dimension.html.

Micro-F1 score is computed using F_1^k while considering the precision as a whole. Specifically, it is defined as follows:

$$\text{Micro-F1} = \frac{2 \sum_{k=1}^{m} \sum_{i=1}^{n} y_i^k \hat{y}_i^k}{\sum_{k=1}^{m} \sum_{i=1}^{n} y_i^k + \sum_{k=1}^{m} \sum_{i=1}^{n} \hat{y}_i^k} \tag{11}$$

Micro-F1 score is largely determined by the common labels, since this measure weights instances evenly.

5.3 Experiment Setup

In the experiments, we use two connectivity based methods: wvRN [3] and SNBC [4]; and two feature based methods: SocDim [1] and MLKNN [2].

We have to construct features for nodes, since all three data sets do not contain nodes' features. We use the ModMax algorithm [1] to construct nodes' features. The dimensionality of feature is set to 50 for DBLP and IMDb data sets and 200 for YouTube data set.

For each data set, the R package, MatchIt [23] is adopted to match nodes, and we choose the Nearest Neighbor method in the package. We set the maximum number of iterations for stochastic gradient descent in SNBC to 100. We set the number of nearest neighbors in MLKNN to 10.

In the code of MLKNN provided by Zhang [2], if the probability of a test instance belonging to a label is more than 0.5, MLKNN assigns this class label to this test instance. We made a minor change. For fair comparison, in all methods, we assume that the number of labels for all unlabeled nodes are already known and we assign the labels based on the classes with highest probability. Such a scheme has been adopted for multi-label evaluation in social network data sets [1, 5] and improve the performance of MLKNN.

We randomly sample a portion of nodes from a data set as training nodes. The fraction of the training data ranges from 40% to 90%. The rest of the nodes are used as testing nodes. Then we report the average performance of the 5 runs in terms of the Macro F1 score and Micro F1 score measures.

In the experiments, we first use our proposed CPP algorithm to calculate the contributions of peer influence and personal preference on each data set for determining the type of classifier. Then we compare the performance of the 4 methods described above on these data sets to evaluate the performance of our proposed framework. The threshold γ is set to 0.4.

5.4 Results

Table 2 shows the classification performance on the DBLP data set. First we have estimated the weights of peer influence, we found that the weights of peer influence w_{PI} are higher than the weights of personal preference w_{PP} for different portions of samples for training. This data set is suitable for using connectivity based methods since $(w_{PI} - w_{PP}) \geq 0.4$. The experimental results support our prediction, wvRN consistently outperforms the others, and connectivity based methods (wvRN and SNBC) achieve better performance than feature based methods (SOCDIM and MLKNN).

Table 2. Classification results on DBLP. The best and the second best results are highlighted. The classification performances consistent with the prediction by our criterion since this data set is a peer influence dominated.

Proportion of labeled nodes		40%	50%	60%	70%	80%	90%
$(w_{PI} - w_{PP})$		0.884	0.886	0.888	0.888	0.896	0.880
Micro-F1 (%)	SOCDIM (feature based)	42.59	42.81	43.39	43.90	44.12	44.48
	MLKNN (feature based)	48.00	49.46	49.75	50.16	51.16	52.25
	wvRN (connectivity based)	**68.51**	**70.29**	**72.52**	**73.27**	**75.33**	**76.85**
	SNBC (connectivity based)	**64.84**	**68.29**	**70.92**	**72.64**	**73.24**	**75.71**
Macro-F1 (%)	SOCDIM (feature based)	32.27	32.53	32.79	32.93	33.13	33.28
	MLKNN (feature based)	38.88	40.73	41.15	41.73	43.26	44.11
	wvRN (connectivity based)	**62.42**	**64.19**	**66.47**	**67.57**	**69.73**	**69.98**
	SNBC (connectivity based)	**59.77**	**62.43**	**64.32**	**65.66**	**65.84**	**67.34**

Table 3 shows the classification performance on the IMDb dataset. The weights of peer influence α are all higher than the weights of personal preference w_{PP}, so this dataset is suitable for using connectivity-based methods since $(w_{PI} - w_{PP}) > 0.4$. From Table 3, we observe that wvRN consistently outperforms the others, and the connectivity-based methods (wvRN and SNBC) achieve better Micro-F1 score and Macro-F1 score than feature-based methods (SOCDIM and MLKNN).

Table 4 shows the classification performance on the YouTube data set. The values of $(w_{PI} - w_{PP})$ are between 0.070 to 0.371 as the proportion of labeled nodes increase, so we cannot predict which type of classifier is suitable. The feature based method SOCDIM has the best performance on both Micro-F1 and Macro-F1. The connectivity based method SNBC has the second best performance on Micro-F1, and the connectivity based method wvRN has the second best performance on Macro-F1.

Table 3. Classification results on IMDb. The best and the second best results are highlighted. The classification performance is largely consistent with the prediction by our criterion since this data set is a peer influence dominated.

Proportion of labeled nodes		40%	50%	60%	70%	80%	90%
$(w_{PI} - w_{PP})$		0.523	0.594	0.611	0.628	0.654	0.668
Micro-F1 (%)	SOCDIM (feature based)	50.17	50.29	50.57	50.48	50.66	52.39
	MLKNN (feature based)	57.51	57.69	59.01	59.34	59.57	60.63
	wvRN (connectivity based)	**74.55**	**74.74**	**75.33**	**75.72**	**76.09**	**76.17**
	SNBC (connectivity based)	**72.08**	**72.76**	**72.89**	**73.28**	**73.96**	**74.56**
Macro-F1 (%)	SOCDIM (feature based)	14.04	13.82	14.27	13.95	13.74	14.69
	MLKNN (feature based)	24.89	25.88	25.36	26.48	26.14	26.20
	wvRN (connectivity based)	**47.35**	**48.56**	**48.74**	**50.07**	**50.12**	**50.17**
	SNBC (connectivity based)	**43.23**	**44.05**	**44.11**	**44.78**	**46.09**	**46.17**

Table 4. Classification results on YouTube. The best and the second best results are highlighted. The classification performance is unpredictable since the data set is mixed, neither personal preference dominated or peer influence dominated

Proportion of labeled nodes		40%	50%	60%	70%	80%	90%
$(w_{PI} - w_{PP})$		0.070	0.146	0.174	0.241	0.327	0.371
Micro-F1 (%)	SOCDIM (feature based)	**45.07**	**45.19**	**45.73**	**46.01**	**46.45**	**46.89**
	MLKNN (feature based)	43.02	43.28	44.15	44.41	44.81	45.56
	wvRN (connectivity based)	41.67	41.43	41.67	41.75	41.57	41.83
	SNBC (connectivity based)	**44.73**	**44.86**	**45.59**	**45.24**	**45.88**	**46.52**
Macro-F1 (%)	SOCDIM (feature based)	**37.27**	**37.35**	**37.53**	**38.92**	**39.43**	**40.43**
	MLKNN (feature based)	34.25	35.07	35.38	36.02	36.46	36.97
	wvRN (connectivity based)	**36.39**	**36.21**	**36.59**	**36.80**	**36.50**	**37.08**
	SNBC (connectivity based)	35.37	34.68	35.38	34.90	34.86	35.81

Table 5 shows the combined classification results on YouTube based on Eq. (8). The combined method SOCDIM+SNBC has the best performance on Micro-F1, and the MLKNN+SNBC has the second best performance on Micro-F1. The combined method SOCDIM+wvRN has the second best performance on Macro-F1, and the SOCDIM +SNBC has the second best performance on Macro-F1.

Table 5. The classification results of combined methods on YouTube. The best and the second best results are highlighted.

Proportion of labeled nodes		40%	50%	60%	70%	80%	90%
$(w_{PI} - w_{PP})$		0.070	0.146	0.174	0.241	0.327	0.371
Micro-F1 (%)	SOCDIM+wvRN	44.66	44.79	45.48	45.67	45.44	46.07
	SOCDIM+SNBC	**45.63**	**45.94**	**46.20**	**46.50**	**47.21**	**47.76**
	MLKNN+wvRN	44.47	44.66	44.90	45.33	45.25	45.97
	MLKNN+SNBC	**45.37**	**45.56**	**45.70**	**46.31**	**46.60**	**47.22**
Macro-F1 (%)	SOCDIM+wvRN	**37.85**	**37.51**	**38.81**	**39.45**	**39.90**	**40.75**
	SOCDIM+SNBC	**37.82**	**37.39**	**38.05**	**38.73**	**39.80**	**39.98**
	MLKNN+wvRN	36.47	36.83	37.67	38.41	39.49	39.84
	MLKNN+SNBC	35.45	35.69	35.90	36.73	37.50	37.57

Comparing Table 5 with the results in Table 4, we found that most combined methods have better performance than single methods. More specifically, MLKNN+SNBC has better performance than MLKNN and SNBC; SOCDIM+SNBC has better performance than SOCDIM and SNBC; MLKNN+wvRN has better performance than MLKNN and SNBC. This work shows that it is promising for a combined approach to deal with the networked data sets which are neither peer influence dominated nor personal preference dominated.

6   Conclusion

In this paper, we have proposed an approach (CCP) which employs the propensity score matching and causal effect estimation to estimate the impact of peer influence and personal preference, to guide the selection of classification methods for a social network.

The CCP shows the characteristic of data set and can be used to selected the type of methods for classification of networked data set. Empirical studies on real-world data sets have demonstrated the CCP can predict the suitability of a method for a data set. The work has shown that the estimation of peer influence and personal preference in a data set is promising for designing integrated methods using two types of classification methods of social network data. Our future work is to develop a deep integration method using both peer influence and personal preference to classify any type of networked data set to achieve the best result.

References

1. Tang, L., Liu, H.: Relational learning via latent social dimensions. In: Proceedings of the 15th ACM SIGKDD International Conference on Knowledge Discovery in Data Mining, pp. 817–826 (2009)
2. Zhang, M.L., Zhou, Z.H.: ML-KNN: a lazy learning approach to multi-label learning. Pattern Recognit. **40**(7), 2038–2048 (2007)
3. Macskassy, S., Provost, F.: A simple relational classifier. In: Proceedings of the Second Workshop on Multi-Relational Data Mining at 9th ACM SIGKDD International Conference on Knowledge Discovery in Data Mining, pp. 64–76 (2003)
4. Nandanwar, S., Murty, M.N.: Structural neighborhood based classification of nodes in a network. In: Proceedings of the 22nd ACM SIGKDD International Conference on Knowledge Discovery in Data Mining, pp. 1085–1094 (2016)
5. Wang, X., Sukthankar, G.: Multi-label relational neighbor classification using social context features. In: Proceedings of the 19th ACM SIGKDD International Conference on Knowledge Discovery in Data Mining, pp. 464–472 (2013)
6. Aral, S., Muchnik, L., Sundararajan, A.: Distinguishing influence-based contagion from homophily-driven diffusion in dynamic networks. Proc. Natl. Acad. Sci. **106**, 21544–21549 (2009)
7. McCallum, A.K.: Multi-label text classification with a mixture model trained by EM. In: Working Notes of the AAAI 1999 Workshop on Text Learning, pp. 1–7 (1999)
8. Boutell, M.R., Luo, J., Shen, X., Brown, C.M.: Learning multi-label scene classification. Pattern Recognit. **37**(9), 1757–1771 (2004)
9. Chen, Z., Chi, Z., Fu, H., Feng, D.: Multi-instance multi-label image classification: a neural approach. Neurocomputing **99**, 298–306 (2013)
10. Zhao, K., Zhang, H., Ma, Z., Song, Y., Guo, J.: Multi-label learning with prior knowledge for facial expression analysis. Neurocomputing **157**, 280–289 (2015)
11. Zhang, M.L., Peña, J.M., Robles, V.: Feature selection for multi-label naive Bayes classification. Inf. Sci. **179**(19), 3218–3229 (2009)
12. Barutcuoglu, Z., Schapire, R.E., Troyanskaya, O.G.: Hierarchical multi-label prediction of gene function. Bioinformatics **22**(7), 830–836 (2006)
13. Neville, J., Jensen, D.: Iterative classification in relational data. In: Proceedings of the Workshop on Learning Statistical Models from Relational Data at the 17th AAAI National Conference on Artificial Intelligence, pp. 42–49 (2000)
14. Heatherly, R., Kantarcioglu, M., Li, X.: Social network classification incorporating link type. In: Proceedings of the 2009 IEEE International Conference on Intelligence and Security Informatics, pp. 19–24 (2009)
15. Lin, F., Cohen, W.W.: Semi-supervised classification of network data using very few labels. In: Proceedings of the International Conference on Advances in Social Networks Analysis and Mining, pp. 192–199 (2010)
16. Rubin, D.: Comment of D. Basu, Randomization analysis of experimental data: the Fisher randomization test. J. Am. Stat. Assoc. **75**, 591–593 (1980)
17. Rubin, D.: Comment of Neyman (1923) and causal inference in experiments and observational studies. Stat. Sci. **5**, 472–480 (1990)
18. Rubin, D.: Causal inference using potential outcomes. J. Am. Stat. Assoc. **100**, 322–331 (2005)
19. Sekhon, J.S.: The Neyman–Rubin model of causal inference and estimation via matching methods. In: The Oxford Handbook of Political Methodology (2007)

20. Rosenbaum, P.R., Rubin, D.: The central role of the propensity score in observational studies for causal effects. Biometrika **70**, 41–55 (1983)
21. Fan, R., Lin, C.: A study on threshold selection for multi-label classification. Technical report, National Taiwan University (2007)
22. Zhang, M.L., Zhou, Z.H.: A review on multi-label learning algorithms. IEEE Trans. Knowl. Data Eng. **26**(8), 1819–1837 (2014)
23. Daniel, H., Imai, K., King, G., Stuart, E.: MatchIt: nonparametric preprocessing for parametric causal inference. J. Stat. Softw. **42**, 1–28 (2011)
24. Hirano, K., Imbens, G., Ridder, G.: Efficient estimation of average treatment effects using the estimated propensity score. Econometrica **71**(4), 1161–1189 (2003)
25. Salemi, B., Noah, S., Aziz, M.: Rfboost: an improved multi-label boosting algorithm and its application to text categorization. Knowl.-Based Syst. **103**, 104–117 (2016)
26. Trohidis, K., Tsoumakas, G., Kalliris, G., Vlahavas, I.: Multi-label classification of music into emotions. In: Proceedings of the 9th International Conference on Music Information Retrieval, pp. 325–330 (2008)
27. Sanden, C., Zhang, J.: Enhancing multi-label music genre classification through ensemble techniques. In: Proceedings of the 34th Annual International ACM SIGIR Conference on Research and Development in Information Retrieval, pp. 705–714 (2011)
28. Tang, L., Wang, X., Liu, H.: Scalable learning of collective behavior. IEEE Trans. Knowl. Data Eng. **24**(6), 1080–1091 (2012)

Diversified and Verbalized Result Summarization for Semantic Association Search

Yu Gu, Yue Liang, Gong Cheng[✉], Daxin Liu, Ruidi Wei, and Yuzhong Qu

National Key Laboratory for Novel Software Technology,
Nanjing University, Nanjing, China
{ygu,yliang,dxliu}@smail.nju.edu.cn, {gcheng,yzqu}@nju.edu.cn,
ruidiwei@gmail.com

Abstract. Semantic association search is to search an entity-relation graph for subgraphs called semantic associations that connect a set of entities specified in a user's query. Recent research on this topic has concentrated on summarizing numerous search results by mining their important patterns to form an abstractive overview. However, top-ranked patterns may have redundancy, and their graph structure may not be comprehensible to non-expert users. To reduce redundancy, we present a novel framework featuring a combinatorial optimization model to select top-k diversified patterns. In particular, we devise a new similarity measure which jointly considers structural and semantic similarity to assess the overlap between patterns. To facilitate non-expert users' comprehension of a pattern, we verbalize its graph structure, transforming it into compact and coherent English text based on a novel method for discourse planning. Extensive experiments demonstrate the effectiveness of our approach compared with existing methods.

Keywords: Semantic association search · Summary generation
Diversity · Graph edit distance · Graph verbalization

1 Introduction

A common type of information needs enabled by the Semantic Web is to search an entity-relation graph (e.g., Fig. 1) for *semantic associations* (SAs) which represent complex relationships between a set of entities specified in a user's query (called query entities). Generally, a SA is a compact subgraph connecting a set of query entities [2–4]. For example, given a query comprising the three people in Fig. 1, two SAs are shown in Fig. 2.

On a large entity-relation graph, numerous SAs can be found and overload users, which can be alleviated by properly summarizing the search results. One established approach is to aggregate SAs using their high-level abstraction called *SA pattern* (SAP), which is obtained by substituting each non-query entity in a SA with one of its types [3,5,7]. For example, the SA x_1 in Fig. 2 matches two

© Springer Nature Switzerland AG 2018
H. Hacid et al. (Eds.): WISE 2018, LNCS 11233, pp. 381–390, 2018.
https://doi.org/10.1007/978-3-030-02922-7_26

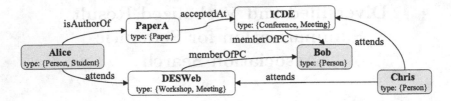

Fig. 1. A running example of entity-relation graph, including the types of each entity and a set of three query entities: Alice, Bob, and Chris.

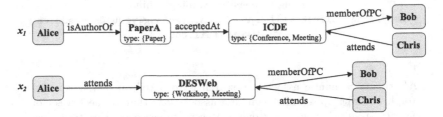

Fig. 2. Two SAs connecting Alice, Bob, and Chris in Fig. 1.

SAPs z_1 and z_2 in Fig. 3. Nonetheless, two problems remain and influence the usability of SAP. First, in a schema-rich entity-relation graph like DBpedia for the encyclopedic domain, the number of aggregated SAPs can still be excessive. Previous research [3] ranks SAPs by their frequency, but simply choosing top-ranked SAPs may suffer from redundancy; e.g., z_1 and z_2 in Fig. 3 are similar. Second, as users may not have expertise in reading or accessing (Semantic Web) graphs, rather than intuitively using node-link diagrams, a more comprehensible way of presenting a SAP is needed for average users.

To meet the above challenges, our research contributions are threefold. (i) We devise a new similarity measure for assessing the overlap between two SAPs based on a variant of graph edit distance, which jointly considers the graph structure and the semantics of graph labels (Sect. 3.2). (ii) To select a set of top-ranked SAPs that are also diversified (i.e., with limited overlap), we model a combinatorial optimization problem and solve it using a greedy algorithm (Sect. 3.1). (iii) To make selected SAPs more comprehensible to non-expert users, we present a verbalization approach that transforms each SAP into compact and coherent English text based on a novel method for discourse planning (Sect. 4).

The reader is also referred to our extended technical report[1] for more details.

2 Preliminaries

We deal with a finite directed labeled *entity-relation graph* denoted by G, where each vertex is labeled with a unique entity, and each arc is labeled with a binary

---

[1] http://ws.nju.edu.cn/association/summ2018/wise18_extended.pdf.

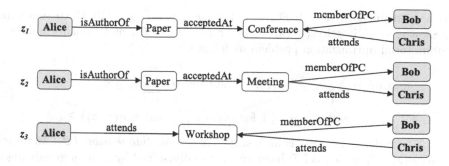

Fig. 3. Three SAPs.

relation on entities, e.g., Fig. 1. An entity has at least one type (i.e., class). Classes are organized into a subsumption hierarchy.

A *query* Q consists of n *query entities* ($n > 1$). All the other entities in G are called *non-query entities*. A result of Q is called a *semantic association* (SA) denoted by x, which is a subgraph of G satisfying: (i) x is connected, (ii) x contains all the query entities, and (iii) x is minimal, i.e., none of its proper subgraphs satisfy (i) and (ii). The minimality of x indicates that its underlying graph is a tree. For example, Fig. 2 illustrates two SAs connecting Alice, Bob, and Chris. The *diameter* of x, denoted by $diam(x)$, is the largest distance between pairs of entities in x, e.g., $diam(x_1) = 3$ in Fig. 2.

A *SA pattern* (SAP) is a directed labeled graph where each vertex is labeled with a query entity or a class, and each arc is labeled with a relation. A SA x *matches* a SAP z if there is a bijection between their vertices such that (i) each query entity in x corresponds to itself in z, (ii) each non-query entity in x corresponds to one of its types in z, and (iii) the bijection is arc-preserving and label-preserving. For example, x_1 in Fig. 2 matches z_1 and z_2 in Fig. 3.

3   Diversified SAP Selection

We present a novel framework featuring a combinatorial optimization model to select top-k SAPs that are diversified, which uses a new similarity measure.

3.1   Framework

We aim to select k top-ranked SAPs that are not similar to each other. For ranking, we follow [3] to consider frequency, which can be replaced by any other ranking criterion. Specifically, for a set of SAs X, the *frequency* of a SAP z, denoted by $f(z, X)$, is the number of SAs in X that match z. Given a threshold $\tau > 0$, z is *frequent* if $f(z, X) \geq \tau$. Let $Z = \{z_1, \ldots, z_m\}$ be the set of all frequent SAPs for X; m is the size of Z. We will select k SAPs from Z.

Given $z_i, z_j \in Z$, let $sim(z_i, z_j) \in [0, 1]$ be the similarity between z_i and z_j, which will be detailed in Sect. 3.2; z_i and z_j are *similar* if $sim(z_i, z_j) \geq \phi$ where

ϕ is a predefined threshold. For $i = 1$ to m, we use $y_i \in \{0,1\}$ to represent whether $z_i \in Z$ is selected ($y_i = 1$) or not ($y_i = 0$), and then formulate a combinatorial optimization problem as follows:

$$\text{maximize} \sum_{i=1}^{m} y_i \cdot f(z_i, X), \quad \text{subject to} \sum_{i=1}^{m} y_i \leq k \tag{1}$$

$$\text{and } y_i + y_j \leq 1 \text{ for every } i \neq j \text{ and } sim(z_i, z_j) \geq \phi.$$

The formulated problem is an instance of the *multidimensional 0-1 knapsack problem* (MKP), which is NP-hard and is usually solved by using a greedy algorithm. The idea is to iteratively select a SAP that satisfies all the constraints and maximizes the following heuristic function:

$$\text{heuristic score of } z_i = \frac{f(z_i, X)}{|\{z_j \in (Z \setminus \{z_i\}) : sim(z_i, z_j) \geq \phi\}| + 1}, \tag{2}$$

until k is reached or Z is exhausted. Here, priority is given to SAPs that are more important (i.e., more frequent) and are similar to fewer other SAPs.

3.2 Similarity Measurement

For $sim(z_i, z_j)$, we devise pGED, a new measure of similarity between SAPs. It is a variant of *graph edit distance* (GED) [8], which is the minimum cost of any sequence of insertion, deletion, and substitution (i.e., relabeling) of vertices and arcs that transforms a source graph into a target graph. Different from GED, pGED disallows any edit operation that involves query entities because they are the focus of a SAP. Only classes and relations can be inserted, deleted, or substituted, for which pGED jointly considers the graph structure and the semantics of graph labels. Specifically, we fix the cost of insertion and deletion to 1. For a substitution which relabels a class or a relation with another, we calculate its cost by one minus the similarity between the old and the new labels: (i) for classes, we use *wpath* [9] to measure semantic similarity; (ii) for relations, we calculate discrete similarity; i.e., 0 for different relations.

Now we define $sim(z_i, z_j)$. For z_i with ν_{z_i} vertices and ϵ_{z_i} arcs, and z_j with ν_{z_j} vertices and ϵ_{z_j} arcs, let $pged(z_i, z_j)$ be the pGED between z_i and z_j, which is calculated by an adapted A* algorithm [8]. By normalizing $pged$, we obtain:

$$sim(z_i, z_j) = 1 - \sqrt{\frac{pged(z_i, z_j)}{\max\{\nu_{z_i}, \nu_{z_j}\} + \epsilon_{z_i} + \epsilon_{z_j}}}. \tag{3}$$

4 SAP Verbalization

We propose a rule-based, domain-independent approach to transform SAP into English text. Each arc is expressed by a subject-predicate-object clause consisting of lexicalized relations, classes, or query entities. Our novel method for discourse planning properly aggregates and orders clauses to improve the compactness and coherence of the generated text.

4.1 Discourse Planning

Discourse planning has two levels. *Sentence level* organizes the clauses within a sentence, and *document level* optimizes the relationship between sentences. At the document level, when a SAP (which is a tree) is large, we decompose it into small subtrees having a diameter of at most 2 ($diam \leq 2$). At the sentence level, from each subtree we generate a simple, a compound, or a complex sentence.

Sentence-Level Planning. We generate a sentence from each (sub-)tree. We assume that some vertex v_{subj} in the tree has been specified as the subject of the sentence by document-level planning which we will introduce later; document-level planning may also require another vertex v_{end} to appear at the end of the sentence. As the tree is small ($diam \leq 2$), we exhaustively plan for all possible cases in the following, to generate a compact sentence.

For $diam = 1$, a tree consists of exactly one arc, as illustrated in Fig. 4(a). We generate a simple sentence consisting of one clause whose subject is v_{subj}.

For $diam = 2$, a tree consists of an internal vertex v_i and at least two leaf vertices v_l, v_{ll}, etc., as illustrated in Fig. 4(b). We distinguish between two cases, depending on the position of v_{subj}.

When v_{subj} is the internal vertex v_i, we generate a compound sentence consisting of clauses joined by commas and the conjunction *and*. These clauses have a common subject v_i, so the subject is omitted from the second clause. Further, we order these clauses such that (i) those having a common predicate are compactly aggregated into one clause, and (ii) the last clause will be the one whose object is v_{end} if it is specified by document-level planning. For example, for z_3 in Fig. 3, if $v_{subj} = $ Workshop and $v_{end} = $ Chris, we will generate the following to-be-lexicalized sentence after aggregation and ordering: ⟨Workshop⟩ ⟨memberOfPC⟩ ⟨Bob⟩, *and* ⟨attends⟩⁻ ⟨Alice⟩ *and* ⟨Chris⟩. We use ⟨···⟩⁻ to represent a relation to be reversely lexicalized.

When v_{subj} is a leaf vertex, say v_l WOLOG, we generate a complex sentence where the main clause expresses the arc between v_l and the internal vertex v_i, and the other arcs form relative clauses with their antecedents referring to v_i. These relative clauses are aggregated and ordered also in the above-mentioned manner. For example, for z_3 in Fig. 3, if $v_{subj} = $ Alice and $v_{end} = $ Chris, we will generate the following to-be-lexicalized sentence: ⟨Alice⟩ ⟨attends⟩ ⟨Workshop⟩, *which* ⟨memberOfPC⟩ ⟨Bob⟩ *and* ⟨attends⟩⁻ ⟨Chris⟩.

Document-Level Planning. To verbalize a SAP, our method allows an arbitrary query entity (denoted by v_Q) to be the subject (i.e., v_{subj}) of the first sentence; in our implementation, v_Q is set to the first query entity. When $diam \leq 2$, we directly generate a sentence using sentence-level planning with $v_{subj} = v_Q$; document-level planning is not needed. When $diam > 2$, we decompose the tree into small subtrees of $diam \leq 2$, and generate one sentence for each subtree; then the goal of our document-level planning is to improve the coherence of these sentences by properly specifying v_{subj} and v_{end} for their sentence-level planning. Here we exhaustively plan for two cases: $diam = 3$ and $diam = 4$, as in

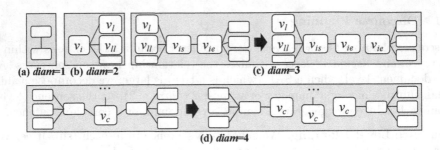

Fig. 4. An illustration of discourse planning.

practice larger ones still cannot be efficiently generated by existing algorithms [3]. However, our method has the potential to be generalized.

For $diam = 3$, a tree consists of two internal vertices v_{is} and v_{ie}, and at least two leaf vertices v_l, v_{ll}, etc., as illustrated in Fig. 4(c). There are two cases.

When v_Q is a leaf vertex, say v_l WOLOG, we decompose the tree into two subtrees as shown on the right side of Fig. 4(c). The first sentence is generated using sentence-level planning with $v_{subj} = v_l$ and $v_{end} = v_{ie}$. The second sentence is generated under $v_{subj} = v_{ie}$, so its subject verbatim repeats the end of the first sentence to improve coherence. For example, for z_1 in Fig. 3, if $v_Q = $ Alice, we will generate the following to-be-lexicalized document: ⟨Alice⟩ ⟨isAuthorOf⟩ ⟨Paper⟩, *which* ⟨acceptedAt⟩ ⟨Conference⟩. ⟨Conference⟩ ⟨memberOfPC⟩ ⟨Bob⟩ *and* ⟨attends⟩⁻ ⟨Chris⟩.

When v_Q is an internal vertex, say v_{is} WOLOG, we decompose the tree in the same way as above and generate two sentences. The only difference is that the first sentence is generated under $v_{subj} = v_Q = v_{is}$.

For $diam = 4$, a tree contains a unique central vertex v_c that is not more than two hops away from other vertices. As illustrated in Fig. 4(d), we decompose the tree at v_c into two or more subtrees. We distinguish between two classes.

When $v_Q \neq v_c$, the first sentence is generated from the subtree containing v_Q under $v_{subj} = v_Q$ and $v_{end} = v_c$. After that, one sentence is generated from each remaining subtree under $v_{subj} = v_c$, so the subjects of these sentences verbatim repeat the end of the first sentence to improve coherence.

When $v_Q = v_c$, we generate one sentence from each subtree under $v_{subj} = v_c$. These sentences form coherent text as they have a common subject.

4.2   Lexicalization

Lexicalization transforms ⟨· · ·⟩ and ⟨· · ·⟩⁻ into natural language. The reader is referred to our technical report for a rule-based, lexicon-free implementation.

As an example, after lexicalization, z_1 in Fig. 3 is finally verbalized as follows when $v_Q = $ Alice: *Alice is author of a Paper, which is accepted at a Conference. The Conference has member of PC Bob and is attended by Chris.*

5 Experiments

Our approach is open-source[2]. We conducted a blind user study involving twenty students majoring in computer science to test three hypotheses.

1. Our proposed top-ranked diversified SAPs form a better summary for SAs than top-ranked undiversified SAPs considered in [3].
2. In particular, our proposed pGED is more effective than a popular similarity measure considered in previous research [6,7].
3. Our proposed discourse planning generates better document structure than the planning considered in a state-of-the-art system [1].

5.1 Datasets and Queries

Our experiments were performed on DBpedia[3]. We used *Mappingbased Objects* for entity-relation graph, *Instance Types* and *Instance Types Transitive* for entity types, and the *DBpedia ontology* for measuring semantic similarity by [9].

We followed [4] to construct queries. We implemented [3] to search for SAs and mine frequent SAPs, which required configuring a diameter constraint λ and a frequency threshold τ. Given a query consisting of n entities, [3] would search the entity-relation graph for all the SAs of $diam \leq \lambda$, and mine their frequent SAPs satisfying $f \geq \tau$. Under different configurations of n, λ, and τ, we constructed a total of 80 queries[4], including

- 10 queries for each $n \in \{2,3,4,5\}$ under $\lambda = 3$ and $\tau = 5$ such that at least 10 frequent SAPs could be mined from the search results of a query, and
- 10 queries for each $n \in \{2,3,4,5\}$ under $\lambda = 4$ and $\tau = 100$ such that at least 50 frequent SAPs could be mined from the search results of a query.

5.2 Experiment on Diversification

Experiment Design. We separately tested the first and the second hypotheses. In these experiments, SAPs were plainly visualized as node-link diagrams.

First, we compared our top-ranked diversified SAPs (denoted by DIV) with top-ranked undiversified SAPs selected by [3] (denoted by FREQ). For DIV, we set $\phi = 0.5$. For each of the 80 queries, DIV and FREQ separately selected five SAPs ($k = 5$) as two summaries to be compared by five users who were invited to decide which summary was preferred and then explain the decision.

Second, we tested the effectiveness of our pGED, compared with a popular similarity measure used in [6,7] which, denoted by JACCARD, measures the Jaccard similarity between sets of graph labels. We used 40 queries due to the availability of users. For each query, three frequent SAPs z, z_i, z_j were selected such that pGED and JACCARD disagreed on their relative similarity; i.e., z_i and z_j were considered more similar to z by pGED and JACCARD, respectively. Five users were invited to decide whether z_i or z_j was more similar to z.

---

[2] http://ws.nju.edu.cn/association/summ2018.

[3] http://wiki.dbpedia.org/dbpedia-dataset-version-2015-10.

[4] http://ws.nju.edu.cn/association/summ2018/query.zip.

Table 1. Average pGED between SAPs Selected by DIV and FREQ

	$\lambda = 3$				$\lambda = 4$			
	$n = 2$	$n = 3$	$n = 4$	$n = 5$	$n = 2$	$n = 3$	$n = 4$	$n = 5$
DIV	0.684	0.596	0.678	0.594	0.565	0.619	0.626	0.588
FREQ	0.229	0.183	0.205	0.171	0.228	0.185	0.204	0.195

Results. In the first experiment, 319 user decisions (80%) preferred the SAPs selected by DIV. One-sample t-test showed that DIV significantly outperformed FREQ under $p < 0.01$, suggesting that our first hypothesis could be accepted.

To characterize the effectiveness of DIV, as shown in Table 1, the average pGED between SAPs selected by FREQ was very small, being much smaller than the one for DIV, showing the necessity of diversification and the effectiveness of DIV. We also analyzed the explanations given by the users for their decisions, and found that the diversity of DIV's selection was recommended in 239 decisions (60%); i.e., diversity was considered in most decisions.

In the second experiment, 154 user decisions (77%) agreed with pGED. One-sample t-test showed that pGED significantly outperformed JACCARD under $p < 0.01$, suggesting that our second hypothesis could be accepted.

5.3   Experiment on Verbalization

Experiment Design. To test the third hypothesis, we compared our verbalization approach (denoted by PaVer) with NaturalOWL [1], which was among the best-performing open-source systems that could process our SAP. We configured NaturalOWL and PaVer to start verbalization from the same query entity. NaturalOWL relied on external templates for lexicalizing relations, and we set it to use ours. Therefore, the main difference between NaturalOWL and PaVer was discourse planning, which was exactly one of our technical contributions.

For each of the 80 queries, we randomly selected a frequent SAP and used NaturalOWL and PaVer to separately generate English text to be evaluated by five users, who had access to the original SAP visualized as a node-link diagram for reference. For each text, each user was asked to complete a questionnaire consisting of four statements about the quality of verbalization:

Correctness. The text is syntactically correct.
Comprehensibility. The text is easy to comprehend.
Conciseness. The text is concise and compact.
Accuracy. The text precisely describes the information contained in the graph.

The user's level of agreement on each statement was to be responded on a five-level Likert item: from 1 (strongly disagree) to 5 (strongly agree).

Results. As shown in Table 2, the mean of all the 400 paired questionnaire results showed that PaVer outperformed NaturalOWL in all the four aspects

Table 2. Mean response to each statement in the questionnaire

Statement	NaturalOWL	PaVer	p-value
Correctness	3.79	3.95	2.0e−3
Comprehensibility	3.58	3.80	1.2e−4
Conciseness	3.36	3.68	6.2e−8
Accuracy	3.95	4.03	4.8e−2

of quality. Paired t-tests suggested that the differences were statistically significant under $p < 0.01$ in correctness, comprehensibility, and conciseness, whereas the difference in accuracy was not significant. That is, NaturalOWL and PaVer generated equally accurate descriptions for SAPs, but the text generated by PaVer was syntactically more correct, easier to comprehend, and more concise and compact, suggesting that our third hypothesis could be accepted.

We attributed the superiority of PaVer in correctness to its proper use of indefinite and definite articles when lexicalizing classes, and attributed its superiority in conciseness and comprehensibility to its use of relative clauses in sentence-level planning and its coherence-oriented ordering in document-level planning. All these features were not considered in NaturalOWL.

6 Related Work

To avoid overloading users with numerous results in SA search, [2,4] present top-ranked results. A complementary paradigm which we follow in this paper is to mine and select a few frequent SAPs [3,5,7]. In [3], top-ranked SAPs are selected, which are ranked by their frequency. Considering there may be redundancy in the information provided by top-ranked SAPs, [5] proposes to select dissimilar SAPs to improve diversity, but its similarity measure can only handle path-structured SAPs. Other measures [6,7] process more general SAPs in a heuristic way. By comparison, our pGED measure systematically considers the cost of transforming one SAP into another, which captures both the graph structure and the semantics of graph labels.

As to verbalization, whereas our lexicalization and aggregation build on common practice in this research field, our domain-independent discourse planning is novel. Compared with [1] which mainly generates simple and compound sentences, our sentence-level planning also supports complex sentences for compactness, and our document-level planning properly orders sentences and clauses for coherence, thereby showing superiority over [1] in the experiments.

7 Conclusion

Towards more usable SA search, we improve its result summarization by (i) diversifying selected SAPs based on a new GED measure which jointly considers structural and semantic similarity, and (ii) verbalizing SAPs based on a novel method for discourse planning to generate compact and coherent English text. Potential applications of our approach are not restricted to SA search. Our framework for diversification can be adapted to diversify the results of keyword queries on graph data [6]. Our verbalization approach can be used to verbalize graph-structured query answers, graph representation of ontologies, etc.

Acknowledgement. This work is supported in part by the NSFC under Grants 61572247 and 61772264, and in part by the Qing Lan and Six Talent Peaks Programs of Jiangsu Province.

References

1. Androutsopoulos, I., Lampouras, G., Galanis, D.: Generating natural language descriptions from OWL ontologies: the naturalowl system. J. Artif. Intell. Res. **48**, 671–715 (2013). https://doi.org/10.1613/jair.4017
2. Chen, C., Wang, G., Liu, H., Xin, J., Yuan, Y.: SISP: a new framework for searching the informative subgraph based on PSO. In: Proceedings of the 20th ACM Conference on Information and Knowledge Management, CIKM 2011, Glasgow, UK, 24–28 October 2011, pp. 453–462 (2011). https://doi.org/10.1145/2063576.2063645
3. Cheng, G., Liu, D., Qu, Y.: Efficient algorithms for association finding and frequent association pattern mining. In: Groth, P., et al. (eds.) ISWC 2016, Part I. LNCS, vol. 9981, pp. 119–134. Springer, Cham (2016). https://doi.org/10.1007/978-3-319-46523-4_8
4. Cheng, G., Shao, F., Qu, Y.: An empirical evaluation of techniques for ranking semantic associations. IEEE Trans. Knowl. Data Eng. **29**(11), 2388–2401 (2017). https://doi.org/10.1109/TKDE.2017.2735970
5. Cheng, G., Zhang, Y., Qu, Y.: Explass: exploring associations between entities via top-k ontological patterns and facets. In: Mika, P., et al. (eds.) ISWC 2014, Part II. LNCS, vol. 9981, pp. 422–437. Springer, UK (2014). https://doi.org/10.1007/978-3-319-11915-1_27
6. Dass, A., Aksoy, C., Dimitriou, A., Theodoratos, D., Wu, X.: Diversifying the results of keyword queries on linked data. In: Cellary, W., Mokbel, M., Wang, J., Wang, H., Zhou, R., Zhang, Y. (eds.) WISE 2016, Part I. LNCS, vol. 10041, pp. 199–207. Springer, Cham (2016). https://doi.org/10.1007/978-3-319-48740-3_14
7. Pirrò, G.: Explaining and suggesting relatedness in knowledge graphs. In: Arenas, M. (ed.) ISWC 2015, Part I. LNCS, vol. 9366, pp. 622–639. Springer, Cham (2015). https://doi.org/10.1007/978-3-319-25007-6_36
8. Riesen, K., Bunke, H.: Approximate graph edit distance computation by means of bipartite graph matching. Image Vision Comput. **27**(7), 950–959 (2009). https://doi.org/10.1016/j.imavis.2008.04.004
9. Zhu, G., Iglesias, C.A.: Computing semantic similarity of concepts in knowledge graphs. IEEE Trans. Knowl. Data Eng. **29**(1), 72–85 (2017). https://doi.org/10.1109/TKDE.2016.2610428

Information Extraction

Main Content Extraction
from Heterogeneous Webpages

Julian Alarte[✉], David Insa, Josep Silva, and Salvador Tamarit

Universitat Politècnica de València, Camí de Vera s/n, 46022 Valencia, Spain
{jalarte,dinsa,jsilva,stamarit}@dsic.upv.es

Abstract. Besides the main content, webpages often contain other complementary and noisy data such as advertisements, navigational information, copyright notices, and other template-related elements. The detection and extraction of main content can have many applications, such as web summarization, indexing, data mining, content adaptation to mobile devices, web content printing, etc. We introduce a novel site-level technique for content extraction based on the DOM representation of webpages. This technique analyzes some selected pages in any given website to identify those nodes in the DOM tree that do not belong to the webpage template. Then, an algorithm explores these nodes in order to select the main content nodes. To properly evaluate the technique, we have built a suite of benchmarks by downloading several heterogeneous real websites and manually marking the main content nodes. This suite of benchmarks can be used to evaluate and compare different content extraction techniques.

Keywords: Information retrieval · Content extraction
Template extraction · Block detection

1 Introduction

Extracting information from webpages is useful for humans and for many different systems. The most important information in a webpage is the main content. However, the main content is almost always next to other noisy elements such as banners, footers, main menus, advertisements, etc. The task of extracting the main content from a webpage consists in isolating the useful information from the noise, removing the elements that do not contain useful knowledge for the user (see, e.g., Fig. 1).

This work has been partially supported by the EU (FEDER) and the *Spanish Ministerio de Ciencia, Innovación y Universidades*/AEI under grant TIN2016-76843-C4-1-R and by the *Generalitat Valenciana* under grant PROMETEO-II/2015/013 (SmartLogic). Salvador Tamarit was partially supported by the *Conselleria de Educación, Investigación, Cultura y Deporte de la Generalitat Valenciana* under the grant APOSTD/2016/036.

© Springer Nature Switzerland AG 2018
H. Hacid et al. (Eds.): WISE 2018, LNCS 11233, pp. 393–407, 2018.
https://doi.org/10.1007/978-3-030-02922-7_27

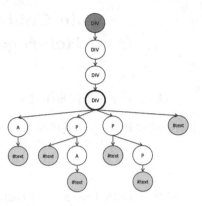

Fig. 1. Main content of WISE 2018's 'keynote-speakers' webpage (extracted with our web content extraction tool).

Fig. 2. Main content DOM nodes example

Our method inputs an arbitrary webpage (the key page) and it outputs a set of DOM nodes representing its main content. Our approach is site-level, so it loads and analyzes several other webpages to detect the main content. This allows it to increase accuracy, because it can detect template (repeated) content (Fig. 2).

This approach is divided into three phases:

1. An algorithm selects a set of webpages that belong to the same website of the key page.
2. For each webpage in the set, an algorithm maps its DOM nodes with the DOM nodes of the key page. If it finds that a node of the key page is repeated in another webpage, the algorithm updates a counter, so that it can know how many times each node appears in other webpages.
3. The set of DOM nodes in the key page that are not repeated in any other page are added to a set of *candidate nodes*. These nodes are analyzed in the following way:
   - An algorithm selects only those DOM nodes of the set that do not have ancestors in the set. They form the *reduced set of candidate nodes*.
   - If there is only one node in the *reduced set of candidate nodes*, that node and all its descendants correspond to the main content. However, if there are several candidate nodes in the set:
     • Each candidate node is analyzed to detect the branch of the DOM tree that more likely contains the main content.
     • Finally, the algorithm selects the candidate nodes (DOM nodes that only appear in the key page) that belong to the main content branch.

2   Main Content Extraction

This section explains the three phases followed in our method to extract the main content of a webpage.

2.1   Set of Webpages Selection

This section proposes an algorithm to identify a set of webpages (in the following n-SET) from the same website of the key page that very likely share the same template of the key page. The process of selecting the n-SET is:

1. The algorithm analyzes the key page and extracts a set containing all the hyperlinks that point to webpages in the same domain.
2. Then, the algorithm sorts the hyperlinks of the set.
3. Finally, the algorithm selects the first n hyperlinks of the sorted set and collects the set of webpages pointed by these hyperlinks.

The first step is trivial. The DOM tree of the key page is traversed and those hyperlink nodes (with the HTML A tag) that point to webpages in the same domain as the key page, and that do not point to the key page are collected.

Once the set of valid hyperlinks of the key page is created, it is necessary to establish an order of relevance. Thereby, the most related hyperlinks to the key page will be positioned in the first places. For this, we use a combination of the two metrics proposed in [3]: the *hyperlink distance* and the *DOM distance*.

The final set of webpages (the n-SET) returned in this phase contains those hyperlinks with a *hyperlink distance* as closer as possible to zero. Webpages with the same *hyperlink distance* are sorted according to their *DOM distance*, selecting hyperlinks as far as possible (among all of them) in the DOM tree (see [2] for details).

2.2   Webpages Mapping

In order to identify the nodes of the key page repeated in other webpages of the n-SET, we use a tree mapping algorithm called *equal top-down mapping* (ETDM) [3]. This mapping analyzes two DOM trees and establishes a correspondence between their nodes.

After we have built the *set of webpages* (n-SET), we compute an ETDM between the key page and each webpage in the n-SET. We have implemented an algorithm that performs each comparison by traversing the DOM trees top-down. Specifically, starting at the root, the algorithm tries to map each node of the key page with the nodes in the other webpage that are at the same depth. If two nodes can be mapped, it means that both nodes are equal (represented with $n_1 \triangleq n_2$), and the algorithm updates an attribute (it works as a counter) called *occurrences* on the node in the key page. This attribute indicates the number of times a node is repeated in the webpages of the n-SET. Then, the algorithm recursively continues with the children of both mapped nodes. When one node cannot be mapped, the algorithm stops exploring the descendants of this node.

Algorithm 1 inputs a key page and a set of webpages (n-SET), and it outputs the same key page including a counter for each node. This counter specifies the number of times a node of the key page appears in the webpages of the n-SET. This counter is used to identify template nodes of the key page: those where the counter is greater than zero (i.e., they are repeated in other webpages of the

---

Algorithm 1. Compute the number of occurrences of each node in the key page

Input: A key page $p_k = (N_1, E_1)$ and a set of n webpages P.
Output: The key page p_k equipped with a variable *occurrences* for each node.
Initialization: $\forall n \in N_1$. $n.occurrences = 0$.

```
begin
    r₁ = root(pₖ);
    foreach (p = (N₂, E₂) in P)
        r₂ = root(p);
        if (r₁ ≜ r₂)
            r₁.occurrences = r₁.occurrences + 1;
            assignOccurrences(r₁, r₂);
    return pₖ;
end

procedure assignOccurrences(node r₁ ∈ N₁, node r₂ ∈ N₂)
    foreach (n₁ ∈ N₁, n₂ ∈ N₂ . n₁ ≜ n₂, (r₁, n₁) ∈ E₁ and (r₂, n₂) ∈ E₂)
        n₁.occurrences = n₁.occurrences + 1;
        assignOccurrences(n₁, n₂);
end procedure
```

---

website). The main content of the webpage will be probably formed by nodes with *occurrences* $= 0$.

The output of this phase is a set of *candidate nodes*. For instance, Fig. 3 shows an example of a key page where grey nodes represent the candidate nodes. Candidate nodes do not appear in other webpages of the n-SET and, thus, some of them can be used to represent the main content. This is explained in the following section.

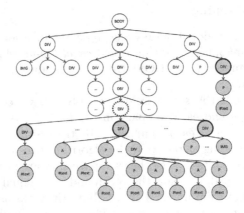

Fig. 3. Set of candidate nodes (excerpt from Fig. 1)

2.3   Candidate Set Reduction

The set of candidate nodes can be simplified by representing subtrees of candidate nodes with their roots. For instance, in Fig. 3, all grey nodes are represented with the four bold-line nodes. This new set is called the *reduced set of candidate nodes*, and it can be computed by exploiting the following theorem.

Algorithm 2. Candidate set reduction

Input: A set of DOM nodes *candidatesSet*
Output: A reduced set of DOM nodes *reducedSet*.
Initialization: *reducedSet* = {}.

begin
   foreach (*node* in *candidatesSet*)
     cand = *node*;
     while (*parent*(*cand*) ∈ *candidatesSet*)

      cand = *parent*(*cand*);
      candidatesSet = *candidatesSet* \
               subtree(*cand*);
    reducedSet = *reducedSet* ∪ *cand*;
   return *reducedSet*;
end

Algorithm 3. Main content branch detection

Input: A key page p_k, and a set *reducedSet* of DOM nodes in p_k.
Output: A DOM node *branch*.

begin
   count = 0;
   foreach (*n* in *reducedSet*)
     node = *parent*(*n*);
     if |*subtree*(*node*)| > *count*
      branch = *node*;
      count = |*subtree*(*node*)|;
   return *branch*;
end

Theorem 1 (parent-child relation of candidate nodes). *Let $P = (N, E)$ be a webpage and let* candidates $\subseteq N$ *be the set of all candidate nodes of P. Then, $n \in$ candidates $\implies \forall n', (n, n') \in E^*$. $n' \in$ candidates.*

Proof. First, if $|descendants(n)| = 0$ the claim follows trivially. We prove the case when $|descendants(n)| \geq 0$ by contradiction. We assume that $n \in$ *candidates* $\land \exists n', (n, n') \in E^*$. $n' \notin$ *candidates*. Because $n' \notin$ *candidates*, there must exist a webpage P' with an ETDM mapping M and $(n', n'') \in M$ (for some n''). Moreover, according to the top-down property of ETDM (see [3]), all ancestors of n' also belong to the ETDM mapping M, and therefore, all ancestors of n' (including n) are not candidates: $n \notin$ *candidates*. But this is a contradiction with the premise $n \in$ *candidates*.

Theorem 1 states an important property that is used by Algorithm 2 to compute the *reduced set of candidate nodes*. This algorithm inputs the set of candidate nodes and, for each node, it explores its ancestors. The ancestor with lower depth on the DOM tree that belongs to the set of candidate nodes, is added to a new set of nodes. Finally, the new set of nodes is returned as the *reduced set of candidates*.

2.4   Main Content Branch Detection

If the reduced set of candidates only contains one node, then this node corresponds to the main content of the webpage. Therefore, it is not necessary to execute the following phases and the algorithm returns that node as the main content of the webpage. If, on the contrary, it contains more than one node, then we need to further analyze the DOM tree to remove those nodes that are not main content.

For each parent of a node in the *reduced set of candidates*, the algorithm counts its number of descendants and stores the value. The node with more descendants represents the root node of the main content branch. Considering the number of nodes in a DOM tree, a draw is very difficult, but possible. If it happens, the node selected is the first one in a deep first traversal because it is more likely to appear in the webpage without scrolling. The reason of selecting the parent of each DOM node in the *reduced set of candidates* is that those parent nodes appear in other webpages, so they likely belong to the template of the website or they are probably located in the border between the template and the main content (according to Theorem 1 all the descendants of these parent nodes are candidates and, therefore, they do not appear in other webpages).

Algorithm 4. Candidate set reduction

> **Input:** A set of DOM nodes *reducedSet* and the branch node *branch*
> **Output:** A set of DOM nodes *finalReducedSet* only including nodes that belong to the main content branch
>
> **begin**
>   **foreach** (*node* in *reducedSet*)
>   **if** (*branch* $\notin$ *ancestors*(*node*))
>   *reducedSet* = *reducedSet* \ {*node*}
>   **return** *reducedSet*;
> **end**

Algorithm 5. Main content selection

> **Input:** A set of DOM nodes *reducedSet*
> **Output:** A set of DOM nodes *mainCont*.
>
> **begin**
>   *mainCont* = *reducedSet*;
>   **foreach** (n_1, n_2 in *mainCont* with $parent(n_1) == parent(n_2)$)
>   *mainCont* = (*mainCont* \ {n_1, n_2}) $\cup$ {$parent(n_1)$}
>   **return** *mainCont*;
> **end**

Algorithm 3 selects the main content branch node of a webpage.

The main content branch is a node used to select a branch of the DOM tree that contains the main content. All nodes outside this branch are discarded (see Sect. 2.5). Of course, it is possible that the selected branch contains several candidates, and thus they should be further processed to extract the final set of main content nodes (this is explained in Sect. 2.6).

For instance, Fig. 3 shows that the node selected as the root node of the main content branch is the dotted-line "DIV". This "DIV" node is the parent node of 3 nodes that belong to the reduced set of candidates. Therefore, it is the root node of the main content branch.

2.5   Discarding Candidates

Once the branch that contains the main content is selected, the nodes that do not belong to that branch can be removed from the *reduced set of candidate nodes*. Therefore, for each node in the *reduced set of candidate nodes*, an algorithm checks whether the node belongs to the main content branch. If the node does not belong to that branch, it is removed. In practice, this means that whenever we find different separate groups of candidate nodes in the DOM tree, we try to join as many groups as possible by selecting the *branch* node with Algorithm 3, and then we discard the other groups.

Example 1. In Fig. 3, once we have computed the branch node (the dotted-line node), Algorithm 4 discards the three grey nodes at the top-right of the tree (because they are not descendants of the branch node).

2.6 Main Content Selection

Once the candidates that do not belong to the main content branch have been discarded, all the remaining nodes in the *reduced set of candidates* are considered main content. However, sometimes, these nodes can be grouped, e.g., when two of them are sibling nodes (as in Fig. 3).

Therefore, the main content is formed by all nodes in the *reduced set of candidates* except for sibling nodes, which are recursively replaced by their parent. This is computed with Algorithm 5.

Example 2. In Fig. 3, after having removed the three grey nodes at the top right side, only three nodes remain in the *reduced set of candidates*: the three bold-line sibling nodes. According to Algorithm 5, the final main content of the webpage is the dotted-line node, because it is replaced by its three children.

3 Empirical Evaluation

We implemented this technique and integrated all the algorithms proposed as a Firefox's addon, publicly available at: http://www.dsic.upv.es/~jsilva/retrieval/Web-ConEx/. For the evaluation, we used the template detection and content extraction benchmark suite (TECO)[1]. Traditionally, most authors of content extraction techniques have measured the recall, precision, and F1 (computed as $(2 * P * R)/(P + R)$, being P the precision and R the recall) of the retrieved words. This implicitly means that they measure the quality of their techniques considering that the main content is text. One important advantage of our technique is that it not only measures the recall, precision, and F1 of the retrieved words, but also of the retrieved DOM nodes. Therefore, we do not only consider the main content as text, it can also contain images, video, and other types of content.

3.1 Precision, Recall, and F1 Evaluation

To evaluate the precision and recall of our technique, we produced a version of our Firefox addon that automatically executes our content extraction algorithm with all the webpages of the benchmark suite. It displays for each benchmark the recall, precision, and F1 of the retrieved words and the retrieved DOM nodes, and the total execution time.

The only parameter we needed to determine was the size of the set of webpages (the n value of the n-SET) needed by Algorithm 1. In order to develop our technique and determine the optimum size of the n-SET, we measured the

---

[1] http://users.dsic.upv.es/~jsilva/retrieval/teco/.

recall, precision and F1 of the retrieved text words and DOM nodes for different n-SET sizes.

Table 1 summarizes the results of the performed evaluation experiments, with a n-SET size from 2 to 8, and with a training set of 15 benchmarks. Each row is the average of repeating all the experiments in the evaluation subset of 15 benchmarks with a different value for n in the n-SET. Column `Size` represents the size of the n-SET. In addition, the table shows, for the retrieved DOM nodes and the retrieved text words, the average `Recall`, `Precision`, and `F1`.

Table 1. Determining the optimal size of the n-SET

Size	DOM nodes			Words			Runtime
	Recall	Precision	F1	Recall	Precision	F1	
2	76,97%	68,15%	71,63%	78,88%	72,52%	74,54%	16,32 s
3	86,25%	85,80%	83,53%	89,24 %	87,33%	86,62%	23,59 s
4	92,92%	92,59%	90,26%	95,90%	94,76 %	93,71%	33,14 s
5	85,21%	97,91%	86,30%	88,50%	99,59%	91,61%	41,31 s
6	85,21%	98,22%	86,44%	88,50%	99,91%	91,76%	50,41 s
7	85,21%	98,22%	86,44%	88,50%	99,91%	91,76%	59,36 s
8	84,80%	98,50%	86,37%	88,50%	99,91%	91,76%	68,15 s

We determined that a set of webpages of size 4 (4-SET) is the best option because it keeps the best F1 value, both in retrieved DOM nodes (90,26%) and in retrieved words (93,71%). Table 1 reveals that sets of 2 webpages (2-SET) obtain low F1 values (around 70%). On the other hand, sets of webpages with 5 or more webpages obtain similar values of F1. Note that sets of 5 or more webpages do increase the precision up to almost 100%, but the recall decreases and their F1 values are lower than the F1 value of the 4-SET. It is also important to highlight that the size of the set directly affects the performance, because as the size is increased, more webpages must be loaded, and more ETDM mappings must be calculated. This is another good reason to select the 4-SET.

In order to evaluate our main content extraction technique, we selected 30 benchmarks from our benchmark suite as the evaluation subset. For the 30 benchmarks, we computed the `Recall`, `Precision`, and `F1` of the retrieved DOM nodes and the retrieved words. In addition, we computed the `Runtime` in seconds. The results (computed with a 4-SET) are shown in Table 2.

The experiments reveal an average F1 around 88% for retrieved DOM nodes, and an average F1 over 91% for retrieved words. To the best of our knowledge, these values are the highest F1 in the state of the art for benchmarks formed by heterogeneous websites. On the one hand, similar techniques as ours that also use heterogeneous websites, produce the following results: Insa et al. obtain an F1 of 74% [10], Gottron et al. 77% [9], and Wu et al. 82% [22]. On the other hand, there are techniques based on evaluating prepared datasets such as Cleaneval [5], BIG5, MYRIAD40, MSS, etc. Other techniques evaluate RSS feeds, or prepared

Table 2. Evaluation of the precision, recall, F1, and runtime

Benchmark	DOM nodes			Words			Runtime
	Rec.	Prec.	F1	Rec.	Prec.	F1	
wise2018.connect.rs	100,00%	95,63%	97,76%	100,00%	98,69%	99,33%	33,57 s
www.javiercelaya.es	100,00%	84,59%	91,65%	100,00%	100,00%	100,00%	27,01 s
www.trendencias.com	99,91%	71,18%	83,13%	100,00%	82,29%	90,28%	189,32 s
www.turfparadise.com	98,59%	92,51%	95,45%	100,00%	100,00%	100,00%	36,89 s
www.u-tokyo.ac.jp	100,00%	92,38%	96,04%	100,00%	100,00%	100,00%	12,61 s
www.savethechildren.net	16,67%	100,00%	28,57%	21,90%	100,00%	35,93%	38,90 s
college.harvard.edu	100,00%	84,29%	91,47%	100,00%	88,21%	93,74%	89,64 s
www.raspberrypi.org	100,00%	82,94%	90,67%	100,00%	86,36%	92,68%	11,82 s
www.annmalaspina.com	100,00%	100,00%	100,00%	100,00%	100,00%	100,00%	10,04 s
dublin.ie	100,00%	94,59%	97,22%	100,00%	100,00%	100,00%	29,20 s
www.amateurgourmet.com	100,00%	90,09%	94,79%	100,00%	97,09%	98,52%	708,39 s
www.museodelprado.es	98,43%	97,67%	98,05%	99,32%	100,00%	99,66%	18,89 s
www.rfet.es	99,90%	96,97%	98,41%	99,73%	97,64%	98,68%	128,88 s
www.centralparknyc.org	71,23%	72,22%	71,72%	100,00%	58,59%	73,89%	121,43 s
manytools.org	100,00%	66,19%	79,65%	100,00%	91,49%	95,56%	32,49 s
clotheshor.se	100,00%	100,00%	100,00%	100,00%	100,00%	100,00%	7,47 s
www.unicef.org	100,00%	99,48%	99,74%	100,00%	97,56%	98,77%	61,82 s
www.news-medical.net	99,59%	77,71%	87,30%	100,00%	88,85%	94,09%	104,72 s
teachreal.wordpress.com	99,29%	67,57%	80,41%	100,00%	90,70%	95,12%	37,90 s
www.grandcentralterminal.com	100,00%	97,95%	98,96%	100,00%	100,00%	100,00%	77,00 s
www.cleanclothes.org	100,00%	88,12%	93,69%	100,00%	91,94%	95,80%	58,37 s
riotimesonline.com	99,74%	63,46%	77,57%	100,00%	65,90%	79,45%	140,09 s
www.ox.ac.uk	96,93%	100,00%	98,44%	100,00%	100,00%	100,00%	80,31 s
www.filmaffinity.com	99,80%	100,00%	99,90%	100,00%	100,00%	100,00%	72,31 s
www.coiicv.org	97,70%	100,00%	98,84%	99,47%	100,00%	99,73%	43,38 s
www.thelawyer.com	91,31%	82,66%	86,77%	94,47%	86,92%	90,53%	281,72 s
www.toureiffel.paris	95,37%	100,00%	97,63%	99,71%	100,00%	99,86%	38,55 s
institute-events.mit.edu	88,76%	96,34%	92,40%	94,35%	100,00%	97,09%	83,40 s
www.w3schools.com	99,76%	100,00%	99,88%	100,00%	100,00%	100,00%	734,78 s
stackoverflow.com	5,97%	98,05%	11,26%	2,76%	94,92%	5,37%	1237,59 s
Average - using metrics	92,08%	89,53%	87,91%	93,72%	93,90%	91,14%	151,62 s
Average - random pages	83,80%	75,70%	73,54%	84,18%	81,41%	79,47%	113,68 s

websites (collections of automatically generated webpages that share the same template). These techniques usually obtain high F1 values: Adam et al. obtain an F1 value of 93% [1], Li et al. 88% [14], Pasternack et al. 95% [15], and Qureshi et al. 94% [16]. Obviously, the comparison of different techniques should not be done using different datasets, thus, we have compared the techniques against the same benchmarks suite. Results are shown in Table 3, and explained in Sect. 4.

We also wanted to evaluate the usefulness of the analyses performed in Sect. 2.1. If we select the pages in the n-SET randomly (instead of using the *hyperlink distance* and *DOM distance* metrics), then our technique decreases its F1 value more than 10% (see the last two rows of Table 2). "using metrics" is the average when using these metrics, while "random pages" do not use the metrics and it selects the pages randomly. Experiments reveal that, in retrieved nodes, the F1 decreases more than about 15%; being about 12% in retrieved words. Regarding the algorithm runtime, the random selection of pages is 38 s faster (on average).

3.2 Performance Evaluation

Column Runtime in Table 2 and Fig. 4 show the time needed to extract the main content from different webpages. 50% of the benchmarks took less than 60 s but, as we expected, usually, the larger (in terms of DOM nodes) the webpages are, the more time the algorithm needs to process them. For larger webpages (those with thousands of DOM nodes) the algorithm takes minutes. We observed that the runtime not only depends on the number of DOM nodes of the website, but it also depends on their structure and features, such as their number of children, their relative position in the DOM tree, their HTML attributes, etc. For that reason, the chart in Fig. 4 is irregular, and the runtime does not strictly grow with the number of DOM nodes.

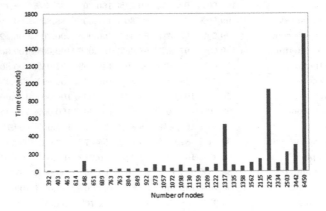

Fig. 4. Relation between webpage size and runtime

4   Comparison with Other Algorithms

We wanted to compare our technique with other well-known algorithms [3, 18–20, 23]. Obviously, it is not possible to compare different content extractors by just comparing the results reported in the bibliography, because each report uses a different measure (e.g., words, characters, DOM nodes...) and also because their evaluations were done with different benchmarks. Unfortunately, some implementations are proprietary or not available, so we decided to reimplement them from scratch. All of them are now available as open-source at: http://users.dsic.upv.es/~jsilva/retrieval/Web-TemEx/.

We compared the performance and accuracy of the following techniques:

SST (2003) [23]: This algorithm introduced a new data structure called *Site Style Tree* (SST) to represent a collection of webpages. The SST stores information about the DOM nodes in the selected set of webpages. The most repeated nodes or groups of nodes are more likely to belong to the template of the key page.

RTDM-TD (2006) [19]: This algorithm inputs a set of webpages and compares their DOM trees using a top-down variant of the tree edit distance (TED) algorithm. The intersection of the DOM trees (the nodes repeated in all webpages) is considered as the template.

IWPTD (2008) [20]: This algorithm divides the DOM trees of the webpages into a set of subtrees whose root nodes are associated with concrete HTML tags (i.e., TABLE, DIV, UL, etc.). Then, it compares the text segments inside the subtrees of all the webpages. The text segments that appear in 5 or more webpages are considered template segments. Finally, it computes a ratio to decide whether a subtree belongs to the template or not.

RBMTD (2009) [18]: This algorithm is similar to RTDM-TD but, in this case, it uses a bottom-up variant of the tree edit distance (TED) algorithm. While comparing the DOM trees, it introduces a restriction to classify a common subtree as template: those subtrees that appear in all webpages must be exactly in the same position.

TemEx (2015) [3]: This algorithm selects a set of webpages from the website and uses a mapping between their DOM trees to determine the number of times a node is repeated across the webpages in the set. A node is considered to belong to the template if it is repeated in t webpages, being t a threshold determined empirically.

We extracted the main content from the test set of 30 benchmarks using all algorithms (see Table 3). Regarding the `recall`, `precision`, and `F1`, the experiments reveal that our algorithm obtains the best F1 values for both, retrieved DOM nodes and retrieved words. The second best F1 values for both retrieved DOM nodes and words is achieved by TemEx. RBMTD is a very conservative

Table 3. Empirical evaluation of six web content extraction algorithms

Algorithm	DOM nodes			Words			Runtime
	Rec.	Prec.	F1	Rec.	Prec.	F1	
SST	66,40%	42,28%	48,08%	70,25%	50,22%	54,25%	554,26 s
RTDM-TD	99,81%	40,74%	53,64%	100,00%	55,25%	68,72%	92,55 s
IWPTD	68,33%	62,53%	61,24%	81,19%	68,87%	72,55%	21,03 s
RBMTD	100,00%	44,20%	56,46%	100,00%	58,94%	71,55%	151,26 s
TemEx	97,17%	76,19%	82,52%	98,59%	82,17%	87,97%	63,37 s
Our algorithm	92,08%	89,53%	87,91%	93,72%	93,90%	91,14%	151,62 s

algorithm. It achieved 100% recall in all experiments. So, it is a good choice if retrieving the whole main content is critical. Another interesting observation is that, in all cases, the F1 values are better if they are measured in retrieved words instead of retrieved DOM nodes. This means that these algorithms are more oriented to retrieve text, but they sometimes miss some image or container (e.g., a `DIV` or a `TABLE`) that do belong to the main content. This fact can be observed because we have used both measures to compare the benchmarks. In the case of RBMTD this difference is higher than 15%.

Given these results, those systems that need a high `Recall` should use RBMTD or RTDM-TD, those systems that need a high `Precision` should use our algorithm or TemEx, and those systems that need high performance should use IWPTD.

With respect to the computation time, it is significantly different for each algorithm. IWPTD is the quickest one, it takes an average of about 21 s per benchmark. In contrast, TemEx takes an average of about one minute per benchmark. Our algorithm is slow compared to the others, it takes more than 2,5 min per benchmark. This is due to the fact that a 4-SET was used in the configuration (it had to find 4 webpages in the website and compare the key page with them). Reducing this number to 3 would speed up the algorithm potentially reducing the runtime around 30% at the cost of 7% F1 (see Table 1). The slowest algorithm is SST, whose computation time is extremely high compared to the others, above 9 min per benchmark on average.

5   Related Work

Besides the techniques explained in the previous section, there are other interesting techniques related to our work. We overview some of them in this section. Content extraction, template extraction, menu detection, etc. are interesting topics due to their relation to web mining, searching, indexing, and web development. There are many different approaches that try to face these problems (see, e.g., [7,9,10,21,22]). Some of these techniques were presented in the CleanEval competition [5], which proposed a collection of examples to be analyzed with a gold standard. This collection of examples was prepared for boilerplate removal and content extraction.

Content Extraction and *template extraction* are very close disciplines. While content extraction tries to isolate the main content pagelets[2] of the webpage, template extraction tries to isolate the template of the webpage. These disciplines are considered an instance of a more general discipline called *Block Detection*, which tries to detect all pagelets that exist in a webpage. In the area of block detection, researchers use three main different approaches to solve the problem: **(1)** Using the textual information of the webpage (i.e., the HTML code). The main idea is that the main content on a webpage has more density of text with less labels. For instance, Ferraresi et al. [8] proposed the main content as

---

[2] Bar-Youssef et al. [4] defined a pagelet as a self-contained logical region with a well defined topic of functionality. Accordingly, webpages are composed of pagelets.

the largest contiguous text area with the least amount of HTML tags. In the same way, Weninger et al. [21] defined the Content Extraction via Tag Ratios (CETR). This method analyzes the HTML code and computes a ratio (CETR) by counting the number of characters and labels inside each label. Kohlschütter et al. [11,13] proposed the exploitation of densitometric features based on the observation of the more common terms in webpage templates.

(2) Using a rendered image of the webpage on the browser. Burget et al. [6] proposed an approach based on the idea that the main content of a webpage is often located in the central part and it is often visible without scrolling (see, e.g., Fig. 1). Kohlschütter et al. [12] concluded that this kind of techniques are not so widespread as others because rendering webpages for classification is a computational expensive operation.

(3) Using the representation of the webpage as a DOM tree. Bar-Yossef et al. [4] proposed a method for template detection based on the analysis of the DOM tree. This approach counts the frequent pagelet item sets. Vieira et al. [19], Yi et al. [23] and Alarte et al. [3] proposed techniques based on finding common subtrees in the DOM trees of a set of webpages from the same website. These common subtrees between webpages of the same website are defined as noisy information or template. Yi et al. [23] introduced a data structure called Site Style Tree (SST). The SST summarizes a set of DOM trees of different webpages of the same website. Every DOM node in the webpages is represented in a single tree (the SST), and the repeated nodes (those appearing in different webpages) are identified by using counters in the SST nodes. The most repeated nodes in the SST (those with highest counter values) are more likely to belong to the template of the website.

Vieira et al. [19] proposed the use of optimal mappings between DOM trees. This mapping (RTDM-TD) finds duplicated nodes across webpages from the same website. RTDM-TD (*restricted top-down mapping*) algorithm was proposed by Reis et al. [17].

6   Conclusions

Our work presents a new technique for content extraction from heterogeneous websites. In contrast to other content extraction techniques, our approach is based on DOM nodes, and it allows for extracting not only text as main content, but also images, videos, animations, and other types of content. As it is a site-level technique, it uses the information of several webpages to extract the main content. In particular, it analyzes the key page and extracts its hyperlinks. Once extracted, the hyperlinks are sorted in order to select the webpages that can provide more information about the main content of the key page. We measured that a set of 4 webpages (4-SET) obtains the best values of F1 for both, retrieved DOM nodes and retrieved words. To select the webpages that should be analyzed we propose the use of two metrics (*hyperlink distance* and *dom distance*). The use of these metrics produce (as an average) an increase of 10% F1. To compare the DOM nodes of the webpages we use an ETDM mapping. Once the webpages are compared, we have information about the number of times each DOM node appears in them. In our algorithms, we consider that a DOM node that only appears in the key page more likely belongs to its main content. This could be relaxed. However, we repeated our experiments with other numbers (i.e., considering that the key page's nodes could be repeated 2, 3, or 4 times in other webpages) and the results were worst in all cases.

The idea of mapping the DOM nodes in order to infer the branch of the DOM tree that probably contains the main content is simple but effective. Usually, the main content of a webpage is not repeated on other webpages of the same website, so identifying the non-repeated nodes can lead us to find the main content. Moreover, the main content usually concentrates a large amount of information that is structurally close. The identification of the main content branch follows this idea.

References

1. Adam, G., Bouras, C., Poulopoulos, V.: CUTER: an efficient useful text extraction mechanism. In: 2009 International Conference on Advanced Information Networking and Applications Workshops, pp. 703–708, May 2009
2. Alarte, J., Insa, D., Silva, J., Tamarit, S.: Automatic detection of webpages that share the same web template. In: ter Beek, M.H., Ravara, A. (eds.) Proceedings of the 10th International Workshop on Automated Specification and Verification of Web Systems (WWV 2014). Electronic Proceedings in Theoretical Computer Science, vol. 163, pp. 2–15. Open Publishing Association, July 2014
3. Alarte, J., Insa, D., Silva, J., Tamarit, S.: Site-level web template extraction based on DOM analysis. In: Mazzara, M., Voronkov, A. (eds.) PSI 2015. LNCS, vol. 9609, pp. 36–49. Springer, Cham (2016). https://doi.org/10.1007/978-3-319-41579-6_4
4. Bar-Yossef, Z., Rajagopalan, S.: Template detection via data mining and its applications. In: Proceedings of the 11th International Conference on World Wide Web (WWW 2002), pp. 580–591. ACM, New York (2002)
5. Baroni, M., Chantree, F., Kilgarriff, A., Sharoff, S.: Cleaneval: a competition for cleaning web pages. In: Proceedings of the International Conference on Language Resources and Evaluation (LREC 2008), pp. 638–643. European Language Resources Association, May 2008
6. Burget, R., Rudolfova, I.: Web page element classification based on visual features. In: Proceedings of the 1st Asian Conference on Intelligent Information and Database Systems (ACIIDS 2009), pp. 67–72. IEEE Computer Society, Washington, DC (2009)
7. Cardoso, E., Jabour, I., Laber, E., Rodrigues, R., Cardoso, P.: An efficient language-independent method to extract content from news webpages. In: Proceedings of the 11th ACM Symposium on Document Engineering (DocEng 2011), pp. 121–128. ACM, New York (2011)
8. Ferraresi, A., Zanchetta, E., Baroni, M., Bernardini, S.: Introducing and evaluating ukWaC, a very large web-derived corpus of English. In: Proceedings of the 4th Web as Corpus Workshop (WAC-4), pp. 47–54 (2008)
9. Gottron, T.: Content code blurring: a new approach to content extraction. In: Proceedings of the 2008 19th International Conference on Database and Expert Systems Application, DEXA 2008, pp. 29–33. IEEE Computer Society, Washington, DC, September 2008

10. Insa, D., Silva, J., Tamarit, S.: Using the words/leafs ratio in the DOM tree for content extraction. J. Log. Algebr. Program. **82**(8), 311–325 (2013)
11. Kohlschütter, C.: A densitometric analysis of web template content. In: Quemada, J., León, G., Maarek, Y.S., Nejdl, W. (eds.) Proceedings of the 18th International Conference on World Wide Web (WWW 2009), pp. 1165–1166. ACM, April 2009
12. Kohlschütter, C., Fankhauser, P., Nejdl, W.: Boilerplate detection using shallow text features. In: Davison, B.D., Suel, T., Craswell, N., Liu, B. (eds.) Proceedings of the 3rd International Conference on Web Search and Web Data Mining (WSDM 2010), pp. 441–450. ACM, February 2010
13. Kohlschütter, C., Nejdl, W.: A densitometric approach to web page segmentation. In: Shanahan, J.G., et al. (eds.) Proceedings of the 17th ACM Conference on Information and Knowledge Management (CIKM 2008), pp. 1173–1182. ACM, October 2008
14. Li, Z., Ng, W.K., Sun, A.: Web data extraction based on structural similarity. Knowl. Inf. Syst. **8**(4), 438–461 (2005)
15. Pasternack, J., Roth, D.: Extracting article text from the web with maximum subsequence segmentation. In: Proceedings of the 18th International Conference on World Wide Web, WWW 2009, pp. 971–980. ACM, New York (2009)
16. Qureshi, P.A.R., Memon, N.: Hybrid model of content extraction. J. Comput. Syst. Sci. **78**(4), 1248–1257 (2012)
17. Reis, D.d.C., Golgher, P.B., Silva, A.S., Laender, A.H.F.: Automatic web news extraction using tree edit distance. In: Proceedings of the 13th International Conference on World Wide Web (WWW 2004), pp. 502–511. ACM, New York (2004)
18. Vieira, K., da Costa Carvalho, A.L., Berlt, K., de Moura, E.S., da Silva, A.S., Freire, J.: On finding templates on web collections. World Wide Web **12**(2), 171–211 (2009)
19. Vieira, K., da Silva, A.S., Pinto, N., de Moura, E.S., Cavalcanti, J.a.M.B., Freire, J.: A fast and robust method for web page template detection and removal. In: Proceedings of the 15th ACM International Conference on Information and Knowledge Management (CIKM 2006), pp. 258–267. ACM, New York (2006)
20. Wang, Y., Fang, B., Cheng, X., Guo, L., Xu, H.: Incremental web page template detection. In: Proceedings of the 17th International Conference on World Wide Web (WWW 2008), pp. 1247–1248. ACM, New York (2008)
21. Weninger, T., Henry Hsu, W., Han, J.: CETR: Content Extraction via Tag Ratios. In: Rappa, M., Jones, P., Freire, J., Chakrabarti, S. (eds.) Proceedings of the 19th International Conference on World Wide Web (WWW 2010), pp. 971–980. ACM, April 2010
22. Wu, S., Liu, J., Fan, J.: Automatic web content extraction by combination of learning and grouping. In: Proceedings of the 24th International Conference on World Wide Web, WWW 2015, pp. 1264–1274. International World Wide Web Conferences Steering Committee, Republic and Canton of Geneva, Switzerland (2015)
23. Yi, L., Liu, B., Li, X.: Eliminating noisy information in web pages for data mining. In: Proceedings of the 9th ACM SIGKDD International Conference on Knowledge Discovery and Data mining (KDD 2003), pp. 296–305. ACM, New York (2003)

Bootstrapped Multi-level Distant Supervision for Relation Extraction

Ying He[1], Zhixu Li[1(✉)], Guanfeng Liu[2], Fangfei Cao[3],
Zhigang Chen[4], Ke Wang[5], and Jie Ma[5]

[1] School of Computer Science and Technology, Soochow University, Suzhou, China
yhe94@stu.suda.edu.cn, zhixuli@suda.edu.cn
[2] Department of Computing, Macquarie University, Sydney, NSW 2122, Australia
guanfeng.liu@mq.edu.au
[3] CFETS Information Technology (Shanghai) Co., Ltd., Shanghai, China
caofangfei@chinamoney.com
[4] IFLYTEK Co., Ltd., Hefei, China
zgchen@iflytek.com
[5] Migu Co., Ltd., Beijing, China
{wangke,majie}@migu.cn

Abstract. Distant supervised relation extraction has been widely used to identify new relation facts from free text. However, relying on a single-node categorization model to identify relation facts for thousands of relations simultaneously inevitably accompanies with serious false categorization problem. Also to the best of our knowledge, no previous efforts has yet considered to update the categorization model with the new identified relation facts, which wastes the chance to further improsve the extraction precision and recall. In this paper, we novelly propose a multi-level distant supervision model for relation extraction, which divides the original categorization task into a number of sub-tasks in multiple levels of a constructed tree-like categorization structure. With the tree-like structure, an unlabelled relation instance would be categorized step by step along a path from the root node to a leaf node. Beyond that, we propose to do bootstrapped distant supervision to update the distant supervision model with new learned relation facts iteratively to further improve the extraction precision and recall. Experimental results conducted on two real datasets prove that our approach outperforms state-of-the-art approaches by reaching more than 10% better extraction quality.

Keywords: Relation extraction · Information extraction
Distant supervision

1 Introduction

Distant supervised Relation Extraction (RE) aims at predicting semantic relations between pairs of entities in texts supervised by a Knowledge Base (KB).

© Springer Nature Switzerland AG 2018
H. Hacid et al. (Eds.): WISE 2018, LNCS 11233, pp. 408–423, 2018.
https://doi.org/10.1007/978-3-030-02922-7_28

Given a KB with a set of relation instances (or entity pairs) under a number of semantic relations respectively, the method detects mentions of these relation instances in a large text depository, based on which a relation instance categorizer could be trained. Then, those unlabelled relation instances detected in the depository could be labelled with different relations by the categorizer [1]. Distant supervision has been proved as an effective way in labelling training data automatically for many Natural Language Processing (NLP) applications such as automatic knowledge completion and question-answering [2].

Distant supervision for RE can scale the task to very large corpora with thousands of relations, however, relying on a single-node categorization model to identify relation facts for thousands of relations simultaneously inevitably accompanies with serious false categorization problem due to: (1) instances of closely-related relations can be easily mixed up; (2) the imbalance of training data between different relations may easily cause the categorizer to take instances of relations with less training data as those of relations with larger training data. Although some efforts have been made towards the problem, no satisfied improvement was achieved so far. For instance, Ji et al. considered to alleviate the training data imbalance problem by using some extra data such as the entity descriptions from Wikipedia pages as a supplement [3], but only limited effect could be reached. Han et al. proposed a global distant supervision model to exploit the consistency between relations and perform joint inference across relation instances, which need to accumulate all evidence from many kinds of weak supervision knowledge [4]. However, the model's consistent decisions are so strict that the extraction recall is still low.

In this paper, we novelly propose a multi-level distant supervision model for relation extraction, which divides the original categorization task into a number of sub-tasks in multiple levels of a constructed tree-like categorization structure. Each node in this tree-like structure is a sub-classifier trained with distant supervision. With the tree-like structure, an unlabelled relation instance would be categorized step by step along a path from the root node to a leaf node. While each classifier at the root node or a non-leaf node will put an entity pair into one of the node's children nodes, the classifier at a leaf node will finally label the entity pair with a relation. In this way, the original classification problem can be divided into a number of many small ones in different levels, which is expected to reach a much higher categorization accuracy.

Furthermore, we propose to do bootstrapped distant supervision to update the distant supervision model with new learned relation facts iteratively to further improve the extraction precision and recall. In this way, the problem of insufficient training data could be further addressed with the selected new-generated relation instances. So far, no previous efforts has yet considered to update the categorization model with the new identified relation facts, which wastes the chance to further improve the extraction precision and recall.

Our contributions are summarized as follows:

- We novelly propose a multi-level distant supervision model for relation extraction, which divides the original categorization task into a number of sub-tasks in multiple levels of a constructed tree-like categorization structure.
- We also propose bootstrapped distant supervision to update the distant supervision model with new learned relation facts iteratively to further improve the extraction precision and recall.
- For experimental evaluation, two real-world datasets are adopted. While one data set is the widely used Riedel data set for RE, the other one is created by us based on the ClueWeb09 data set. Our experimental results conducted on these two datasets prove that our approach outperforms state-of-the-art approaches by reaching more than 10% better extraction quality.

The remainder of the paper is organized as follows: We cover the related work in Sect. 2, and then define the task in Sect. 3. After introducing our proposed approach in Sect. 4, we present our experiments in Sect. 5. Finally, we present the conclusions and future work in Sect. 6.

2   Related Work

Relation Extraction (RE) has been studied extensively in recent years, and the study focus has shifted from a single relation to a variety of relations. Within the realm of information extraction, the RE approaches can be roughly divided into three aspects: supervised approaches, semi-supervised approaches and distant supervised approaches.

The supervised approaches work on building a classification model with a training data set. Some work used kernel methods, such as subsequence kernel and dependency tree kernel [5–7] to extract relations from unstructured natural language sources. Some other work [8,9] explored a set of features (lexical, syntactic and semantic) that are selected by performing textual analysis. Zhou et al. converted the features into symbolic IDs and fed them into a SVM classifier [9], while Zhou et al. constructed a rich semantic relation tree structure to integrate both syntactic and semantic information [8]. Conversely, Zeng et al. exploited a convolutional deep neural network (CNN) to extract lexical and sentence level features [10]. Based on the CNN model, Santos et al. proposed to do classification by Ranking CNN (CR-CNN) model [11]. The main drawback of the supervised approaches lies on that they require a great deal of labelled data, which is not always available.

The semi-supervised approaches use a small number of seed examples to generate extraction patterns, which in turn results in new tuples being extracted from the document collection. In the later iterations, new involved tuples would help generate more extraction patterns, while new generated patterns would also bring in more new tuples. Given the effectiveness of this bootstrapping mechanism, several well-known open information systems are developed such as Snowball [12], KnowItAll [13], NELL [14], and Probase [15]. However, this kind of approaches relies heavily on the quality of learned patterns, which can hardly reach a balance between high precision and recall.

To address the issues of the two kinds of approaches above, Mintz et al. first proposed to align plain text with Freebase by distant supervision [1], that is, let the existing KB provide training data for building the classification model, which saves plenty of time consuming and human labor. However, distant supervision inevitably accompanies with wrong labelling problem. To alleviate the problem, Riedel et al. modeled distant supervision for relation extraction as a multi-instance single-label problem [16], while Hoffmann et al. and Surdeanu et al. adopted multi-instance multi-label learning in relation extraction [17,18]. Multi-instance multi-label learning considers the reliability of the labels for each instance, and learns features from sentences containing the instance. But all feature-based methods depend strongly on the quality of the features generated by NLP tools, which will suffer from error propagation problem. Later, Zeng et al. proposed the use of Piecewise Convolutional Neural Networks (PCNNs) with multi-instance learning to automatically learn features without complicated NLP preprocessing, and then devise a piecewise max pooling layer to address the wrong labelling problem [19]. However, the PCNNs approach still makes mistakes given that all the instances of closely-related relations can be easily mixed up and instances of a relation with insufficient training data can be easily taken as those of a relation with larger training data. To alleviate the training data imbalance problem, Han et al. proposed a global distant supervision model to exploit the consistency between relations and perform joint inference across relation instances [4]. Ji et al. considered to use some extra data such as the entity descriptions from Wikipedia pages as a supplement [3], but only limited effect could be reached.

Hierarchical structure over classification has been extensively studied in several fields such as machine learning and vision communities. A large number of hierarchical Bayesian models for transfer learning [20–22] are presented in the field of visual recognition. Hierarchical topic model for image features learning [21] can discover visual taxonomies in an unsupervised fashion from large datasets. A hierarchical CNN models [22] have shown same or even better performance than standard CNN models in the field of visual recognition. Besides, many models [23,24] based on hierarchical Dirichlet processes are also used for transfer learning. However most of these models use MCMC(Markov Chain Monte Carlo) approaches for inference so that they are difficult to scale to large datasets.

3 Task Definition

Let $t = \langle e_1, e_2, r \rangle$ denote a relation triple representing that an entity pair $p = \langle e_1, e_2 \rangle$ is an instance of a relation r, where e_1 and e_2 are two entities and r is a particular relation between the two entities. Let $KB = (R, T)$ denote an existing knowledge base, where R contains a number of concerned relations and T contains a set of relation triples of the concerned relations.

Given a knowledge base $KB = (R, T)$, and a large text depository containing a set of sentences S, each of them has mentioned an entity pair. All sentences in S mentioning the same entity pair constitute a bag. Suppose there are N labelled bags $\{B_1^l, B_2^l, \ldots, B_N^l\}$, each of which corresponds to a relation triple in T, the task of distant supervised RE is to build a prediction model based on these labelled bags, which could be used to predict a relation label (from R) for each of the left M unlabelled bags $\{B_1^u, B_2^u, \ldots, B_M^u\}$.

4   Methodology

The workflow of our approach is described in Fig. 1. Basically, we train a *Tree-like categorization structure* with the labelled bags to categorize those unlabelled ones. The model performs distant supervision in a *Bootstrapping* manner. As we adopt PCNNs at each node, we call our approach as Bootstrapped Multi-Level PCNNs (or *BM-PCNNs* for short).

In the left of this section, we first introduce how we build the *Tree-like categorization structure*, and then give the details of the PCNNs model at each tree node. We finally present the bootstrapping strategy of the approach.

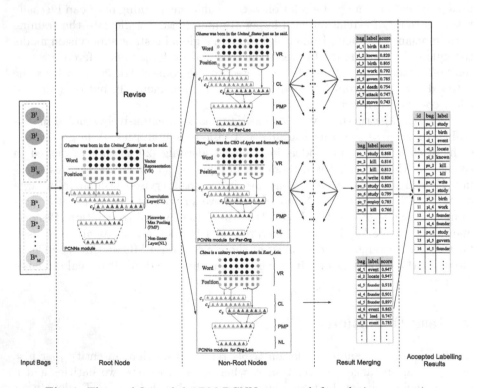

Fig. 1. The workflow of the BM-PCNNs approach for relation extraction

4.1 Building the Tree-Like Structure

The tree-like categorization structure (*tree* for short) is constructed from the root node to every of its leaf nodes. Different from the existing hierarchical structures over classification, which use Bayes theorem and Dirichlet processes, we propose our own set of steps to create the *tree*.

According to the experimental results of existing methods, we found that lots of wrong labels are caused by entity types. For example, there exist two relations: */person/organization/memberOf* and */location/organization/memberOf*, both representing *membership*. But they have different types of entity pairs, one is person-organization, and the other is location-organization. And the two relations are always confused in existing models. To solve the problem, we consider types of entities at the root node of *tree*. For instance, assume that there are three types of entities: person (Per), organization (Org) and location (Loc), then we could have three different combinations of entities, i.e., Org-Loc, Per-Loc, Per-Org. Thus the root node would have three children nodes as shown in Fig. 1.

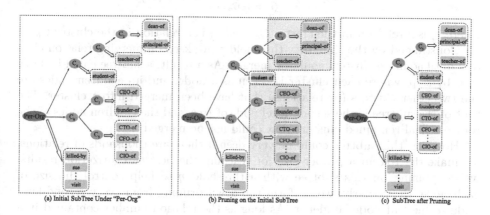

(a) Initial SubTree Under "Per-Org"   (b) Pruning on the Initial SubTree   (c) SubTree after Pruning

Fig. 2. Building a subtree under "Per-Org" for the tree-like structure

To further divide the classification task into smaller sub-tasks, we build subtrees for every second-level node in the *tree*. There are mainly two steps as described in Fig. 2: Firstly, we adopt agglomerative hierarchical clustering to build an initial subtree for a second-level node (Fig. 2(a)). Secondly, we prune the subtrees (Fig. 2(b) and (c)) to solve the problem of overfitting, which helps to improve its performance.

More specifically, in the first step, for all the relations under a second-level node, we calculate the similarity between every pair of relations with the equation below.

$$rSim(r_i, r_j) = \frac{\sum_{(p_{i_1}, p_{j_2}) \in P_i \times P_j} pSim(p_{i_1}, p_{j_2})}{\min(|P_i|, |P_j|)} \tag{1}$$

where P_i is a collection of all instances associated with the relation r_i, $|P_i|$ is the number of elements in the set, and $pSim(p_i, p_j)$ calculates the similarity between two relation instances as given below:

$$pSim(p_i, p_j) = \frac{\sum_{r \in |R_i \cap R_j|} f(r)}{\sum_{r \in |R_i \cup R_j|} f(r)} \qquad (2)$$

where R_i is a set of relations that are connected to the instance p_i in the given KB and $f(r)$ indicates the time that the relation r appears in the dataset.

Among all the relation pairs, the one with the highest similarity would be chosen. As long as the similarity is larger than a predefined similarity threshold τ, the two relations would be merged into a cluster.

Next, we would calculate the similarity between the new generated cluster and every of the existing relations or clusters. Here we quantify the similarity between two clusters with the relations with the highest similarity between the two clusters as follows:

$$Sim(C_a, C_b) = \max_{(r_i, r_j) \in C_a \times C_b} rSim(r_i, r_j) \qquad (3)$$

where r_i is a relation in the cluster C_a, and r_j is a relation in the cluster C_b.

Constrained by the similarity threshold τ, we keep on merging relation clusters until no more clusters could be merged. As a result, a subtree could be built step by step, where each cluster is taken as a node, and its children nodes are the two other clusters (or relations) that have been merged into a cluster. The root of the subtree is the second-level node of *tree*, and the children nodes of the root are final remained clusters that could not be merged.

However, this subtree could be very deep if there are thousands of relations. To make the *tree* in a proper size for reaching the best categorization results, we do pruning on this subtree with a threshold η to help control the size of each node. Particularly, we perform top-down pruning from every second-level node to the leaf nodes under it. As long as the relation number contained in a node is less than η, the subtree of the node will be pruned directly to let this node become a leaf node. Until then, a tree-like structure for multi-level distant supervision is built.

4.2   PCNNs Model

We now present the details of the PCNNs model that is used at each node of the *tree*. Basically, the model is used to extract features from a sentence, and then merge all features into one vector for a bag. Basically, PCNNs is composed by *Vector Representation*, *Convolution*, *Piecewise Max-pooling*, *Selective Attention* and *Training Objective*. We describe these parts in details below.

4.2.1   Vector Representation

The inputs of our model are raw word tokens gotten from those unlabelled bags. We first transform words into low-dimensional vectors. Here, we transform each input word into a vector via word embedding matrix. Besides, we add the position information of each word into the vector by looking up position embeddings.

Word Embeddings. Word embeddings transform words into distributed representations, which can capture both semantic and syntactic information about words very well. Much work has shown the power in many NLP tasks [25]. In the past years, many models for training word embeddings have been proposed [26–28]. In this paper, we use the Skip-gram model proposed by [28] to train word embeddings.

Fig. 3. An example of relative distances.

Position Embeddings. As shown in [10], position features are important in RE. The closer the words is to the target entities, the more useful information is available for determining the relation between entities. Each word has two relative distances. Figure 3 shows an example of the relative distance, and position feature is the combination of the two relative distance. The position embedding matrixes are randomly initialized. Then we transform the relative distances into real valued vectors by looking them up.

In the example shown in each tree node in Fig. 1, combining word embeddings and position embeddings, we can transform a sentence into a vector sequence $x = \{x_1, x_2, \cdots, x_m\}$, where m is the length of the input sentence. Assume that the size of word representation is k_w and that of position representation is k_d, then the size of x is $k = k_w + 2k_d$.

4.2.2 Convolution, Piecewise Max-pooling

The input of the network is a set of sentences of different lengths in RE. And the important information for predicting relation can appear in any area of the sentence. Thus, we should utilize all local features and perform relation prediction globally, which can be done by a convolution layer. First, the convolution layer extracts local features with a sliding window of length l over the sentence. In the example shown in Fig. 1, the length of the sliding window l is 3. Then it combines all local features via a max-pooling operation to obtain a fixed-sized vector for the input sentence.

Here, convolution is an operation between a vector of inputs that are treated as a sequence $\mathbf{X}$, and a convolution matrix $\boldsymbol{W} \in \mathbb{R}^{d^c \times (l \times k)}$, where d^c is the sentence embedding size. We use $\mathbf{X}_{j-l+1:j}$ to present the concatenation of a sequence of l word embeddings within the j-th window, where $1 \leq j \leq m+l-1$. If the window is outside of the sentence boundaries, we regard all out-of-range input vectors x_i ($i < 1$ or $i > m$) as zero vector. Hence, the convolution operation can obtain another sequence $c \in \mathbb{R}^{m+l-1}$:

$$c_j = \boldsymbol{W} \otimes \mathbf{X}_{j-l+1:j} \quad 1 \leq j \leq m + l - 1 \tag{4}$$

To capture different features, we use multiple filters in the convolution. Therefore, we also need n convolution matrices $\widehat{W} = \{W_1, W_2, \cdots, W_n\}$, so the convolution operation can be expressed as follows:

$$c_{ij} = W_i \otimes X_{j-l+1:j} \quad 1 \le j \le m+l-1, 1 \le i \le n \tag{5}$$

We obtain a matrix $C = \{c_1, c_2, \cdots, c_n\} \in \mathbb{R}^{n \times (m+l-1)}$ through the convolution layer. Figure 1 shows an example that we use 3 different filters in the convolution layer.

Using max-pooling operation can make features extracted by the convolution layer independent of the sentence length. The idea is to capture the most significant features in each feature map. In RE, an input sentence can be divided into three segments based on the two selected entities. Therefore, we do max-pooling operation on each segment [19]. As shown in Fig. 1, the result vector $\mathbf{x}_i$ of the convolution layer is divided into three parts $\{c_{i1}, c_{i2}, c_{i3}\}$ by *Obama* and *United_States*. Then the piecewise max-pooling procedure is:

$$p_{ij} = \max c_{ij} \quad 1 \le i \le n, 1 \le j \le 3 \tag{6}$$

Then we can concatenate all 3-dimensional vectors $p_i = \{p_{i1}, p_{i2}, p_{i3}\}$ to obtain a vector $p \in \mathbb{R}^{3n}$. Finally, we apply a non-linear function, such as the hyperbolic tangent to get a feature vector for a sentence:

$$f = \tanh p_{1:n} \tag{7}$$

where the size of f is fixed, $3n$, and is not related to the sentence length m.

4.2.3   Selective Attention

Adopted from the selective attention [29], suppose there is a bag B containing n sentences for an entity pair $p\langle e_1, e_2 \rangle$. The vector b of B is then computed as a weighted sum of these sentence vector f_i:

$$b = \sum_i \alpha_i f_i \tag{8}$$

where α_i is the weight of each sentence vector f_i, and α_i is further defined as:

$$\alpha_i = \frac{\exp(score(f_i, r))}{\sum_k \exp(score(f_k, r))} \tag{9}$$

where $score(f_i, r)$ is referred as a score that how well the input sentence f_i matches the predict relation r, and k is the number of sentences in a bag. Details of the function $score$ can be found in [29].

And the final output of the neural network is a score vector $o \in \mathbb{R}^{n_r}$, where n_r is the number of relations. o shows how well the entity pair p is associated to all relations and it is defined as follows:

$$o = Mb + d \tag{10}$$

where M is the representation matrix of relations and d is a bias vector.

Finally, the conditional probability $p(r|B,\theta)$ through a softmax layer is defined as follows:

$$p(r|B,\theta) = \frac{\exp(o_r)}{\sum_{k=1}^{n_r} \exp(o_k)} \tag{11}$$

where o_r is referred as a score that how well the input bag B matches the predict relation r, and n_r indicates the number of all relations. And θ indicates all parameters of our model.

4.2.4 Training Objective

In multi-instance learning, all sentences labeled by a triple constitute a bag. The objective of multi-instance learning is to predict the labels of the unseen bags. We define the objective function using cross-entropy as follows:

$$O(\theta) = \sum_{i=1}^{N} \log p(r_i|B_i,\theta) \tag{12}$$

where N indicates the number of bags. We adopt stochastic gradient descent to minimize the objective function, and employ dropout [30] to prevent over-fitting.

4.3 The Bootstrapping Strategy

BM-PCNNs performs distant supervision in a *Bootstrapping* manner. It accepts a number of top-ranked labelled bags with a satisfying labelling confidence, and keeps the left bags unlabelled. Then, the accepted labelled bags will be added into the training data to help adjust all the classifiers at every node of the *tree*. After that, the left unlabelled bags will be labelled again with the updated tree-like structure in later iterations. This process is performed iteratively until no more bags could be labelled with a satisfied labelling confidence.

To identify satisfied newly-labelled instances from the others at each iteration, we need a proper way to estimate a labelling confidence for an unlabelled bag, as well as a predefined confidence threshold, say δ. Without using human intervention, the quality of the labelling result to a bag, say B^u, can be obtained from the score vector o output by BM-PCNNs as described in Eq. 10. More specifically, from the set of scores in a vector o, the highest score, denoted by $maxScore(B^u, o)$, would be taken as the confidence of the labelling result to the bag. As long as $maxScore(B^u, o) \geq \delta$, then the labelling result to the bag B^u would be accepted.

5 Experiments

The experiments are intended to demonstrate that BM-PCNNs model can separate instances of close-related relations well and learn new features to further improve the extraction precision and recall. To this end, we first introduce the datasets and evaluation metrics used in the experiments. Next, we use cross-validation to determine the parameters of our model. Finally we compare the performance of our method with several state-of-the-art methods.

5.1   Datasets and Evaluation Metrics

We evaluate our model on two data sets: Riedel dataset[1] and ClueWeb09 dataset[2].

- The **Riedel** data set was developed by [16], and has been widely used in many previous studies [3,18,29]. The data set is generated by aligning Freebase relations with the New York Times corpus (NTY) and contains 377,948 relation instances with 695,059 sentences. The dataset has 53 possible relations, such as */people/person/ethnicity*, */time/event/locations*, */people/family/member*, involving so many kinds of entities: people, ethnicity, event, location and so on.
- Given that the similarities between relations in the Riedel data set are very low, using Riedel only can not fully reflect our advantages in distinguishing relations with high similarity. Thus, we created a larger data set for this task based on the **ClueWeb09** data set. This one contains about 1 billion web pages, and we successfully extract 44,915 relation instances with 254,665 sentences under 1,322 relations between three concerned entity types, i.e., organization, person and location.

To build the *BM-PCNNs* model, about 70% relation instances under each relation are randomly selected for training, while the left 30% are used for testing. Each reported result is the average of several evaluations, that is, several different training data will be generated with different random seeds, and the experimental results we present are the average results based on the generated testing data.

To reduce the consumption of artificial time, we evaluate our method in the held-out evaluation, similar to [1], which compares the *precision* and *recall* of the labelled relation instances in the testing data set.

5.2   Experimental Setting

Word Embeddings. In this paper, we use the word2vec of Skip-gram model[3] to train the word embeddings on the two datasets. First, words that appear more than 100 times make up a vocabulary. And we concatenate the words of an entity if it has more than one word. Since the dimension of a word embedding is not the focus of this paper and learned by previous works, we directly set the dimension of the vector k_w to 50.

Parameter Settings. Following previous work, we tune all of the models using three-fold validation on the training set. We adopt grid search to determine the optimal parameters, including sentence embedding size k, learning rate λ, batch size B, window size l, labelling confidence δ, similarity threshold τ and relation number contained in a node η.

---

[1] http://iesl.cs.umass.edu/riedel/ecml/.

[2] http://lemurproject.org/clueweb09/.

[3] http://code.google.com/archive/p/word2vec/.

Table 1. Parameter settings for *BM-PCNNs*

Sentence embedding size k		$150, 200, \mathbf{230}, 250$
Word dimension k_w		$\mathbf{50}$
Position dimension k_d		5
Learning rate λ		$0.001, \mathbf{0.01}, 0.1$
Batch size B		$100, \mathbf{160}, 220, 280$
Dropout probability p		$\mathbf{0.5}$
Window size l	Riedel	$\mathbf{3}, 4, 5, 6, 7$
	ClueWeb09	$3, 4, \mathbf{5}, 6, 7$
Labelling confidence δ	Riedel	$0.65, 0.7, 0.75, 0.8, 0.85, \mathbf{0.9}, 0.95$
	ClueWeb09	$0.65, \mathbf{0.7}, 0.75, 0.8, 0.85, 0.9, 0.95$
Similarity threshold τ	Riedel	$0.325, 0.35, \mathbf{0.375}, 0.4, 0.425, 0.45, 0.475$
	ClueWeb09	$0.5, 0.55, 0.6, \mathbf{0.65}, 0.7$
Relation number in a node η	Riedel	$\mathbf{5}, 10, 15, 20, 25$
	ClueWeb09	$25, \mathbf{30}, 35, 40, 45, 50$

For most of our parameters, since they have little effect on the results, we follow the setting used in [10]. As shown in the first 6 lines of Table 1, where the best choice for each parameter are boldfaced.

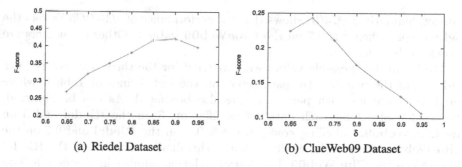

(a) Riedel Dataset   (b) ClueWeb09 Dataset

Fig. 4. Estimating the effect of δ to the F-score of *BM-PCNNs*.

But for the three hyper-parameters, say δ, τ and η, which only exist in our proposed method, we perform 6 group of experiments to estimate their influence on the experimental results as reported in Figs. 4, 5, and 6. Each time we do experiments for one parameter, we just set the other two parameters to a certain value. It can be seen from the Fig. 4(a), as δ increases from 0.65 to 0.9, the performance of *BM-PCNNs*, i.e., F-score, increases steadily from 0.268 to 0.422. The F-score reaches at its peak score when $\delta = 0.9$, and then begins to fall after that. Thus we choose 0.9 as the best setting of δ for the **Riedel**

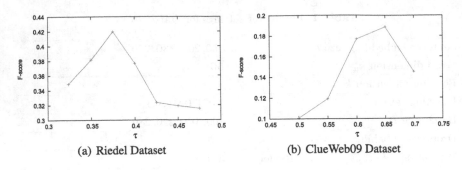

Fig. 5. Estimating the effect of τ to the F-score of *BM-PCNNs*.

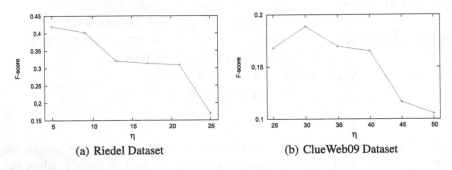

Fig. 6. Estimating the effect of η to the F-score of *BM-PCNNs*.

dataset. Similarly, Fig. 4(b) shows that the performance of *BM-PCNNs* gets the highest score when δ is 0.7 on the **ClueWeb09** dataset. Other parameters are chosen in the same way.

We list all the possible values we have tried for the three parameters (δ, τ and η) and the window size parameter l in the last 4 lines of Table 1, where the best choice for each parameter are also boldfaced. As can be observed, the window size l for each data set is different, 3 on the **Riedel** and 5 on the **ClueWeb09**. Labelling confidence δ is 0.9 on the **Riedel** and 0.7 on the **ClueWeb09** respectively, while similarity threshold τ is 0.375 on the **Riedel** and 0.65 on the **ClueWeb09**. In addition, relation number in a node η is 5 on the **Riedel** and 30 on the **ClueWeb09**.

5.3   Experimental Results and Analysis

We compare our *BM-PCNNs* method with four previous methods. *Mintz* proposed by [1] extracts features from all instances. *MultiR* proposed by [17] uses multi-instance learning. *MIML* proposed by [18] is a multi-instance multi-label model. *PCNNs* proposed by [29] adds attention system based on CNNs model proposed by [19]. We implement them with the source codes released by the authors.

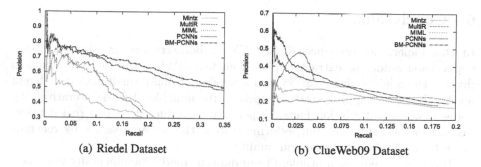

(a) Riedel Dataset   (b) ClueWeb09 Dataset

Fig. 7. P-R comparison of the five methods

Table 2. Evaluating the bootstrapping strategy

Dataset		turn0	turn1	turn2	turn3	turn4
Riedel	P	0.4145	0.4175	0.4195	0.4215	0.4262
	R	0.4251	0.4282	0.4303	0.4324	0.4371
	F-score	**0.4197**	**0.4228**	**0.4248**	**0.4268**	**0.4316**
ClueWeb09	P	0.1908	0.2458	0.2749	0.3040	0.3327
	R	0.1867	0.2404	0.2690	0.2975	0.3256
	F-score	**0.1887**	**0.2431**	**0.2719**	**0.3007**	**0.3291**

Figure 7 shows the P-R (precision-recall) curves for each method. In Fig. 7(a), we can observe that both in precision and recall, *BM-PCNNs* outperforms all the other evaluated methods, and it enhances the precision to approximately 51.0% without any loss of recall. In Fig. 7(b), *MultiR* performs better than our method when recall varies from 0.02 to 0.045. Since *MultiR* takes relations with high similarities into account, it could reach a high precision. But on the other hand, the features the method used are defined in advance, which results in a low recall. Besides, *MIML* also has a line beyond our method. But after reaching the highest point, the line sharply decreases, leading to the final precision as same as ours. By contrast, our multi-level structure could automatically learn features from original words and distinguish instances of closely-related relations.

We also evaluate the effectiveness of our bootstrapping strategy. As could be observed in Table 2, updating the model with new learned relation facts could further improve the extraction performance. As can be observed, the F-score could be improved by 14% on ClueWeb09. However, the improvement is not that obvious on Riedel. Through observations to the Riedel dataset, we find that more than 70% instances are labelled NA, and the training data has a lot of overlap with the testing data. Thus the bootstrapping strategy could not improve a lot on this dataset. But still, our experimental results could prove the effectiveness of the bootstrapping strategy in *BM-PCNNs*.

6   Conclusions

In this paper, we introduce *BM-PCNNs*, a bootstrapped multi-level distant supervision model to extract triples from texts. While the multi-level model divides the original classification into a number of many small ones in different levels, the bootstrapping strategy updates the multi-level model iteratively by using new learned relation facts. Experiments conducted on two real datasets prove that our approach outperforms state-of-the-art approaches by reaching more than 10% better extraction quality.

However, it remains a problem to update our model incrementally when new class and their instances come in. In our future work, we would consider to use more semantic information between instances and more features like syntax tree to better integrate a new class into the current hierachical classification structure.

Acknowledgments. This research is partially supported by National Natural Science Foundation of China (Grant No. 61632016, 61402313, 61472263), and the Natural Science Research Project of Jiangsu Higher Education Institution (No. 17KJA520003).

References

1. Mintz, M., Bills, S., Snow, R., Dan, J.: Distant supervision for relation extraction without labeled data. In: ACL-IJCNLP, pp. 1003–1011 (2009)
2. Savenkov, D., Agichtein, E.: When a knowledge base is not enough: question answering over knowledge bases with external text data. In: SIGIR, pp. 235–244 (2016)
3. Ji, G., Liu, K., He, S., Zhao, J.: Distant supervision for relation extraction with sentence-level attention and entity descriptions. In: AAAI, pp. 3060–3066 (2017)
4. Han, X., Sun, L.: Global distant supervision for relation extraction. In: AAAI (2016)
5. Culotta, A., Sorensen, J.: Dependency tree kernels for relation extraction. In: ACL, pp. 423–429 (2004)
6. Bunescu, R.C., Mooney, R.J.: A shortest path dependency kernel for relation extraction. In: HLT-EMNLP, pp. 724–731 (2005)
7. Zhou, G., Zhang, M., Ji, D.H., Zhu, Q.: Tree kernel-based relation extraction with context-sensitive structured parse tree information. In: EMNLP-CoNLL, pp. 728–736 (2007)
8. Zhou, G., Qian, L., Fan, J.: Tree kernel-based semantic relation extraction with rich syntactic and semantic information. Inf. Sci. **180**(8), 1313–1325 (2010)
9. Zhou, G., Su, J., Zhang, J., Zhang, M.: Exploring various knowledge in relation extraction. In: ACL, pp. 419–444 (2002)
10. Zeng, D., Liu, K., Lai, S., Zhou, G., Zhao, J., et al.: Relation classification via convolutional deep neural network. In: COLING, pp. 2335–2344 (2014)
11. Santos, C.N.D., Xiang, B., Zhou, B.: Classifying relations by ranking with convolutional neural networks. Comput. Sci. **86**(86), 132–137 (2015)
12. Agichtein, E., Gravano, L.: Snowball: extracting relations from large plain-text collections. In: ACMDL, pp. 85–94 (2000)

13. Etzioni, O., et al.: Web-scale information extraction in knowItAll. J. Korean Med. Sci. **27**(2), 146–52 (2004)
14. Carlson, A., Betteridge, J., Kisiel, B., Settles, B., Hruschka Jr., E.R., Mitchell, T.M.: Toward an architecture for never-ending language learning. In: AAAI, pp. 1306–1313 (2010)
15. Wu, W., Li, H., Wang, H., Zhu, K.Q.: Probase: a probabilistic taxonomy for text understanding. In: SIGMOD, pp. 481–492 (2012)
16. Riedel, S., Yao, L., Mccallum, A.: Modeling relations and their mentions without labeled text. In: ECML-PKDD, pp. 148–163 (2010)
17. Hoffmann, R., Zhang, C., Ling, X., Zettlemoyer, L., Weld, D.S.: Knowledge-based weak supervision for information extraction of overlapping relations. In: ACL-HLT, pp. 541–550 (2011)
18. Surdeanu, M., Tibshirani, J., Nallapati, R., Manning, C.D.: Multi-instance multi-label learning for relation extraction. In: EMNLP-CoNLL, pp. 455–465 (2012)
19. Zeng, D., Liu, K., Chen, Y., Zhao, J.: Distant supervision for relation extraction via piecewise convolutional neural networks. In: EMNLP, pp. 1753–1762 (2015)
20. Srivastava, N., Salakhutdinov, R.: Discriminative transfer learning with tree-based priors. In: Advances in Neural Information Processing Systems, pp. 2094–2102 (2013)
21. Bart, E., Porteous, I., Perona, P., Welling, M.: Unsupervised learning of visual taxonomies. In: CVPR, pp. 1–8 (2008)
22. Yan, Z., Zhang, H., Piramuthu, R., Jagadeesh, V., Decoste, D., Di, W., Yu, Y.: HD-CNN: hierarchical deep convolutional neural networks for large scale visual recognition. In: ICCV, pp. 2740–2748 (2016)
23. Xue, Y., Liao, X., Carin, L., Krishnapuram, B.: Multi-task learning for classification with dirichlet process priors. J. Mach. Learn. Res. **8**(1), 35–63 (2007)
24. Salakhutdinov, R.R., Tenenbaum, J., Torralba, A.: Learning to learn with compound HD models. In: Advances in Neural Information Processing Systems, pp. 2061–2069 (2012)
25. Socher, R., Huval, B., Manning, C.D., Ng, A.Y.: Semantic compositionality through recursive matrix-vector spaces. In: EMNLP-CoNLL, pp. 1201–1211 (2012)
26. Bengio, Y., Schwenk, H., Sencal, J.S., Morin, F., Gauvain, J.L.: Neural probabilistic language models. J. Mach. Learn. Res. **3**(6), 1137–1155 (2006)
27. Collobert, R., Weston, J., Karlen, M., Kavukcuoglu, K., Kuksa, P.: Natural language processing (almost) from scratch. J. Mach. Learn. Res. **12**(1), 2493–2537 (2011)
28. Mikolov, T., Chen, K., Corrado, G., Dean, J.: Efficient estimation of word representations in vector space. arXiv preprint arXiv:1301.3781 (2013)
29. Lin, Y., Shen, S., Liu, Z., Luan, H., Sun, M.: Neural relation extraction with selective attention over instances. In: ACL, pp. 2124–2133 (2016)
30. Srivastava, N., Hinton, G., Krizhevsky, A., Sutskever, I., Salakhutdinov, R.: Dropout: a simple way to prevent neural networks from overfitting. J. Mach. Learn. Res. **15**, 1929–1958 (2014)

On the Discovery of Continuous Truth: A Semi-supervised Approach with Partial Ground Truths

Yi Yang[1(✉)], Quan Bai[1], and Qing Liu[2]

[1] Auckland University of Technology, Auckland, New Zealand
{yi.yang,quan.bai}@aut.ac.nz
[2] Software and Computational Systems, Data61, CSIRO, Hobart, Australia
Q.Liu@data61.csiro.au

Abstract. In many applications, the information regarding to the same object can be collected from multiple sources. However, these multi-source data are not reported consistently. In the light of this challenge, truth discovery is emerged to identify truth for each object from multi-source data. Most existing truth discovery methods assume that ground truths are completely unknown, and they focus on the exploration of unsupervised approaches to jointly estimate object truths and source reliabilities. However, in many real world applications, a set of ground truths could be partially available. In this paper, we propose a semi-supervised truth discovery framework to estimate continuous object truths. With the help of ground truths, even a small amount, the accuracy of truth discovery can be improved. We formulate the semi-supervised truth discovery problem as an optimization task where object truths and source reliabilities are modeled as variables. The ground truths are modeled as a regularization term and its contribution to the source weight estimation can be controlled by a parameter. The experiments show that the proposed method is more accurate and efficient than the existing truth discovery methods.

Keywords: Truth discovery · Source relabilities
Semi-supervised learning

1 Introduction

In many applications, the information regarding to the same object can be collected from multiple sources. However, these multi-source data are not reported consistently. For example, different stations may provide different daily high temperature for a city; the stock data provided by different websites may be conflicting. Without resolving conflicts among multi-source data, the data quality cannot be guaranteed, and these data cannot be used in analytic tasks to extract useful information. In the light of this challenge, truth discovery is emerged to resolve conflicts among multi-source data and it has become a hot topic in the community.

© Springer Nature Switzerland AG 2018
H. Hacid et al. (Eds.): WISE 2018, LNCS 11233, pp. 424–438, 2018.
https://doi.org/10.1007/978-3-030-02922-7_29

Truth discovery resolves conflicts by estimating the truths, i.e. the most trust-worthy information, of the objects. Different from the native approaches, such as majority voting (for categorical data) and mean (for continuous data), truth discovery estimates object truths by estimating source reliabilities. The general principle of truth discovery is that sources which frequently provide trustworthy information are reliable, and the data supported by reliable sources is trustworthy and selected as truth of an object. As the source reliabilities are unknown a priori, most existing truth discovery methods explore unsupervised approaches to jointly estimate object truths and source reliabilities, and they assume that the ground truths of objects are entirely unknown. Obtaining the entire set of ground truths from highly reliable sources is usually expensive, but acquiring a small set of ground truths is usually practical. For example, part of the objects' truths may be available from government websites, information released by governments is usually real and we can use it as ground truths. If we can use these partial objects' ground truths and add some supervisions in the truth discovery process, we believe the accuracy of truth discovery can be improved.

A semi-supervised truth discovery method is studied in [19]. However, their method is originally designed for processing categorical object truths. It requires that the ground truths must be among the observations provided by sources. This is impractical for many real world applications, especially when both ground truth and observations are real numbers. As demonstrated in Sect. 5, the existing methods perform poorly on datasets when object truths are continuous data instead of categorical data.

In this paper, we study the semi-supervised truth discovery problem for continuous object truths. We propose an *Op*timization based *S*emi-supervised *T*ruth *D*iscovery (OpSTD) method for discovering continuous object truths, in which the truth discovery problem is formulated as an optimization task where both object truths and source reliabilities are modeled as variables, and the ground truth is modeled as a regularization term to propagate its trustworthiness to the estimated truths. Furthermore, we present theoretical analysis for the proposed method and conduct a series of experiments on both real datasets and synthetic dataset to demonstrate its effectiveness.

In summary, we make the following contributions:

- We formulate a semi-supervised truth discovery framework, OpSTD, for continuous object truths.
- An iterative based algorithm to estimate source weights and object truths is developed that can converge to an optimal solution.
- We theoretically prove the convergence and analyze the time complexity of OpSTD.
- The experiment results on both real world datasets and synthetic dataset show that the proposed method outperforms the existing methods significantly.

The rest of this paper is organized as follows. In Sect. 2, we review existing work of truth discovery. In Sect. 3, we present the optimal framework and the iterative solution. In Sect. 4, we prove the convergence property of the proposed method and analyze time complexity. In Sect. 5, it shows the experiments to evaluate the performance of the proposed method. Finally, we conclude in Sect. 6.

2   Related Work

Truth discovery has received a lot of attentions in the community of data quality, data mining and trust management [2,8,9,11,20,23]. Being different from the traditional approaches such as majority voting and mean, truth discovery estimates object truths by taking the source reliabilities into consideration. Most truth discovery methods follows the principle that sources which frequently provide trustworthy observations are reliable, and observations that are supported by reliable sources are trustworthy. Guided by this principle, truth discovery was first proposed by Yin et al. [18]. Since then, various truth discovery methods are proposed to solve various aspects of truth discovery problems. In terms of data type, [3,7,14,15] are developed to process categorical data while [12,13,21] are specifically designed for continuous data. In order to resolve conflicts among heterogeneous data, Li et al. [10] proposed a general truth discovery method in which various distance functions can be plugged in to capture the difference between observations. Most existing truth discovery methods assume that each object has only one property. This assumption is relaxed in [5,6,16,22] and these methods are develop to solve multi-truth truth discovery problem. In [3,4,17], the authors assume that the sources are not independent. These methods identify object truths by considering the copying relationship among sources.

The existing truth discovery methods can be generally categorized into three ways to formulate the truth discovery task. The *iterative based methods* [3,7,18] treat object truths and source weights dependently, and the object truths and source weights are updated iteratively until algorithms converge. The *probabilistic graphical model based methods* [21,22] model the object truth, source weight and observation as random variables. The dependencies among the random variables are usually captured by a Bayesian network, then the object truths are inferred by probabilistic inference techniques. The *optimization based methods* [10,12,13] formulate the truth discovery as an optimization task where the object truths and source reliabilities are modeled as variables. An objective function that involves these variables needs to be optimized and the optimal object truths that minimize the objective function are selected as truths.

The truth output of most truth discovery methods can be categorized into the following two methods. The *scoring method* [18,19] assigns a trustworthy score to each observation. Then it requires a post decision making process to select the observation as the truth of an object based on the scores of the observations. Usually the observation with the highest score for an object is selected as the truth. The *labeling method* [10,12,21] directly assigns a label or a truth to an object. The labeling method is especially helpful when the truth discovery deals with continuous data. In this scenario, the truth of an object might not be observed by any source. For example, the ground truth of temperature of Auckland on June 5th is 26.3. The observations provided by three sources claiming the temperature are 25.6, 26.1 and 26.5, and the ground truth is not among the observations. Thus, the scoring method may fail to work in this case.

There is some work that share similarities with ours. In [15], source reliabilities are modeled as latent variables, its EM based solution can incorporate a small set of ground truths to help truth inference. But it is limited to work with categorical only. The work that is closet to ours is the semi-supervised truth discovery SSTF [19]. SSTF is originally designed for categorical data, and it uses a graph based semi-supervised technique, label propagation, to propagate the trustworthiness of ground truths to the observations. SSTF is limited that it uses scoring technique to output object truths. Therefore, it requires the ground truths are among observations, which is not suitable for the truth discovery applications in which the data is continuous. SSTF also uses a predefined similarity function to capture the relations among observations. This similarity function is application specific and usually hard to define in practice. In contrast, the proposed method OpSTD is specially designed for semi-supervised truth discovery over continuous data and the setting of OpSTD is much simpler. The experiments in Sect. 5 also demonstrates that OpSTD outperforms SSTF to find continuous object truths.

3   Semi-supervised Truth Discovery on Continuous Data

In this section, we will formulate the problem of the semi-supervised truth discovery for continuous object truths first. Then the framework and an optimization based method are presented.

3.1   Problem Formulation

We first define the notations for the truth discovery problem, which are also summarized in Table 1.

Definition 1. *Object, Source and Observation: An object, o, is a thing or an event that has a continuous property. A source, s, is an information provider which can observe and report the property value of object o. An observation, $v_o^s \in \mathbb{R}$, is the continuous property value of object o reported by source s.*

Definition 2. *Ground truth and estimated truth: The ground truth, $\bar{v}_o^* \in \mathbb{R}$, of object o is the fractal truth that correctly describes the property value of o. It is usually unknown a priori. The Estimated truth, $v_o^* \in \mathbb{R}$, of object o, is the estimated most trustworthy information describing the property value of o, it is the output of a given truth discovery method.*

Definition 3. *Source Weight: The source weight, $w^s \in \mathbb{R}^+$, reflects the reliability of source s. The information provided by sources with high source weights is usually more trustworthy and closer to the truth.*

Table 1. Notations

Notation	Description
O	Set of all the objects
O_u	Set of objects whose ground truths are unknown
O_g^s	Set of objects whose ground truths are available
O_u^s	Set of objects observed by s, and the objects' ground truths are unknown
O_g	Set of objects observed by s, and the objects' ground truths are available
S	Set of all the sources
S_o	Set of sources that observe object o
V	Set of all the observations
V_u^*	Set of estimated truths for objects in O_u
W	Set of all the source weights
v_o^s	The observation for object o reported by source s
w^s	Weight of source s
$\bar{v}_o^*$	The ground truth of object o
v_o^*	The estimated truth of object o

In this paper, we study the semi-supervised truth discovery for continuous object truths, in which we use some partially available ground truths to supervise the truth discovery process. Let S be the set of all the sources and O be the set of all the objects. We split O into two sets O_g and O_u where O_g and O_u are disjoint and $O_g \cup O_u = O$. O_g is the set of objects whose ground truths are available, and O_u is the set of objects whose ground truths are unknown. Usually $|O_g| << |O_u|$. Next, we formally define the truth discovery task.

Problem Definition. Given the observations V where $V = \{v_o^s\}_{o \in O, s \in S}$ and a set of ground truths $\{\bar{v}_o^*\}_{o \in O_g}$, semi-supervised truth discovery for continuous object truths aims at resolving conflicts among multi-source data and estimating the truths $V_u^* = \{v_o^*\}_{o \in O_u}$ with the help of available ground truths.

3.2   The OpSTD Framework

In this section, we present the OpSTD framework. We formulate the semi-supervised truth discovery as an optimization problem. The intuitions are (1) objects that provide observations closer to the ground truth can be inferred as reliable sources; and (2) the observations reported by reliable sources should be close to the estimated truth. Based on this intuition, we use ground truths to guide the source weight estimation that can in turn impact on the truths estimation for the objects whose ground truths are unknown. Following, we present

the optimization framework that can incorporate the available ground truths for
truth discovery.

$$\min_{V_u^*, W} f(V_u^*, W) = \sum_{o \in O_u} \left\{ \sum_{s \in S_o} w^s (v_o^* - v_o^s)^2 \right\} + \theta \sum_{o \in O_g} \left\{ \sum_{s \in S_o} w^s (\bar{v}_o^* - v_o^s)^2 \right\}$$

$$\sum_{s \in S} \exp(-w^s) = 1 \tag{1}$$

In Eq. (1), S_o is the set of sources that observe object o. In the first term
$\sum_{o \in O_u} \{ \sum_{s \in S_o} w^s (v_o^* - v_o^s)^2 \}$, for source s, $(v_o^* - v_o^s)^2$ models the estimated error
made by s on the observation for object o, and it computes the discrepancy
between the observation provided by sources and the estimated object truths.
This term itself estimates the source weights and object truths in an unsupervised manner. In order to minimize f, the optimization process will assign high
weights to sources which make small estimated errors. Similarity, if the estimated error is large, it will assign a low weight to w^s to minimize the error's
contribution in the objective function.

The second term $\sum_{o \in O_g} \{ \sum_{s \in S_o} w^s (\bar{v}_o^* - v_o^s)^2 \}$ introduces supervision into
the objective function to supervise source weight and object truth estimation
process. For a source s, $(\bar{v}_o^* - v_o^s)^2$ models the discrepancy between the ground
truth and the source's observation for object o. It is the real error made by s for
object o. To minimize the objective function, it penalizes the unreliable sources
and assigns low weights to them if the real error is large. θ is a hyper parameter
which balances these two terms in the objective function. Combining these two
terms makes the proposed framework semi-supervised. The source weights are
determined by both estimated errors and real errors, and the ground truths
supervises object truths and source weights estimation. This will be further
discussed in Sect. 3.3.

The constraint function, $\sum_{s \in S} \exp(-w^s) = 1$ is required mathematically to
constrain the source weights between 0 and 1. Otherwise the source weights can
be set to $-\infty$ to minimize the objective function.

3.3 The Iterative Solution

The object truths and source weights shall be learned jointly to minimize the
objective function, and the optimal values learned after the optimization process
will be selected as the object truths and source weights. In order to minimize
the objective function f, we choose to use block coordinate descent [1] in which
it iteratively updates one set of variables while fixing the other set to keep
reducing the value of f until reaching convergence. There are two steps involved
to minimize function f. Step one is to update the estimated truths V_u^* while fixing
the source weights W. Step two is to update the source weights W while fixing
the estimated truths V_u^*. These two steps can be mathematically formulated by
Formulas (2) and (3). Next we discuss in details on how to derive the rules to
update source weights and estimated truths.

$$V_u^* \leftarrow \arg \min_{V_u^*} f(V_u^*, W) \tag{2}$$

$$W \leftarrow \arg\min_{W} f(V_u^*, W) \qquad s.t. \qquad \sum_{s \in S} \exp(-w^s) = 1 \qquad (3)$$

Object Truth Update Rule: In this step, we update the set of estimated object truths V_u^* while fixing W. By setting $\frac{df_W(V_u^*)}{dv_o^*} = 0$ for the object $o \in O_u$, we get the following update rule for each estimated object truth:

$$v_o^* = \frac{\sum_{s \in S_o} w^s v_o^s}{\sum_{s \in S_o} w^s} \qquad (4)$$

Source Weight Update Rule: We use Lagrange multiplier approach to solve Formula (3). The Lagrangian can be formulated as:

$$\mathcal{L}(W, \lambda) = f(V_u^*, W) + \lambda(\sum_{s \in S} \exp(-w^s) - 1) \qquad (5)$$

where λ is the Lagrange multiplier. By setting $\frac{d\mathcal{L}(W,\lambda)}{dw^s} = 0$, from the constraint we can derive that

$$\lambda \exp(-w^s) = \sum_{o \in O_u^s} (v_o^* - v_o^s)^2 + \theta \sum_{o \in O_g} (\bar{v}_o^* - v_o^s)^2 \qquad (6)$$

where O_u^s and O_g^s are both observed by source s, but their ground truths are unknown and available respectively. Combined with the constraint equation $\sum_{s \in S} \exp(-w^s) = 1$, we can compute the Lagrange multiplier as:

$$\lambda = \sum_{s \in S} \left\{ \sum_{o \in O_u^s} (v_o^* - v_o^s)^2 + \theta \sum_{o \in O_g^s} (\bar{v}_o^* - v_o^s)^2 \right\} \qquad (7)$$

Plugging Eq. (7) back to Eq. (6), we can derive the source weight update rule in Eq. (8).

$$w^s = -\log\left(\frac{\sum_{o \in O_u^s} (v_o^* - v_o^s)^2 + \theta \sum_{o \in O_g^s} (\bar{v}_o^* - v_o^s)^2}{\sum_{s \in S} \left\{ \sum_{o \in O_u^s} (v_o^* - v_o^s)^2 + \theta \sum_{o \in O_g^s} (\bar{v}_o^* - v_o^s)^2 \right\}} \right) \qquad (8)$$

Discussion: From Eq. (8) of the source weight update rule, we can see that a source has higher weight if it makes few errors among all the sources. Specifically, the errors are determined by the estimated errors and real errors, and the proportion can be adjusted by controlling θ. If we increase θ, the source weight will be computed mostly by the real errors. In the extreme case where $\theta = \infty$, the term $\sum_{o \in O_u^s} (v_o^* - v_o^s)^2$ is ignored and the source weight is totally determined by objects with the ground truths. Conversely, if we decrease θ, the source weight will be computed mostly by the estimated errors. If $\theta = 0$, this is equivalent to the truth discovery in an unsupervised setting where the ground truths do not contribute to the truth discovery process and we estimate source weights solely from the observations. In Sect. 5.3, we will experimentally show the effect of θ to the performance of OpSTD.

From Eq. (4) we can see that the estimated object truth is computed by weighted aggregation in which all the observations for object $o \in O_u$ contribute to the estimated truth, but the contribution is discounted by the weights of the sources which provide these observations. As a result, the estimated truth will be close to the observations from sources with high weights. Furthermore, the source weights are partially computed by ground truths as in Eq. (8). Thus, the ground truths also impact the truths estimation for the objects whose ground truths are unknown.

The algorithm flow of the OpSTD is summarized in Algorithm 1. First, the source weights are initialized. If no prior knowledge is available about the reliabilities of the sources, the source weights can be initialized uniformly, i.e. $w^s = -\log(\frac{1}{|S|})$. Otherwise the source weights can be changed accordingly to reflect the initial belief of the source reliability. Then the algorithm iteratively update object truths and source weights by Eqs. (4) and (8) until convergence.

Algorithm 1. OpSTD Algorithm Flow

Input : Observations V, ground truths V_l^* for O_l
Output: Inferred object truths V_u^*
1 Initialize source weights;
2 **repeat**
3   **for** $o \in O_u$ **do**
4 | Update v_o^* by Equation (4);
5   **end**
6   **for** $s \in S$ **do**
7 | Update w^s by Equation (8);
8   **end**
9 **until** *Convergence*
10 **return** V_u^*

---

4 Theoretical Analysis

In this section, we theoretically analyze the convergence property of the OpSTD algorithm and its time complexity.

4.1 Convergence Analysis

We prove the following theorem to show the convergence of OpSTD algorithm, and it is valid to use block coordinate descent to minimize the objective function given in Eq. (1).

Theorem 1. *The iterative process in OpSTD algorithm converges, and the optimal solutions, V_u^* and W, is a stationary point for the objective function in Eq. (1) to attain minimum.*

Proof. There are two blocks of variables, V_u^* and W, involved in the objective function f. We use $\mathcal{Y}$ to denote the union of the two blocks of variables, i.e. $\mathcal{Y} = \{V_u^*, W\}$. Let the size of $\mathcal{Y}$ be l where $l = |V_u^*| + |W|$. Then the optimization problem can be rewritten as:

$$\text{minimize} \quad f(y), \quad \text{s.t.} \quad y \in \mathcal{Y}$$

According to [1], let $\{y^z\}$ be the sequence generated by the following rule:

$$y_i^{z+1} = \arg\min_{\xi \in \mathcal{Y}_i} f(y_1^{z+1}, \ldots, y_{i-1}^{z+1}, \xi, y_{i+1}^z, \ldots, y_l^z) \quad \text{for} \quad i = 1, 2, \ldots, l$$

where z is the current iterate index, then every limit point of y^z is a stationary point and $f(\{y^z\})$ is the global minimum of f if f satisfies the following two conditions:

1. f is continuously differentiable over $\mathcal{Y}$.
2. For each $y_i \in \mathcal{Y}_i$, $f(y_1, y_2, \ldots, y_{i-1}, \xi, y_{i+1}, \ldots, y_l)$, viewed as a function of ξ while the other variables are fixed, attains a unique minimum $\bar{\xi}$ over $\mathcal{Y}_i$, and is monotonically non-increasing in the interval from y_i to $\bar{\xi}$.

Next, we show the objective function f satisfies the two above conditions in the following two scenarios:

- Scenario 1: Update V_u^* while fixing W. In this case, $f_W(V_u^*)$ is a combination of quartic functions $w^s(v_o^* - v_o^s)^2$ where $w^s > 0$. Hence, $f_W(V_u^*)$ is strictly convex and continuously differentiable and attains a unique minimum.
- Scenario 2: Update W while fixing V_u^*. In this case, $f_{V_u^*}(W)$ is a combination of linear functions w.r.t w^s, which is affine, strictly convex and continuous differentiable. In addition, the exponential function is strictly convex, the constraint in the objective function is also strictly convex. Thus, $f_{V_u^*}(W)$ is continuously differentiable and attains a unique minimum while fixing V_u^*.

Therefore, Algorithm 1 converges when f attains its minimum $f(y^z)$, and $\{V_u^*, W\} = \{y^z\}$ is the stationary point. □

4.2   Time Complexity Analysis

We analyze the time complexity of OpSTD algorithm by analyzing the computational complexity of each iteration in Algorithm 1. In the object truth update step, each object can be observed by up to $|S|$ sources. The cost of updating object truths is $O(|O_u| \times |S|)$ since this step computes the sum of observations weighted by source weights. In the source weight update step, each source can observe up to $|O|$ objects. The cost of updating source weight is $O(|O| \times |S|)$ since this step computes the squared error between each source's observation and truths. Therefore, the computational complexity of each iteration in OpSTD algorithm is $O(|O| \times |S|)$. In the truth discovery application, there are at most $|O| \times |S|$. Hence, computational complexity of each iteration is also linear with the number of observations.

5   Experiments

In this section, we experimentally compare the proposed method with the state-of-art truth discovery methods on both real and synthetic datasets. All the experiments are conducted on a PC with Intel i7 processor and 16 GB RAM.

5.1   Experiment Setup

In this subsection, we describe the baseline methods, datasets and performance metrics used to evaluate OpSTD.

Baseline Methods. OpSTD is state-of-art truth discovery methods:

- GTM [21]: A probabilistic graph model based method for resolving conflicts on continuous data.
- CRH [10]: Finding truth of heterogeneous data by using various loss functions.
- SSTF [19]: A semi-supervised truth discovery method adopts graph based semi-supervised learning method to learn object truths with a small set of ground truths.
- Mean: This method does not consider source reliabilities. It simply aggregates the object truth by taking the mean of the observations.

Datasets. We use two real world datasets and one synthetic dataset to evaluate the proposed method. The ground truths of all the datasets are available for evaluation.

- **Weather**[1]: It contains daily weather information for 30 cities over 6 months. The daily temperature property for each city is adopted in the experiments.
- **Stock** (see footnote 1): It records data for 1000 stocks collected from 55 sources over 21 working days in 2011. The open price property for each stock is adopted in the experiments.
- **Gas Price**: In order to compare OpSTD against SSTF, we generate a synthetic dataset that is suitable for SSTF. In this dataset, regular gas prices of 500 gas stations in US from Gasbuddy are collected for one day as ground truth. We generate 30 sources with different reliabilities. For each object, we select a random source with high reliability and let it provide ground truth as the observation to the object. The observations provided by the rest of the sources are generated by adding different levels of Gaussian noise based on their reliabilities to the ground truth. Thus, the ground truth of each object is among the observations and it satisfies the condition of SSTF.

---

[1] http://lunadong.com/fusionDataSets.htm.

Table 2. Accuracy comparison

Method	Dataset					
	Weather		Stock		Gas price	
	MAE	RMSE	MAE	RMSE	MAE	RMSE
OpSTD	**0.7274**	**1.1546**	**0.0038**	**0.0002**	**0.2264**	**0.0781**
SSTF	N/A	N/A	N/A	N/A	0.2613	0.1057
GTM	0.8196	1.5074	0.0044	0.0004	0.2502	0.0946
CRH	0.7829	1.4518	0.0046	0.0004	0.2525	0.0987
Mean	0.9524	2.2517	0.0128	0.004	0.3156	1.1514

Performance Metrics. The data in these datasets are continuous, the difference between estimated truth and ground truth can be measured by their numerical distance. Therefore, we use the following two metrics to evaluate the accuracy of OpSTD:

- Mean Absolute Error (MAE): it measures the mean of the overall absolute error between estimated truth and ground truth.
- Root Mean Square Error (RMSE): it measure the root of the mean squared error between estimated truth and the ground truth.

Both MAE and RMSE measure the discrepancy between estimated truth and ground truth. The lower the measure, the more accurate the method is. Being different from MAE, RMSE penalizes heavily on large errors.

We use running times of OpSTD to evaluate its efficiency.

5.2   Performance Comparison

In this section, we report the performance evaluation for OpSTD against the baseline methods on the three datasets. For weather and stock datasets, since the ground truths are not among the observations, it does not satisfy the condition of SSTF, SSTF is not able to estimate object truths for these two datasets. For each dataset, we randomly choose 20% objects and use the ground truths of these objects in the truth discovery process, the ground truths of the rest objects are used for evaluation. For all the baseline methods, we use the best parameters that results in the best performance.

Accuracy Comparison. The experiment results conducted on the three datasets in terms of accuracy are summarized in Table 2. As shown in the table, OpSTD consistently achieves the best accuracy in terms of MAE and RMSE. Among all the methods, Mean performs worst because it simply takes the average of observations for each object as truth, which does not take source reliabilities into consideration. Compared with GTM and CRH, OpSTD's error is reduced ranging from 7%–14% in terms of MAE and 17%–50% in terms of RMSE over the three datasets. The reason is that these two methods explore an unsupervised approach which does not use ground truths in the truth discovery process. Therefore, their errors are larger compared to our method. OpSTD also outperforms the semi-supervised method SSTF. Note that SSTF's accuracy is even lower than GTM and CRH even if it uses ground truths to estimate object truths. This is because its algorithm is designed for handling categorical data and it runs poorly on continuous data scenarios.

Table 3. Running times (second(s))

Method	Dataset		
	Weather	Stock	Gas price
OpSTD	0.245	0.371	0.125
SSTF	N/A	N/A	7.129
GTM	0.277	0.453	0.173
CRH	0.283	0.409	0.151
Mean	0.031	0.04	0.019

Efficiency. The experiment results conducted on the three datasets in terms of running times are summarized in Table 3. From this table we can see that Mean achieves the optimal efficiency. This is because Mean ignores source reliabilities estimation and it outputs mean of observations as truths directly. Among the baseline methods, OpSTD runs about 10% faster than GTM and CRH over the three datasets. The reason is that OpSTD uses 20% ground truths as its input and it estimates the truths for the rest 80% objects, while GTM and CRH discovers truths for the whole dataset. Compared with SSTF, OpSTD runs 57 times faster, which demonstrates the superiority of OpSTD for truth finding with ground truths.

5.3   Sensitivity Analysis

We study the effect of ground truth size and θ on our method. We first evaluate the effect of ground truth size to the accuracy of OpSTD. The θs are fixed at 20, 15 and 35 for weather, stock and gas price datasets, respectively. We vary the available ground truth size over the whole dataset from 0 to 0.8 with the step of 0.2. Since the accuracy of SSTF is also sensitive to the ground truth size, we also use different ground truth sizes to test SSTF on gas price dataset in this experiment. The experiment result is plotted in Fig. 1. From Fig. 1, on one hand, we can see that both MAE and RMSE are high for all the three datasets when ground truth size is 0. This is the case when no ground truth is used in our method and its accuracy is the same as CRH. As the ground truths are involved in our truth discovery process, the errors begin to drop; on the other hand, we can also see that the errors are inverse proportional to the size of ground truths.

This demonstrates that the ground truth indeed benefits the truth estimation in OpSTD. On the other hand, from Figs. 1(c) and (f) we can find that OpSTD outperforms SSTD in terms of MAE and RMSE under all ground truth sizes. This shows that OpSTD can utilize ground truths better for truth discovery tasks with continuous object truths.

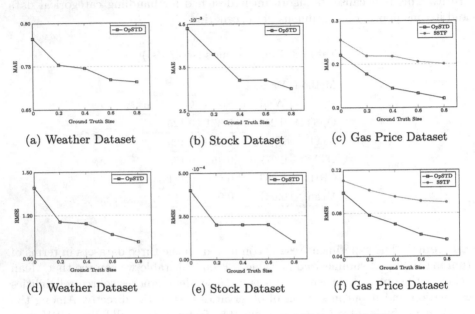

(a) Weather Dataset   (b) Stock Dataset   (c) Gas Price Dataset

(d) Weather Dataset   (e) Stock Dataset   (f) Gas Price Dataset

Fig. 1. Effects of ground truth size to MAE and $RMSE$

The effect of θ to the accuracy of our method is plotted in Fig. 2. In this experiment, we fix ground truth size at 0.2 and vary θ from 0 to 50. From this figure we can see the errors begin to decrease when θs begin to increase from 0 and reach the optimal error very soon. Being different from the ground truth size, as we keep increasing θ, the errors begin to increase after it reaches the optimal ones. The reason is that as we increase θ, the real errors become significant and it dominates the estimated errors in Eq. (8). This may cause the estimated source weights overfit the objects whose ground truths are available, but less general to the rest 80% objects whose object truths are estimated. Given different datasets having different distribution and characteristics, θ is sensitive to OpSTD and we use the best θ to achieve the optimal performance.

In summary, ground truth, even a small set of ground truth, are beneficial for truth discovery. By effectively incorporating ground truths into our method, the accuracy can be improved significantly. When ground truth size is small, theta is sensitive to different datasets and can be tuned to achieve optimal results.

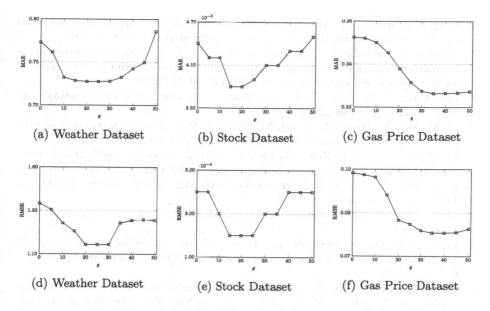

(a) Weather Dataset   (b) Stock Dataset   (c) Gas Price Dataset

(d) Weather Dataset   (e) Stock Dataset   (f) Gas Price Dataset

Fig. 2. Effects of θ to MAE and $RMSE$

6   Conclusion

In this paper, we investigate semi-supervised truth discovery method for continuous object truths. We formulate the truth discovery problem as an optimization task in which object truths and source weights are modeled as variables, and the ground truths is formulated as a regularization term to reinforce the source weights. An iterative solution is developed to estimate object truths and source weights and its convergence property and time complexity are analyzed. We also conduct a series of experiments to demonstrate that the proposed method outperforms the existing truth discovery methods in terms of both accuracy and efficiency.

References

1. Bertsekas, D.P.: Nonlinear Programming. Athena Scientific, Belmont (1999)
2. Cho, J.H., Swami, A., Chen, R.: A survey on trust management for mobile ad hoc networks. IEEE Commun. Surv. Tutor. **13**(4), 562–583 (2011)
3. Dong, X.L., Berti-Equille, L., Srivastava, D.: Integrating conflicting data: the role of source dependence. Proc. VLDB Endow. **2**(1), 550–561 (2009)
4. Dong, X.L., Berti-Equille, L., Srivastava, D.: Truth discovery and copying detection in a dynamic world. Proc. VLDB Endow. **2**(1), 562–573 (2009)
5. Fang, X.S.: Truth discovery from conflicting multi-valued objects. In: Proceedings of the 26th International Conference on World Wide Web Companion, pp. 711–715. International World Wide Web Conferences Steering Committee (2017)

6. Fang, X.S., Sheng, Q.Z., Wang, X., Ngu, A.H.: SmartMTD: a graph-based approach for effective multi-truth discovery. arXiv preprint arXiv:1708.02018 (2017)

7. Galland, A., Abiteboul, S., Marian, A., Senellart, P.: Corroborating information from disagreeing views. In: Proceedings of the Third ACM International Conference on Web Search and Data Mining, pp. 131–140. ACM (2010)

8. Lee, Y.W., Pipino, L.L., Funk, J.D., Wang, R.Y.: Journey to Data Quality. The MIT Press, Cambridge (2009)

9. Li, M., Sun, X., Wang, H., Zhang, Y., Zhang, J.: Privacy-aware access control with trust management in web service. World Wide Web 14(4), 407–430 (2011)

10. Li, Q., Li, Y., Gao, J., Zhao, B., Fan, W., Han, J.: Resolving conflicts in heterogeneous data by truth discovery and source reliability estimation. In: Proceedings of the 2014 ACM SIGMOD International Conference on Management of Data, pp. 1187–1198. ACM (2014)

11. Li, Y., et al.: A survey on truth discovery. ACM SIGKDD Explor. Newsl. 17(2), 1–16 (2016)

12. Li, Y., et al.: On the discovery of evolving truth. In: Proceedings of the 21st ACM SIGKDD International Conference on Knowledge Discovery and Data Mining, pp. 675–684. ACM (2015)

13. Meng, C., et al.: Truth discovery on crowd sensing of correlated entities. In: Proceedings of the 13th ACM Conference on Embedded Networked Sensor Systems, pp. 169–182. ACM (2015)

14. Pasternack, J., Roth, D.: Knowing what to believe (when you already know something). In: Proceedings of the 23rd International Conference on Computational Linguistics, pp. 877–885. Association for Computational Linguistics (2010)

15. Pasternack, J., Roth, D.: Latent credibility analysis. In: Proceedings of the 22nd International Conference on World Wide Web, pp. 1009–1020. ACM (2013)

16. Pochampally, R., Das Sarma, A., Dong, X.L., Meliou, A., Srivastava, D.: Fusing data with correlations. In: Proceedings of the 2014 ACM SIGMOD International Conference on Management of Data, pp. 433–444. ACM (2014)

17. Qi, G.J., Aggarwal, C.C., Han, J., Huang, T.: Mining collective intelligence in diverse groups. In: Proceedings of the 22nd International Conference on World Wide Web, pp. 1041–1052. ACM (2013)

18. Yin, X., Han, J., Philip, S.Y.: Truth discovery with multiple conflicting information providers on the web. IEEE Trans. Knowl. Data Eng. 20(6), 796–808 (2008)

19. Yin, X., Tan, W.: Semi-supervised truth discovery. In: Proceedings of the 20th International Conference on World Wide Web, pp. 217–226. ACM (2011)

20. Zhang, J., Tao, X., Wang, H.: Outlier detection from large distributed databases. World Wide Web 17(4), 539–568 (2014)

21. Zhao, B., Han, J.: A probabilistic model for estimating real-valued truth from conflicting sources. In: Proceedings of QDB (2012)

22. Zhao, B., Rubinstein, B.I., Gemmell, J., Han, J.: A Bayesian approach to discovering truth from conflicting sources for data integration. Proc. VLDB Endow. 5(6), 550–561 (2012)

23. Zheng, Y., Li, G., Li, Y., Shan, C., Cheng, R.: Truth inference in crowdsourcing: is the problem solved? Proc. VLDB Endow. 10(5), 541–552 (2017)

Web Page Template and Data Separation for Better Maintainability

Chenxu Zhao, Rui Zhang(✉), and Jianzhong Qi

School of CIS, The University of Melbourne, Parkville, Australia
chenxuz@student.unimelb.edu.au,
{rui.zhang,jianzhong.qi}@unimelb.edu.au

Abstract. Separating a web page into template code and data records populated into the template is an important problem. This problem has a wide range of applications in web page compression and information extraction. We study this problem with the aim to separate a web page into easily maintainable template code and data records. We show that this problem is NP-hard. We then propose a heuristic algorithm to solve the problem. The main idea of our algorithm is to parse a web page into a tree and then to process it recursively in a bottom-up manner with three steps: splitting, folding, and alignment. We perform experiments on real datasets to evaluate the performance of our proposed algorithms in maximizing the maintainability of the template code produced. The experimental results show that our proposed algorithms outperform the baseline algorithms by 25% in the maintainability measure.

Keywords: Web page template extraction · Maintainability index
Dual teaching and learning based optimization

1 Introduction

Separating a web page into template code and data records populated into the template is an important problem. This problem has a wide range of applications in web page compression and data record extraction. We study this problem and focus on the maintainability of the template code generated, since easily maintainable template code is reliable, and it will simplify further developments on top of the template code, e.g., to update the web templates. Figure 1 shows an example of the web page separation problem. In particular, Fig. 1(a) shows the HTML source of a web page which contains a list of items (cf. the "$\langle li \rangle$" tags). Figure 1(b) shows the template code separated from the HTML source, which effectively says that the HTML page can be generated by a for-loop (cf. the "$\langle for1 \rangle$" tag) to produce a list of "$\langle li \rangle$" items. The attributes and data records to be populated into this list of items are represented by variables "r1" and "r2", the values of which are stored in a data record file as illustrated in Fig. 1(c). Note that, in this example HTML source code, the second list item

© Springer Nature Switzerland AG 2018
H. Hacid et al. (Eds.): WISE 2018, LNCS 11233, pp. 439–449, 2018.
https://doi.org/10.1007/978-3-030-02922-7_30

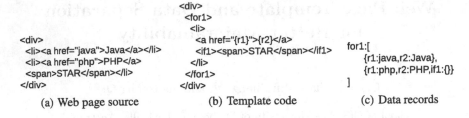

(a) Web page source   (b) Template code   (c) Data records

Fig. 1. An example of the web page separation problem

contains an additional "$\langle span \rangle$" tag. This is reflected in the "$\langle if1 \rangle$" tag in the generated template in Fig. 1(b).

Making such a separation has important applications. For example, we can improve the maintainability of web page source code by writing "for-loop" to produce repeating data records rather that writing a duplicate code segment for each data record. Besides, we may reduce the size of the web page by storing it in its template code form, e.g., a long list in the HTML source can be replaced by a simple "for-loop" tag in the template code.

SYNTHIA [7] is the state-of-the-art web page separation algorithm that aims at compressing web pages. It processes the *DOM tree* of a web page hierarchically and utilizes pairwise similarity to determine which *siblings* should be *folded*, i.e., replaced by a for-loop in the template code. It uses a local *alignment* algorithm to capture the differences between the siblings folded together, i.e., to add the "if" tags in a for-loop in the template code. After separating a web page into its template and data records, SYNTHIA will use the template code and data records to replace the HTML source only if they are shorter. For example, the number of characters of the HTML source in Fig. 1(a) is 90, while the total number of characters in the template code and data records in Figs. 1(b) and (c) is 131. Thus, SYNTHIA will keep the HTML source as it is shorter.

In comparison, our work focuses on the maintainability of the web page source. In the example shown in Fig. 1, the HTML source has a maintainability score of 110.106, while the template code has a maintainability score of 110.132 which indicates that the template code is easier to maintain (detailed in Sect. 3). Intuitively, replacing all those list items with a for-loop makes it easier to update the web page, e.g., if we want to change the style of each list item, we only need to change it once in the template for-loop rather than changing every list item in the HTML source.

In this paper, we study how to separate a web page into easily maintainable template code and data records. We evaluate the maintainability of a separation by the *maintainability index* (MI) [1]. MI is an important software metric to measure the maintainability of software source code. A higher MI score suggests a higher maintainability and is more preferable. Intuitively, a piece of code would have a higher MI score if it is shorter, having few variable and functions, and having few branches (detailed in Sect. 3). To the best of our knowledge, we are the first to formulate a web page separation problem based on the MI score.

We analyze the complexity of our separation problem and show that it is NP-hard. We then adapt the SYNTHIA algorithm to solve our problem. In SYN-THIA, the tree sibling splitting procedure compares pairwise sibling similarity and uses a splitting threshold to determine whether the siblings should be folded together or be separated into different chunks. This threshold is heuristically defined to obtain short template code. We adapt the algorithm by replacing this threshold with ours that aims to obtain a high MI score.

We further develop a population-based optimization algorithm, named *dual teaching and learning based optimization* (dual-TLBO), to optimize sibling splitting. The algorithm considers a splitting plan as an individual and evaluates the maintainability of a population of different individuals (splitting plans) globally under the subtrees being considered, which addresses the limitation of SYNTHIA that only considers pairwise similarity. We also develop a global alignment algorithm with dynamic costs of alignment operations, which achieves a high-quality separation. In summary, we make the following contributions:

- We define a novel web page separation problem and show that it is NP-hard.
- We propose a population-based algorithm named dual-TLBO to help select the siblings in the web page source code tree that should be folded together with the aim to achieve a high MI score.
- We propose a global alignment algorithm to align siblings to be folded together with the aim to achieve a high MI score.
- We perform a experimental study to evaluate our algorithms in both max-imizing the MI score and minimizing the length of the generated template code and data records.

2   Related Work

We review closely related work on data record extraction and optimization. MDR [5] uses XPath to identify the text nodes. It presents an XPath matching algorithm to detect template and text nodes with similar XPaths. The partial tree alignment algorithm [11] aligns similar sub-DOM trees. FiVaTech [4] applies alignment algorithms and template detection algorithms on multiple web pages. It defines a new pattern tree named "fixed/variant pattern tree". These techniques aim to extract the data records only. The web page templates are simply discarded. The original pages cannot be reconstructed from the extraction results because of the loss of template information. TLBO [9] is a population-based optimization algorithm that every solution learns from the current best solution and representative solutions towards achieving the best solution. The number of nodes of a DOM tree may change when deduplicating repetitive sub-DOM trees. Pang et al. [8] present an algorithm for constructing a minimal dominating set on digraphs when the number of nodes is changing.

3   Maintainability Index of Template

The *maintainability index* (MI) [1] is a software metric to evaluate the maintainability of programs. We use it to evaluate the quality of a separation. MI is formed by three important software metrics: Halstead Volume (V), Cyclomatic Complexity [6] (G), and Source Lines of Code (SLOC). The function of MI is given as $MI = 171 - 5.2 \times ln(V) - 0.23 \times (G) - 16.2 \times ln(SLOC)$.

We evaluate our work using the MI score and aim to separate the web page into data records and template code that has a high MI score (i.e., better maintainability). We adapt the model given in SYNTHIA [7]. To compute the maintainability index score for a separation, we define the maintainability index of layout trees as summarized in Table 1. Every text node adds one operand and one SLOC. The text node is identified by its value. An element node adds one operator and its starting tag and an ending tag adds one SLOC, respectively. They are identified by its tag string. A condition/iteration node adds one operator, one Cyclomatic Complexity, and two SLOCs. A reference node adds one operand. It further adds one SLOC when it refers text nodes.

Table 1. MI of layout trees

Node type	Operand	Operator	G	SLOC
Text node	1	0	0	1
Element node	0	1	0	2
For/If node	0	1	1	2
Reference node (attrs)	1	0	0	0
Reference node (others)	1	0	0	1

4   Problem Definition and Complexity Analysis

4.1   Problem Definition

We parse the web pages into a DOM tree. There exist many possible separations of a given DOM tree as formed by different folding combinations of the siblings in the DOM tree. We define a new concept called the *variation* to represent a specific folding combination. A variation of a given DOM tree is defined as a set S of subsets of siblings of the DOM tree that satisfies the following three conditions:

1. The intersection of any two elements of S (i.e., subsets of siblings) is empty.
2. The union of all elements of S contains all nodes of the DOM tree.
3. The siblings of an element of S are continuous, i.e., they are all adjacent to each other in the DOM tree.

The maintainability index of a variation of a given DOM tree is the maintainability index of the layout tree that is generated by the variation. We aim to find a variation that has the maximum maintainability index. A formal definition of our problem is as follows.

Definition 1. Maintainability based web page separation (optimization version): *Given a DOM tree T and the maintainability index function, find a variation that has the maximum maintainability index.*

4.2 Complexity Analysis

To analyze the complexity of our separation problem, we first recast it as a decision problem as follows:

Definition 2. Maintainability based web page separation (decision version): *Given a DOM tree T and the maintainability index function, determine whether there exists a variation whose maintainability index is less than or equal to a given constant c.*

Then, we reduce the *exact cover* problem, which has been proved to be an NP-complete problem [3], to our separation problem. As the solution of a general exact cover problem does not have a weight (which is to be reduced to our maintainability index score c), we define that all valid solutions of an exact cover problem have a constant weight c^*, while non-valid solutions have a negative infinity weight. The definition of an exact cover problem is as follows:

Definition 3. EXACTCOVER: *Given a set X, a set S of subsets of X, and a cost function F, determine whether there is a subcollection S^* of S such that the intersection of any two distinct subsets in S^* is empty, the union of the subsets in S^* is X, and $F(S^*) = c^*$. Here, $F(S*)$ is a function that returns a value of $c*$ if $S*$ is an exact cover of X, and $-\infty$ otherwise.*

Theorem 1. *The decision version of the maintainability based web page separation problem is NP-hard.*

Proof. In the following, we construct an instance ms of the decision version of the maintainability based web page separation and form a mapping from $EXACTCOVER$ to ms. Given any X of $EXACTCOVER$, we define the nodes of T of ms as the elements of X. A subcollection of S consists of a number of subsets of S, denoted by S'. If we view each subset of S as a subset of siblings of T in ms, then each subcollection corresponds to a variation in ms. We map each subcollection S' that covers all elements of X to its corresponding variation and define the cost of the variation as the cost of all subsets in S'. The cost of the variation corresponding to any other subsets of X than those in S is defined as $-\infty$. For any instance of $EXACTCOVER$, if there exists a subcollection of S that covers all elements of X and has the cost of c^*, then its corresponding variation also has the cost of c^*. Therefore, $EXACTCOVER$ is reduced to the decision version of the maintainability based web page separation problem, and the decision version of the maintainability based web page separation problem is NP-hard.

5 Our Methods

5.1 Overall Algorithm Procedure

Our algorithm traverses the DOM tree in a bottom-up manner to generate a layout tree with the aim to maximize the maintainability index of every sub-layout tree. For every node in the DOM tree, our algorithm uses a splitting algorithm (i.e., dual-TLBO, detailed in Sect. 5.3) to split its child nodes (siblings) into chunks. For the chunks that contain multiple siblings, our algorithm folds them together and uses an alignment algorithm (detailed in Sect. 5.2) to capture the differences among the siblings. The splitting algorithm adds iteration nodes, and the alignment algorithm adds condition nodes to the layout tree generated. When the root node of the DOM tree is reached, the algorithm terminates and returns the layout tree generated.

5.2 Alignment Algorithm

Our alignment algorithm aligns siblings in the same chunk and captures the differences among the siblings. For every two siblings (which have been converted to layout trees already since our algorithm works in a bottom-up manner) in the same chunk, our algorithm aligns their children globally and recursively. Let the two layout trees be A and B. The reference nodes, which refer to attributes, are aligned as the attributes of element nodes. Besides, the reference nodes, which replace text nodes, are kept. Therefore, reference nodes which refer to attributes are not included in A or B. The score of aligning A and B is the difference between the MI score of after and before aligning A and B. We consider the following alignment operations: (1) Aligning a layout tree with null means adding a condition as the root of the layout tree. If the root of the layout tree is already a condition node, the layout tree does not change. We denote the layout trees of align A and B with null by A_c and B_c respectively. (2) If at least one of A and B has an instruction node as the root, the alignment begins from their child, which is not an instruction node and the instruction node is kept as the root of the layout tree after aligning. (3) If the roots of A and B are element nodes, and their attribute names and tag string are the same, then they become one element node and their children are aligned recursively. (4) If A and B are reference nodes, which refer to text nodes, they become one reference node.

Let the number of children of A and B be m and n and the ith child of A and the jth child of B be $A[i]$ and $B[j]$. The algorithm fills a matrix $ALIGN$ of size $(m+1) \times (n+1)$. $ALIGN[i,j]$ is the maximum value of (1) $ALIGN[i-1,j-1] + score(A[i], B[j])$, (2) $ALIGN[i-1,j] + score(A[i])$ and (3) $ALIGN[i,j-1] + score(B[j])$. The alignment algorithm fills $ALIGN$ from $ALIGN[0,0]$ to $ALIGN[m+1,n+1]$ row by row. When the filling is done, $ALIGN[m+1,n+1]$ stores the maximum score of an alignment. We trace back from $ALIGN[m+1,n+1]$ to $ALIGN[0,0]$ to obtain the alignment. Let T_a be the root of the new layout tree. During tracing back, we add children to T_a to generate an aligned layout tree. If $ALIGN[i,j] = ALIGN[i-1,j+1] + score(A[i], B[j])$, $A[i]$ and $B[j]$ are matched, we align them and add the generated layout tree to T_a. If $ALIGN[i,j] = A[i-1,j] + score(A[i])$, we add a condition node as the parent of $A[i]$ to generate $A_c[i]$ and add $A_c[i]$ to T_c. If $ALIGN[i,j] = A[i,j-1] + score(B[j])$, we add a condition node as the parent of $B[j]$, which is $B_c[j]$, and add $B_c[j]$ to T_c. Besides, the data records are also aligned to adapt to the changes of the layout tree. The alignment algorithm returns T_c.

Table 2. Score

		A_2	B_2	D_2	F_2
	0	-2	-5	-3	-4
E_1	-4	-7	-3	2	1
A_1	-1	6	-1	-3	-5
C_1	-6	-2	-1	-1	-3
D_1	-4	-1	-1	5	-2
F_1	-5	-5	-2	-4	6

Table 3. Alignment

		A_2	B_2	D_2	F_2
	0	←-2	←-7	←-10	←-14
E_1	↑-4	←↑-6	↖-5	↖-5	←↖-9
A_1	↑-5	↖2	←-3	←↑-6	←↖↑-10
C_1	↑-11	↑-4	↖1	←-2	←-6
D_1	↑-15	↑-8	↑-3	↖6	←2
F_1	↑-20	↑-13	↑-8	↑1	↖12

To help understand our alignment algorithm, we give a running example. In Tables 2 and 3, where $[A_1, B_1, D_1, F_1]$ and $[A_2, C_2, D_2, E_2, F_2]$ are two lists of siblings. Table 2 shows the score of matching every branch with others from the other list. We fill the Table 3 from left to right in a top-down manner. Finally, we trace back from the last element of Table 3 which is (F_2, F_1). The cells in orange show the matching path. The alignment result is $(E_1), (A_1, A_2), (C_1, B_2), (D_1, D_2), (F_1, F_2)$ and it increases the maintainability index by 12. Our alignment adds a condition node as the parent of E_2. Although $score(C_1, B_2)$ is smaller than 0, aligning C_1 and B_2 with null gets a lower score than $score(C_1, B_2)$.

5.3 Dual Teaching and Learning Based Optimization Algorithm

We model the problem of splitting siblings into chunks as a problem of finding boundaries to separate the siblings. We represent the boundary of N siblings with a list of 0 or 1 whose length is $N - 1$, where 1 stands for a boundary (i.e., in different chunks), while 0 stands for none boundary (i.e., in the same chunk). We design a population-based splitting algorithm which is named *dual teaching and learning based optimization* (dual-TLBO). Figure 2 shows the key steps of our dual-TLBO. First, the algorithm initializes the population and a termination criterion. The termination criterion is the maximum number of loops. Each individual in the population is a list of 0's and 1's, which represents a solution B of the "Finding Chunk Boundaries" problem. It is initialized randomly. The mean of all individuals in the population is computed. Then a teacher phase starts. In this phase, we get a new for every individual X_{old}, we create a new individual X_{new} based on the teacher and the mean value, which is denoted by $X_{new} = X_{old} + r(Teacher - Mean)$, where r is a system parameter between 0 and 1 that represents the learning rate. The computed individual X_{new} may contain non-integer numbers or negative numbers. We round the number to its nearest non-negative integer. If X_{new} represents a chunking that yields a higher MI score, we replace X_{old} by X_{new}. Otherwise, X_{old} is kept and X_{new} is discarded. Then, it comes to the student phase. We randomly select two individuals X_i and X_j and for every individual X_{old}, we create a new individual $X_{new} = X_{old} + r(X_i - X_j)$. Similar to the teacher phase, we compare X_{old} with

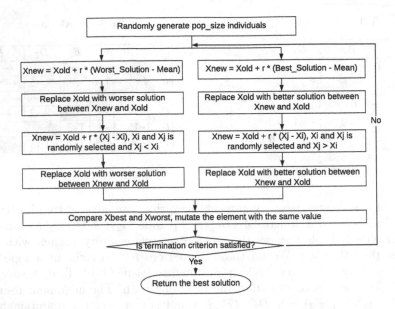

Fig. 2. Workflow of dual-TLBO

X_{new} and keep the one producing a higher MI score. We add it a dual step to the algorithm, where we run another teacher phase and another student phase but learn from the worst (rather than the best) individual of the original population. After learning from the worst and the best individuals, we can compare the new best individual and the new worst individual element-wise. If the element in the same position is the same, we consider it as a not well-learned element because the contribution to improving maintainability index of the element is not clear. For these positions, we mutate their values and keep the new best solution for the next iteration. If the termination criterion is met, the algorithm returns the best solution, otherwise found so far.

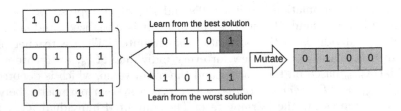

Fig. 3. Running example of dual-TLBO

To help understand the dual-TLBO algorithm, we give a running example next. We assume 5 siblings. The workflow is shown in Fig. 3. First, we randomly generate three individuals: $[0, 1, 1, 1], [1, 0, 1, 1]$ and $[0, 1, 0, 1]$. We assume that

$[0, 1, 0, 1]$ is the current best individual. The mean of the population is $[1/3, 2/3, 2/3, 1]$ and every individual can generate three new individuals which are $[2/3, 1/3, 1/3, 1]$, $[-1/3, 4/3, -2/3, 1]$ and $[-1/3, 4/3, 1/3, 1]$. After rounding to the nearest non-negative integers, these three new individuals become $[1, 0, 0, 1]$, $[0, 1, 0, 1]$ and $[0, 1, 0, 1]$. In this step, we only replace the old individuals that are worse than the new individuals. After learning from the teacher, the population becomes $[1, 0, 0, 1]$, $[0, 1, 0, 1]$ and $[0, 1, 0, 1]$. Then, we randomly select two students from the population and the students are $[1, 0, 0, 1]$, $[0, 1, 0, 1]$. Every student learns from these two students, and the population becomes three $[1, 0, 0, 1]$. Then, we use the original population: $[0, 1, 1, 1]$, $[1, 0, 1, 1]$ and $[0, 1, 0, 1]$ to learn from the worst individual. In this step, we aim to minimize the maintainability index of individuals. We assume that $[1, 0, 1, 1]$ is the current worst individual. After learning from them, the bad population becomes three $[1, 0, 1, 1]$. Compare the best individual $[0, 1, 0, 1]$ and the worst individual $[1, 0, 1, 1]$, we find that the last element in these two individuals is the same. This is a non-well learned element. We mutate it and update the best individual to be $[0, 1, 0, 0]$.

6   Experiments

We evaluate the empirical performance of our algorithms on real datasets and report the results in this section. We show that our algorithms can produce separations with high MI scores than those by the baseline algorithms. Meanwhile, our algorithms also produce separations with small sizes, which are comparable to those produced by the SYNTHIA algorithm that is designed to produce separations with small sizes.

We use two real data sets. The first dataset is from the SYNTHIA paper [7]. We denote the dataset by SYNTHIA-DATA. SYNTHIA-DATA contains 200 web pages from 40 websites (5 web pages per website). We use it to compare the performance of our algorithms with the SYNTHIA algorithm. As the code of SYNTHIA is not available, we do our best to implement the algorithm following the SYNTHIA paper [7] for our experiments. The result is shown in Table 4.

Table 4. SYNTHIA-DATA

	MI	CR	Time (s)
Original page	−3.46 (0%)	100%	−
SYNTHIA	−3.06 (+0.7%)	63.34%[a]	310
MAX-MI-LCS	6.97 (+19.8%)	66.31%	315
MI-Dual-TLBO	*9.38 (+24.4%)*	67.27%	5943

[a]The compression ratio reported in the original SYNTHIA paper is 62.8%.

In the table, MI denotes the maintainability index score. CR denotes the compression ratio, i.e., the output file size over the original file size, and time denotes the algorithm response time. We also report CR because SYNTHIA is designed to achieve a small CR value. We report the performance of five algorithms, SYNTHIA is our implementation of the SYNTHIA algorithm [7]; MAX-MI-LCS adds the alignment algorithm described in Sect. 5.2; MI-Dual-TLBO is our proposed algorithm as described in Sect. 5.3.

We can see from the table that the proposed algorithm MI-Dual-TLBO outperforms all the baseline algorithms in terms of the MI score, while its compression ratio is very close to that of SYNTHIA which is designed to optimize the compression ratio. We notice that MI-Dual-TLBO is slower than SYNTHIA. We argue that web page separation is usually an offline task which can allow a slower algorithm. Further, if the algorithm response time is critical, then our adapted algorithm MAX-MI and MAX-MI-LCS can be used instead of SYNTHIA as they are as fast as SYNTHIA while obtaining larger MI scores.

Table 5. TBDW

Method	Dataset	MI	CR	Time (s)
Original page	TBDW	17.80 (0%)	100%	–
SYNTHIA	TBDW	23.02 (+13.45%)	74.21%	379
MI-Dual-TLBO	TBDW	27.50 (+25.0%)	75.13%	2944

Table 6. UW-CAN

Method	Dataset	MI	CR	Time (s)
Original page	UW-CAN	41.69 (0%)	100%	–
SYNTHIA	UW-CAN	45.04 (+6.1%)	81.16%	127
MI-Dual-TLBO	UW-CAN	48.30 (+12.1%)	83.92%	1610

We also test our algorithm on other datasets. The result is shown in Tables 5 and 6. TBDW [10] is a data set that contains 253 web pages from 51 websites. UW-CAN [2] is a data set that contains 314 web pages from the University of Waterloo and other Canadian websites. The result is shown in Table 6. Our proposed algorithm again outperforms SYNTHIA in terms of the MI score, and the advantage is 25.0% and 12.1% on these two datasets, respectively.

7   Conclusions

We proposed and studied a web page separation problem that aims to extract easily maintainable template code and data records from web pages. We showed the NP-hardness of the problem and presented a heuristic algorithm to solve the problem. We proposed a dual-teaching and learning based optimization algorithm to detect siblings generated by the same template. The experimental results show that our algorithms outperform state-of-the-art web page separation techniques. Our algorithms extracts easily maintainable template code from web pages, Web developers can compare their web page source code with the template code generated by our algorithms and identify ways to improve the maintainability of their source code. In the future, we plan to work on extracting template code on the website level and data record extraction.

Acknowledgment. This work is supported by Australian Research Council (ARC) Future Fellowships Project FT120100832 and Discovery Project DP180102050.

References

1. Counsell, S., et al.: Re-visiting the 'maintainability index' metric from an object-oriented perspective. In: SEAA, pp. 84–87 (2015)
2. Hammouda, K.M., Kamel, M.S.: Phrase-based document similarity based on an index graph model. In: ICDM, pp. 203–210 (2002)
3. Karp, R.M.: Reducibility among combinatorial problems. In: Miller, R.E., Thatcher, J.W., Bohlinger, J.D. (eds.) Complexity of Computer Computations. The IBM Research Symposia Series, pp. 85–103. Springer, Boston (1972). https://doi.org/10.1007/978-1-4684-2001-2_9
4. Kayed, M., Chang, C.H.: FiVaTech: page-level web data extraction from template pages. IEEE Trans. Knowl. Data Eng. **22**(2), 249–263 (2010)
5. Liu, B., Grossman, R., Zhai, Y.: Mining data records in web pages. In: KDD, pp. 601–606 (2003)
6. McCabe, T.J.: A complexity measure. IEEE Trans. Softw. Eng. **4**, 308–320 (1976)
7. Omari, A., Kimelfeld, B., Yahav, E., Shoham, S.: Lossless separation of web pages into layout code and data. In: KDD, pp. 1805–1814 (2016)
8. Pang, C., Zhang, R., Zhang, Q., Wang, J.: Dominating sets in directed graphs. Inf. Sci. **180**(19), 3647–3652 (2010)
9. Rao, R.V., Savsani, V.J., Vakharia, D.: Teaching-learning-based optimization: a novel method for constrained mechanical design optimization problems. Comput.-Aided Des. **43**(3), 303–315 (2011)
10. Yamada, Y., Craswell, N., Nakatoh, T., Hirokawa, S.: Testbed for information extraction from deep web. In: WWW, pp. 346–347 (2004)
11. Zhai, Y., Liu, B.: Web data extraction based on partial tree alignment. In: WWW, pp. 76–85 (2005)

Text Mining

Combining Contextual Information by Self-attention Mechanism in Convolutional Neural Networks for Text Classification

Xin Wu[1], Yi Cai[1(✉)], Qing Li[2], Jingyun Xu[1], and Ho-fung Leung[3]

[1] School of Software Engineering, South China University of Technology, Guangzhou, China
ycai@scut.edu.cn
[2] Department of Computer Science, City University of Hong Kong, Kowloon, Hong Kong SAR, China
[3] Department of Computer Science and Engineering, The Chinese University of Hong Kong, Kowloon, Hong Kong SAR, China

Abstract. Convolutional neural networks (CNN) are widely used in many NLP tasks, which can employ convolutional filters to capture useful semantic features of texts. However, convolutional filters with small window size may lose global context information of texts, simply increasing window size will bring the problems of data sparsity and enormous parameters. To capture global context information, we propose to use the self-attention mechanism to obtain contextual word embeddings. We present two methods to combine word and contextual embeddings, then apply convolutional neural networks to capture semantic features. Experimental results on five commonly used datasets show the effectiveness of our proposed methods.

Keywords: Convolutional neural networks · Text classification
Attention mechanism · Word representation

1 Introduction

Text classification is an essential task in many natural language processing (NLP) applications, such as web searching, sentiment analysis, and information filtering [1]. Traditional text classification methods mainly focus on human designed features and different types of machine learning algorithms [26]. The most widely used feature is the bag-of-words (BoW) feature, which represents a word as a high-dimension and sparse vector with only one non-zero value, resulting in the poor performance of representation of semantic and syntax. More complex features such as POS tagging and tree kernel [18] are designed to capture more semantic features. Classifier such as support vector machine and logical regression can be used for classification with these features [26]. However, such handcrafted features are time-consuming due to the extensive feature engineering.

© Springer Nature Switzerland AG 2018
H. Hacid et al. (Eds.): WISE 2018, LNCS 11233, pp. 453–467, 2018.
https://doi.org/10.1007/978-3-030-02922-7_31

Recently, deep neural networks demonstrate their effectiveness of automatic feature extraction in text classification tasks. The commonly used neural networks text classification tasks include Recursive Neural Network (RecursiveNN), Convolutional Neural Network (CNN), Recurrent Neural Network (RNN) and more complex networks architecture. One influential work is proposed by Kim [7], where one single convolution layer is employed to capture semantic features. Despite its simple structure, the model shows its effectiveness on many text classification datasets. Some advanced models are proposed based on this standard one-layer CNN [9,29,32].

However, most CNN methods tend to use simple convolutional kernels with a fixed size window. Though these kernels can effectively capture local useful features, they may perform poorly when some important features are far away from current filter window. For example, in the sentence *"Who is the head of the World Bank?"*, without global context information *"Who is the head of"*, the phrase *"World Bank"* may be misunderstood as a famous scenic spot, which may lead to the wrong classification of this sentence. With the help of global context information, the phrase *"World Bank"* can be easily distinguished as the meaning of a financial institution. Though global context information is very essential for text classification, it can appear anywhere in a sentence and it is hard to catch. Although simply increasing the size of filters can capture more information of texts, but the parameters of large filters are hard to train. There can be two approaches to tackle this problem.

One approach is to use the recurrent neural network to capture global context information [8] and use CNN to capture features from global context information and perform classification. RNN has been proven to be efficient in terms of constructing variable-length texts, it handles input texts word by word and stores the semantic information of previous words, which means RNN can be used to learn dependencies between words in a sentence. However, when RNN is used to capture the context information of a whole sentence, the forward and backward signals of these long-distance dependencies have to traverse in the networks, such long-distance information may lose during the propagation [24]. Moreover, RNN is time-consuming due to their poor ability in parallel computation.

Another approach is text conceptualization technique, which can represent sentences as a set of concepts [25]. These concepts contain the general information of sentences. However, such a conceptualization technique relies on the external knowledge base and may perform poorly when the intersection between the knowledge base and the corpus is small.

To address the limitation of the above-mentioned methods. We propose to use the self-attention mechanism to capture global context information of words. The self-attention mechanism is an attention mechanism for modeling the dependencies without regarding their distance in a sentence, which can be used to compute the representation of the sentence [24]. As compared to RNN, self-attention is not affected by the distance between words, it is easy for self-attention to capture important features of a whole sentence. Moreover, self-attention is suitable for parallel computation and is easy to implement without complex networks architecture or external knowledge base.

The main contributions of this paper are summarized as follows:

- We apply the self-attention mechanism to capture context information of each word in sentences. The additional training cost of self-attention is relatively small, and it is task-independent without requiring any external knowledge base.
- We compare two score functions to calculate the attention weights and give an intuitive explanation for how to choose the score function. We present two methods to combine word embedding and context information we obtained, which are designed from two different perspectives.
- We conduct extensive experiments on five commonly used text classification datasets. The results show that our methods further improve the performance of standard one-layer CNN and give competitive results against some state-of-the-arts models without any external knowledge or logical rules.

The remainder of this paper is organized as follows. Section 2 introduces the related work of text classification and attention mechanism. In Sect. 3, we present our method in details. Section 4 presents experimental results on five commonly used datasets. We discuss the results and analyze our methods in Sect. 4.3 and draw a conclusion in Sect. 5.

2 Related Work

Traditional text classification works mainly use machine learning algorithms with the handcrafted feature. The most widely used feature is the bag-of-words (BoW) feature, which represents a word as a high-dimension sparse vector with only one non-zero value. The most critical problem of BoW feature is it ignores the relationship between words. More complex features such as part-of-speech tags and tree kernels [18] are designed to capture more semantic features. Traditional methods often use classifiers such as logistic regression (LR), Naive Bayes (NB), and support vector machine (SVM) to perform classification [26].

Word embedding, which first proposed in Rumelhart et al. [19], have been successfully used in language models [13,14], named entity recognition [23], parsing [20,21] and other NLP applications. The pre-trained word embeddings on large corpus can be used in text classification task [7,29,32] to improve the performance because word embedding methods represent words with more semantic and syntactic information as low-dimension dense vectors.

Most recently, CNN with pre-trained word embeddings is becoming increasingly popular in text classification. Collobert et al. [4] first used CNN with pre-trained word embedding for text classification. Kim [7] further improved the performance by using multi-channel embedding. Multiple pre-trained word embeddings are exploited to enhance the model in Yin and Schütze [29] and Zhang et al. [32]. Li et al. [9] proposed a novel initialization technique for CNN, which encode semantic features into convolutional layers by initializing them with important n-grams. Moreover, other external knowledge also is used to improve performance. In Hu et al. [5], neural models are harnessed by logic rules.

Wang et al. [25] proposed to use concepts extracted from the external knowledge base and character level features to enhance the embedding of short text.

In addition to convolutional neural networks, other deep neural networks are also widely used in text classification. Socher et al. [22] introduced the recursive neural tensor network to obtain representations of phrases and sentences for sentiment analysis. Irsoy and Cardie [6] proposed deep recursive neural network to modeling sentences. Zhou et al. [33] proposed a bidirectional Long short-Term Memory network with attention mechanism for relation classification. Yang et al. [28] introduced hierarchical attention networks (HAN) for classifying documents.

Attention mechanisms have been widely used in sequence to sequence model [2]. Self-attention is one variation of attention mechanisms, which focus on obtain representation of a sentence. Self-attention has been used successfully in reading comprehension [3], textual entailment [17], machine translation [24] and other NLP tasks.

3   Model

In this section, we first present how to capture words' global context information by self-attention mechanism. Then we propose two methods to combine context information with word embedding. Finally, we employ convolution neural networks to extract semantic features among words and their context information.

3.1   Capturing Global Context Information

Since CNN remains unsatisfactory for capturing long distance features, it is crucial to take these global context information into consideration. Due to the limitations of RNN and conceptualization technique, we propose to use self-attention mechanism to capture global context information. Self-attention, also known as intra-attention is an attention mechanism modeling of dependencies in a single sentence without regarding the distance between them.

In the self-attention mechanism, we select a target word at each time, and then each of other words in this sentence will compute a dependency weight

Table 1. Two input sentences with the same target word *"bank"* from two categories, weight vector is obtained by self-attention respectively.

Sentence	Target word	Weight vector	Category
Who is the head of the World Bank?	Bank	$\{<Who, 0.1>, <is, 0.005>, ...,$ $<head, 0.05>, <of, 0.005>,$ $<World, 0.2>, <Bank, 0.6>\}$	*Human*
What city boasts Penn's Landing, on the banks of the Delaware river?	Banks	$\{<What, 0.1>, <city, 0.05>, ...,$ $<landing, 0.005>, <banks, 0.6>,$ $<of, 0.005>, <river, 0.2>\}$	*Location*

with the target word, resulting those most relevant words get higher dependency weight. We believe these strong dependency words are the global context information of the target word, and we use them and their word embeddings to compute a contextual word embedding of the target word. Let's consider an example.

Example 1. Considering the first sentence *"Who is the head of the World Bank?"* in Table 1, we select *"Bank"* as the target word. According to self-attention mechanism, we obtain a weight vector $\{<Who, 0.1>, <is, 0.005>, ..., <head, 0.05>, <of, 0.005>, <World, 0.2>, <Bank, 0.6>\}$. The weight vector shows that such large weight words *"Bank"*, *"head"* and *"World"* have strong dependency with target word *"Bank"* in this sentence. We then obtain the contextual word embedding of *"Bank"* by weighted sum the word embedding of each word and their corresponding weight in the weight vector. Similarly, we can obtain the contextual word embedding of *"banks"* in the second sentence in Table 1. Though *"Bank"* and *"banks"* share the same one word embedding, they have discriminative contextual word embeddings.

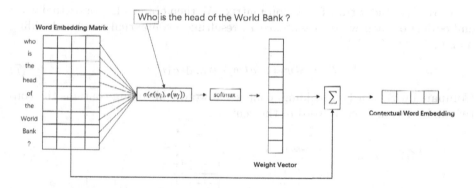

Fig. 1. Self-Attention. *"who"* is the target word at this time.

Figure 1 shows the overall architecture of self-attention mechanism. Given a sentence L with s words, $e(w_i)$ represent the word embedding of the ith word in the sentence, resulting in an embedding matrix $E \in R^{s \times d}$ for sentence L:

$$E = e(w_1) \oplus e(w_2) \oplus ... \oplus e(w_s) \tag{1}$$

Here d is the dimension of word embedding and $\oplus$ is the concatenation operator. Given a target word w_i with word embedding $e(w_i)$, we obtain contextual word embedding $c(w_i) \in R^{1 \times d}$ of target word w_i by self-attention mechanism:

$$c(w_i) = \sum_{j=1}^{s} \alpha_{ij} e(w_i) \tag{2}$$

The weight α_{ij} of each word embedding $e(w_i)$ is computed by:

$$\alpha_{ij} = \frac{exp(e_{ij})}{\sum_{k=1}^{s} exp(e_{ik})} \tag{3}$$

$$e_{ij} = S(e(w_i), e(w_j)) \tag{4}$$

where e_{ij} is the score between word embeddings $e(w_i)$ and $e(w_j)$, which represents the intensity of dependency between them. Here S is a score function to compute the intensity of dependency between two input word embeddings. There are several alternatives [11] of S, such as dot product operation or cosine similarity. We use two S functions (*dot* and *general*) in our experiments, *dot* function:

$$S(e(w_i), e(w_j)) = e(w_i)^{\top} e(w_j) \tag{5}$$

and *general* function:

$$S(e(w_i), e(w_j)) = e(w_i)^{\top} W_a e(w_j) \tag{6}$$

Here, W_a is a linear transformation matrix. We then obtain the contextual word embedding of each word in sentence L, resulting a contextual word embedding matrix $C \in R^{s \times d}$ of L:

$$C = c(w_1) \oplus c(w_2) \oplus ... \oplus c(w_s) \tag{7}$$

Compared with word embedding, contextual word embedding contains the context information of each word in the sentence.

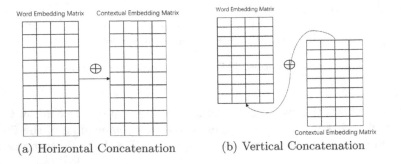

(a) Horizontal Concatenation   (b) Vertical Concatenation

Fig. 2. Two methods to combine contextual embedding with word embedding.

3.2   Combination of Word and Contextual Embedding

To make use of context information, we propose two methods to combine word embedding and contextual word embedding from two different perspectives. Figure 2 shows the details of our methods.

Horizontal Concatenation (HC). One word may have multiple senses in different contexts, and these senses should have discriminative representations. However, most previous works mainly use one single word embedding vector as word representation for one word. In this method, we use word embedding and contextual word embedding together to represent words, thus one word in different contexts have discriminative representations. we combine word embedding matrix and contextual embedding matrix by concatenating in horizontal orientation to obtain the final embedding matrix $X \in R^{s \times 2d}$:

$$X = [E; C] \tag{8}$$

Here, $[;]$ is the horizontal concatenation operator.

Vertical Concatenation (VC). Since horizontal concatenation method use word embedding and contextual word embedding together to represent words, this kind of word representation may suffer from more sparsity problem as compared to single word embedding representation. Moreover, as the dimension of word representation expanded, more parameters of filters need to be trained. In this method, we utilize contextual word embedding as expansion of sentences to make use of contextual information. We concatenate contextual word embedding matrix after word embedding matrix as. Then we obtain final embedding matrix $X \in R^{2s \times d}$:

$$X = E \oplus C \tag{9}$$

Here, $\oplus$ is a vertical concatenate operator.

3.3   Convolutional Neural Networks

We employ convolution layer to capture semantic features, and use max pooling operation to extract the most important feature. We then feed the obtained feature vector into output layer to perform classification. Figure 3 shows the details of CNN.

Convolution Layer. Given the final embedding matrix X as the input of convolutional layer, let $w_{f_i} \in R^{h \times d}$ (for horizontal concatenation method, $w_{f_i} \in R^{h \times 2d}$) be the parameters of the ith filter with window size h in convolutional layer. We apply a convolution operation between a filter and a sub-matrix of X to extract a higher level feature o_i:

$$o_i = f\left(w_{f_i} \cdot [X(w_i) : X(w_{i+h-1})] + b\right) \tag{10}$$

Here, $b \in R$ is a bias term, f is a non-linear function, $[X(w_i) : X(w_{i+h-1})] \in R^{h \times d}$ (for horizontal concatenation method, $[X(w_i) : X(w_{i+h-1})] \in R^{h \times 2d}$) is a sub-matrix of X with height h. This filter is applied to each possible sub-matrix

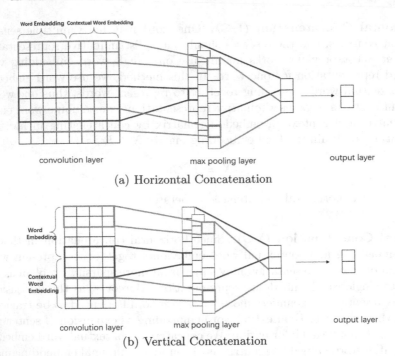

(a) Horizontal Concatenation

(b) Vertical Concatenation

Fig. 3. CNN for our two methods, red lines and blue lines represent different window size filter (red for 2 height, blue for 3 height). (Color figure online)

in X to generate a feature map vector $o \in R^{2s-h+1}$ ($o \in R^{s-h+1}$ for horizontal method):

$$o = \begin{cases} [o_1, o_2, ..., o_{2s-h+1}] & VC\ method \\ [o_1, o_2, ..., o_{s-h+1}] & HC\ method \end{cases} \tag{11}$$

Pooling Layer. A max-over-time pooling operation is applied to capture the most importance feature by extracting the largest number from a particular feature map vector. We apply multiple filters with different heights to capture different features. With max-over-time pooling operation, we obtain a fixed-length feature vector which contains the important features from each filters.

Output Layer. We feed the fixed-length feature vector into a fully connected softmax layer to perform classification. We employ dropout operation and l_2-norm penalty for regularization: (1) We randomly set elements of the fixed-length feature vector to zero. (2) We additionally constrain l_2-norm of the weight vectors of output layer. we rescale the weight vectors if their l_2-norm is larger than a preset value, which considered as a hyper parameter.

4   Experiments

4.1   Datasets

To demonstrate the effectiveness of our proposed method, we test our model on several commonly used datasets. Table 2 shows the summary statistics of the datasets.

Table 2. Summary statistics of datasets. Class: Number of classes. Length: Average sentence length. Train: Train set size. Test: Test set size (CV means there was no standard train/test split and thus 10-fold CV was used)

Datasets	Class	Length	Train	Test
TREC	6	10	5952	500
MR	2	20	10662	CV
Subj	2	23	10000	CV
AG	4	7	120000	7600
MPQA	2	3	10606	CV

Movie Reviews. This dataset[1] is a movie reviews dataset consisting of 10,662 movie-review snippets (a striking extract usually one sentence long), each snippet was labeled with its source reviews label positive or negative [16].

TREC. TREC question dataset[2] contains 6452 questions about 6 discriminative question types (abbreviation, description, entity, human, location, numeric value), each question was labeled only one type [10].

Subjectivity Dataset. Subjectivity dataset[3] - task involves classifying a sentence into subjective or objective. Subjective sentences consist of 5000 movie review snippets, and 5000 objective sentences were taken from plot summaries available from the Internet Movie Database [15].

AG News. AG[4] is a collection of more than 1 million news articles. News articles have been gathered from more than 2000 news sources. Following the setting of Wang et al. [25], we only use the title of the 4 largest classes from this corpus for classification, each class contains 30000 samples of train set and 1900 of test set.

---

[1] https://www.cs.cornell.edu/people/pabo/movie-review-data/.
[2] http://cogcomp.cs.illinois.edu/Data/QA/QC/.
[3] http://www.cs.cornell.edu/home/llee/data/search-subj.html.
[4] http://www.di.unipi.it/~gulli/AG_corpus_of_news_articles.html.

MPQA. This corpus[5] was collected and annotated as part of the summer 2002 NRRC Workshop on Multi-Perspective Question Answering (MPQA), which contains 10606 sentences of opinion. The task is to classify a sentences as positive or negative [27].

Table 3 shows the hyperparameters of our model, we set them empirically. We use the pre-trained word2vec vectors that were trained on 100 billion tokens from Google News to initialize the embedding layer and keep it fine-tune during the training. We use Adadelta [30] update rule to optimize the training process. For the score function in the self-attention mechanism, we use *dot* for TREC and MPQA, and use *general* for AG, MR, and Subj.

Table 3. Hyper parameters

Parameter	Values
Filter sizes	[3, 4, 5]
Filters num	[100, 100, 100]
Dropout rate	0.5
Learning rate	1.0
l2 norm penalty	3

4.2 Baselines

We compared our models with traditional methods, standard one-layer CNN, and some state-of-the-art models. We use accuracy as the metric for all datasets [12].

BoW+SVM. Bag of words SVM used in Pang and Lee [15,26], this method use support vector machine with unigram and bigram as features to perform classification. The weight of each feature is the term frequency. We select this method as a baseline of traditional method in text classification. We reprint the results from Wang et al. [25] and Wang and Manning [26].

CNN-Non-static. The standard one-layer CNN, which employs only one simple convolution layer with little hyperparameter tuning. we reprint the results of the non-static model proposed by Kim [7], which the pre-train word embeddings on Google News via word2vec toolkit were used to initialize the embedding layer and keep it fine-tuning during the training.

CharCNN. We also selected a convolutional neural network with only character level features as the input, which proposed by Zhang et al. [31]. CharCNN

---

[5] http://www.cs.pitt.edu/mpqa/.

is an effective method for text classification without the need of words. Abnormal character combinations such as misspellings and emoticons may be learned when using CharCNN. The results of our experiments are reprinted from Wang et al. [25].

KPCNN. A joint model called Knowledge Powered Convolutional Neural Network proposed by Wang et al. [25]. KPCNN use Probase to perform text conceptualization and employs two convolutional neural networks to extract features from word-concept and character-level input respectively. Finally, the combination of feature vectors is passed into the output layer for classification. We reprint the results for comparison of our methods.

CNN-Non-static+UNI. An one-layer CNN with a novel weight initialization method proposed by Li et al. [9]. It first selected useful semantic features (i.e. n-grams) by Naive Bayes (NB) weighting and then used this semantic features to initialize the weights of filters. Here we reprint the results of the initialization method with unigram features from Li et al. [9].

4.3   Results and Discussion

The results of all datasets are shown in Table 4.

Table 4. Experiment results

Model	TREC	MR	Subj	AG	MPQA
Bow+SVM	52.8	60.44	90.0	61.56	-
CNN-non-static	93.6	81.52	93.4	86.11	89.6
CharCNN	76	44.96	-	85.64	-
KPCNN	93.46	**83.77**	-	88.36	-
CNN-non-static+UNI	94.4	82.1	93.7	-	89.3
CNN-HE	94.8	82.12	93.8	**88.92**	**90.54**
CNN-VE	**95.0**	81.89	**93.91**	88.43	90.5

Compared with traditional models, convolutional neural networks approaches achieve better performance of text classification. It proves that CNN can capture richer semantic features than traditional models and may suffer from the data sparsity problem less.

Compared with standard one-layer CNN, our methods improve the accuracies on all datasets, which demonstrates the effectiveness of the self-attention mechanism for capturing global context information in text classification task.

Compared with some state-of-the-arts models, our methods give competitive results without requiring any extra knowledge or complex networks architectures, it can be easily applied to other NLP tasks. Moreover, A amount of computation can be parallelized in the self-attention mechanism, which means the additional computation is small.

In Table 4, experiments show that our two concatenate methods-i.e., horizontal concatenation and vertical concatenation do not always perform better than the other one. We believe that horizontal concatenation method uses word embedding and contextual embedding together to represent a word, which allows convolutional neural networks pay more attention to the relation and interaction between word and its context information. However, horizontal concatenation method suffers from more sparsity issue as compared to single word embedding method. Moreover, in the horizontal concatenation method, the dimension of word representation is twice as large as single word embedding, larger parameters are more difficult to train. In vertical concatenation method, though the number of parameters does not increase, the convolution layer captures word and context features respectively, which means it does not consider the interaction between them.

Table 5. Comparison experiment results of different score functions in the self-attention mechanism.

Model	TREC	MR	Subj	AG	MPQA
Self-attention (*dot*)	94.8	81.87	93.57	88.10	90.54
Self-attention *general*	94.8	82.12	93.8	88.92	90.52

Table 5 shows the comparison experiment results of two score function a according to Eqs. 5 and 6 (i.e. *dot* and *general*). Particularly, these comparison experiment base on our horizontal concatenation method. Overall, there is a small gap between two score functions except in AG news corpus. The following is our assumption: Compared to the rest four datasets, the size of AG news is much bigger. It makes sense that additional parameters may improve the performance when learning from big data.

5   Conclusion

In this paper, we proposed two methods based on convolutional neural networks with the self-attention mechanism for text classification. We improve the performance of CNN for text classification by considering the context information of texts. We evaluate our methods on five commonly used text classification datasets with some state-of-the-art models. The experiments results demonstrate the effectiveness of our methods. Furthermore, without any external knowledge or human-designed rules, our method also gives competitive results against more complex models. Our future work will focus on taking more attention mechanism into consideration and addressing the huge computation problem in long sentence or document classification.

Acknowledgement. This work was supported by the Fundamental Research Funds for the Central Universities, SCUT (No. 2017ZD048), the Tiptop Scientific and Technical Innovative Youth Talents of Guangdong special support program (No. 2015TQ01X633), the Science and Technology Planning Project of Guangdong Province (No. 2016A030310423, 2017B050506004), the Science and Technology Program of Guangzhou International Science & Technology Cooperation Program (No. 201704030076) and partially supported by a CUHK Direct Grant for Research (Project Code EE16963) and an internal grant from City University of Hong Kong (project no. 9610367).

References

1. Aggarwal, C.C., Zhai, C.: A survey of text classification algorithms. In: Aggarwal, C., Zhai, C. (eds.) Mining Text Data, pp. 163–222. Springer, Boston (2012). https://doi.org/10.1007/978-1-4614-3223-4_6
2. Bahdanau, D., Cho, K., Bengio, Y.: Neural machine translation by jointly learning to align and translate. arXiv preprint arXiv:1409.0473 (2014)
3. Cheng, J., Dong, L., Lapata, M.: Long short-term memory-networks for machine reading. arXiv preprint arXiv:1601.06733 (2016)
4. Collobert, R., Weston, J., Bottou, L., Karlen, M., Kavukcuoglu, K., Kuksa, P.: Natural language processing (almost) from Scratch. J. Mach. Learn. Res. **12**, 2493–2537 (2011)
5. Hu, Z., Ma, X., Liu, Z., Hovy, E., Xing, E.: Harnessing deep neural networks with logic rules. arXiv preprint arXiv:1603.06318 (2016)
6. Irsoy, O., Cardie, C.: Deep recursive neural networks for compositionality in language. In: Advances in Neural Information Processing Systems, pp. 2096–2104 (2014)
7. Kim, Y.: Convolutional neural networks for sentence classification. arXiv preprint arXiv:1408.5882 (2014)
8. Lai, S., Xu, L., Liu, K., Zhao, J.: Recurrent convolutional neural networks for text classification. In: AAAI, vol. 333, pp. 2267–2273 (2015)
9. Li, S., Zhao, Z., Liu, T., Hu, R., Du, X.: Initializing convolutional filters with semantic features for text classification. In: Proceedings of the 2017 Conference on Empirical Methods in Natural Language Processing, pp. 1884–1889 (2017)
10. Li, X., Roth, D.: Learning question classifiers. In: Proceedings of the 19th International Conference on Computational Linguistics-Volume 2, pp. 1–7. Association for Computational Linguistics (2002)
11. Luong, M.T., Pham, H., Manning, C.D.: Effective approaches to attention-based neural machine translation. arXiv preprint arXiv:1508.04025 (2015)
12. Manning, C.D., Schütze, H.: Foundations of Statistical Natural Language Processing. MIT Press, Cambridge (1999)
13. Mikolov, T., Sutskever, I., Chen, K., Corrado, G.S., Dean, J.: Distributed representations of words and phrases and their compositionality. In: Advances in Neural Information Processing Systems, pp. 3111–3119 (2013)
14. Mnih, A., Hinton, G.E.: A scalable hierarchical distributed language model. In: Advances in Neural Information Processing Systems, pp. 1081–1088 (2009)
15. Pang, B., Lee, L.: A sentimental education: Sentiment analysis using subjectivity summarization based on minimum cuts. In: Proceedings of the 42nd annual meeting on Association for Computational Linguistics, p. 271. Association for Computational Linguistics (2004)

16. Pang, B., Lee, L.: Seeing stars: Exploiting class relationships for sentiment categorization with respect to rating scales. In: Proceedings of the 43rd Annual Meeting on Association for Computational Linguistics, pp. 115–124. Association for Computational Linguistics (2005)

17. Parikh, A.P., Täckström, O., Das, D., Uszkoreit, J.: A decomposable attention model for natural language inference. arXiv preprint arXiv:1606.01933 (2016)

18. Post, M., Bergsma, S.: Explicit and implicit syntactic features for text classification. In: Proceedings of the 51st Annual Meeting of the Association for Computational Linguistics (Volume 2: Short Papers), vol. 2, pp. 866–872 (2013)

19. Rumelhart, D.E., Hinton, G.E., Williams, R.J.: Learning representations by back-propagating errors. Nature **323**(6088), 533 (1986)

20. Socher, R., Bauer, J., Manning, C.D., et al.: Parsing with compositional vector grammars. In: Proceedings of the 51st Annual Meeting of the Association for Computational Linguistics (Volume 1: Long Papers), vol. 1, pp. 455–465 (2013)

21. Socher, R., Lin, C.C., Manning, C., Ng, A.Y.: Parsing natural scenes and natural language with recursive neural networks. In: Proceedings of the 28th International Conference on Machine Learning (ICML 2011), pp. 129–136 (2011)

22. Socher, R., et al.: Recursive deep models for semantic compositionality over a sentiment treebank. In: Proceedings of the 2013 Conference on Empirical Methods in Natural Language Processing, pp. 1631–1642 (2013)

23. Turian, J., Ratinov, L., Bengio, Y.: Word representations: a simple and general method for semi-supervised learning. In: Proceedings of the 48th Annual Meeting of the Association for Computational Linguistics, pp. 384–394. Association for Computational Linguistics (2010)

24. Vaswani, A., et al.: Attention is all you need. In: Advances in Neural Information Processing Systems, pp. 6000–6010 (2017)

25. Wang, J., Wang, Z., Zhang, D., Yan, J.: Combining knowledge with deep convolutional neural networks for short text classification. In: Proceedings of the 26th International Joint Conference on Artificial Intelligence, pp. 2915–2921. AAAI Press (2017)

26. Wang, S., Manning, C.D.: Baselines and bigrams: simple, good sentiment and topic classification. In: Proceedings of the 50th Annual Meeting of the Association for Computational Linguistics: Short Papers-Volume 2, pp. 90–94. Association for Computational Linguistics (2012)

27. Wiebe, J., Wilson, T., Cardie, C.: Annotating expressions of opinions and emotions in language. Lang. Resour. Eval. **39**(2–3), 165–210 (2005)

28. Yang, Z., Yang, D., Dyer, C., He, X., Smola, A., Hovy, E.: Hierarchical attention networks for document classification. In: Proceedings of the 2016 Conference of the North American Chapter of the Association for Computational Linguistics: Human Language Technologies, pp. 1480–1489 (2016)

29. Yin, W., Schütze, H.: Multichannel variable-size convolution for sentence classification. arXiv preprint arXiv:1603.04513 (2016)

30. Zeiler, M.D.: ADADELTA: an adaptive learning rate method. arXiv preprint arXiv:1212.5701 (2012)

31. Zhang, X., Zhao, J., LeCun, Y.: Character-level convolutional networks for text classification. In: Advances in Neural Information Processing Systems, pp. 649–657 (2015)

32. Zhang, Y., Roller, S., Wallace, B.: MGNC-CNN: a simple approach to exploiting multiple word embeddings for sentence classification. arXiv preprint arXiv:1603.00968 (2016)
33. Zhou, P., Shi, W., Tian, J., Qi, Z., Li, B., Hao, H., Xu, B.: Attention-based bidirectional long short-term memory networks for relation classification. In: Proceedings of the 54th Annual Meeting of the Association for Computational Linguistics (Volume 2: Short Papers), vol. 2, pp. 207–212 (2016)

Cpriori: An Index-Based Framework to Extract the Generalized Center Strings

Shuhan Zhang[1,2]([✉]) [iD], Shengluan Hou[1,2] [iD], and Chaoqun Fei[1,2] [iD]

[1] Key Laboratory of Intelligent Information Processing of Chinese Academy of Sciences, Institute of Computing Technology, Chinese Academy of Sciences, Beijing, China
zhangshuhan@ict.ac.cn
[2] University of Chinese Academy of Sciences, Beijing, China

Abstract. The common approximate substring (CAS) problem is to extract CAS in all sequences of a large sequence set. The restriction of requesting the exact match results in losing a large amount of useful information in sequential pattern mining. Instead of extracting the exact substrings, it is more significant to extract the generalized center string (GCS). The GCS is the string that can produce all other exact substrings through limited mutation. The GCS problem can be used for accurate reasoning on mutation in real-world applications (e.g. biological sequence analysis). However, this task is very challenging due to the exponentially increasing complexity after loosening the constraints. In this paper, we propose an index-based framework to solve the GCS problem using the divide-and-conquer strategy. Particularly, we propose an efficient algorithm, named CValidating, that converts the problem of pattern extracting to the problem of query processing. Moreover, a heuristic filtering strategy is devised to reduce the search space. Experimental results show that our algorithm outperforms the existing algorithms.

Keywords: Sequential pattern mining
Common approximate substring · Generalized center string · Index

1 Introduction

Sequential pattern mining (SPM) is a central task in data mining [1], which has a large number of applications, including bioinformatics [23], market basket analysis [25], log analysis [28] and webpage click analysis [8]. The common approximate substring (CAS) problem is one of the SPM problems, which aims to extract center strings appearing in all sequences [6,17,24]. Due to the commonly existing noises in real bioinformatics data, Lu et al. proposed the generalized center string (GCS) problem [19]. The GCS problem relaxes the constraint on the CAS problem that at least q sequences occuring in all sequences have a CAS.

Supported by National Key Research and Development Program of China.

H. Hacid et al. (Eds.): WISE 2018, LNCS 11233, pp. 468–482, 2018.
https://doi.org/10.1007/978-3-030-02922-7_32

The relaxed constraint is more consistent with the real applications, but brings more severe combinatorial explosion problem.

Inspired by the Apriori algorithm [2] and the FP-growth algorithm [11,12], Lu et al. proposed three algorithms to solve the GCS problem [19]. To achieve the completeness of the results, they enumerate strings in a brute force way based on downward closure property, with two strategies: query expansion and table expansion. The query expansion method follows the generate-and-test strategy to enumerate all possible patterns [18]. This method is time-consuming, due to the enumerating paradigm, especially in testing whether a candidate string is a center string or not. The table expansion method costs massive memory to store all possible patterns in advance [22], and therefore it has high space complexity. However, these approaches inevitably have the following drawbacks: (1) Time-consuming, due to enumerating paradigm. (2) High space complexity, due to the enormous search space. Shang et al. proposed an indexing technique to deal with frequent substrings in a single sequence, which reached a rational time complexity [23].

Motivated by [23], in this paper, we propose a novel framework, named Cpriori to solve the GCS problem. The Cpriori framework adopts an indexing technique by using the divide-and-conquer strategy. To efficiently extract the GCS, we propose an algorithm, named CValidating, that converts the pattern extracting problem to a querying problem. Moreover, we devise a heuristic filtering strategy to reduce the search space when generating candidates. The main contributions of Cpriori are:

- We propose an index-based algorithm to extract the GCS, that increases the efficiency in the searching step.
- We develop a heuristic filtering strategy that reduces the search space significantly.
- We conduct comprehensive experiments using real datasets, demonstrating the performance of the proposed algorithm.

The rest of this paper is organized as follows. The next section discusses the related works. Section 3 presents preliminaries and illustrates the GCS problem. Section 4 introduces the methodology of Cpriori framework, including an index-based algorithm Indexing, CValidating algorithm and a heuristic filtering algorithm. Section 5 presents the experimental analysis and Sect. 6 concludes the paper.

2 Related Work

SPM has a long history [1–3,25]. Numerous methods are proposed to solve the SPM problem. For instance, Wang et al. proposed the BIDE+ algorithm to mine frequent closed sequences, which are the sequences that contain no super-sequences with the same support [26]. Garcia-Hernandez et al. proposed an approximate algorithm DIMASP for maximal sequential pattern mining [9]. However, these two constraints are incompatible with our goal.

The SPM has many extensions that focus on various patterns [3, 5, 7, 10, 13, 16, 21]. Fuzzy sequential pattern mining considers items in sequences which have multiple values. Weighted sequential pattern mining aims to find sequential patterns that have the minimum weight. High-utility sequential pattern mining considers the quality of items rather than support [29]. Uncertain sequential pattern mining considers data uncertainty in database, such as noise, inaccuracy, and data missing. The key idea behind these above works is that the quantity of patterns is more important than the quality of patterns.

Extracting center strings belongs to the SPM [4], which aims to discover frequent subsequences in a single or multiple sequences. This problem is NP complete, recently, many efficient algorithms [14, 20, 22, 23] are proposed. Hufsky et al. [14] developed an advanced preprocessing method to filter out unsolvable instances efficiently. They introduce data reduction techniques by inferring redundant instances, and describe an iterative search strategy to speed up search tree algorithms. However, this method focuses on the binary strings and still enumerates strings naively. Sahli et al. presented ACME [22], a parallelized method for extracting common motifs from a gigabyte-long single sequence. ACME follows the table expansion way which is adopted by Bpriori2 [19]. Considering the parallel and distribute computation, Miliaraki et al. proposed an algorithm called MG-FSM [20], which is used for distributed frequent sequence mining in MapReduce. They extend the item-based partitioning to gap constrained frequent sequence mining. However, they do not take the mismatching situation into consideration. In the purpose of mining maximal consecutive pattern, Shang et al. presented an algorithm called MACFP [23]. They take advantage of indexing all strings by Suffix Array [15], which inspired us to adopt this technique. However, MACFP is used to detect frequent sub-patterns appearing in single DNA sequences. Our algorithm, on the contrary, is used to detect the frequent sub-patterns in the individuals of a community. GCS can be considered as a generalization of MACFP. However, some techniques used by MACFP such as divide-and-conquer may be useful for solving the GCS problem.

3   Preliminaries

Frequently used notations are listed in Table 1.

Definition 1. 1-mutated copy. *Let $a = a_1a_2...a_n$ and $b = b_1b_2...b_n$ be two strings, $a, b \in \Sigma^n$. The Hamming distance $d_H(a,b)$ between a and b is defined as:*

$$d_H(a, b) = \sum_{i=1}^{n} \varepsilon(a_i, b_i), where \ \varepsilon(x, y) = \begin{cases} 0, & x \neq y \\ 1, & otherwise \end{cases}$$

If $d_H(a,b) = 1$, we call a is a 1-mutated copy of b, and we also call b is a 1-mutated copy of a.

Definition 2. Common Approximate Substring (CAS) problem [24]. *Given $S = \{s_1, s_2, ..., s_N\}$, such that $|s_i| \leq L$, $1 \leq i \leq N$ and a positive integer*

Table 1. Frequently used notations

Notation	Description
Σ	Alphabet
Σ^+	The set of non-empty strings over Σ
Σ^l	The set of all strings $\subseteq \Sigma^+$ with length l
S	A set of input sequences over Σ
$\|a\|$	a can be a sequence or a set, $\|a\|$ means the length of sequence, the size of set respectively
d	Hamming distance threshold, $d \geq 0$
q	The support threshold, $q \geq 0$
$s_i \subseteq s_j$	s_i is a substring of s_j
$d_H(s_i, s_j)$	Hamming distance between s_i and s_j

l such that $1 \leq l \leq N$ and $l > d$. Find a string $g \in \Sigma^l$ such that for any $s_i \in S$, there exists a substring g' of s_i, which satisfy $|g'| = |g|$ and $d_H(g,g') \leq d$. g is often called a **center string**.

Definition 3. Mutated Instance String (MIS). *Let g be a sequence with $|g| \geq d$, we call g' is a mutated instance string(MIS) of g if $|g'| = |g|$ and $d_H(g,g') \leq d$. g' is also called a consensus string, if g is a center string.*

Definition 4. Generalized Center String (GCS) problem [19]. *Given $S = \{s_1, s_2, \ldots, s_N\}$, such that $|s_i| = L$, $1 \leq i \leq N$, q is a positive integer such that $0 < q \leq N$. For any l in range $(d, L]$ and any $g \in \Sigma^l$,if s_i has a substring g', which is a MIS of g, we say g and s_i are **related**, denoted as $g \sim s_i$. What's more, if $|\{i | g \sim s_i\}| \geq q$, we call g is a GCS of S. GCS has downward closure property which leads to two statements in Sect. 4.*

Property 1. **Downward closure property.** If g is a center string of S, then each substring of g is also a center string, but not vice versa.

Example 1. Given $S = \{0101101, 1101011, 1011100, 0110110\}$, which means $N = 4, L = 7, |\Sigma| = 2$. We set $q = 4, d = 0$, all center strings are: $\{0, 1, 01, 10, 11, 011, 101, 110, 1011\}$

4   Methodology

In this section, we propose a novel and efficient framework, named Cpriori, which is inspired by [18,23,27]. The Cpriori framework consists of the following algorithms, the Indexing, the CValidating and the Filtering. The Indexing algorithm works in a divide-and-conquer way to search efficiently. The Heuristic Filtering algorithm can significantly reduce the search space.

4.1   Cpriori Framework

Definition 5. Consensus tree [19]. *Consensus tree is a TRIE-like data structure to store all substrings in input sequences. Each path represents a consensus string $b_1 \ldots b_i$. Each node of the consensus tree is denoted by $S_{b_1 \ldots b_i}$, which is a set containing all positions of $b_1 \ldots b_i$, each position (j, k) means the consensus string starts at the k-th position of the j-th sequence.*

Note that the boolean variable $R_{b_1 \ldots b_i}$ records the $b_1 \ldots b_i$ that have contributed to at least one center string. Consensus strings with $R_{b_1 \ldots b_i} = 0$ are removed. An example of consensus tree is shown in Fig. 1.

Algorithm 1. Cpriori Framework

---

1 **for** $j = 1$ *to* N **do**
2   **for** $k=1$ *to* L **do**
3   **if** *the k-th element of the j-th sequence is $b_1 \in \Sigma$* **then**
4   put (j, k) into S_{b_1}, put b_1 into CES_1, put j into T_{b_1}, set $R_{b_1} = 1$
             and $CAS_2 = \Sigma \times \Sigma$

5 $i = 1$
6 **while** $i < L$ **do**
7   **foreach** $R_{b_1 \ldots b_i} = 1$ **do**
8   **foreach** (j, k) *in* $S_{b_1 \ldots b_i}$ **do**
9   **if** $k < L - i + 1$ *and the $(k + i)$-th element of the j-th sequence is*
             $b_{i+1} \in \Sigma$ **then**
10   put (j, k) into $S_{b_1 \ldots b_{i+1}}$ and j in $T_{b_1 \ldots b_{i+1}}$
11   Indexing()
12   remove all $S_{b_1 \ldots b_i}$ and $T_{b_1 \ldots b_i}$
13   **if** $CAS_{i+2} = \emptyset$ **then**
14   exit
15      $i = i + 1$

By using the consensus tree and level-wise search strategy, Cpriori converts the input sequences into a consensus tree. Let $T_{b_1 \ldots b_i}$ denote the set of sequences number which contain at least one MIS of $b_1 \ldots b_i$. Let CES_l denote the set of all center strings with the length of l. Let CAS_l denote the set of all candidate strings with the length of l. Note that in Line 4 (Algorithm 1), CAS is initialized based on *property 1* of GCS, the operator "×" denote Cartesian product of two sets. First, Cpriori scans the sequences sequentially and initializes S_{b_1}, CES_1, T_{b_1}, R_{b_1} (Line 1–4). Then, it extends nodes in the consensus tree and constructs the position sets further (Line 7–10). If the CAS which returned by Filtering algorithm (Sect. 4.4) is null, then exit this procedure (Line 13–14).

Example 2. When $l = 1$, substrings $0, 1$ appear in all sequences, their positions such as $(1, 1)$ $(1, 2)$ are inserted into nodes 0 resp. 1.

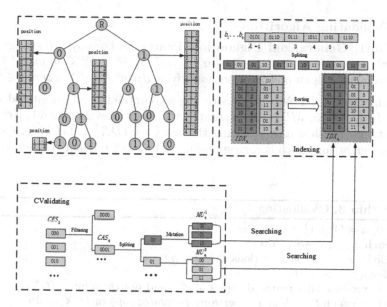

Fig. 1. An Example of Cpriori with $d = 2, \Sigma = \{0,1\}$ and $S = \{0101101, 1101011, 1011100, 0110110\}$. Let $q = 4$.

4.2 An Index-Based Algorithm

Divide-and-conquer is an efficient paradigm for SPM under Hamming distance and Edit distance [18,23]. Inspired by above works, we build indices of consensus string in a divide-and-conquer way, as shown in Algorithm 2.

Let λ denote the indexed variable of consensus string. Let IDX_{i+1} denote the index set of $b_1 \ldots b_{i+1}$, i.e. $IDX_{i+1} = \{IDX_{i+1}^h, 1 \le h \le m\}$. For each consensus string, it is divided into m partitions, then put them as several key-value pairs $< [b_1 \ldots b_{i+1}]^h, \lambda >$ into IDX_{i+1}^h (Line 5). The function "$sort(IDX_{i+1}^h)$" in Line 7 means that the elements of IDX_{i+1}^h are sorted according to $[b_1 \ldots b_{i+1}]^h$'s lexicographical order.

Algorithm 2. Indexing

1 $m = \lfloor \frac{d}{2} + 1 \rfloor, IDX_{i+1} = \emptyset,$
2 **for** $h=1$ to m **do**
3 $\quad \lambda = 1$
4 $\quad$ **foreach** $S_{b_1 \ldots b_{i+1}} \ne \emptyset$ **do**
5 $\quad\quad$ build IDX_{i+1}^h
6 $\quad\quad \lambda = \lambda + 1$
7 $\quad$ sort(IDX_{i+1}^h)
8 $\quad$ put IDX_{i+1}^h into IDX_{i+1}
9 CValidating(IDX_{i+1})

4.3   CValidating Algorithm

Theorem 1. Generalized Pigeonhole Principle. For any two strings $a, b \in S$ such that $d_H(a,b) \leq d$, $a^1 \ldots a^p \ldots a^m \subseteq a$, $b^1 \ldots b^p \ldots b^m \subseteq b$, $m = \lfloor \frac{d}{2} + 1 \rfloor$, there must exist at least one position p satisfying $d_H(a^p, b^p) \leq 1$, where $p \in [1, m]$.

In Algorithm 3, let MU_{i+1}^h denote the set containing all 1-mutated copies of c_{i+1}^h. In Line 4–5, MU_{i+1}^h is generated which satisfies Theorem 1. Next, the CValidating algorithm starts probing the index set IDX_{i+1}^h. Let $b_1' \ldots b_l'$ denote the consensus string which is corresponding with λ. Let P denote the set contains $< \lambda, d_H(\cdot) >$.

Algorithm 3. CValidating

---

1 $P = \emptyset, m = \lfloor \frac{d}{2} + 1 \rfloor$
2 **foreach** $c_{i+1} \in CAS_{i+1}$ **do**
3   divide c_{i+1} into m partitions $c_{i+1}^h, 1 \leq h \leq m$
4   **for** $h = 1$ *to* m **do**
5   produce all 1-mutated copies of c_{i+1}^h and put them into MU_{i+1}^h
6   **foreach** $b_1' \ldots b_{i+1}'$ *by searching 1-mutated copy in* IDX_{i+1} **do**
7   **if** $d_H(c_{i+1}, b_1' \ldots b_{i+1}') \leq d$ **then**
8   insert $< \lambda, d_H(\cdot) >$ into P
9   Filtering(P)

Example 3. Suppose a candidate string "0001" is splitted into "00" and "01", CValidating produces all 1-mutated copies "00", "01", "10" and "00", "01", "11" via enumerating "00" and "01" respectively. Then, it traverses IDX_4^1, IDX_4^2 respectively and finds several consensus strings $b_1' \ldots b_{i+1}'$ according to the 1-mutated copies. At last, it computes the Hamming distance between $b_1' \ldots b_{i+1}'$ and "0001", and check whether $d_H(\text{"0001"}, b_1' \ldots b_{i+1}') \leq 2$ or not. If true, save the indexed variable of $b_1' \ldots b_{i+1}'$, i.e. $\lambda = 1$ as $d_H(\text{"0001"}, \text{"1"})$ into P.

Average-case Time Complexity Analysis. Let us make an assumption that all consensus strings are generated randomly. The size of a partition of consensus string is $\lfloor \frac{i}{m} \rfloor$, there are $|\Sigma|^{\lfloor \frac{i}{m} \rfloor}$ distinct partitions in the i-th level. In the worst case, $N \times (L - i + 1)$ nodes are produced in the i-th level, therefore the expectation of occurrences of a partition is $\frac{N \times (L-i+1)}{|\Sigma|^{\lfloor \frac{i}{m} \rfloor}}$. As shown in Algorithm 3, it finally makes the expected time complexity of validating algorithm in $O(\frac{NL}{|\Sigma|^{\lfloor \frac{i}{m} \rfloor}})$, which is lower than the expected time complexity of validating step in Bpriori3, which is $O(NL)$. The improvement in performance is shown in Fig. 2.

Space Complexity Analysis. Since we build indices with linear memory, it costs $O(mNL)$. However, we need $O(\sum_{j=0}^{min\{d,i\}} \binom{i}{j}(|\Sigma| - 1)^j \times \frac{N \times (L-i+1)}{q})$ to store the center strings in each level. Therefore, the total space complexity of Cpriori is $O(NL \times \frac{i^d}{q \times d!} \times |\Sigma|^d)$.

4.4   A Heuristic Filtering Strategy

Algorithm 4 tests whether each of $(n-1)$-candidate strings is a center string, if true, extend it further.

Let $\mathcal{I}$ denote the number set of sequences which contain at least one $b_1 \ldots b_{i+1}$, $\mathcal{N}$ denotes the number set of sequences which contain at least one $b'_1 \ldots b'_{i+1}$ which satisfied $d_H(\cdot) < d$, $\mathcal{N} \subseteq \mathcal{I}$. Let T'_δ denote a number set which contains at least one δ in $(b'_1 \ldots b'_{i+1} \circ \delta)$. For each element in P, we operate these three sets above (Line 3–8). If the number of sequences which contain the $(d-1)$-mutated copies have met the q requirement, then generate the candidate strings and put them into CAS_{i+2}. Otherwise, if the number of sequences which contain the mutated copies which satisfy $d_H(\cdot) = d$ meet the requirement of $q - |\mathcal{N}|$, then generate the candidate strings and put them into CAS_{i+2} (Line 15–18). Note that the operator "$\circ$" denote string concatenation, for example, 'a' $\circ$ 'b' = 'ab'.

Example 4. Suppose there exists a GCS "000", a MIS "110", $d_H($"000", "110"$)=$ $2 = d$. Check the number of different sequences in T'_0 which contain "1100" or T'_1 which contain "1101" respectively. Since $|T'_1|$ is $4 > (q - |\mathcal{N}|)$, "0001" can be generated to a candidate.

Algorithm 4. Filtering

---

1 $\mathcal{I} = \emptyset, \mathcal{N} = \emptyset, T'_\delta = \emptyset$
2 **foreach** $< \lambda, d_H(\cdot) >\in P$ **do**
3 $\quad \mathcal{I} = \mathcal{I} \bigcup T_{b'_1 \ldots b'_{i+1}}$
4 $\quad$ **if** $d_H(\cdot) < d$ **then**
5 $\quad\quad$ put $T_{b'_1 \ldots b'_{i+1}}$ into $\mathcal{N}$
6 $\quad$ **else**
7 $\quad\quad$ **foreach** $\delta \in \Sigma$ **do**
8 $\quad\quad\quad$ put the sequences' number which contain at least one δ in $(b'_1 \ldots b'_{i+1} \circ \delta)$ into T'_δ

9 **if** $|\mathcal{I}| \geq q$ **then**
10 $\quad$ output the key-value pair $< c_{i+1}, all\ S_{b'_1 \ldots b'_{i+1}} >$ and sct all $R_{b'_1 \ldots b'_{i+1}} = 1$
11 $\quad$ **if** $|\mathcal{N}| \geq q$ **then**
12 $\quad\quad$ **foreach** $\delta \in \Sigma$ **do**
13 $\quad\quad\quad$ put $(c_{i+1} \circ \delta)$ into CAS_{i+2}
14 $\quad$ **else**
15 $\quad\quad$ **foreach** $T'_\delta \neq \emptyset$ **do**
16 $\quad\quad\quad$ **if** $|T'_\delta| \geq q - |\mathcal{N}|$ **then**
17 $\quad\quad\quad\quad$ put $(c_{i+1} \circ \delta)$ into CAS_{i+2}

18 **else**
19 $\quad$ exit

---

5   Experiment

We implemented Cpriori in C++ to evaluate the performance of our proposed approach, and explored the speedup techniques used in our algorithm under different parameter settings.

Datasets. DNA sequence is one of application scenarios of our method. To evaluate the performance of our approach, we used two datasets with different DNA string length from Eukaryotic Promoter Database (EPD)[1] and Human Genome DNA Sequence chr6 in NCBI[2]. The alphabet set for DNA sequence is $\Sigma = \{A, C, G, T\}$.

- Small Dataset, $N = $ **30K**, $L = $ **300**, culled from 300–nucleotide human promoter sequences in EPD;
- Large Dataset, $N = $ **1M**, $L = $ **100**, culled from chr6.

As analyzed before, Cpriori outperforms other algorithms under different parameter settings, due to our Indexing techniques and Heuristic Filtering strategy. We measured the average running time of each algorithm, averaged over 5 runs of the same workload of each experiment. Cpriori supports extracting all motifs of any length, and the positions of all their mutations in input sequences. However, considering the applications of DNA, we measured the algorithm of extracting maximal GCS.

Experiment Setting. All experiments were performed on a local machine with Intel(R) Core(TM) i7-6700 CPU @3.40 GHZ and 32 GB memory. To evaluate the algorithm efficiency, we used a single thread during experiments.

We compared five algorithms as introduced below.

- **Bpriori2** extracts all GCSs and their MISs of input sequences S at any length $\leq l$ for all possible mismatches $\leq d$ via table expansion paradigm. Its worst case time complexity is $O(lNL \times v(d, l))$, where $v(d, l)$ is the number of MISs of a GCS of length l with a Hamming distance at most d [19].
- **Cpriori-without-indexing-filtering** turns off Indexing technique and Heuristic Filtering strategy (our baseline algorithm). The way of generating candidates follows the expansion way which is introduced in Sect. 4.1.
- **Cpriori-without-indexing** applies Heuristic Filtering strategy to accelerate the baseline algorithm.
- **Cpriori-without-filtering** applies Indexing technique to accelerate the baseline algorithm.
- **Cpriori** applies Indexing technique and Heuristic Filtering strategy in our Cpriori algorithm.

---

[1] http://epd.vital-it.ch/index.php.
[2] https://www.ncbi.nlm.nih.gov.

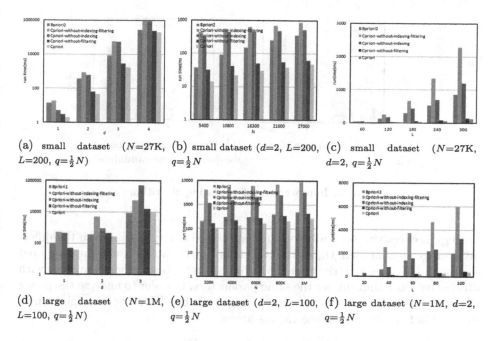

(a) small dataset (N=27K, (b) small dataset (d=2, L=200, (c) small dataset (N=27K, L=200, $q=\frac{1}{2}N$) $q=\frac{1}{2}N$ d=2, $q=\frac{1}{2}N$

(d) large dataset (N=1M, (e) large dataset (d=2, L=100, (f) large dataset (N=1M, d=2, L=100, $q=\frac{1}{2}N$) $q=\frac{1}{2}N$ $q=\frac{1}{2}N$

Fig. 2. Comparison with alternative algorithms on small dataset and large dataset

5.1   Comparison with Alternative Algorithms

Figure 2 depicts the running time for all algorithms with varying d, N, and L on small dataset and large dataset, respectively. We observe that our Cpriori algorithm outperforms existing algorithms on these two datasets for the following techniques: Indexing technique and Filtering strategy.

For **Indexing technique**, we compared Cpriori with Cpriori-without-indexing algorithm, it makes Cpriori algorithm approximately achieve 10 times improvement over the baseline algorithm on both datasets. It is because that Indexing technique omits some irrelevant consensus strings in CValidating algorithm (in step 6 of Algorithm 1). In addition, we compare our Indexing technique used in Cpriori with the Fast Chunk Indexing technique used in [23] and show the performance in Fig. 3. We observe that our Indexing technique substantially reduces the runtime of CValidating algorithm over Fast Chunk Indexing technique. In particular, for each single candidate, the speed of our Indexing technique is approximately two times over the Fast Chunk Indexing technique with fixed parameters N, L and q. These two Indexing techniques are all based on Pigeonhole principle, the difference between them is the number of partitions. In Fast Chunking Indexing, it is $\lfloor d+1 \rfloor$, in Cpriori, it is $\lfloor \frac{d}{2}+1 \rfloor$. The results show that the number of partitions $\lfloor \frac{d}{2}+1 \rfloor$ is more suitable for extracting GCS from a large sequence set.

For **Filtering strategy**, we compared Cpriori algorithm with Cpriori-without-filtering algorithm, the advantage shows that our Heuristic Filtering

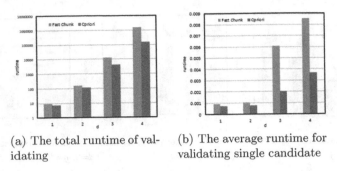

(a) The total runtime of val-   (b) The average runtime for
idating   validating single candidate

Fig. 3. Improvement of validating algorithm

strategy is effective, especially for long patterns (i.e. larger L). Given the different average sequence length, and data distribution of two datasets, the compared
results also demonstrate that our algorithm is most robust, when compared with
alternative algorithms under the same parameters. In order to analyze the power
of our Heuristic Filtering strategy, we compared three filtering strategies which
are used in five alternative algorithms above.

- **No filtering** adopted by Bpriori2 algorithm [19].
- **Enumerate filtering** generate n-candidates by expanding the $(n-1)$-GCSs
  with a character in Σ. This strategy is adopted by baseline algorithm.
- **Heuristic filtering** our filtering strategy, as described in Sect. 4.4.

We measure the filtering strategies by *candidate ratio*, defined as $\frac{|CAS|}{|GCS|}$ where
$|CAS|$ is the number of candidates and $|GCS|$ is the number of GCSs, the
ratio and filtering power are inversely proportional. The results are shown in
Table 2, which indicates that Heuristic Filtering strategy outperforms other filtering strategies for varying d, N, L on these two datasets. This strategy can
reduce 2.5 times search space than the other filtering strategies. For instance,
on a large dataset, the candidate ratios of Heuristic Filtering is still acceptable
(nearly 1.6), while the candidate ratios of all other filtering strategies are larger
than Heuristic Filtering (exactly 4). We observe that the candidate ratios of the
No filtering and Enumerate filtering strategies are exactly $|\Sigma|$, since they follow
the way of enumerating the alphabet to generate the candidates.

5.2 Scalability

To test the scalability of our algorithm, we randomly sampled 20%, 40%, 60%
and 80% of two datasets. In this experiment, we measured the *relative time ratio*
which reflects the performance of each algorithm on a sampled dataset over that
on the whole dataset. From Fig. 4(a) and (c), we observe that the relative time
ratios of all algorithms scale approximately linearly with the increasing of N.

Table 2. Candidate ratios

Datasets	d	No filtering ($\frac{|CAS|}{|GCS|}$)	Enumerate filtering ($\frac{|CAS|}{|GCS|}$)	Heuristic filtering ($\frac{|CAS|}{|GCS|}$)
Small dataset	1	4	4	1.6
	2	4	4	1.69
	3	4	4	1.7
	4	4	4	1.65
Large dataset	1	4	4	1.83
	2	4	4	1.66
	3	4	4	1.7
	4	4	4	1.8
Datasets	N	No filtering ($\frac{|CAS|}{|GCS|}$)	Enumerate filtering ($\frac{|CAS|}{|GCS|}$)	Heuristic filtering ($\frac{|CAS|}{|GCS|}$)
Small dataset	5.4K	4	4	1.6
	10.8K	4	4	1.61
	16.2K	4	4	1.6
	21.6K	4	4	1.6
	27K	4	4	1.62
Large dataset	200K	4	4	1.61
	400K	4	4	1.64
	600K	4	4	1.65
	800K	4	4	1.65
	1M	4	4	1.66
Datasets	L	No filtering ($\frac{|CAS|}{|GCS|}$)	Enumerate filtering ($\frac{|CAS|}{|GCS|}$)	Heuristic filtering ($\frac{|CAS|}{|GCS|}$)
Small dataset	60	4	4	1.6
	120	4	4	1.3
	180	4	4	1.4
	240	4	4	1.3
	300	4	4	1.5
Large dataset	20	4	4	1.9
	40	4	4	1.8
	60	4	4	1.45
	80	4	4	1.5
	100	4	4	1.6

The most time-consuming step in these five algorithms is the computation of Hamming distance between two strings, it depends on two factors:

(1) The number of GCSs ($|GCS|$).
(2) The data size of input sequences, including the number of sequences (N) and the length of sequences (L).

We found that the change of $|GCS|$ is not obvious when varying N. Therefore, the runtime is consistent with the theoretical time complexity, scales linear with the increasing of N. From Fig. 4(b) and (d), we observe that all algorithms are sensitive to the length of sequences L. Since $|GCS|$ scales nearly linear with the increasing of L, the relative time ratios scale approximately quadratic with the increasing of L.

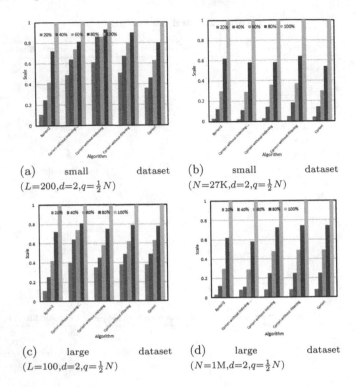

(a)   small   dataset
$(L=200,d=2,q=\frac{1}{2}N)$

(b)   small   dataset
$(N=27\text{K},d=2,q=\frac{1}{2}N)$

(c)   large   dataset
$(L=100,d=2,q=\frac{1}{2}N)$

(d)   large   dataset
$(N=1\text{M},d=2,q=\frac{1}{2}N)$

Fig. 4. Comparison with alternative algorithms varying data size

6   Conclusion

Extracting the generalized center string (GCS) has a great use in various applications, such as bioinformatics, market basket analysis, web log analysis, and etc. In this paper, we introduce Cpriori, an index-based combinatorial framework to mine the GCS from a large set of long sequences. Cpriori adopts the Indexing algorithm in the divide-and-conquer way to reduce the computation. Cpriori has no preference on the length of motifs, and it retains the positions of all mutations of GCSs in input sequences. Therefore, Cpriori can be used for accurately reasoning on mutation in biological sequence analysis. Moreover, Cpriori adopts Heruistic Filtering strategy that can more effectively generate the candidates with the user-defined support threshold. Comprehensive experimental results demonstrate the effectiveness and efficiency of Cpriori.

In the future, we will further devise a generic framework which includes more parallel combinatorial search algorithms to support longer sequences with limited memory.

Acknowledgement. This work is supported by National Key Research and Development Program of China under grant 2016YFB1000902, National Natural Science Foundation of China (No. 61232015, 61472412, 61621003).

References

1. Agrawal, R., Srikant, R.: Mining sequential patterns. In: 1995 Proceedings of the Eleventh International Conference on Data Engineering, pp. 3–14. IEEE (1995)
2. Agrawal, R., Srikant, R., et al.: Fast algorithms for mining association rules. In: Proceedings of the 20th International Conference on Very Large Data Bases, VLDB, vol. 1215, pp. 487–499 (1994)
3. Alkan, O.K., Karagoz, P.: CRoM and HuspExt: improving efficiency of high utility sequential pattern extraction. IEEE Trans. Knowl. Data Eng. **27**(10), 2645–2657 (2015)
4. Cao, H., Mamoulis, N., Cheung, D.W.: Mining frequent spatio-temporal sequential patterns. In: Fifth IEEE International Conference on Data Mining, pp. 82–89. IEEE (2005)
5. Chang, J.H.: Mining weighted sequential patterns in a sequence database with a time-interval weight. Know. Based Syst. **24**(1), 1–9 (2011)
6. Evans, P.A., Smith, A.D., Wareham, H.T.: On the complexity of finding common approximate substrings. Theoret. Comput. Sci. **306**(1–3), 407–430 (2003)
7. Fiot, C., Laurent, A., Teisseire, M.: From crispness to fuzziness: three algorithms for soft sequential pattern mining. IEEE Trans. Fuzzy Syst. **15**(6), 1263–1277 (2007)
8. Fournier-Viger, P., Gueniche, T., Tseng, V.S.: Using partially-ordered sequential rules to generate more accurate sequence prediction. In: Zhou, S., Zhang, S., Karypis, G. (eds.) ADMA 2012. LNCS (LNAI), vol. 7713, pp. 431–442. Springer, Heidelberg (2012). https://doi.org/10.1007/978-3-642-35527-1_36
9. García-Hernández, R.A., Martínez-Trinidad, J.F., Carrasco-Ochoa, J.A.: A new algorithm for fast discovery of maximal sequential patterns in a document collection. In: Gelbukh, A. (ed.) CICLing 2006. LNCS, vol. 3878, pp. 514–523. Springer, Heidelberg (2006). https://doi.org/10.1007/11671299_53
10. Ge, J., Xia, Y., Wang, J.: Mining uncertain sequential patterns in iterative mapreduce. In: Cao, T., Lim, E.-P., Zhou, Z.-H., Ho, T.-B., Cheung, D., Motoda, H. (eds.) PAKDD 2015. LNCS (LNAI), vol. 9078, pp. 243–254. Springer, Cham (2015). https://doi.org/10.1007/978-3-319-18032-8_19
11. Han, J., Pei, J., Yin, Y.: Mining frequent patterns without candidate generation. In: ACM SIGMOD Record, vol. 29, pp. 1–12. ACM (2000)
12. Han, J., Pei, J., Yin, Y., Mao, R.: Mining frequent patterns without candidate generation: a frequent-pattern tree approach. Data Min. Knowl. Disc. **8**(1), 53–87 (2004)
13. Hong, T.P., Lin, K.Y., Wang, S.L.: Mining fuzzy sequential patterns from multiple-item transactions. In: 2001 Joint 9th IFSA World Congress and 20th NAFIPS International Conference, vol. 3, pp. 1317–1321. IEEE (2001)
14. Hufsky, F., Kuchenbecker, L., Jahn, K., Stoye, J., Böcker, S.: Swiftly computing center strings. In: Moulton, V., Singh, M. (eds.) WABI 2010. LNCS, vol. 6293, pp. 325–336. Springer, Heidelberg (2010). https://doi.org/10.1007/978-3-642-15294-8_27
15. Kärkkäinen, J., Sanders, P.: Simple linear work suffix array construction. In: Baeten, J.C.M., Lenstra, J.K., Parrow, J., Woeginger, G.J. (eds.) ICALP 2003. LNCS, vol. 2719, pp. 943–955. Springer, Heidelberg (2003). https://doi.org/10.1007/3-540-45061-0_73
16. Lan, G.C., Hong, T.P., Tseng, V.S., Wang, S.L.: Applying the maximum utility measure in high utility sequential pattern mining. Exp. Syst. Appl. **41**(11), 5071–5081 (2014)

17. Lanctot, J.K., Li, M., Ma, B., Wang, S., Zhang, L.: Distinguishing string selection problems. Inf. Comput. **185**(1), 41–55 (2003)
18. Liu, A.X., Shen, K., Torng, E.: Large scale hamming distance query processing. In: 2011 IEEE 27th International Conference on Data Engineering (ICDE), pp. 553–564. IEEE (2011)
19. Lu, R., Jia, C., Zhang, S., Chen, L., Zhang, H.: An exact data mining method for finding center strings and all their instances. IEEE Trans. Knowl. Data Eng. **19**(4), 509–522 (2007)
20. Miliaraki, I., Berberich, K., Gemulla, R., Zoupanos, S.: Mind the gap: large-scale frequent sequence mining. In: Proceedings of the 2013 ACM SIGMOD International Conference on Management of Data, pp. 797–808. ACM (2013)
21. Muzammal, M.: Mining sequential patterns from probabilistic databases by pattern-growth. In: Fernandes, A.A.A., Gray, A.J.G., Belhajjame, K. (eds.) BNCOD 2011. LNCS, vol. 7051, pp. 118–127. Springer, Heidelberg (2011). https://doi.org/10.1007/978-3-642-24577-0_12
22. Sahli, M., Mansour, E., Kalnis, P.: Parallel motif extraction from very long sequences. In: Proceedings of the 22nd ACM International Conference on Information and Knowledge Management, pp. 549–558. ACM (2013)
23. Shang, J., Peng, J., Han, J.: Macfp: Maximal approximate consecutive frequent pattern mining under edit distance. In: Proceedings of the 2016 SIAM International Conference on Data Mining, pp. 558–566. SIAM (2016)
24. Smith, A.D.: Common approximate substrings. Ph.D. thesis. Citeseer (2004)
25. Srikant, R., Agrawal, R.: Mining sequential patterns: generalizations and performance improvements. In: Apers, P., Bouzeghoub, M., Gardarin, G. (eds.) EDBT 1996. LNCS, vol. 1057, pp. 1–17. Springer, Heidelberg (1996). https://doi.org/10.1007/BFb0014140
26. Wang, J., Han, J., Li, C.: Frequent closed sequence mining without candidate maintenance. IEEE Trans. Knowl. Data Eng. **19**(8), 1042–1056 (2007)
27. Yang, X., Wang, Y., Wang, B., Wang, W.: Local filtering: improving the performance of approximate queries on string collections. In: Proceedings of the 2015 ACM SIGMOD International Conference on Management of Data, pp. 377–392. ACM (2015)
28. Yen, T.-F., Reiter, M.K.: Traffic aggregation for malware detection. In: Zamboni, D. (ed.) DIMVA 2008. LNCS, vol. 5137, pp. 207–227. Springer, Heidelberg (2008). https://doi.org/10.1007/978-3-540-70542-0_11
29. Yin, J., Zheng, Z., Cao, L.: USpan: an efficient algorithm for mining high utility sequential patterns. In: Proceedings of the 18th ACM SIGKDD International Conference on Knowledge Discovery and Data Mining, pp. 660–668. ACM (2012)

Topic-Net Conversation Model

Min Peng[1(✉)], Dian Chen[1], Qianqian Xie[1], Yanchun Zhang[2], Hua Wang[2],
Gang Hu[1], Wang Gao[1,3], and Yihan Zhang[4]

[1] Computer School, Wuhan University, Wuhan, China
{pengm,dchen,xieq,hoogang,gaowang}@whu.edu.cn
[2] Centre for Applied informatics, University of Victoria, Melbourne, Australia
{yanchun.zhang,hua.wang,}@vu.edu.au
[3] College of Sports Science and Technology, Wuhan Sports University, Wuhan, China
[4] Computer School, National University of Singapore, Singapore, Singapore
e0261914@u.nus.edu

Abstract. Most sequence-to-sequence neural conversation models have
a ubiquitous problem that they tend to generate boring and safe
responses with almost none useful information, such as "I don't know" or
"I'm OK". In this paper, we study the response generation problem and
propose a topic-net conversation model (TNCM) via incorporating topic
information into the sequence-to-sequence framework. TNCM generates
every response word using not only its word embedding hidden state but
also the topic embedding, which guides the model to form more interesting and informative responses in conversation. The model increases
the possibility of the topic word appearing in the response further via
a mixed probabilistic model of two modes, namely the generate-mode
and the topic-mode. Moreover, we improve the process of beam search
during the test, which enhances the performance with better efficiency.
Evaluation results on large scale dataset indicate that our model can significantly outperform state-of-the-art methods on response generation of
the conversation system.

Keywords: Conversation model · Response generation
Sequence-to-sequence

1 Introduction

With the development of artificial intelligence technology, natural language processing (NLP) is widely used in many fields, such as data anonymisation [21–23],
electronic business [8], access control [7,26] and bioinformatics [27]. Conversation systems, also known as dialogue systems and sometimes chatbots, aim at
generating relevant and fluent responses in free-form natural language, which is
a challenging task in AI and NLP. Conversation systems can be divided into the
goal-driven systems [31] and open-domain chatbots. The former such as technical support services goals to help people to complete a specific task, while the
latter focuses on talking like human chit-chat in the open domains [16,25] such

© Springer Nature Switzerland AG 2018
H. Hacid et al. (Eds.): WISE 2018, LNCS 11233, pp. 483–496, 2018.
https://doi.org/10.1007/978-3-030-02922-7_33

as language learning tools or computer game characters. Previous researches on conversation focused on goal-driven systems. Recently, with the large amount of conversation data available on the Internet, open-domain chatbots are drawing more and more attention in both academia and industry [28].

Sequence-to-sequence model (seq2seq) [1,24], as a data-driven method mapping between the two sequences of arbitrary length, has achieved remarkable success in various natural language processing (NLP) tasks [5], including but not limited to conversation systems [25] which is referred to the neural conversational model. Seq2seq is essentially an encoder-decoder model, in which the encoder first transforms the input sequence to a certain representation which can then transformed into the output sequence by the decoder. [5] These methods have become mainstream for capturing semantic and syntactic relationships between messages and responses in a scalable and end-to-end manner. However, in practice, the neural conversational model tends to generate trivial or non-committal responses, often involving in high-frequency general responses such as "I don't know" [18], which is boring and frustrating with almost none useful information.

Furthermore, in seq2seq model, during the test period, seq2seq model can only perceive the information of the partially-formed sequence that has been generated. However, the training process maximizes the generation probability of the fully-formed word sequence conditioned on the input sequence and the history of target words, which misses the information of the partially-formed sequence. The difference between the training and the testing process also involves in high-frequency general responses. In practice, during the test period, for lack of the information of the fully-formed word sequence, the commonly used method is that the response generation is accomplished by searching over output sequence greedily with beam-search. However, in this method, the problem of the high-frequency general responses still exists.

In this paper, we study the response generation problem of open-domain chatbots. Notably our goal is to generate responses which are more interesting, diverse and informative. We observe that the people often subconsciously extract the theme of the input message and then generate a follow-up response with the extracted theme. Inspired by this observation, we extract the topic of the input message to guide the generation of the response, thus increasing the diversity of responses and reducing the probability of high-frequency general responses. Moreover, the topic of the input message can be used as the extracted prior knowledge to increase the controllability.

We propose a topic-net conversation model (TNCM). TNCM is based on the sequence-to-sequence framework. It contains the topic generation model and the topic-net seq2seq model. The topic generation model is a convolutional neural network to obtain the topic words. As for the topic-net seq2seq model, it represents the input message as the hidden vector and extracts the topic embedding in encoder. In decoder, the model generates every response word using not only its word embedding hidden state but also the topic embedding. Furthermore, the model increases the possibility of the topic word appearing in the response based on a mixed probabilistic model of two modes, namely the generate-mode and the topic-mode, that the former is the traditional probabilistic model and the latter picks words from the embedding of the topic. This mechanism makes the word in the response not only related to the input message but also to the topic information of the message and further increases the possibility that the topic word appears in the response. Moreover, to solve the problem of the difference between the training and the testing process, we improve the process of beam search during the test stage, which enhances the performance with better efficiency. Evaluation results on large scale test data indicate that our model can significantly outperform state-of-the-art methods for response generation of the conversation system.

Our contributions in this paper include: (1) a proposal of topic-net conversation model that naturally incorporates topic information into the encoder-decoder structure; (2) a proposal of a method of beam search optimization which enhance the performance with better efficiency; (3) an empirical verification of the effectiveness of topic-net conversation model for response generation.

The rest of this paper is arranged as follows. We survey the related literature in Sect. 2. In Sect. 3, we provide background on sequence-to-sequence model and attention mechanism. Section 4 gives the detail of our model. We report experimental results in Sects. 5 and 6 concludes the paper.

2   Related Work

The early traditional dialogue system relied on heuristic rules for response generation even there was a statistical component, which was inefficient and could only generate very limited responses [15, 30].

[17] deemed response generation as a statistical machine translation (SMT) problem. It inspired attempts to extend neural language models in SMT to response generation [24], which [24] present a data-driven approach to generating responses to Twitter status posts, based on phrase-based SMT. [20] improved [17] with a seq2seq model and represented the utterances in previous turns as a context vector and incorporate the context vector into response generation, which became the mainstream method in this area [19, 25], that [19] propose Neural Responding Machine (NRM), a neural network-based response generator for short-text conversation. However, the neural conversational model tends to generate high-frequency general responses. To solve this issue, various methods have been proposed. [10] propose using Maximum Mutual Information (MMI) as the objective function in neural models. [11] present persona-based models for handling the issue of speaker consistency in neural response generation. There were also attempts to model complex conversational structures [18] or to seek better optimization strategies [12, 13].

There have been studies that introduce the topic information to the conversation system. [28] combined the topic with the tensor network, which the message vector, the response vector, and the two topic vectors are fed to neural tensors to calculate a matching score, but the message-response matching is limited to data sets and non-scalable. [29] leveraged the topic information obtained from a pre-trained LDA model in response generation by a joint attention mechanism and a biased generation probability. However, the probabilistic topic is not suitable for seq2seq model and the topic is only introduced between the encoder and the decoder.

3   Background: Sequence-to-Sequence Model and Attention Mechanism

Before introducing our model, let us first briefly review the sequence-to-sequence model and the attention mechanism.

3.1   Sequence-to-Sequence Model

In seq2seq, given an input sequence (message) $X = \{x_1, x_2, \ldots, x_{N_x}\}$ and the output sequence (response) $Y = \{y_1, y_2, \ldots, y_{N_y}\}$, the model can be expressed in a probabilistic view as maximizing the generation probability of observing the Y conditioned on $X : p\left(y_1, y_2, \ldots, y_{N_y} \mid x_1, x_2, \ldots, x_{N_x}\right)$. Seq2seq is essentially an encoder-decoder model.

The encoder first transforms X to a context vector c through a recurrent neural network (RNN), i.e.

$$h_t = f\left(x_t, h_{t-1}\right); c = \phi\left(\{h_1, \ldots, h_T\}\right) \tag{1}$$

where $\{h_t\}$ is the RNN hidden state at time t, f is the dynamics non-linear function, and ϕ summarizes the hidden states. In practice, it is found that gated RNN alternatives such as LSTM [6] or GRU [2] often perform much better than vanilla ones [5]. In this work, the source sentence is encoded with a Bi-directional RNN, making each hidden state h_t aware of the contextual information from both ends.

The decoder then estimates the generation probability of Y with the context vector c as input, through the following dynamics and prediction model:

$$s_t = f\left(y_{t-1}, s_{t-1}, c\right); p\left(y_t | y_{<t}, X\right) = g\left(y_{t-1}, s_t, c\right) \tag{2}$$

where $\{s_t\}$ are the RNN hidden state at time t, and $\{y_{<t}\}$ denoting the history $\{y_1, \ldots, y_{t-1}\}$.

3.2   Attention Mechanism

The traditional seq2seq model takes input as a complete sequence X and compresses all information into a fixed-length vector. In practice, however, different words in Y could be semantically related to different parts of X. To address this issue, the attention mechanism was introduced to seq2seq [1]. It allows seq2seq model to inspect all the information in the input sequence X, then generate the Y according to the current word and context. Each y_i in Y corresponds to a dynamically changing context vector c_i, which is a weighted average of all hidden states $\{h_t\}$ of the encoder, i.e.

$$c_t = \sum_{\tau=1}^{T} \alpha_{t\tau} h_\tau; \alpha_{t\tau} = \frac{e^{\eta(s_{t-1}, h_\tau)}}{\sum_{\tau'} e^{\eta(s_{t-1}, h_{\tau'})}} \tag{3}$$

where η is usually implemented as a multi-layer perceptron (MLP) with tanh as the activation function.

4   Topic-Net Conversation Model

Let X denotes an input message $X = \{x_1, x_2, \ldots, x_{N_x}\}$, where N_x denotes the number of words in X, and K denotes the topic representation of the message X. Let Y denotes a sequence in response to the message X, where $Y = \{y_1, y_2, \ldots, y_{N_y}\}$ and N_y is the length of the response. Our goal is to learn a generation model to generate response candidates for X with the topic representation K.

To learn the model, we need to deal with two questions: (1) how to obtain topic words? (2) how to perform learning? In the following sections, we first present our method on topic word generation, then we give details of our seq2seq model.

4.1   Topic Generation Model

Our topic generation model is based on neural language model TDLM [9]. TDLM is a topically driven neural language model with a convolutional neural network topic model, which is concise and efficient. We do some improvements on the model to make the topic more suitable for the conversational model, as shown in Fig. 1.

We assume that the input message is $X = \{x_1, x_2, \ldots, x_i, \ldots, x_{N_x}\}$, where $x_i \in \mathbb{R}^e$ is the e-dimensional word vector for the i-th word in the message. Then we can use a number of convolutional filters to process the word vectors. Let $W_v \in \mathbb{R}^{eh}$ be a convolutional filter which is applied to a window of m words to generate a feature. A feature u_i for a window of words $x_{i:i+m-1}$ is given as follows:

$$u_i = \delta \left(W_v^T x_{i:i+m-1} + b_v \right) \tag{4}$$

where b_v is a bias term and δ is the activation function, which is generally the Relu function. Then we apply a max-pooling operation, yielding the message vector d:

$$d = \max_i u_i \tag{5}$$

The topic vectors are stored in two lookup tables $A \in \mathbb{R}^{k \times a}$ (input vector) and $B \in \mathbb{R}^{k \times b}$ (output vector), where k is the number of topics, a and b are the dimensions of the topic vectors. The attention vector p can be written as

$$p = \gamma (Ad) \tag{6}$$

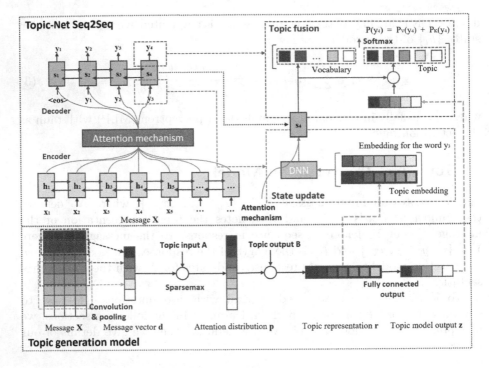

Fig. 1. The structure of the topic-net conversation model.

where $p \in \mathbb{R}^k$ and γ is usually the softmax function. But here we use sparsemax function [14], which is similar to the traditional softmax and outputs sparse probabilities.

Then, we calculate the topic representation r by

$$r = B^T p \tag{7}$$

where $r \in$. Intuitively, r is a weighted mean of topic vectors, with the weighting given by the attention p.

At last, r is connected to a dense layer to generate the topic word, and the model is optimized by using categorical cross-entropy loss.

In addition, considering that the conversation sequences are generally short, we introduce a penalty item similarity penalty to ensure that the similarity between topics is small enough. The output vector $B \in \mathbb{R}^{k \times b}$ determines the similarity of the topic representation. A row in B represents a topic. Therefore, we normalize B by

$$\tilde{B}_{ij} = \frac{B_{ij}}{\parallel B_i \parallel^2} \tag{8}$$

The cosine similarity of the topic representations is $\tilde{B}\tilde{B}^2$. Therefore, we can penalize the similarity by

$$Penality = \beta max \left(\tilde{B}\tilde{B}^T - E \right)^2 \tag{9}$$

where E is the identity matrix; and β is the penalty parameter.

4.2   Topic-Net Seq2Seq Model

As illustrated in Fig. 1, topic-net seq2seq model is built on the sequence-to sequence framework, which is still an encoder-decoder model.

In encoder, as same as [1], we use a bi-directional RNN to convert the input sequence X to a series of hidden states $\{h_t\}$ with equal length, specifically each hidden state h_t corresponding to word x_t. Then these hidden states $\{h_t\}$ are transformed to context vectors.

In decoder, a RNN predicts the target sequence with the context vector, which is similar with [1]. But, there are two important differences: the state update module and the topic fusion module. The state update module integrates the topic into the hidden states $\{s_t\}$. The topic fusion module predicts words based on a mixed probabilistic model of two modes, namely the generate-mode and the topic-mode, where the former is the traditional probabilistic model and the latter picks words from the embedding of the topic. It increases the possibility that the topic word appears in the response.

State Update. The state update module optimizes the update process of the hidden states $\{s_t\}$ with the topic. Specifically, it updates each decoding state s_t with the previous state s_{t-1}, the previous symbol y_{t-1}, the context vector c_{t-1} and the topic embedding r, so the Eq. (2) can be rewritten as

$$s_t = f\left(y_{t-1}, s_{t-1}, c, r\right) \tag{10}$$

where r is the topic representation extracted by the topic generation model. Compared to the traditional Eq. (2), the state update module makes each word generated in the response not only relevant to the input message, but also to the topic information.

Topic Fusion. In the traditional decoder, the words of the response y_t is predicted from the vocabulary $V = \{v_1, v_2, \ldots, v_{N_v}\}$. In addition, in the topic fusion module as in Fig. 2, we have another set of topic words K, for all the words extracted by the topic generation model $K = \{k_1, k_2, \ldots, k_{N_k}\}$, where K may contain words not in V.

In the topic fusion module, every word in response is predicted based on a mixed probabilistic model of two modes, namely the generate-mode for the word in V and the topic-mode for the word in K. Given the hidden state s_t at time t, we define the mixture generation probability of the response word y_t as

$$p\left(y_t|y_{t-1}, s_{t-1}, c_t, K\right) = p_V\left(y_t|y_{t-1}, s_{t-1}, c_t\right) + p_K\left(y_t|y_{t-1}, s_{t-1}, c_t, K\right) \tag{11}$$

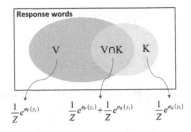

Fig. 2. The illustration of the decoding probability $p(y_t|\cdot)$ in the topic fusion module. V is the vocabulary of the predicted words. And K the vocabulary of the topic words.

where p_V is the generate-mode and p_K is the topic-mode. The probability of the word y_t in these two modes are defined by

$$p_V\left(y_t|y_{t-1}, s_t, c_t\right) = \begin{cases} \frac{1}{Z}e^{\varphi_V(y_t)}, y_t \in V \\ 0, y_t \notin V \end{cases}$$
$$p_K\left(y_t|y_{t-1}, s_t, c_t, K\right) = \begin{cases} \frac{1}{Z}\sum_{j:k_j=y_t} e^{\varphi_K(y_t)}, y_t \in K \\ 0, y_t \notin K \end{cases} \tag{12}$$

where φ_V and φ_K are the measure functions for the generate-mode and the topic-mode and Z is the normalization term $Z=\sum_{v\in V} e^{\varphi_V(v)} + \sum_{k\in K} e^{\varphi_K(k)}$. These two modes are basically competing through one softmax function. Therefore, they share the same normalization term.

As shown in Fig. 2, the measure functions of the generate-mode and topic-mode are calculated as follow.

For the generate-mode, the measure function is the same as [1]:

$$\varphi_V\left(y_t = v_i\right) = v_i^T W_o s_t \tag{13}$$

where $W_o \in \mathbb{R}^{(N+1)\times d_s}$ and v_i is the one-hot indicator vector for the response word in V.

For the topic-mode, the measure function is defined by

$$\varphi_K\left(y_t = k_i\right) = \delta\left(k_i^T W_c\right) s_t \tag{14}$$

where $W_c \in \mathbb{R}^{d_h \times d_s}$, k_i is the one-hot indicator vector for the response word in K and δ is a non-linear activation function, which is the role of mapping k_i to the semantic space of s_t.

With the topic fusion, the generation probability of response word is partial to the topic words. For the non-topic words, only the generate-mode can be activated, and the measure functions of topic-mode does not work. For the topic words, topic-mode further increases the possibility of the topic words appearing in responses.

4.3   Beam-Search Optimization

As illustrated in Fig. 3, to solve the problem of the difference between the training and the testing process, we reconstruct the response sequence of the train

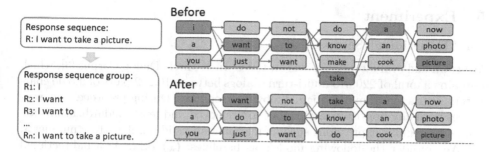

Fig. 3. Beam-search optimization. The left is the schematic diagram of the sequence group. The right is the optimization of the argmax. Right top: possible $y_{1:t}$ formed in training with a beam of size $K = 3$ and with target sequence $y1:6 =$ "i want to take a picture". Note that in time-step $t = 4$ the predicted prefixes involve in margin violations. Right bottom: after grouping, margin violations rarely appear.

dataset, which makes the training process perceive the information of partially-formed sequences.

Specifically, for train dataset, we split each response sequence into a sequence group order by the word, as shown in Fig. 3. The sequence group contains the information of partially-formed sequences, which exposes partially-formed sequences to the training process. In practice, in order to simplify the experiment process, we directly weight each word according to word order in the loss function to achieve the effect of splitting. For each partially-formed sequence in sequence group, the loss can be written as

$$loss_p = - \sum_{t=1}^{N_R} \log p\left(y_t | y_{<t}, X\right) \tag{15}$$

where N_R is the sequence length of every partially-formed sequence. So, the loss of the sequence group is

$$loss = - \sum_{N_R=1}^{N_R=N_y} \sum_{t=1}^{N_R} \log p\left(y_t | y_{<t}, X\right)$$
$$= - \sum_{t=1}^{N_y} (N_y - i + 1) \log p\left(y_t | y_{<t}, X\right) \tag{16}$$

Furthermore, we replace beam-search with argmax to enhance the performance with better efficiency. In traditional beam-search, the model is never exposed to its own errors during the test process, and so the inferred histories in the test do not resemble the training histories. In Fig. 3, the partially decoded words could not be included in the beam-size of the beam-search before the ending of the test process. However, after exposing partially-formed sequences to the training process, we discover that almost all the partially decoded words emerged in the beam-size of the beam-search have the relatively high scores. Therefore, to enhance efficiency, we substitute the beam-search with argmax, which narrows the search space and simplifies the test process.

5   Experiment

5.1   Experiment Setup

We train and evaluate the models on the Cornell Movie Dialog Corpus [3], which contains a total of 220,579 multi-turn dialogs between 10,292 pairs of movie characters, extracted from 617 original movie screenplays. During preprocessing, sentences with any non-Roman alphabet are removed and few standardizations are made via regular expressions such as mapping all valid numbers to <number>.

We consider the following models as baselines: **(1) bS2SA:** a bidirectional sequence-to-sequence model with attention; **(2) S2SA-MMI:** a sequence-to-sequence model using MMI as the objective function [10]; **(3) bS2SA-Topic Attention (bS2SA-TA):** a bS2SA with topic attention, which synthesizes topic vectors from a pre-trained LDA [29]; **(4) bS2SA-State update(bS2SA-Su):** a bS2SA with state update module, to verify the effectiveness of the topic fusion module of TNCM; **(5) bS2SA-State update & Topic fusion(bS2SA-ST):** a bS2SA with state update module and the topic fusion module, to verify the effectiveness of the beam-search optimization of TNCM.

The accurate evaluation of a non-goal-driven dialogue system is an open problem [4] but not the focus of the paper. We follow the existing work and employ the automatic evaluation metrics include perplexity and Distinct-1 & Distinct-2 [10].

Perplexity. For probabilistic language models word perplexity is a well-established performance metric and has been suggested for generative dialogue models previously [18]. Perplexity is defined by

$$PPL = \exp\left(-\frac{1}{N}\sum_{n=1}^{N}\log\left(p\left(Y_i\right)\right)\right)$$

Perplexity measures how well the model generates a response. It explicitly measures the model?s ability to account for the syntactic structure of the dialogue and the syntactic structure of each utterance [18]. A lower perplexity score indicates better generation performance.

Distinct-1 and Distinct-2. Following [10], we calculate Distinct-1 and Distinct-2. Distinct-1 and Distinct-2 are respectively the number of distinct unigrams and bigrams divided by total number of generated words. The two metrics measure how informative and diverse the generated responses are. High numbers and high ratios mean that there is much content in the generated responses, and high numbers further indicate that the generated responses are long [29].

In addition to automatic evaluation metrics, we also recruit human annotators to determine the quality of responses generated by different models [29]. Five volunteers with rich experience are invited to do this. The volunteers judge the quality of the response according to the following criteria: Good (+2): the response was not only relevant and natural, but also interesting and informative; Medium (+1): the response can be used as a reply to a message, but it is too common, such as "Yes, I see", "Me too" and "I don't know"; Bad (+0): the response cannot be used as a reply to the message.

Table 1. Results on automatic metrics.

Model	Perplexity	Distinct-1	Distinct-2
bS2SA	87.24	.095	.199
S2SA-MMI	87.24	.148	.295
bS2SA-TA	81.62	.122	.273
bS2SA-Su	81.03	.160	.297
bS2SA-ST	74.85	**.172**	.382
TNCM	**73.96**	.170	**.383**

5.2 Evaluation Results

Table 1 shows the results of automatic metrics. It is clear that our model has achieved the best performance. On perplexity, bS2SA-State update, bS2SA-State update & Topic fusion and TNCM beat all the baselines, where TNCM is the best. S2SA-MMI is an after-processing mechanism on the responses generated by S2SA, therefore, following [29], we report the perplexity of S2SA to approximately represent the generation ability of S2SA-MMI. It is worth mentioning that bS2SA-TA and our model perform better than bS2SA and S2SA-MMI, which verifies our claim that the topic information does have an effect on conversation models. However, bS2SA-TA is worse than bS2SA-Su and bS2SA-ST; probably because bS2SA-TA leverages the LDA topic, where the probabilistic topic may not be suitable for seq2seq model and the topic information is only introduced between the encoder and decoder. The bS2SA-ST shows better performance than bS2SA-Su, indicating the importance of the topic fusion module. Compared to bS2SA-ST, the perplexity of TNCM reduces a little, suggesting that the beam-search optimization contributes to enriching the response in some extent. On distinct-1 and distinct-2, bS2SA-State update & Topic fusion outperforms all the baseline models, which further illustrates the state update module and the topic fusion module have positive effects. Note that bS2SA-ST performs better than TNCM in term of the number of distinct unigram, which results from the beam-search optimization narrowing the search space of the response words.

Table 2. Results on human evaluation.

Model	Good	Medium	Bad	Ave
bS2SA	20.8%	39.6%	39.6%	.812
S2SA-MMI	32.7%	36.5%	30.8%	1.019
bS2SA-TA	38.4%	29.6%	32.0%	1.064
bS2SA-Su	39.3%	28.1%	32.6%	1.067
TNCM	41.6%	28.3%	30.1%	**1.115**

Table 2 shows the results of human evaluation. Obviously, TNCM are better than all the baseline models. Compared with the S2SA-MMI, bS2SA-TA and bS2SA-Su all generates more "Good" responses, confirming the topic information takes effect on generating interesting and informative responses. However, generic responses ("Medium") of bS2SA-Su reduces and the "Bad" responses of bS2SA-Su increases. It is because that noise in the topic is brought to generation without the topic fusion module. This shows that the topic fusion module is indispensable.

Table 3. Results on the efficiency improvement of the beam-search optimization.

Model	BLEU	Speedup
bS2SA-ST	3.59	10.4x
TNCM	3.62	3.38x

Table 3 shows the results of the efficiency improvement of the beam-search optimization. Beam-search is applied to generating natural language of the response. Therefore, we use BLEU [4] to measure the quality of the natural language. On the one hand, it's obvious that TNCM performs a little bit better than the bS2SA-ST with regards to BLEU. On the other hand, our model has almost 3 times speedup in test inference. This suggests that the beam-search optimization enhances the effect with better efficiency.

6   Conclusions

In this paper, we study the response generation problem and propose a topic-net conversation model (TNCM) to incorporate topic information into the sequence-to-sequence framework. TNCM generates every response word using not only its word-embedding hidden state but also the embedding of the topic and increases the possibility that the topic word appears in the response. Moreover, we improve the process of beam search during the test stage, which enhances the effect with better efficiency. Experimental results show that TNCM can significantly outperform state-of-the-art models and generate more interesting and more informative responses.

References

1. Bahdanau, D., Cho, K., Bengio, Y.: Neural machine translation by jointly learning to align and translate. arXiv preprint arXiv:1409.0473 (2014)
2. Cho, K., Van Merriënboer, B., Gulcehre, C., Bahdanau, D., Bougares, F., Schwenk, H., Bengio, Y.: Learning phrase representations using rnn encoder-decoder for statistical machine translation. arXiv preprint arXiv:1406.1078 (2014)

3. Danescu-Niculescu-Mizil, C., Lee, L.: Chameleons in imagined conversations: a new approach to understanding coordination of linguistic style in dialogs. In: Proceedings of the 2nd Workshop on Cognitive Modeling and Computational Linguistics, pp. 76–87. Association for Computational Linguistics (2011)
4. Galley, M., Brockett, C., Sordoni, A., Ji, Y., Auli, M., Quirk, C., Mitchell, M., Gao, J., Dolan, B.: deltaBLEU: a discriminative metric for generation tasks with intrinsically diverse targets. arXiv preprint arXiv:1506.06863 (2015)
5. Gu, J., Lu, Z., Li, H., Li, V.O.: Incorporating copying mechanism in sequence-to-sequence learning. arXiv preprint arXiv:1603.06393 (2016)
6. Hochreiter, S., Schmidhuber, J.: Long short-term memory. Neural Comput. **9**(8), 1735–1780 (1997)
7. Kabir, M.E., Wang, H., Bertino, E.: A role-involved purpose-based access control model. Inf. Syst. Front. **14**(3), 809–822 (2012)
8. Khalil, F., Li, J., Wang, H.: An integrated model for next page access prediction. Int. J. Knowl. Web Intell. **1**(1–2), 48–80 (2009)
9. Lau, J.H., Baldwin, T., Cohn, T.: Topically driven neural language model. arXiv preprint arXiv:1704.08012 (2017)
10. Li, J., Galley, M., Brockett, C., Gao, J., Dolan, B.: A diversity-promoting objective function for neural conversation models. arXiv preprint arXiv:1510.03055 (2015)
11. Li, J., Galley, M., Brockett, C., Spithourakis, G.P., Gao, J., Dolan, B.: A persona-based neural conversation model. arXiv preprint arXiv:1603.06155 (2016)
12. Li, J., Monroe, W., Ritter, A., Galley, M., Gao, J., Jurafsky, D.: Deep reinforcement learning for dialogue generation. arXiv preprint arXiv:1606.01541 (2016)
13. Li, J., Monroe, W., Shi, T., Jean, S., Ritter, A., Jurafsky, D.: Adversarial learning for neural dialogue generation. arXiv preprint arXiv:1701.06547 (2017)
14. Martins, A., Astudillo, R.: From Softmax to Sparsemax: a sparse model of attention and multi-label classification. In: International Conference on Machine Learning, pp. 1614–1623 (2016)
15. Nio, L., Sakti, S., Neubig, G., Toda, T., Adriani, M., Nakamura, S.: Developing non-goal dialog system based on examples of drama television. In: Mariani, J., Rosset, S., Garnier-Rizet, M., Devillers, L. (eds.) Natural Interaction with Robots, Knowbots and Smartphones. Springer, New York (2014)
16. Perez-Marin, D.: Conversational Agents and Natural Language Interaction: Techniques and Effective Practices: Techniques and Effective Practices. IGI Global (2011)
17. Ritter, A., Cherry, C., Dolan, W.B.: Data-driven response generation in social media. In: Proceedings of the Conference on Empirical Methods in Natural Language Processing, pp. 583–593. Association for Computational Linguistics (2011)
18. Serban, I.V., Sordoni, A., Bengio, Y., Courville, A.C., Pineau, J.: Building end-to-end dialogue systems using generative hierarchical neural network models. In: AAAI 2016, pp. 3776–3784 (2016)
19. Shang, L., Lu, Z., Li, H.: Neural responding machine for short-text conversation. arXiv preprint arXiv:1503.02364 (2015)
20. Sordoni, A., Galley, M., Auli, M., Brockett, C., Ji, Y., Mitchell, M., Nie, J.Y., Gao, J., Dolan, B.: A neural network approach to context-sensitive generation of conversational responses. arXiv preprint arXiv:1506.06714 (2015)
21. Sun, X., Li, M., Wang, H., Plank, A.: An efficient hash-based algorithm for minimal k-anonymity. In: Proceedings of the Thirty-First Australasian Conference on Computer Science, vol. 74, pp. 101–107. Australian Computer Society, Inc. (2008)
22. Sun, X., Wang, H., Li, J., Zhang, Y.: Injecting purpose and trust into data anonymisation. Comput. Secur. **30**(5), 332–345 (2011)

23. Sun, X., Wang, H., Li, J., Zhang, Y.: Satisfying privacy requirements before data anonymization. Comput. J. **55**(4), 422–437 (2012)
24. Sutskever, I., Vinyals, O., Le, Q.V.: Sequence to sequence learning with neural networks. In: Advances in neural information processing systems, pp. 3104–3112 (2014)
25. Vinyals, O., Le, Q.: A neural conversational model. arXiv preprint arXiv:1506.05869 (2015)
26. Wang, H., Cao, J., Zhang, Y.: Ticket-based service access scheme for mobile users. In: Australian Computer Science Communications, vol. 24, pp. 285–292. Australian Computer Society, Inc. (2002)
27. Wang, H., Zhang, Y., et al.: Detection of motor imagery eeg signals employing naïve bayes based learning process. Measurement **86**, 148–158 (2016)
28. Wu, Y., Wu, W., Li, Z., Zhou, M.: Response selection with topic clues for retrieval-based chatbots. arXiv preprint arXiv:1605.00090 (2016)
29. Xing, C., Wu, W., Wu, Y., Liu, J., Huang, Y., Zhou, M., Ma, W.Y.: Topic aware neural response generation. In: AAAI 2017, pp. 3351–3357 (2017)
30. Young, S., Gašić, M., Keizer, S., Mairesse, F., Schatzmann, J., Thomson, B., Yu, K.: The hidden information state model: a practical framework for pomdp-based spoken dialogue management. Comput. Speech Lang. **24**(2), 150–174 (2010)
31. Young, S., Gašić, M., Thomson, B., Williams, J.D.: POMDP-based statistical spoken dialog systems: a review. Proc. IEEE **101**(5), 1160–1179 (2013)

A Hybrid Model Reuse Training Approach for Multilingual OCR

Zhongwei Xie, Lin Li(✉), Xian Zhong(✉), Luo Zhong,
Qing Xie, and Jianwen Xiang

School of Computer Science and Technology,
Wuhan University of Technology, Wuhan, China
{kevinsnest,cathylilin,zhongx,zhongluo,felixxq,jwxiang}@whut.edu.cn

Abstract. Nowadays, there is a great demand for multilingual optical character recognition (MOCR) in various web applications. And recently, Long Short-Term Memory (LSTM) networks have yielded excellent results on Latin-based printed recognition. However, it is not flexible enough to cope with challenges posed by web applications where we need to quickly get an OCR model for a certain set of languages. This paper proposes a Hybrid Model Reuse (HMR) training approach for multilingual OCR task, based on 1D bidirectional LSTM networks coupled with a model reuse scheme. Specifically, Fixed Model Reuse (FMR) scheme is analyzed and incorporated into our approach, which implicitly grabs the useful discriminative information from a fixed text generating model. Moreover, LSTM layers from pre-trained networks for unilingual OCR task are reused to initialize the weights of target networks. Experimental results show that our proposed HMR approach, without assistance of any post-processing techniques, is able to effectively accelerate the training process and finally yield higher accuracy than traditional approaches.

Keywords: Multilingual OCR · Model reuse · Parameter transfer

1 Introduction

Optical character recognition (OCR), as a major application of machine learning, has been widely used in industrial field to convert text from scanned document images into digitally editable texts, which can be an important data source for subsequent text mining. However, the number of multilingual data including texts and images over the Internet is increasing continually with active exchanges between international Web users. While there are a lot of practical applications

Supported by the National Social Science Foundation of China (Grant No: 15BGL048), the Hubei Province Science and Technology Support Project (Grant No: 2015BAA072), the National Natural Science Foundation of China (Grant No. 61672398), the Hubei Provincial Natural Science Foundation of China (Grant No: 2017CFA012), the Fundamental Research Funds for the Central Universities (WUT: 2017II39GX).

H. Hacid et al. (Eds.): WISE 2018, LNCS 11233, pp. 497–512, 2018.
https://doi.org/10.1007/978-3-030-02922-7_34

(e.g., UniClip [16]) consisting of OCR module, such as image search engine and end-to-end image documents processing based information systems, still today most OCR systems just specialize in building a unilingual OCR model for one particular language.

Multilingual OCR (MOCR) is a difficult task and it presents several unique challenges (e.g., multiple languages on a page). The traditional OCR process for dealing with such documents is a two-step pipeline. The first step is to identify individual languages and the second is to apply mature unilingual OCR technique for each language. By doing so, the problem is reduced to unilingual OCR. However, errors from language identification will propagate through the pipeline and pose great difficulties for latest stage applications. The goal of transcribing multilingual documents completely and accurately, under moderate degradation is still far. Recent studies on Latin script using LSTM networks [18] have shown that reliable MOCR results can be obtained, without language identification or any post-processing steps. But an OCR system with a unified recognition model for all languages is still difficult to achieve. In this paper we study how to train a general MOCR model for several selected languages.

Moreover, the training process of an MOCR system from scratch would be very time-consuming due to various character arrangement in different language context. Web applications applied in different countries, like extracting texts from screenshots of smartphone and landmark recognition module in the self-driving system, have to be confronted with challenges in particular multilingual scenarios. It is practically significant for scalability of web applications to accelerate the training process. Considering this, model reuse can be a nice choice to help reduce the time cost, data amount and expertise required [20].

There have been efforts [9,11,21,22] reported to leverage model reuse or transfer learning to construct a model by utilizing existing available models, mostly trained for other tasks, rather than building a model from scratch. They directly use the weights from pre-trained deep network as initial values for the target model. However, during training the new deep model, they kick off the pre-trained model that in fact can help model training. Currently, Fixed Model Reuse (FMR) is proposed in [20] to incorporate the helpful information in the fixed model into a new convolutional model training.

Reusing the weights from pre-trained networks and fixed model reuse approach both have their strengths in accelerating the training process and reducing the time cost, which motivate us to combine these methods to further raise the convergence rate of MOCR models. In this paper, we propose a Hybrid Model Reuse training approach (HMR) which focuses on how to effectively train and optimize a bidirectional LSTM networks for multilingual OCR to accelerate the training process. Firstly, we incorporate the FMR approach from the traditional ConvNets scheme into our LSTM based model. A fixed multilingual text generating model based on LSTM networks [14] is reused to provide valuable contextual information for our target networks. In addition, our HMR approach takes advantage of parameter transfer learning. Before training our networks, a pre-trained model for unilingual OCR task on a small training data is exploited to set a good basis for the target network. Weight propagation, knockdown and weight propagation after knockdown steps are used to utilize the deep information or features in the fixed text-generating model, accelerating the training procedure and improving OCR accuracy. Experimental results show that our HMR approach can speed up the convergence rate of multilingual OCR model and yield higher recognition accuracy than traditional approaches.

The rest of this paper starts from introduction on related work in Sect. 2. Then we elaborate on the hybrid model reuse training approach (HMR) in Sect. 3, followed by experiments in Sect. 4 and conclusions in Sect. 5.

2 Related Work

The usual approach to address multilingual OCR problem is to somehow combine two or more separate classifiers [12], as it is believed that a reasonable OCR output for a single script can not be obtained without sophisticated postprocessing steps such as language modeling, use of dictionary to correct OCR errors, font adaptation.

More specifically, approaches to multilingual OCR can be broadly divided into segmentation-based and segmentation-free approaches, depending on whether text-line images are segmented into smaller units prior to recognition. As segmentation based MOCR systems, open-source Tesseract engine [15] concentrates effort on enabling generic multilingual operation. Philip et al. proposed a SVM based bilingual OCR system [13] on Malayalam scripts. In [5], Firmani et al. proposed a convolutional neural network based OCR component to recognize Latin characters. In our experiences, segmentation error is the limiting factor for the performance of segmentation-based OCR systems.

In contrast, segmentation-free MOCR methods implicitly incorporate character segmentation in producing a globally optimized character/word sequence as the recognition result using methods such as the left-to-right Hidden Markov Model (HMM) [10], which avoids many difficulties of segmentation-based MOCR systems. However, unsegmented MOCR methods still require careful choices of model structures, and face issues in heuristic modifications of their cost functions to achieve overall good performance.

Recently, Recurrent Neural Networks have shown a great success in many sequence learning tasks, due to the Long Short Term Memory architecture, which differs significantly from the previous RNN and appears to overcome many limitations and problems of earlier architectures. In [18], LSTM networks have been used in MOCR for printed document images, yielding excellent results. In this paper, an 1D bidirectional Long Short Term Memory network is directly trained on the gray-level text lines as a basic model to explore a better training approach to multilingual OCR.

With regard to accelerating the training process, model reuse approaches can be of great help. Model reuse [20] usually attempts to construct a model by utilizing existing available models, mostly trained for other tasks, rather than from scratch, which offers a great potential to reduce the required amount of training examples and training time cost, because the exploitation of existing models may help set a good basis for the training of a new model. In the deep learning community, there are several pieces of studies [9,11,21,22] trying to reuse the layers in deep structures, e.g., by initializing a network with weights from pre-trained networks [21]; by proposing a new network architecture to transfer features from pre-trained networks [9]. These models are generally based on the strategies of re-training on dataset B with trained deep networks on dataset A, i.e., they mainly focus on the transfer of information of the latent weights in deep networks.

Recently, Fixed Model Reuse approach [20], a more thorough model reuse scheme has been proposed, which arranges the convolution layers and the model or features for general tasks in parallel, fully connecting to the output layer nodes and gradually reducing the dependencies between the fixed model/features and the output layer nodes.

In this paper, based on the 1D bidirectional LSTM networks for MOCR, FMR scheme is incorporated from the ConvNets scheme to implicitly grab the useful discriminative information from a fixed multilingual text model. Moreover, LSTM layers are reused by initializing the target network with weights from pre-trained network for unilingual OCR, which helps set a good basis for the training process of multilingual OCR.

3 Our Proposed HRM Approach

3.1 Overview

In this paper, inspired by FMR scheme and parameter transfer approach, we propose a hybrid model reuse training approach to accelerate training process and improve the accuracy for multilingual machine-printed OCR task. The structure of our hybrid model reuse approach is shown in Fig. 1. We use a multilingual OCR system for Chinese and Japanese for example. Prior to the training, we initialize the weights in the bidirectional LSTM layers with those from a pre-trained unilingual (Chinese or Japanese) OCR model. Then, a fixed multilingual (Chinese and Japanese) text generating model built on corpora in corresponding languages, is directly connected to the 1D bidirectional LSTM network in parallel to the output layer nodes. Finally, weight propagation, knockdown, weight propagation after knockdown steps (introduced in Sect. 3.3) are adopted to train the deep model and retain consistency of the networks. All-round introductions of each part will be given in the following.

3.2 Pre-trained LSTM Multilingual OCR

As shown in Fig. 1, an 1D bidirectional LSTM architecture on the text-line images is constructed for our proposed HMR approach. Recurrent Neural Networks have had somewhat of a renaissance due to the Long Short Term Memory (LSTM) architecture [7]. The LSTM architecture differs significantly from

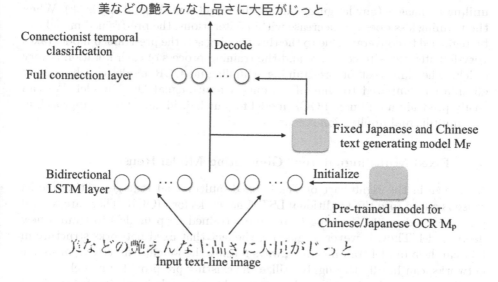

Fig. 1. Network structure of our hybrid model reuse approach

the architecture of previous recurrent neural networks and appears to overcome many of the limitations and problems of those earlier architectures, like vanishing gradient and exploding gradient problems [3].

Recurrent Neural Networks are considered good at context-aware processing and recognizing patterns occurring in time series. However, traditional recurrent neural networks have not shown competitive performance in large scale tasks like OCR or speech recognition, perhaps due to the vanishing gradient problem. The Long Short Term Memory architecture is designed to overcome this problem. It is a highly non-linear recurrent network with multiplicative gates and additive feedback. Ul-Hasan et al. [18] introduced bidirectional LSTM architectures for accessing context in both forward and backward directions. Both layers are then connected to a single output layer. To avoid the requirement of segmented training data, Graves et al. [6] then use a forward backward algorithm to align transcripts with the output of the neural network and yield very good results.

We have tried several LSTM network structures, whose results show that 1D architecture outperforms their 2D or higher dimensional siblings for printed OCR tasks. Apart from that, in our preliminary experiments, we find that adding convolution layers to extract the deep features from text-line images cannot obviously improve the performance either. Therefore, in this paper, we directly train an 1D bidirectional LSTM architecture on the multilingual text-line images as basic model. Both left-to-right and right-to-left LSTM layers contain 128 LSTM memory blocks.

As for the pre-trained unilingual OCR models, we train an 1D bidirectional LSTM networks based OCR model [18] (identical network structure with the target network) on a very small set (the size of training set we use is 200) of unilingual images (any language among the target multilingual sets is ok). When the training loss does not decrease within 5 iterations, the pre-trained model can be regarded to converge. Due to the tiny training set, the pre-trained model takes very few iterations to converge and the training process of each iteration is very quick. The time cost of pre-training a unilingual OCR model can be simply eliminated compared to that of training a multilingual OCR model. We can easily pre-train a unilingual OCR model for our hybrid model reuse approach in any multilingual application scenarios.

3.3 Fixed Multilingual Text Generating Model Reuse

As shown in the right part of Fig. 1, fixed multilingual text generating model reuse is added into the traditional LSTM networks for MOCR. There are several pieces of studies [9,21] trying to reuse pre-trained deep models to train a new deep model. They, however, neglecting the fact that fixed networks structure in fact can help model training and always reusing the weights directly from source networks, can hardly throughly utilize the existing pre-provided model.

Thus, a complete novel model reuse technique with deep structures, Fixed Model Reuse approach, has been proposed [20], which directly substitutes the sophisticated fixed model/features used in general tasks with a deep network rather than transfers the pre-trained weights or learns with source/target examples. A new operator knockdown is used for eliminating the connections between the sophisticated model/features and the deep structure, and it seems similar to dropout but they are completely different in purpose and effects. Dropout [17] is proposed to reduce the overfitting problem by randomly setting hidden unit activities to zero during training a deep network. In contrast, different to the connections can be reset and updated with another trial of iteration in dropout, knockdown strategy vanishes the dependencies between the basic model and the fixed model/features. Once the dependencies is disconnected by a knockdown operation, the related features will not be functional anymore.

Here, we mainly introduce the concrete steps on how to reuse the fixed model in our HMR approach. In the FMR part, the basic model and the fixed model are directly arranged in parallel to the output layer nodes (Fig. 2).

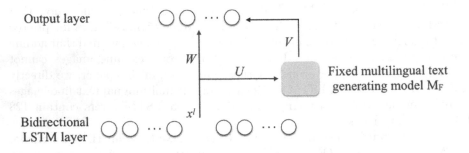

Fig. 2. Reuse the fixed model in HMR scheme

The left side in Fig. 2 is our basic model. Specifically, the raw features x can be calculated through a bidirectional LSTM layer and can be finally represented as x^l, which is eventually fully connected with the output layer. And the fully connected weights can be organized as a linear mapping matrix W together with a nonlinear soft-max function.

Besides, we fully connect the fixed multilingual text generating model M_F between the bidirectional LSTM layers and output layers. The LSTM networks based text generating model M_F is built on multilingual corpora [14], which contains contextual information of multilingual texts. The weights between the fixed model and basic model can be denoted as U and V respectively (refer to Fig. 2). It is obvious that the provided fixed features and V compose a traditional linear prediction model which could be useful in conventional application.

3.4 Model Training

Training Process. During the training process, knockdown strategy facilitates the information transfer from fixed model M_F to the basic networks, and eventually ensure that deep networks possesses the representation or discriminative information of the fixed models M_F. The whole training procedure of FMR part can be largely divided into three iterative steps: weight propagation, knockdown, weight propagation after knockdown, as shown in Fig. 3.

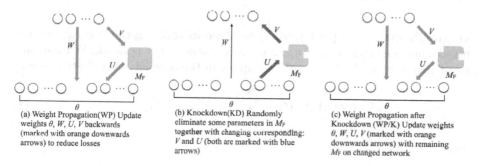

(a) Weight Propagation(WP) Update weights θ, W, U, V backwards (marked with orange downwards arrows) to reduce losses

(b) Knockdown(KD) Randomly eliminate some parameters in M_F together with changing corresponding: V and U (both are marked with blue arrows)

(c) Weight Propagation after Knockdown (WP/K) Update weights θ, W, U, V (marked with orange downwards arrows) with remaining M_F on changed network

Fig. 3. Training process for FMR part: WP → WP/K iteration. It is notable that in KD step, corresponding connections $V_{i,*}$, $U_{*,i}$ in weight matrix V, U are set to zero while the output feature z_i in the model M_F are eliminated (marked with white blanks in plot(b))

The detail training procedure is shown in Algorithm 1. At first, arrange the basic deep network and the fixed model built on multilingual texts M_F in parallel to the output layer. Then prepare the training batch and start the training process of fixed model reuse, which contains WP, KD and WP/K steps (introduced in the following). After the network has converged or reached the max iteration, the corresponding character sequence towards each text-line image by leveraging extra information from fixed model can be obtained.

Algorithm 1. Training Algorithm for Hybrid Model Reuse

Input:

Weights of the 1D bidirectional LSTM layer from pre-trained network for unilingual OCR: W_J;

Train data $\widehat{D}$ containing text-line images x_i together with corresponding text labels y_i: $\widehat{D} = \{x_i, y_i\}_{i=1}^{N}$;

The number of elements in each batch: n;

Eliminate number of features in each iteration: m;

Max-iter: k;

Fixed model built on multilingual texts: M_F

Output:

Corresponding character sequence towards each text-line image

1: Arrange the deep network and M_F in parallel to the output layer nodes;

2: Initialize Weight: Initialize the LSTM layer with W_J;

3: **repeat**

4:    Create Batch: Randomly pick up n examples from $\widehat{D}$ with replacement;

5:    Input the examples to the networks and calculate the loss L;

6:    Weight propagation(for the first iteration) or Weight propagation after knock-down step(weight consist): Obtain the derivative $\partial L/\partial W$, $\partial L/\partial U$, $\partial L/\partial V$, $\partial L/\partial \theta$. Update remaining parameters W, V, U, θ;

7:    Knockdown step(model transfer): Randomly eliminate m fixed features without replacement and set corresponding connections to zero, i.e., $U_{*,j1}, U_{*,j2}, \cdots, = 0, V_{j1,*}, V_{j2,*}, \cdots = 0$;

8: **until** converge or reach the max iteration k

Weight Propagation (WP). As general model training, the WP step focuses on reducing the errors made in the current status of the network. Without any loss of generalities, the loss function implied in the parallel network structure is:

$$L(\Theta, W, U, V) = \sum_{i=1}^{N} l(x_i, z_i, y_i) + \lambda L_{reg} \tag{1}$$

where

$$l(x_i, z_i, y_i) = \tilde{l}(x_i^l, z_i, y_i) + \hat{l}(x_i^l, z_i)$$
$$\tilde{l}(x_i^l, z_i, y_i) = y_i \log p(f(x_i^l) + g(z_i))$$
$$\hat{l}(x_i^l, z_i) = \frac{1}{2}\|z_i - x_i^l U\|_F^2$$

Here suppose we have N examples, denoted by $D = \{(x_1, y_1), (x_2, y_2), \cdots, (x_i, y_i), \cdots, (x_N, y_N)\}$, where x_i is the raw image input and y_i is the label of x_i, Θ are the parameters in the bidirectional LSTM layer and z_i is the output of fixed model M_F. $\tilde{l}(x_i^l, z_i, y_i)$ is the label prediction loss function ($\tilde{l}$ actually can be with any convex loss functions), in which $f(x_i^l)$ is the prediction of x_i^l (we define as linear function $f(x_i^l) = x_i^l W + b_x^l$ for simplicity, b_x^l is the bias for pre-dictors of x^l), p is a soft-max operator and $g(z_i)$ is the predictor of the provided features z_i (we also define as linear function $g(z_i) = z_i V + b_z$, b_z is the bias of fixed features z). L_{reg} can be any convex regularization, while in order to facilitate the WP step, in this paper, we choose L_{reg} as: $L_{reg} = \|W\|_2^2 + \|U\|_2^2 + \|V\|_2^2$. The parameter λ controls the trade-off between the loss and regularization. In WP step, the derivatives are taken to a portion of parameters, i.e., Θ, W, U, V, with the help of Back Propagation technique as shown in Fig. 3(a).

Knockdown (KD). In order to eliminate the influence of the fixed features z during the training procedure, we need to remove those connected parts corresponding to features z gradually and finally vanish all related components as shown in Fig. 3(b). Suppose there are additional d_f dimensional sophisticated features for each instance in fixed model M_F. We randomly remove several components of features saying $z_{i,j_1}, z_{i,j_2}, \cdots$, where $j_1, j_2 \in [1, d_f]$, and consequently, the corresponding connections, i.e., $U_{*,j_1}, U_{*,j_2}, \cdots$ in matrix U, $V_{j_1,*}, V_{j_2,*}, \cdots$ in matrix V are restricted to zero as well. These KD steps will be carried out for several trials during the early training process. In each iteration, the KD step randomly eliminates components without replacement, and finally will cause the fixed prediction model M_F disconnected with the deep networks. After removing all features or the whole fixed model, the trained model is the same structure as the traditional deep network.

Weight Propagation After Knockdown (WP/K). This step is generally the same as WP step. However, we here emphasize that the KD step could break the structure of the originally consistent network. Therefore, in each adjustment on removing parts of features, additional steps are required for making the whole deep structure self-consistent which can further reduce the prediction errors. After the knockdown, WP/K step is used to harmonize the remaining weights and makes the network re-consistent, finally reducing the errors. It is notable that in the WP/K step, the parameters Θ, U, V are depended only on those fixed features remaining on the network changed by the KD step as shown in Fig. 3(c).

4 Experiments

We conduct two sets of experiments on two multilingual application scenarios. One is for Han script OCR (Chinese and Japanese), and the other is for Latin script OCR (English, French and German). In each scenario, we conduct experiments with the following approaches. In some segmented approaches (SVM and CNN), vertical projection histogram analysis is used to divide text lines into character units.

- **Basic model:** The simple 1D bidirectional LSTM network [18] with ground-truth alignment using a forward-backward algorithm (Connectionist Temporal Classification, CTC [6]).
- **Parameter transfer:** We use the parameter transfer approach [21] by initializing the weights in target networks with those from pre-trained unilingual OCR model.

- **Fixed model reuse**: We adopt the weight propagation, knockdown and weight propagation after knockdown steps [20] to train the target network with a fixed model built on multilingual texts.
- **SVM**: We train a classifier based on support vector machine [13] for multilingual character units.
- **CNN**: Convolutional networks are trained as a classifier [5] to recognize the multilingual characters.
- **Tesseract**: We directly use the Tesseract open source OCR engine [15] to recognize multilingual text-line images.
- **Our HMR approach**: We incorporate the parameter transfer and fixed model reuse method. That is to say, before beginning training, weights of the target multilingual networks are initialized by those from pre-trained unilingual OCR models. And during the early training process, weight propagation, knockdown and weight propagation after knockdown steps are used to train the target networks together with a fixed multilingual text model.

4.1 Dataset

Our datasets for multiple languages are developed by OCRopus [4] based on multilingual corpora. This utility only requires a bunch of UTF-8 encoded text files and a set of true-type fonts. With these two things available, one can artificially generate any number of text-line images. This utility also provides control to induce scanning defects such as distortion, jitter, and other degradations. These images are degraded using degradation models [2] to reflect scanning defects. There are four degradation parameters, namely elastic elongation, jitter, sensitivity and threshold. Some text-line image samples in our dataset are shown in Fig. 4. Datasets are further divided into training, test, and validation subsets, the numbers of which are 9000, 1000 and 1000 respectively. The text-line images are normalized to 32 pixels in height and 540 pixels in width in the pre-processing step.

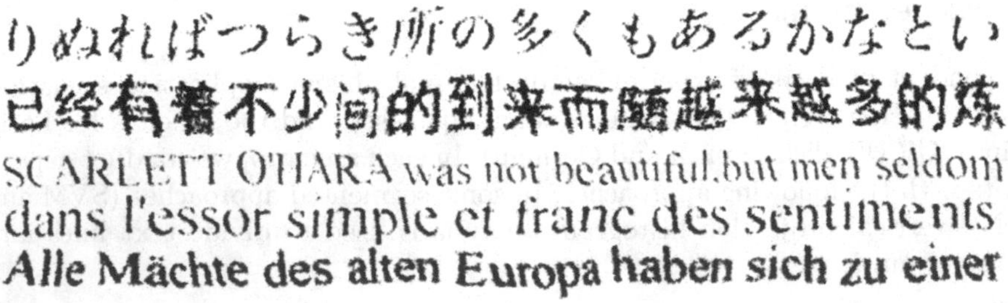

りぬれ ばつらき所の多くもあるかなとい
已经有着不少间的到来而随越来越多的炼
SCARLETT O'HARA was not beautiful,but men seldom
dans l'essor simple et franc des sentiments
Alle Mächte des alten Europa haben sich zu einer

Fig. 4. Some text-line image samples from our dataset

4.2 Evaluation

We run the following experiments with Keras framework implementation on an Intel(R) E5-2680 v3 CPU server. As for the deep weight propagation in our implementation, we use gradient descent with momentum. The hyperparameters are the same as used by [8]: momentum 0.9; weight decay $5*10^{-4}$; initial learning rate 10^{-2}, which is decreased by a factor of 10. To evaluate the models, we report the iteration number that each model needs until convergence, which measures the time cost that each model takes. In our experiments, even though our HMR approach connects fixed structure and increases complexity, the parameters in the fixed model are untrainable and the additional weights (U, V), of which the number is much less than that in the basic model, will be eliminated by the knockdown step during the early training process. Thus, the time cost of each model can be simply presented by the iteration number which networks need to converge (when the validation loss doesn't decrease within 5 iterations, we consider the model gets into convergence).

And we are also interested in the following accuracy measure: let S be a set of training examples drawn from a fixed distribution $D_{X \times Z}$, given a test set $\bar{S} \subset D_{X \times Z}$ disjoint from S, where label error rate (LER) of a temporal classifier h is defined as the mean normalized edit distance between its classifications and the targets on $\bar{S}$, i.e.

$$LER(h, \bar{S}) = \frac{1}{|\bar{S}|} \sum_{(x,z) \in \bar{S}} \frac{ED(h(x), z)}{|z|} \tag{2}$$

$$Accuracy = 1 - LER(h, \bar{S}) \tag{3}$$

where $ED(p, q)$ is the edit distance between two sequences p and q, i.e. the minimum number of insertions, substitutions and deletions required to change p into q. This is a natural measure for tasks (such as speech or handwriting recognition) where the aim is to minimize the rate of transcription mistakes. Then the corresponding accuracy can be calculated by the label error rate. The accuracy results in this paper are all obtained by 5-fold cross-validation method.

Table 1. Results of basic model and parameter transfer model

Han-based MOCR			Latin-based MOCR		
Model	Iteration	Accuracy (%)	Model	Iteration	Accuracy (%)
Basic model [18]	63	91.67	Basic model [18]	34	93.27
Chinese→Han	**40**	**92.31**	English→Latin	24	93.47
Japanese→Han	43	92.11	French→Latin	**21**	**93.88**
			German→Latin	22	93.71

4.3   Experiment 1: Parameter Transfer Approach *vs* Basic Model

In this experiment we use the parameter transfer approach [21] by initializing the weights of 1D bidirectional LSTM network [18] (basic model) for multilingual OCR with those from pre-trained network for unilingual OCR task. We initialize networks for the Han-based multilingual OCR in turn with the pre-trained unilingual model for Chinese (Chinese→Han) and Japanese (Japanese→Han). Then we run the same experiments on Latin script. The results of these experiments can be found in Table 1. The results indicates that models using parameter transfer method dramatically outperforms basic model in term of convergence rate (e.g. Iteration number 63→40 in Han-based MOCR and 34→21 in Latin-based MOCR) and also can yield somewhat improvement in accuracy. It seems that exploitation of the pre-trained unilingual OCR model can be of assistance to help set a good basis for the training of multilingual OCR model and strengthen the generalization ability. This gives the reason why transferring the parameter can accelerate the training process and improve the performance.

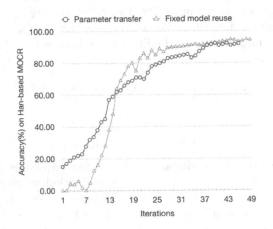

Fig. 5. Han-based results of fixed model reuse and parameter transfer method

4.4   Experiment 2: Fixed Model Reuse *vs* Parameter Transfer

In this experiment, Fixed Model Reuse scheme is adopted to train the target network with a fixed multilingual text-generating model. The experimental results are compared with those of model using parameter transfer method (we select the best experimental results according to the results in experiment 1, i.e., Chinese→Han and French→Latin), which can be found in Figs. 5 and 6.

The line charts show that Parameter transferring approach can get better results at the beginning, but they are excelled by fixed model reuse approach at 15th iteration on Han script and 12th iteration on Latin script. Besides, parameter transfer approach can help the networks converge faster than FMR approach. In contrast, FMR approach can achieve higher accuracy when models converge

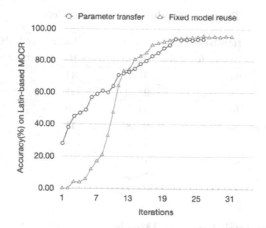

Fig. 6. Latin-based results of fixed model reuse and parameter transfer method

(2.29% higher on Han-based MOCR and 1.82% higher on Latin-based MOCR). It seems that basic networks can learn useful information from the fixed model M_F during the training process and reduce the recognition errors.

4.5   Experiment 3: Our HMR Approach *vs* Baselines

As we can see from the experiment 1 and 2, parameter transfer approach can obviously accelerate the training process, while fixed model reuse method can also raise the convergence rate and accuracy. They both have respective advantages in accelerating the training process and improving the accuracy, which motivate us to combine these two techniques and propose our hybrid model reuse approach. In this experiment, our hybrid model reuse approach is used to train the target network by transferring the parameters from the pre-trained unilingual OCR model plus exploiting the information from the multilingual text-generating model. The results of the aforementioned approaches together with some segmented approaches (SVM [13] and CNN [5] as classifiers, and open-source Tesseract engine [15]) are presented in Table 2.

On the basis of the previous experiments, we assume that better results can be obtained by using our Hybrid Model Reuse approach. And the experimental results showed in Table 2 validates our assumption. The fastest convergence rate and the highest accuracy can be obtained by using the Hybrid Model Reuse approach. Generally, sufficient training samples on every single character are essential to the results of character classifiers. Our datasets may not provide enough samples for each character, which accounts for the unsatisfactory accuracy of the segmentation-based methods. And the Tesseract engine using the off-the-shelf multilingual model seems not to be able to handle the degradation of the images, which also leads to low accuracy.

Table 2. Results of different models

Model	Han-based MOCR		Latin-based MOCR	
	Iteration number	Accuracy (%)	Iteration number	Accuracy (%)
Basic model [18]	63	91.67	34	93.27
Parameter transfer [21]	40	92.31	21	93.88
Fixed model reuse [20]	44	94.6	26	95.7
Our hybrid model reuse	**35**	**95.87**	**18**	**96.92**
SVM [13]	-	71.2	-	75.3
CNN [5]	-	79.8	-	85.2
Tesseract [15]	-	73.68	-	81.33

5   Conclusion and Future Work

The typical approach to multilingual OCR task is to identify the language category first and then leverage the specific unilingual OCR system to continue the recognition task. But errors from language identification often propagate through the pipeline and restrain the final performance.

In this paper, we focus on the end-to-end multilingual OCR task under moderate degradation and propose a hybrid model reuse approach to pursue the aim of shorter training process and higher recognition accuracy. Different from mainly reuse the pre-trained network weights, our approach exploits the structure of mature models which have been widely used in various applications and adopts parameter transfer method to help set a good basis for the target model. The experimental results indicate that our hybrid training approach has promising potentials in coping with these kinds of complicated multilingual OCR tasks to accelerate the training process and yield satisfactory accuracy.

In the future, attention mechanism [1] can be explored and incorporated into the current LSTM networks as a new training scheme. Besides, region-based convolutional neural networks (R-CNNs) [19], which allow for parametrically efficient and effective image feature extraction, are also worth noticed.

References

1. Ablavatski, A., Lu, S., Cai, J.: Enriched deep recurrent visual attention model for multiple object recognition. In: Applications of Computer Vision, pp. 971–978 (2017)
2. Baird, H.S.: Document image defect models and their uses. In: 2nd International Conference Document Analysis and Recognition, ICDAR 1993, Tsukuba City, Japan, 20–22 October 1993, pp. 62–67 (1993)
3. Bengio, Y., Simard, P., Frasconi, P.: Learning long-term dependencies with gradient descent is difficult. IEEE Trans. Neural Netw. **5**(2), 157–166 (2002)
4. Breuel, T.M.: The OCRopus open source OCR system. In: Proceedings of the Document Recognition and Retrieval XV, Part of the IS&T-SPIE Electronic Imaging Symposium, San Jose, CA, USA, 29–31 January 2008, p. 68150F (2008)

5. Firmani, D., Merialdo, P., Nieddu, E., Scardapane, S.: In codice ratio: OCR of handwritten Latin documents using deep convolutional networks. In: International Workshop on Artificial Intelligence for Cultural Heritage, pp. 9–16 (2017)
6. Graves, A., Gomez, F.: Connectionist temporal classification: labelling unsegmented sequence data with recurrent neural networks. In: International Conference on Machine Learning, pp. 369–376 (2006)
7. Hochreiter, S., Schmidhuber, J.: Long short-term memory. Neural Comput. **9**(8), 1735–1780 (1997)
8. Krizhevsky, A., Sutskever, I., Hinton, G.E.: ImageNet classification with deep convolutional neural networks. In: International Conference on Neural Information Processing Systems, pp. 1097–1105 (2012)
9. Long, M., Cao, Y., Wang, J., Jordan, M.I.: Learning transferable features with deep adaptation networks. In: International Conference on International Conference on Machine Learning, pp. 97–105 (2015)
10. Naz, S., Umar, A.I., Shirazi, S.H., Ajmal, M.M., Salahuddin: The optical character recognition for cursive script using HMM: a review. Res. J. Appl. Sci. Eng. Technol. **8**(19), 2016–2025 (2014)
11. Pan, S.J., Yang, Q.: A survey on transfer learning. IEEE Trans. Knowl. Data Eng. **22**(10), 1345–1359 (2010)
12. Peng, X., Cao, H., Setlur, S., Govindaraju, V., Natarajan, P.: Multilingual OCR research and applications: an overview. In: International Workshop on Multilingual OCR, pp. 1–8 (2013)
13. Philip, B., Samuel, R.D.S.: A novel bilingual OCR system based on column-stochastic features and SVM classifier for the specially enabled. In: Second International Conference on Emerging Trends in Engineering & Technology, pp. 252–257 (2009)
14. Shi, Z., Shi, M., Li, C.: The prediction of character based on recurrent neural network language model. In: IEEE/ACIS International Conference on Computer and Information Science, pp. 613–616 (2017)
15. Smith, R., Antonova, D., Lee, D.S.: Adapting the tesseract open source OCR engine for multilingual OCR. In: International Workshop on Multilingual OCR, p. 1 (2009)
16. Song, R., Umemoto, K., Nie, J., Xie, X., Tanaka, K., Rui, Y.: UniClip: leveraging web search for universal clipping of articles on mobile. Data Sci. Eng. **1**(2), 101–113 (2016)
17. Srivastava, N., Hinton, G., Krizhevsky, A., Sutskever, I., Salakhutdinov, R.: Dropout: a simple way to prevent neural networks from overfitting. J. Mach. Learn. Res. **15**(1), 1929–1958 (2014)
18. Ul-Hasan, A., Breuel, T.M.: Can we build language-independent OCR using LSTM networks? In: International Workshop on Multilingual OCR, p. 9 (2013)
19. Yang, B., Zhang, Y., Cao, J., Zou, L.: On road vehicle detection using an improved faster RCNN framework with small-size region up-scaling strategy. In: Satoh, S. (ed.) PSIVT 2017. LNCS, vol. 10799, pp. 241–253. Springer, Cham (2018). https://doi.org/10.1007/978-3-319-92753-4_20
20. Yang, Y., Zhan, D., Fan, Y., Jiang, Y., Zhou, Z.: Deep learning for fixed model reuse. In: Proceedings of the Thirty-First AAAI Conference on Artificial Intelligence, San Francisco, California, USA, 4–9 February 2017, pp. 2831–2837 (2017)

21. Yosinski, J., Clune, J., Bengio, Y., Lipson, H.: How transferable are features in deep neural networks? In: Advances in Neural Information Processing Systems 27: Annual Conference on Neural Information Processing Systems 2014, Montreal, Quebec, Canada, 8–13 December 2014, pp. 3320–3328 (2014)
22. Zhou, Z.H.: Learnware: On the Future of Machine Learning. Springer, New York (2016)

Author Index

Aamir, Tooba II-178
Alarte, Julian I-393
Alghamdi, Bandar I-189
Almars, Abdulqader I-319
Alsadie, Deafallah II-167
Al-Shammari, Ahmed II-121
Alshammari, Ahmed II-167
Alzahrani, Eidah J. II-167
Ansah, Jeffery I-281

Bagozi, Ada II-487
Bai, Quan I-424
Balasubramaniam, Thirunavukarasu II-285
Baravalle, Andres II-502
Beheshti, Amin I-161
Benatallah, Boualem I-199, II-301
Benharkat, Aïcha Nabila I-68
Bianchini, Devis II-361, II-487
Bononi, Luciano II-209
Bouguettaya, Athman II-151, II-178
Boukadi, Khouloud I-68

Cai, Xiangrui II-193
Cai, Yi I-453
Cao, Buqing II-19
Cao, Fangfei I-408
Cao, Jinli II-111
Chandra, Anita II-438
Chang, Elizabeth I-127
Chao, Wenhan I-297
Chen, Aaron II-135
Chen, Dian I-483
Chen, Jinjun II-19
Chen, Zhigang I-408, II-35
Cheng, Gong I-381
Christensen, Helen II-100
Cristea, Alexandra I. II-395

De Antonellis, Valeria II-361, II-487
de Guzmán, Pablo Chico I-81
Dillon, Tharam I-127
Ding, Xiaoke II-193
Dong, Hai II-178
Dong, Manqing I-199

Fang, Junhua I-333
Fang, Yuan II-240
Fattah, Sheik Mohammad Mostakim II-151
Fei, Chaoqun I-468
Fogelman-Soulié, Françoise II-312
Fu, Hao I-178
Fu, Yumeng I-178

Gao, Neng I-247, II-335
Gao, Rong II-51
Gao, Wang I-483
Garda, Massimiliano II-361
Georgakopoulos, Dimitrios II-209
Getoor, Lise II-410
Ghafari, Seyed Mohssen I-161
Ghedira-Guegan, Chirine I-68
Gorostiaga, Felipe I-81
Gu, Xiwu I-262
Gu, Yu I-381
Guo, Jinwei II-225

Han, Fengling I-18
Han, Yuehui I-213
Hao, Maoxiang II-345
Hartmann, Sven II-135
He, Jinyuan II-85
He, Qiang II-269
He, Ying I-408
Hong, Xiaoguang II-3
Hou, Shengluan I-468
Hu, Gang I-483
Hu, Yupeng II-3
Huang, Feitao II-253
Huang, Hao I-199
Huang, Jianwei II-225
Huang, Xinyi I-111
Huo, Yingxiang II-457

Ibrahim, Ibrahim A. I-319
Insa, David I-393

Jasberg, Kevin II-422
Jayaraman, Prem Prakash II-209

Jia, Weijia   II-325
Jiang, Qingshan   I-348
Jiang, Xin   I-297

Kang, Wei   I-281
Kayes, A. S. M.   I-127
Khalil, Ibrahim   I-111
Kotagiri, Rao   II-269
Kudo, Michiharu   I-3

Largeron, Christine   II-312
Larsen, Mark E.   II-100
Le, Trung   II-100
Leckie, Christopher   II-269
Lee, Sin Wee   II-502
Leung, Ho-fung   I-453
Li, Jiuyong   I-281, I-364
Li, Lin   I-497
Li, Peiyao   I-145
Li, Qing   I-453, II-253
Li, Ruixuan   I-262
Li, Xiang   II-335
Li, Xue   I-319
Li, Yuhua   I-262
Li, Zhixu   I-213, I-408, II-35, II-345
Liang, Tianan   I-262
Liang, Yue   I-381
Liao, Chang   II-379
Liao, Kewen   II-209
Lin, Zehang   II-253
Liu, An   I-213, II-35
Liu, Chengfei   II-121
Liu, Daxin   I-381
Liu, Guanfeng   I-408
Liu, Jixue   I-281, I-364
Liu, Lin   I-281, I-364
Liu, Qing   I-424
Liu, Wenyin   II-253
Lou, Jiong   II-325
Loukil, Faiza   I-68
Lu, Jie   II-67
Lu, Lingjiao   I-333
Luo, Yonghong   II-193
Luo, Yun   II-269
Luo, Zhunchen   I-297
Luong, Khanh   II-285
Lv, Jianming   II-253
Lyu, Tianshu   I-308

Ma, Hui   II-135
Ma, Jiangang   II-85
Ma, Jie   I-408
Ma, Wenjia   I-297
Mahbub, Syed   I-127
Maiti, Abyayananda   II-438
Marini, Alessandro   II-487
Melchiori, Michele   II-361
Mistry, Sajib   II-151
Miyamoto, Kohtaroh   I-3
Mo, Jingjie   I-247
Montori, Federico   II-209
Muzammal, Muhammad   I-36, I-52, I-348

Nakamura, Hiroaki   I-3
Naseriparsa, Mehdi   II-121
Nasrulin, Bulat   I-52
Nayak, Richi   II-285
Nepal, Surya   I-18, I-111
Nguyen, Duc Thanh   II-100
Nguyen, Hung   II-100
Nguyen, Thin   II-100
Nguyen, Van   II-100
Nguyen, Vanh Khuyen   II-472
Ning, Xiaodong   II-301
Nobari, Sadegh   I-348
Nurgaliev, Ildar   I-36

O'Dea, Bridianne   II-100
Orgun, Mehmet   I-161

Pang, Yanxia   II-379
Pardede, Eric   I-127
Pei, Yang   I-247
Peng, Min   I-483
Peng, Zhaohui   II-3
Phung, Dinh   II-100

Qi, Jianzhong   I-439
Qian, Weining   II-225
Qin, Hongchao   I-230
Qu, Qiang   I-36, I-52, I-348
Qu, Yuzhong   I-381

Rahayu, Wenny   I-127
Ramesh, Arti   II-410
Ren, Xiaoxuan   I-308
Rong, Jia   II-85

Salehi, Mahsa   II-269
Sánchez, César   I-81
Sellis, Timos   II-209
Sheng, Quan Z.   I-199, II-472
Shi, Lei   II-395
Silva, Josep   I-393
Situ, Runwei   II-253
Sizov, Sergej   II-422
Song, Chunyao   II-193
Song, Huan   II-457
Su, Yijun   II-335
Sun, Le   II-85
Sun, Lili   II-111

Tamarit, Salvador   I-393
Tan, Wenan   II-379
Tang, Wei   II-335
Tari, Zahir   II-167
Teng, Luyao   II-457
Teng, Shaohua   II-457

Venkatesh, Svetha   II-100
Vimalachandran, Pasupathy   II-111
Vo, Bao Quoc   II-121

Wang, Can   II-51
Wang, Chen   II-135
Wang, Feiran   II-19
Wang, Guoren   I-230
Wang, Hao   I-364
Wang, Hua   I-96, I-483, II-51, II-85, II-457
Wang, Jiong   I-247
Wang, Ke   I-408
Wang, Lizhen   II-240
Wang, Wei   I-178
Wang, Xianzhi   I-199, II-301
Wang, Xiaoxuan   II-240
Wang, Yi   I-96
Watson, Jason   I-189
Wei, Ruidi   I-381
Wen, Yiping   II-19
Wu, Dianshuang   II-67
Wu, Xin   I-453
Wu, Yueping   II-379

Xiang, Ji   II-335
Xiang, Jianwen   I-497
Xie, Qianqian   I-483
Xie, Qing   I-497

Xie, Zhongwei   I-497
Xiong, Xiaoqing   I-262
Xu, Jiajie   I-333
Xu, Jingyun   I-453
Xu, Yang   II-3
Xu, Yue   I-189

Yakhchi, Shahpar   I-161
Yang, Jiali   II-35
Yang, Jian   I-145
Yang, Peizhong   II-240
Yang, Wu   I-178
Yang, Xu   I-111
Yang, Xuechao   I-18
Yang, Yi   I-424
Yang, Zhenguo   II-253
Yao, Lina   I-199, II-301
Yi, Xun   I-18, I-111
Yin, Dan   I-178
Yin, Hongzhi   I-213, I-333, II-35
Yong, Jianming   II-111
Yu, Yonghong   II-51
Yuan, Xiaojie   II-193
Yuan, Ye   I-230

Zeng, Yali   I-111
Zhang, Fuyong   I-96
Zhang, Guangquan   II-67
Zhang, Jianyu   II-312
Zhang, Li   II-51
Zhang, Qian   II-67
Zhang, Rui   I-439
Zhang, Shaohua   II-325
Zhang, Shuai   II-301
Zhang, Shuhan   I-468
Zhang, Wei Emma   II-472
Zhang, Xiang   II-301
Zhang, Xuyun   II-269
Zhang, Yan   I-308
Zhang, Yanchun   I-483, II-85, II-111, II-457
Zhang, Yihan   I-483
Zhang, Ying   II-193
Zhang, Zan   I-364
Zhang, Zhao   II-225
Zhao, Chenxu   I-439
Zhao, Lei   I-213, I-333, II-35
Zhao, Pengpeng   I-333, II-35
Zhao, Weiliang   I-145
Zhao, Xin   I-319
Zhao, Yan   II-345

Zheng, Kai   II-345
Zhong, Luo   I-497
Zhong, Xian   I-497
Zhou, Aoying   II-225
Zhou, Chunyi   II-379

Zhou, Rui   II-121, II-269
Zhou, Xiaojie   II-325
Zhou, Yiming   I-213
Zhou, Yujing   I-247
Zhu, Feida   I-230

Printed in the United States
By Bookmasters